Spawning Migration of the European Eel

FISH & FISHERIES SERIES

VOLUME 30

For other titles published in this series, go to
http://www.springer.com/series/5973

Guido van den Thillart • Sylvie Dufour
J. Cliff Rankin
Editors

Spawning Migration of the European Eel

Reproduction Index, a Useful Tool for Conservation Management

Springer

Editors
Guido van den Thillart
Institute of Biology
Leiden University
POB 9516, 2300 RA Leiden
The Netherlands

Sylvie Dufour
Museum National d'Histoire Naturelle
DMPA, UMR CNRS 5178 "Biology of Marine Organisms and Ecosystems"
7 rue Cuvier, CP 32
75231 Paris cedex 05
France

J. Cliff Rankin
Department of Chemical and Biological Science
University of Huddersfield
Queensgate, HD1 3DH
UK

Cover picture: European silver eels in swimming tunnels at Leiden University
Photograph by: David Doubilet
Reproduced with kind permission of Undersea Images Inc, New York, USA

ISBN 978-1-4020-9094-3 e-ISBN 978-1-4020-9095-0

Library of Congress Control Number: 2008934776

Printed on acid-free paper

springer.com

Volume Foreword

Freshwater eels are almost infinitely improbable creatures. They spawn and die in the middle of the ocean, often associated with undersea mountains. Their transparent, leaf-like larvae move with ocean currents for months or years until they approach the mouths of freshwater rivers. Then they undergo a dramatic transformation in morphology, physiology and behavior. They move from their planktonic oceanic environment, migrate upstream and live for several years as apex freshwater predators. Then, almost impossibly, as they become sexually mature, they reverse their migration downstream to the ocean and back to spawning grounds to complete their life cycle.

The dramatic changes in their life cycles are incredible. The efforts to unravel the details of their life history have been truly daunting. Much of the past research was the work of dedicated individuals who devoted their lifetime research to these fishes. Freshwater eels merit a separate chapter in almost any textbook dealing with ichthyology, marine biology or animal migration. We know a great deal about some aspects of the biology of freshwater eels. However, our understanding of their biology still resembles a work of art as much as a work of science. To some it appears like the sweeping brush strokes of a Japanese Zen landscape, to others it resembles the work of a French impressionist, and to still others it appears as magic realism. Of course there is art because of the irresistible nature of eels, but there must also be art because there are so many details that science does not yet know. It is a challenge to fit together all the bits and pieces from researchers around the world, on different species, to resolve the details.

This volume is notable because it represents a quantum change in the approach to studies of eel biology. This volume brings together the concerted research efforts of numerous authors, at different institutions in several countries, focused on a single species, the European eel, Anguilla anguilla. For the first time we have an integrated consideration of evidence from the molecular to the ecosystem. The volume goes far beyond the usual collection of papers on a theme or even papers on a species. The research on which all these chapters are based was planned and integrated from the start to address clearly defined questions about the European eel. The research questions were daunting, the approaches had to be creative.

Exactly how do you study the migration of a fish in the depths of the Atlantic Ocean? How can you conduct research on the sexual maturation of eels as they migrate from rivers into the open ocean? What are the impacts of parasites, chemical contaminants and viruses on eel recruitment? In the end, how could you bring all this information together to deal with broader questions of recruitment and conservation of the eel? It is a remarkable tribute to the ingenuity, originality and perseverance of these authors that they have dealt with all these questions, and more. They are able to address such broad questions as the evolution of continental dispersal, and the evolution of migration in freshwater eels. It is a testament to the underlying project that brought all these researchers together that we have such a wonderfully integrated and synthetic volume.

This volume should be a model, for those who will continue to study freshwater eels, for those entrusted with the conservation and management of freshwater eels, and for those looking for a species and study system rich with future possibilities.

Dr. David L. G. Noakes
Editor, Springer Fish and Fisheries Series
Professor of Fisheries and Wildlife
Senior Scientist, Oregon Hatchery Research Center
Oregon State University
Corvallis, Oregon, USA

Preface

The idea for this book was started during the EU project EELREP (October 2001–May 2005), which focused particularly on the capacity of wild silver eels to mature and to contribute to recruitment (EELREP report at: www.fishbiology/silvereel.html). The main goal of this project was to find useful quality parameters for migrating silver eel, indicating their contribution to recruitment. Obviously recruitment is the product of the number and the quality of escaping silver eels. Just counting the numbers, although rather easy, can never be used as a reliable estimate of the number of effective spawners. Remarkably this approach has (and still is) thus far hardly been considered. Quantitative parameters, mainly based on fisheries, showed that the eel populations all over the world were declining since the 1960s, and even more seriously since the glass eel decline in the 1980s. Actually nobody knows the cause for this decline and management measures were thus far only based on quantity and not on the quality of spawners; which means that they are mainly controlling fisheries.

The best quality parameter for silver eels is reproductive capacity. To obtain an impression about reproductive capacity it is not necessary to go through the whole maturation process. This would not have been easy as reproduction of the European eel is still not under our control. It was anticipated that a reasonable estimate for the reproductive capacity would be the responsiveness to hormonal stimulation. Another crucial parameter for recruitment would be the physiological fitness, i.e. the capacity of the European silver eels to swim across the Atlantic Ocean. There are few fish species that swim similar distances to their spawning sites. Salmon, *Oncorhynchus* species, swim up to a few thousand kilometre upstream the rivers, however, the distance to the Sargasso Sea (6,000 km) appears thus far to be the longest distance known for any fish species. Clearly, such a distance requires an enormous effort and in addition a rather large energy store. Evidently migrating eels need to be in great shape, and without such perfect physical condition they would not be able to contribute to recruitment. Clearly poor physical condition can never be compensated by large numbers. Therefore, apart from large numbers of escaping silver eels, we need silver eels that are prone to mature sexually and that are able to swim across the Atlantic Ocean. An integrative approach regarding eel maturation is not a simple one as there appear to be two groups of scientists that hardly communicate with each other; i.e. ecologists and physiologists. While the former mainly focuses on fisheries and population dynamics, the latter has its focus on reproductive physiology and aquaculture. In this book we included both aspects, and we hope that this approach will help to solve the current problems regarding the collapsing eel population.

The support of Erik Burgerhout was crucial for finishing the book in time; the editors are greatly indebted and like to acknowledge his punctuality and endurance.

February 2008

Guido van den Thillart
Sylvie Dufour
Cliff Rankin

Contents

Part I
Introduction

Chapter 1
How to Estimate the Reproductive Success of European Silver Eels

Guido van den Thillart and Sylvie Dufour

1.1 Introduction

Over the past 25 years populations of the European eel have been declining to such a degree that major concerns have been raised for their long-term well being. Adult stocks started to dwindle in the 1940s in major areas of the continent, while recruitment (glass eel arrivals) has collapsed since the early 1980s. Fishing yield has gradually declined by 75% and the abundance of glass eels (young eel migrating from the ocean) is now below 5% of the historical level. The stock is considered outside safe biological limits, and immediate protection measures have been recommended. There is no sign of recovery and the phenomenon seems to occur over the natural range of the European eel (*Anguilla anguilla*). A parallel development is observed in the closely related American eel (*A. rostrata)* (Castonguay et al. 1994). EIFAC and related scientific work groups (Stone 2003) started a campaign to alert politicians, pointing to the rather serious collapse of the European and other eel populations. Also the EU-commission showed its concern by the fisheries policy publication of October 2003. Recently the EU has decided that all countries have to reduce the fishing pressure and to take protective measures such as allowing a 40% escapement of silver eels. Since 2007 the European eel is protected under CITES Appendix II (Annex B of Reg. (EC) 338/97). In addition the European Commission agreed in 2007 on measures for the recovery of the stock of European Eel (COM (2005) 472). Discussions have been going on for many years trying to pinpoint the

G. van den Thillart
Institute Biology, Leiden University, P.O. Box 9516, 2300 RA Leiden, The Netherlands

S. Dufour
Museum National d'Histoire Naturelle, DMPA, UMR CNRS 5178 "Biology of Marine Organisms and Ecosystems", 7 rue Cuvier, CP 32, 75231 Paris cedex 05, France

G. van den Thillart et al. (eds.), *Spawning Migration of the European Eel,*

causes that led to the downfall of a very common fish species. However, possibly due to the complex life cycle we have only indications such as: overfishing, restriction of habitat, migration barriers, and pollution.

The life cycle of the eel is fairly complex: (1) About 1 year as larvae drifting on the oceanic currents. (2) Then a few months as glass eels swimming into the estuaries along the European coasts. (3) This is followed by the longest interval: a feeding interval of 4 to 20 years as yellow eels. (4) Then, finally a second metamorphosis from yellow into the silver form takes place. This last interval has a duration of about 6 months, which includes downstream migration and an unknown oceanic interval of long distance migration to the spawning sites. Although the natural spawning process has never been observed, circumstantial evidence suggests that spawning takes place in the Sargasso Sea, some 6,000-km away (Miller and McCleave 1994; Schmidt 1923; Tesch and Wegner 1990). Further proliferation of the gonads must take place during the migration and/or during their short stay at the spawning site. A few records indeed show a GSI (gonad somatic index) higher than 10 for ocean caught silver eels (Bast and Klinkhardt 1988). The life cycle of the migrating silver eel can thus be characterized by refusal of food; an impressive swimming activity; and final maturation of the gonads.

During the second metamorphosis or 'silvering', eels go through a number of morphological and physiological changes which prepares them for their oceanic migration. All sensory organs become more developed: eyes and nostrils are enlarged, the lateral line is more visible (Barni et al. 1985; Dave et al. 1974; Lewander et al. 1974; Pankhurst and Lythgoe 1983). At the silver stage, eels stop feeding (Fricke and Kaese 1995; Tesch 1977) and their digestive tract regresses, the structure and metabolism of the liver changes too (Hara et al. 1980). Moreover, silver eels are sexually more developed than yellow eels, although the gonads remain in a pre-pubertal stage (Dufour et al. 2003). The silvering process is a complex phenomenon. The actual sequence of events (intermediate phases), the link between external and internal modifications as well as the duration of the silvering process, remain largely unknown. Up to now, the identification of silver eels was generally based on Pankhurst's eye index. An eye index value of 6.5 was supposedly the threshold between yellow and silver eels. But even though the eye index correlates with GSI, a large sized eel may have a high eye index without necessarily being a migrant. Moreover, the 6.5 limit was based on artificially matured individuals and no direct link to downstream migration was made. An eye index of 6.5 therefore overestimates the number of migrants. Recently a detailed description of the silvering process was conducted on a subpopulation of eels from the Grand-Lieu Lake in France. From this five stages (F-I to F-V) for females and two stages for males were based on four easy to measure external parameters (length, weight, eye diameter, and pectoral fin length), predicting the stage with an accuracy of 82% (Durif et al. 2005). The pectoral fin length increases significantly when eels start their downstream migration. These fins do not actually propel the eel in the water, but they contribute to its stabilization in the open water and this may explain the increase in length at the time of migration.

1.1.1 Possible Causes for the Decline

So far, the causes for the decline of the eel population are not known, we have only assumptions without hard evidence. Apart from the most obvious factors such as habitat reduction, and overfishing, also pollution, in particular with PCB's (De Boer et al. 1999), infections with the swimbladder parasite (*Aguillicola crassus*) and the EVEX virus (Van Ginneken et al. 2004) might be causes for the current decline of the eel population. PCB's accumulate in the fat tissue of eels during their feeding interval (yellow eels), even to levels that make them by totally unsuited for human consumption (Dirksen et al. 1995; Reijnders 1986). During their spawning migration the silver eels do not eat, instead they use their fat for swimming. As a result of increasing PCB plasma levels, the animals may not be able to reach their spawning ground. It is likely that PCBs interfere with hormonal regulation and/or gonad development. Furthermore, not only will the adult eels be directly affected by the toxicity of PCBs but also subsequent fertilisation (Spies et al. 1988), hatching (Hose et al. 1982; Von Westernhagen et al. 1987) and viability of the larvae (Von Westernhagen et al. 1981) will be disturbed by high PCB levels. It is unknown to what extent levels of PCBs rise in blood plasma, fat and gonads in migrating silver eels during their journey to the Sargasso Sea, but they may well reach toxic levels. A large proportion of the European eel population is seriously infected with the swimbladder parasite (Moravec et al. 1994; Thomas and Ollevier 1992) and with the EVEX virus (Van Ginneken et al. 2004). Both have been introduced recently in Europe from Asia and are candidates for the rapid decline of the eel populations. Eels infected with the swimbladder parasite have a lower cruising speed (Barni et al. 1985). This may have serious consequences for the adult silver eels when migrating to the spawning grounds. In addition, infected eels loose the functionality of their swimbladder (Haenen 1995; Kleckner 1980), which means that they cannot maintain buoyancy and thus lose depth orientation.

1.1.2 Artificial Reproduction

Maturation and artificial reproduction of eels has been studied since the early work of Fontaine (1936). In Japan this research has been continued since the 1970s, resulting in fertilized eggs and non-feeding larvae in 1997 (Ohta et al.). Research on the induction of maturation and larval rearing of the Japanese eel, *Anguilla japonica*, is being conducted in mainland China and Japan. Most progress has been made at the National Research Institute of Aquaculture (Nansei, Japan). The maturation procedure involves injection of salmon pituitary extract (20 mg per week) for 11 or 12 weeks. At 24-h after the last injection, 17, 20 β-dihydroxy-4-pregnen-3-one was injected (2 μg g^{-1} body weight) to induce ovulation within 15–18 h. Five or six weekly injections of human chorionic gonadotropin (1 IU g^{-1} body weight) induced spermiation in male eels. Up to 62.8% fertility and 54.3% hatchability were obtained in the best case. (Ohta et al. 1997). Currently maturation and fertilisation result in controlled hatching and most larvae commence feeding. Also since 2001 leptocephali from *Anguilla japonica* were reared through metamorphosis, resulting in several yellow eels. This suggests that the eel life cycle may be closed in the near future. The artificially produced larvae are, however, very much smaller than wild leptocephali, when they arrive at the coast of Taiwan (Fricke and Tsukamoto 1998). Despite the enormous investments in time and energy, the Japanese scientists did not fully resolve the basic and applied question of the reproduction of eels. Obviously this stage is not reached with the European eel; only recently some papers have appeared showing successful fertilisation and embryonic development, but hatched larvae die before feeding (Palstra et al. 2005; Pedersen 2004). The fact that many fertilized eggs don't develop and that most larvae die prematurely, indicate that the problems may lie in the quality of eggs and possibly also in the quality of sperm. We assume that a major aspect of this problem may be resolved by trying to find natural stimuli of maturation and reproduction.

1.1.3 Maturation Sensitivity of Wild Eels

For wild silver eels, recruitment depends apart from successful migration also on successful spawning. As both processes occur in the ocean, we need other parameters to estimate reproductive success of the migrating silver eels. Thus far, there are no indicators for recruitment success. While obviously quality is of major importance, only the numbers of escaping silver eels are currently used in protection measures. A possible quality parameter is the sensitivity of silver eels to hormonal stimulation.

Full gonadal development has been successfully and repeatedly induced in female and male silver European eels, using the classical gonadotropic treatments (CPE- carp pituitary extract -in females and hCG- human chorionic gonadotropin -in males). As demonstrated by the ability of exogenous gonadotropic treatment to induce maturation, the insufficient production of endogenous gonadotropins is responsible for the blockade of eel maturation. Eel LH (luteinizing hormone) and FSH (follicle-stimulating hormone) are known to be expressed at a low level in the pituitary of silver eels. These hormones are regulated in an opposite manner during induced maturation, with a large increase in LH synthesis, but a decrease in FSH, likely resulting from differential endogenous steroid feedbacks. It has been shown that LH release is under a dual neurohormonal control, positive by GnRH (gonadotropin releasing hormone) and negative by dopamine.

High variability in the kinetics of oocyte maturation and egg quality has been observed. Maturation performance of female silver eels, as estimated by their gonadal development in response to short term and long term CPE treatments, correlates to individual initial biometric parameters. Inter-individual variability of ovarian development in response to gonadotropic treatment has often been mentioned, but this question was never directly addressed. Clearly it would be very useful if we would be able to predict which silver eel respond quickly and which would respond slowly or not at all.

1.2 Long Term Swimming

An important aspect of the reproduction of European silver eels is the huge distance they have to swim to reach their spawning grounds. After leaving the West European coast they still have to swim 5,000–6,000 km to the Sargasso Sea, the assumed spawning site. To cover this distance female eels must swim continuously for 6 months at 0.5 BL/s, which requires an impressive endurance as well as high energy reserves coupled with low cost of transport. So, obviously long term swimming capacity is a major prerequisite for reproduction. Until recently no long term swimming studies were carried out. It was not known how much energy is required for the crossing of the Atlantic Ocean. Still, this knowledge is crucial for any estimation about recruitment. We need to know at what speed eels swim, whether they can swim continuously, and at what energy costs. The energy requirement of course corresponds fat content, obviously eels must be fat to cross the ocean, but how fat?

The capacity to swim across the Atlantic Ocean can easily be influenced by several environmental factors such as parasites, virus infections, and pollutants. The swimbladder parasite *A. crassus* is known to have infected all European eel populations in the 1980s. Although eels usually don't die from the infection, it affects buoyancy control and therefore swimming mode and navigation in open water. Similarly infection with the eel viruses is not lethal, but those infections might be more severe in combination with strenuous swimming for a long period.

1.2.1 High Pressure

Also very impressive is the requirement of silver eels to navigate at 50–200 atm water pressure. Current knowledge suggests that pressure is an important factor because of its disturbing effects on biophysical and biochemical processes (Kleckner 1980). Previous experiments have shown that the first hours under pressure constitute a critical period during which the mortality could be high (Sebert and Barthelemy 1985). This mortality is related to the fish's fitness. However, pressure has also been suggested as a stimulus for maturation (Fontaine et al. 1985).

1.2.2 Our Goal

This book has been written for only one reason: to provide information about the quality of escaping silver eels by estimating the chance of success in reproduction. The current state of knowledge is such that we can make reasonable estimates of the escaping silver eels for reproduction. That implies that new ways for eel conservation can be implemented; instead of only setting percentages of escapement, it would be wise to include also the protection of habitats that produce high quality spawners. A second implication of the current knowledge is that this knowledge should help us to find solutions for artificial reproduction. The current problem of eel reproduction is the limited fertility of adults and the large scale mortality of larvae; understanding the natural process of silvering and maturation should be understood in order to develop successful protocols.

Acknowledgements This work has been supported by the European contract EELREP N°Q5RS-2001-01836.

References

Barni S, Bernocchi G, Gerzeli G (1985) Morphohistochemical changes during the life cycle of the European eel. Tissue Cell 17: 97–109.

Bast HD, Klinkhardt MD (1988) Fang eines Silberaales (Anguilla anguilla L. 1758) im Iberischen Becken (Nordost Atlantik) (Teleostei: Anguillidae). Zool. Anz. 221: 386–398.

Castonguay M, Hodson PV, Moriarty C, Drinkwater KF, Jessop BM (1994) Is there a role of ocean environment in American and European eel decline? Fish. Oceanogr. 3(3): 197–203.

Dave G, Johansson ML, Larsson A, Lewander K, Lidman U (1974) Metabolic and haemotological studies on the yellow and silver phases of the European eel, Anguilla anguilla L.-11. Fatty acid composition. Comp. Biochem. Physiol. 47B: 583–591.

De Boer J, Pieters H, De Wit M (1999) Verontreinigingen in aal en snoekbaars: Monitorprogramma ten behoeve van de Nederlandse sportvisserij, 1998. (RIVO-DLO report C036/99, The Netherlands).

Dirksen S, Boudewij TJ, Slager LK, Mes RG, Van Schaick MJM, De Voogt P (1995) Reduced breeding success of cormorants (*Phalacrocorax carbo sinensis*) in relation to persistent organochlorine pollution of aquatic habitats in the Netherlands. Environ. Pollut. 88: 119–132.

Dufour S, Burzawa-Gerard E, Le Belle N, Sbaihi M, Vidal B (2003) Reproductive endocrinology of the European eel, Anguilla anguilla. In: Aida K, Tsukamoto K, Yamauchi K (eds) Eel Biology. Springer, Tokyo, pp. 373–386.

Durif C, Dufour S, Elie P (2005) The silvering process of Anguilla anguilla: A new classification from the yellow resident to the silver migrating stage. J. Fish Biol. 66: 1025–1043.

Fontaine M (1936) Sur la maturation complete des organes genitaux de l'anguille et l'emission spontane des produits sexuels. C.R. Acad. Sci. Paris, 252: 1258–1260.

Fontaine YA, Dufour S, Alinat J, Fontaine M (1985) A long immersion in deep sea stimulates the pituitary gonadotropic function of the female European eel (Anguilla anguilla L.). C. R. Acad. Sci. Paris, Ser. III 300: 83–87.

Fricke H, Kaese R (1995) Tracking of artificially matured eels (*Anguilla anguilla*) in the Sargasso Sea and the problem of the Eel's Spawning Site. Naturwissenschaften 82: 32–36.

Fricke H, Tsukamoto K (1998) Seamounts and the mystery of eel spawning. Naturwissenschaften 85: 290–291.

Haenen OLM (1995) *Anguillicola crassus* (Nematoda, Dracunculoidea) infections of European eel (*Anguilla anguilla*) in the Netherlands: Epidemiology, pathogenesis, and pathobiology. Ph.D. thesis. Agricultural University, Wageningen, 127 pp.

Hara A, Yamauchi K, Hirai H (1980) Studies on female specific serum protein (vitellogenin) and egg yolk protein in Japanese eel (*Anguilla japonica*). Comp. Biochem. Physiol. 65B: 315–320.

Hose JE, Hannah JB, DiJulio D, Landolt ML, Miller BS, Iwaoka WT, Felton SP (1982) Effects of benzo(a)pyrene on early development of flatfish. Arch. Environ. Contam. Toxicol. 11: 167–171.

Kleckner RC (1980) Swimbladder volume maintenance related to initial migratory depth in silver-phase Anguilla rostrata. Science 208: 1481–1482.

Lewander K, Dave G, Johansson ML, Larsson A, Lidman U (1974) Metabolic and haemotological studies on the yellow and silver phases of the European eel, *Anguilla anguilla* L.-I. Carbohydrate, lipid, protein and inorganic ion metabolism. Comp. Biochem. Physiol. 47B: 571–481.

Miller M, McCleave JD (1994) Species assemblages of leptocephali in the subtropical convergence zone of the Sargasso Sea. J. Mar. Res. 52: 743–772.

Moravec F, Di Cave D, Orecchia P, Paggi L (1994) Experimental observation on the development of *Anguillicola crassus* (Nematoda: Drancunuloidea) in its definitive host, *Anguilla anguilla* (Pisces). Folia Parasit. 41: 138–148.

Ohta H, Kagawa H, Tanaka H, Okuzawa K, Iinuma N, Hirose K (1997) Artificial induction of maturation and fertilization in the Japanese eel, *Anguilla japonica*. Fish Physiol. Biochem. 17: 163–169.

Palstra AP, Cohen EGH, Niemantsverdriet PRW, Van Ginneken VJT, Van den Thillart GEEJM (2005) Artificial maturation and reproduction of European silver eel: Development of oocytes during final maturation. Aquaculture 249: 533–547.

Pankhurst NW, Lythgoe JN (1983) Changes in vision and olfaction during sexual maturation in the European eel *Anguilla anguilla* (L.). J. Fish Biol. 23: 229–240.

Pedersen BH (2004) Fertilisation of eggs, rate of embryonic development and hatching following induced maturation of the European eel *Anguilla anguilla*. Aquaculture 237: 461–473.

Reijnders PJH (1986) Reproductive failure in common seals feeding on fish from polluted coastal waters. Nature 324: 456–457.

Schmidt J (1923) Breeding places and migration of the eel. Nature 111: 51–54.

Sebert P, Barthelemy L (1985) Effects of high hydrostatic pressure per se, 101 atm, on eel metabolism. Resp. Physiol. 62: 349–357.

Spies RB, Rice DW, Felton JH (1988) Effects of organic contaminants on reproduction of the starry flounder *Platichhys stellatus*, in San Francisco Bay 11. Reproductive success of fish captured in San Francisco Bay and spawned in the laboratory. Mar. Biol. 98: 191–200.

Stone R (2003) Freshwater eels are slip-sliding away. Science 302: 221–222.

Tesch FW, Wegner G (1990) The distribution of small larvae of Anguilla sp. related to hydrographic conditions 1981 between Bermuda and Puerto Rico. Int. Rev. Ges. Hydrobiol. 75: 845–858.

Tesch FW (1977) The eel, Biology and Management of Anguilled Eels, Chapman & Hall, London, 434 pp.

Thomas K, Ollevier F (1992) Population biology of *Anguillicola crassus* in the final host *Anguilla anguilla*. Dis. Aquat. Org. 14: 163–170.

Van Ginneken V, Haenen O, Coldenhoff K, Willemze R, Antonissen E, Van Tulden P, Dijkstra S, Wagenaar F, Van den Thillart G (2004) Presence of virus infections in eel species from various geographic regions. Bull. Eur. Ass. Fish Pathol. 24(5): 268–272.

Von Westernhagen, H., Rosenthal, H., Dethlefsen, V., Ernst, W., Harms, U.L., Hanse, P.D., Cameron, P., Janssen, D. 1981. Bioaccumulating substances and reproductive success in Baltic flounder *Platichthys flesus. Aquatic Toxicology*, 1: 85–99.

Von Westernhagen H, Dethlefsen V, Cameron P, Janssen D (1987) Chlorinated hydrocarbon residues in gonads of marine fish and effects on reproduction. Sarsia 72: 419–422.

Part II
Silvering

Chapter 2
Seasonal Evolution and Individual Differences in Silvering Eels from Different Locations

Caroline M.F. Durif, Vincent van Ginneken, Sylvie Dufour, Tamás Müller, and Pierre Elie

2.1 Introduction

Silvering is a requirement for downstream migration and reproduction. It marks the end of the growth phase and the onset of sexual maturation. This true metamorphosis involves a number of different physiological functions (osmoregulatory, reproductive), which prepare the eel for the long return trip to the Sargasso Sea. Unlike smoltification in salmonids, silvering of eels is largely unpredictable. It occurs at various ages (females: 4–20 years; males 2–15 years) and sizes (body length of females: 50–100 cm; males: 35–46 cm) (Tesch 2003). It is most common when studying eels, to separate individuals into two groups, yellow (resident) and silver (presumably migrant), and to compare the physiological profiles between the two. Basic knowledge was obtained in this way and we will first review what is known about these two stages.

Because of the difficulty of getting individuals while they are in the process of metamorphosing, little is known about the dynamics of the silvering process. Here we present new information on the triggers, duration and succession of events up to

C.M.F. Durif
Center for Ecological and Evolutionary Synthesis, University of Oslo, P.O. Box 1066, Blindern, 0316 Oslo/Institute of Marine Research, 5392 Storebø, Norway

V. van Ginneken
Integrative Zoology, Institute Biology Leiden (IBL), Van der Klaauw Laboratorium, P.O. Box 9516, 2300 RA Leiden, The Netherlands

S. Dufour
USM 401, UMR 7178 CNRS "Biology of Marine Organisms and Ecosystems", Museum National d'Histoire naturelle, 7 rue Cuvier, 75231 Paris Cedex 05, France

T. Müller
Szent István University, Faculty of Agricultural and Environmental Sciences. Gödöllo, Páter K. u. 1., H-2103 Gödöllo/Pannon University Georgikon Faculty of Agriculture, Deák F. u. 16, H-8360 Keszthely, Hungary

C.M.F. Durif and P. Elie
Cemagref Agricultural and Environmental Engineering Research Center Research Unit "Estuarine Ecosystems and Migratory Amphihaline Fish Species" 50 avenue de Verdun Gazinet 33612 Cestas, France

G. van den Thillart et al. (eds.), *Spawning Migration of the European Eel*,

the silver migratory stage. The transitional stages – between the yellow (resident) stage and the silver (migrating) stage – were characterized through exhaustive sampling of both resident and migratory eels. New aspects concerning the endocrinology of the silvering process will be given based on seasonal sampling of eels undergoing metamorphosis. These results will also bring new insights on the triggers and timing of silvering.

2.2 Literature Review

2.2.1 Characteristics of Silver Eel

2.2.1.1 External Changes

The first obvious change in the eel is the colour modification. Migrating eels display counter-shading similar to pelagic fishes and marine mammals. The belly turns silvery white, while the sides and back, initially yellow or dark green, become a dark bronze colour. This colour change is viewed as an adaptation to the pelagic environment. As in salmonids this change occurs before the fish actually enters the marine environment. Other distinctive features of silver eels are their enlarged eyes, their elongated pectoral fins, which become black and the increased thickness of their skin.

2.2.1.2 Energy Stores

Migratory eels must have the necessary energy stores to accomplish their long spawning migration. Lipids are the main source of energy in fishes, especially migratory fishes. In eels, accumulation of fat seems to occur during silvering, since in yellow eels energy stores are mainly glycogen (Barni et al. 1985). Fat stores increase from 8% to 28% between the yellow and silver stage (Bergersen and Klemetsen 1988; Larsson et al. 1990). A small portion of the fat will be used for gonad development while the majority will be used for swimming. These lipids are thus in direct contact with the muscles (Fontaine 1975; Pankhurst 1982a). Fat is also stored under the skin and in the liver. Larsson et al. (1990) hypothesized that a certain amount of fat stores (28%) are a requirement for eels to silver. However many migratory eels display lower percentages (Svedäng and Wickström 1997).

2.2.1.3 Sensory Organs

The size and structure of the eye change during silvering. The volume and consequently the surface of the eye are considerably increased (Bertin 1951; Stramke 1972; Pankhurst 1982b). This makes the eye of silver eels more efficient in absorbing light than that of yellow eels (Carlisle and Denton 1959). In addition, the number of rods increases and cones appear to decrease since the retinal surface area increases (Pankhurst 1982b). Lack of cone cells is common in deep-sea fish where color vision is unnecessary (Locket 1975). Before eels even reach the sea, their retinal pigments shift from porphyrin and rhodopsin to chrysopsin which is for scotopic vision (Carlisle and Denton 1959; Es-Souni and Ali 1986). Maximal absorption of the pigment is shifted towards short wave lengths as it is in deep-sea fishes.

The lateral line is an important sensory organ used by fishes to detect currents in the close environment to avoid obstacles, capture prey or avoid predators, and for schooling. During silvering, the lateral line becomes visible and sensory cells increase (Zacchei and Tavolaro 1988).

Olfaction is particularly well developed in eels. However, Pankhurst and Lythgoe (1983) and Sorensen and Pankhurst (1988) showed that olfactory cells and associated mucus cells degenerate in male and female eels after hormonally induced sexual maturation. This suggests that olfaction will be less important during the marine migratory phase of the life cycle than during the freshwater phase when eels are still feeding. However, later in the life cycle, during sexual maturation, development of sensitivity to sexual pheromones is likely to occur and to be involved in the expression of the spawning behaviour.

2.2.1.4 Alimentary Tract

Silvering eels stop feeding and this continues until the end of their life cycle. This occurs in August according to Bertin (1951). The alimentary tract degenerates, and this is further increased during hormonally induced sexual maturation. However, it still plays a role in osmoregulation (Chapter 6). First, the intestine becomes shorter, and then the thickness of the wall decreases. This partial degeneration is not merely a consequence of fasting but a true degeneration (Pankhurst and Sorensen 1984). This state is reversible if the eel starts feeding again, even after complete sexual maturation (Dollerup and Graver 1985; Le Belle and Fontaine 1987).

2.2.1.5 Swim Bladder

Most fish tissues are denser than water and therefore fishes tend to "sink". The swim bladder of fishes decreases density by secretion and re-absorption of gas. It allows the fish to adjust its buoyancy. Silver eels can secrete more gas than yellow eels. This is due to a considerable development of the capillaries of the *rete mirabilis* which increase in diameter and in length (Kleckner and Krueger 1981; Yamada et al. 2001). Guanine coats the wall of the swim bladder which becomes thicker, thus reducing loss of gas by diffusion (Kleckner 1980). These modifications are considered as an adaptation to life in the deep sea (see chapter 5).

2.2.1.6 The Osmoregulatory System

The eel is euryhaline, meaning it can withstand rapid and large changes in salinity. The immediate passage from freshwater to saltwater causes a loss of water and therefore increases ion concentration in the blood. Osmotic balance is achieved by absorption of water by the alimentary tract. The intestine, the kidneys, and the gills eliminate ions. Mucus is more abundant and the thickness of the skin increases to reduce loss of water (Fontaine 1975).

In contrast with salmonids, the eel adapts quickly to saltwater whatever the stage. Yellow eels can be found both in fresh- and saltwater, so changes in osmoregulatory function are less important. Nevertheless, yellow eels seem to adapt more slowly to saltwater than migratory eels according to Boucher-Firly and Fontaine (1933), but this will be further developed in Chapter 6. As long as silver eels stay in freshwater, they will undergo a progressive demineralization (Fontaine 1975; Dutil et al. 1987). According to Fontaine (1985) this state could stimulate silver eels to migrate and leave freshwater.

2.2.1.7 Gonads

Eels become sexually differentiated when they reach 14 to 35 cm in length (Bienarz et al. 1981; Colombo et al. 1984). In captivity, gonads were differentiated after 2 years (Beullens et al. 1997). In male eels, differentiation occurs at the same time as silvering while gonads of female differentiate well before (Colombo et al. 1984; Durif et al. 2005).

In females, the gonad weight increases during silvering. Nevertheless, the gonadosomatic index (GSI) rarely goes above 1.5%, while it can reach 50% after hormonally induced sexual maturation. The gonadotropic axis is blocked as long as the eel remains in fresh- or coastal water. It is surmised that the eel naturally becomes sexually mature only in the open ocean (although this has never been observed) and under the influence of certain, yet unknown, environmental factors.

2.2.2 Hormonal Control

Silvering is considered to be a secondary metamorphosis because during this phase the eel undergoes many morphological and physiological modifications. The neuroendocrine system regulates these modifications. Analogies with the parr-smolt transformation or smoltification of diadromous salmonids are often made, since these both correspond to a pre-adaptation to the marine environment. Changes in skin colour are similar in smolts and silver eels have counter-shading typical of marine species. In smolts, the most important changes concern the osmoregulatory system. By analogy, the first studies on endocrinology of silvering focused on these hormones and their role in the osmoregulatory changes. However, one must bear in mind that smoltification relates to juvenile fish (smolts) while silvering marks the onset of sexual maturation and the most important changes in silvering eels relate to the reproductive system.

Modifications in the activity of ionic transport in gills have been linked with production of cortisol and thyroid hormones have been implicated in this. According to histology, the thyroid gland is more active in silver eels than in yellow eels (Callamand and Fontaine 1942). The production of thyroid hormones induces hyper-locomotor activity. One can read many anecdotes on the agitation of eels at the onset of migration. Captive eels will try to escape and this particular behaviour has been used by fishermen to predict on which nights migratory runs of eels will occur (Fontaine 1975). Hyperactivity caused by hyperthyroidism could even cause them to leave the water if there is no outlet, as occurs in aquatic amphibian larvae before their adult metamorphosis (Fontaine 1975). Han et al. (2004) found that transcription of the beta subunit of TSH (thyroid stimulating hormone) increases along with serum thyroxine levels during silvering of the Japanese eel, *Anguilla japonica*. However in the European eel, recent results have shown that thyrotropin mRNA levels did not change nor did serum thyroid hormone increase during silvering (Aroua et al. 2005). Aroua et al. (2005) did not find any effect of T4 treatments on yellow eels and they concluded that the thyrotropic axis is not or only moderately involved in silvering.

Adrenocorticotropic hormone (ACTH) secreted by the pituitary gland stimulates the adrenal gland and boosts the synthesis of corticosteroids. Leloup-Hâtey (1964) showed that adrenal cells increase in numbers and serum cortisol level increases during silvering. It has been shown in the North American eel, *Anguilla rostrata*, that cortisol induces a rise in the activity of Na^+, K^+ ATPase of gills and intestine as well as silvery hues in yellow eels (Epstein et al. 1971). Utida et al. (1972) showed that water absorption in the gut was increased in cortisol-treated freshwater eels. Finally, cortisol probably has a role in lipid breakdown in fasting silver eels (Fontaine 1975).

The changes described previously remain slight in comparison with changes related to the beginning of sexual maturation. Silvering marks the onset of sexual maturation. Among all vertebrates, the development and activity of gonads are under the positive control of gonadotropins (GTH: LH: luteinising hormone and FSH = follicular-stimulating hormone) secreted by the pituitary. Gonads will produce gametes and steroid hormones in response to GTH stimulation. It is at this gonadal level that sexual maturation can be artificially induced by injecting carp or salmon pituitary extracts in females and hCG (human chorionic gonadotropins). Steroids will in turn stimulate target organs involved in reproduction. This will allow yolk deposition (vitellogenesis) in oocytes. Vitellogenin protein is synthesized by the liver under the effect of estradiol and incorporated in the oocyte under the control of GTH. Only silver eels have hepatocytes capable of being stimulated for this synthesis via estradiol receptors although large extraphysiological doses of estradiol can stimulate vitellogenin synthesis in yellow female eels as well as in male eels (Burzawa-Gérard et al. 1994). Silver eels also have the structures necessary for endocytosis of vitellogenin.

Gonadotropic function is totally inactive in yellow eels, while it shows some activity in silver eels (Dufour et al. 1983a, b; Durif et al. 2005). The brain controls the production of pituitary LH (luteinising hormone) via GnRH (gonadotropin releasing hormone) produced by neurosecretory cells located in the hypothalamus. The status of these neurons (activation) can depend on the developmental stage of the animal (age, growth, and metabolism) as well as on environmental factors. The blockade of sexual maturation in the silver eel is due to an insufficient production of GnRH and to a dopaminergic inhibition that prevents the production of LH (Dufour et al. 1991). At the yellow stage this blockade is reinforced since gonads do not respond to pituitary extract (Lopez and Fontaine 1990). Combined treatments of GnRH agonist and dopamine receptor antagonist resulted in a significant increase in pituitary LH as well as increased locomotor activity of eels (Dufour et al. 1988).

Growth hormone, another pituitary hormone, stimulates growth and cell reproduction in animals. In addition, it acts on several physiological processes, including reproduction, metabolism and osmoregulation (Evans 1993). GH also stimulates estradiol receptors in the liver (Messaouri et al. 1991) and potentiates estradiol-induced vitellogenin synthesis (Peyon et al. 1996). GH also controls the secretion of IGF-I (Insulin-like Growth Factor 1) by the liver in teleosts as in other vertebrates. Huang et al. (1998) showed that IGF-I may exert a negative feedback on growth by inhibiting pituitary GH secretion but a stimulatory effect on reproduction by activating LH production. Estradiol and testosterone then amplify the activity of the gonadotropic axis by stimulating the production of LH while inhibiting GH.

It is not known what triggers silvering in eels. In vertebrates, sexual maturation occurs when individuals have reached a certain age and size and accumulated enough energy stores to ensure the success or reproduction. In eels, it has been suggested that there is a "critical fat mass" for triggering silvering (Larsson et al. 1990); although other studies did not find any link between fat content and silvering (Svedäng and Wickström 1997).

Huang et al. (2001) have shown that in vivo injections of cortisol and testosterone have a stimulatory effect on LH synthesis. Yellow eels that received this treatment also showed silver coloring, increase of retinal surface area, increase in liver weight, and regression of the gut. Similar results were obtained with 17α-methyltestosterone (11-KT) implants on *Anguilla australis* (Rohr et al. 2001). Six weeks after implantation, treated yellow eels displayed increased eye surface area, darkening of the pectoral fins, thickening of the skin and development of the liver and gonads. It appears that 11-KT is capable of inducing silvering. Recent data highlighted the possible roles of gonadotropins (FSH then LH) in the control of silvering and are reported in Chapter 3 of this book.

2.2.3 Length and Age at Migration

Length and age at migration are extremely variable in female eels. Differences in male and female growth rates and strategies have often been discussed but it is not clear why certain eels choose to maximize size while others opt for a time-minimizing strategy (Vøllestad and Jonsson 1986; Vøllestad 1992; Poole and Reynolds 1996; Svedäng et al. 1996; Holmgren et al. 1997). Age at silvering is negatively correlated with growth rate and some variation is due to habitat differences (Svedäng et al. 1996). A correlation has been reported between length at metamorphosis and longitude as well as distance to the spawning ground (Vøllestad 1992). A clear trend in the proportion of migrating small and large eels throughout the migratory season was observed on the river Loire (Durif and Elie 2008). Those authors showed that the number of large eels increased regularly during migration, indicating that these large eels were located upstream in the watershed. Therefore, eels located further away from the spawning grounds were longer. Also of interest, is the fact that gonad development during hormonally induced sexual maturation increases exponentially with body length and eels over 70 cm show the highest gonad weight/body length ratio (Durif et al. 2006). Although female eels may silver at 50 cm (Fig. 2.1), they certainly benefit from attaining a greater size in terms of fecundity. The time needed to reach that size will vary according to their habitat and growth conditions. The extreme variability in length and age of female silver eels (Figs. 2.1 and 2.2) is

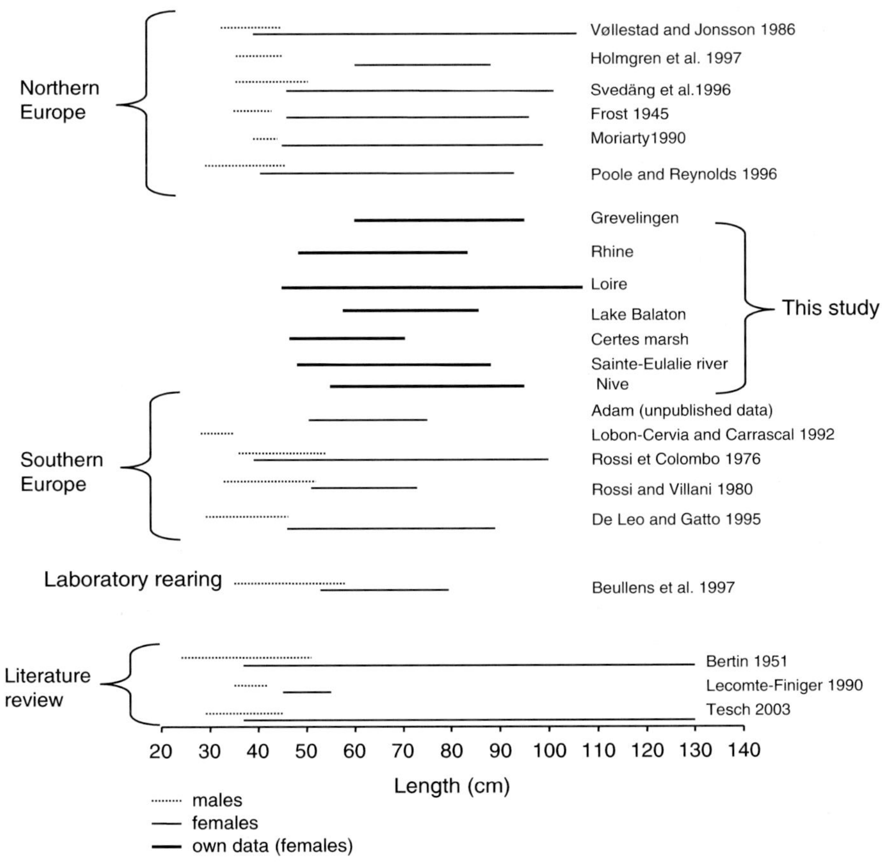

Fig. 2.1 Length of silver eels (*Anguilla anguilla*)

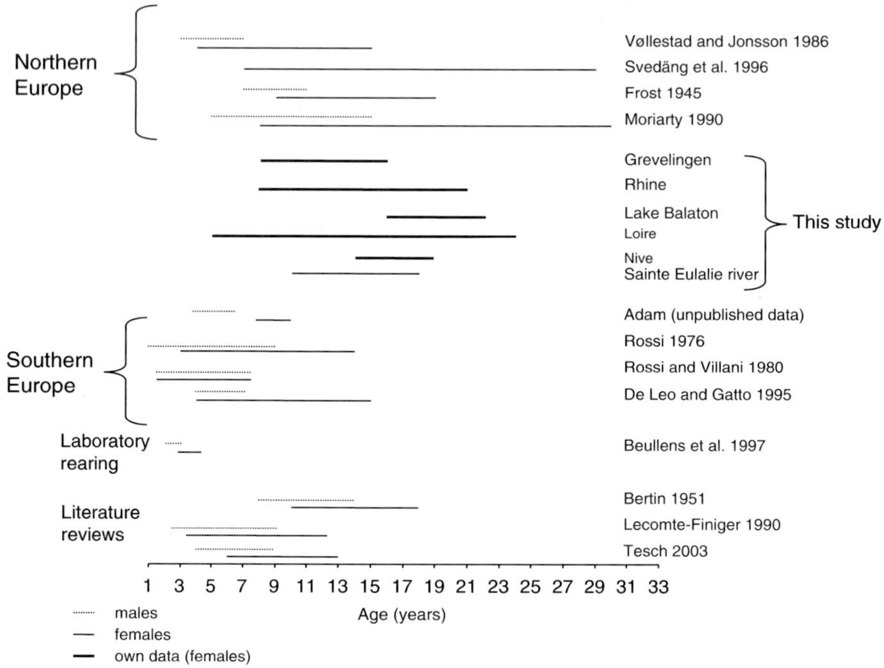

Fig. 2.2 Age of silver eels (*Anguilla anguilla*)

probably a reflection of the variability in habitats and growth conditions of eels. This variability is much reduced in males since they rarely grow over 50 cm (except when reared, Fig. 2.1). In the wild, they can start to silver at 30 cm, but most are around 35–40 cm. Their energetic demand obviously will not be as high as for females during reproduction.

2.3 Silvering Stages and Dynamics

As we have seen in the previous sections the overall characteristics of silver eels have been well documented. The modifications and differences between both yellow and silver stages are clear. Nevertheless, in the field one finds intermediate looking individuals that seem to fit both categories. Tesch (2003) writes that the best indication of whether the animal is physiologically a fully developed silver eel is the method by which it was caught. Thus little is known about the dynamics of silvering. Therefore, several studies were undertaken within the framework of the Eelrep project with the aim to give a fuller description of the silvering process by: (1) describing the events that lead to the migrating stage (2) gathering knowledge on the seasonal occurrence of these stages (onset and duration of silvering) and finally by (3) comparing different sub-population of silver eels from different locations.

2.3.1 Silvering Stages

Ecologists use clusters of individuals displaying similar characteristics to describe continuous phenomena. It then is easier to set boundaries to describe processes even when none are apparent. Thus the yellow and silver stages have been described as two separate stages whereas in fact the silvering process is gradual. Here we will give a description of the different stages eels go through during silvering. Further details of the study are given in Durif et al. (2005) and Durif et al. (in press).

Eels were collected at six locations in France which corresponded to different types of hydrosystems (rivers in both small and large catchments, small coastal river, estuary and marsh). Different types of fishing gear were used targeted on migrating individuals or not. Overall, 1,188 eels were sampled at different times of the year between 1996 and 2002. For each individual a profile was defined based on morphological and physiological characteristics. Table 2.1 shows the different measurements taken to define each profile based on previously described differences between yellow and silver eels.

Principal Component Analysis and Cluster Analysis were used to group individuals according to their profile (Fig. 2.3). Results of the analyses showed that many individuals displayed similar characteristics. The characteristics of the clusters that were defined by the analysis were consistent with a gradual transition from the yellow to the silver stage. Five different stages were determined for females that described the

Table 2.1 List of parameters measured and sampled on eels to define their morpho-physiological profile

Process	Measurement	Index
Growth	Total body length	L
	Pituitary growth hormone	GH expressed as $\mu g\ g^{-1}$ of total body mass
	Age based on otoliths	Age in years
Fat stores	Body mass	Fulton's condition factor K
Increase in retinal surface area	Eye diameter	Pankhurst eye index EI
Increase of the pectoral fin	Pectoral fin length	Fin index: I_F = fin length/L*100
Cessation of feeding: Regression of the alimentary tract	Mass of the digestive tract	Digestive tract index: DTI = mass of the alimentary tract/body mass*100
Development of gonads	Mass of gonads	Gonado-somatic index: GSI = mass of gonads/body mass*100
Development of liver	Mass of liver	Hepato-somatic index: HSI = mass of liver/body mass*100
Onset of puberty	Pituitary gonadotropin (GtH-II or LH-like)	LH expressed as $ng\ g^{-1}$ of total body mass
	Vitellogenin	Log (VTG) expressed in $\mu g\ ml^{-1}$

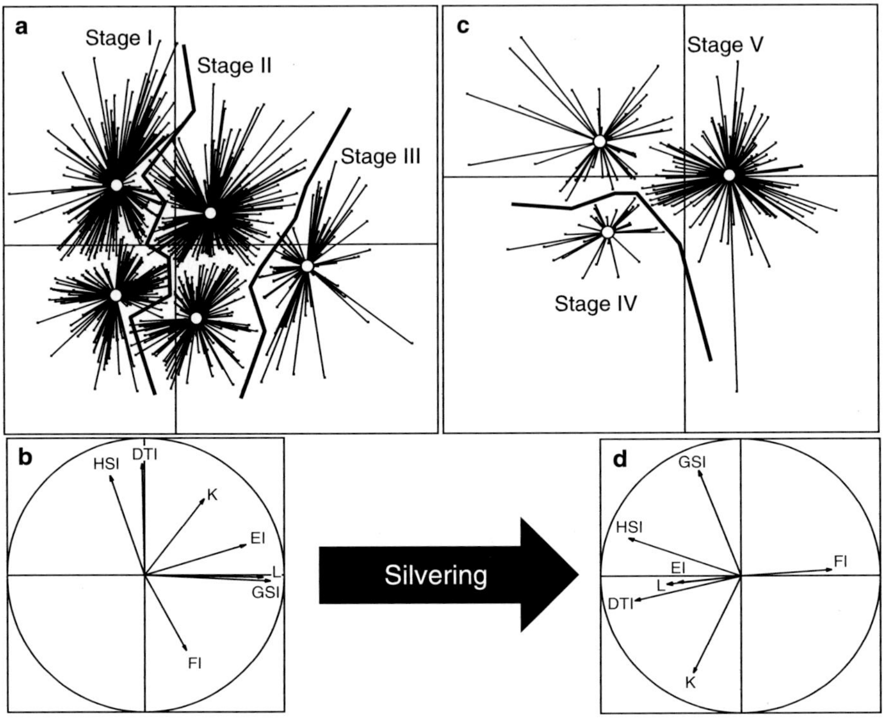

Fig. 2.3 Results of the principal component analysis and cluster analysis on female resident and migratory eels. Stars represent clusters of yellow (**a**) and (**c**) silver eels. (**b**) Correlation circle of the variables for yellow eels. (**d**) Correlation circle of the variables for silver eels. Broken lines in the factorial plots and separating the clusters correspond to the successive stages during the silvering process (Stages I to V). L: body length, K: Fulton's condition factor, EI: Pankhurst eye index, FI: fin index, GSI: gonado-somatic index, HSI: hepato-somatic index, DTI: digestive tract index

progression from small, undifferentiated yellow eels to pre-pubertal migratory eels. Figure 2.4 shows the evolution of the different morphological and physiological parameters for each stage in undifferentiated (stage I) and female (stages II to V) European eel.

Comparisons were made using ANCOVA and body length (for comparison of eye diameter and fin length) or body mass (for comparison of gonads, liver, and digestive tract masses) as a cofactor to remove any size effect. Bonferroni pairwise comparisons were used to detect which stages were different. Differences in vitellogenin, GtH-II (LH), and GH were detected using the Kruskal-Wallis test. Based on these results, the silvering process can be described as follows. Stages I and II correspond to the yellow stage, separating sexually undifferentiated individuals (stage I) from female eels (stage II). Eels at these stages have only slightly developed gonads, GSI is approximately 0.5% and under. There is not yet any production of vitellogenin or gonadotropin. Silvering is initiated at stage III, when vitellogenin shows a significant increase ($p < 0.05$). GtH-II (=LH)

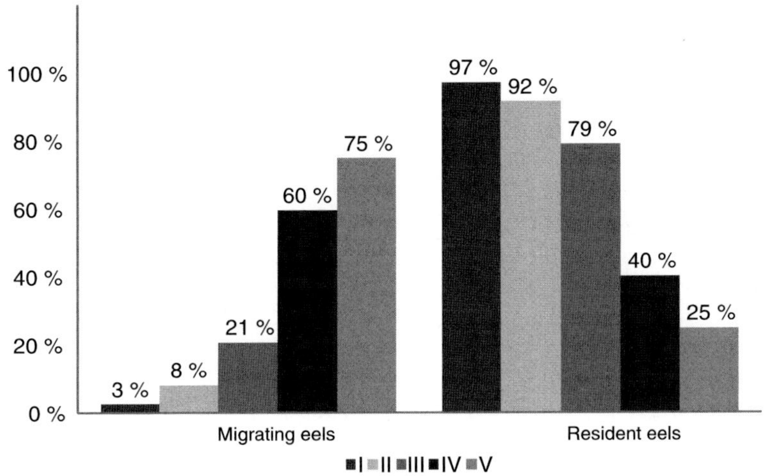

Fig. 2.4 Proportion of downstream migrating eels (caught in a downstream trap or a stow net) and resident eels (caught with fyke nets or eel pots) in each silvering stage

level shows a slight production. GH level reaches a maximum at this stage which suggests high growth rate before or at the onset of silvering. GSI of these pre-silver eels is approximately 0.8%. Eye diameter also increases significantly ($p < 0.05$) and mean eye index is 7.6. By backtracking the type of fishing gear used for each individual we found that most stage III eels in the sample (72%) had not yet begun their downstream movements since only 28% were caught with fishing gear targeting migratory eels (Fig. 2.4).

The silver migratory stage is reached at stages IV and V. Vitellogenin and GtH-II levels reach their maximum value as does the development of gonads (Fig. 2.5). The alimentary tract has significantly ($p < 0.05$) regressed and the eels have probably stopped feeding at this point. Eye index is around 10. Pectoral fin length continues to increase between stages IV and V. GH however decreases significantly from stage IV suggesting that at this stage, eels have switched from somatic growth to starting sexual maturation. The proportions of stage IV and V eels that were caught by specifically targeting migrants (captures at night with stow nets or trap during directed downstream movements) were respectively 68% and 82%.

2.3.2 Comparison of Silver Eels from Different Locations

Do migratory eels from different types of hydrosystems present the same morphological and physiological characteristics? Eight different locations were compared

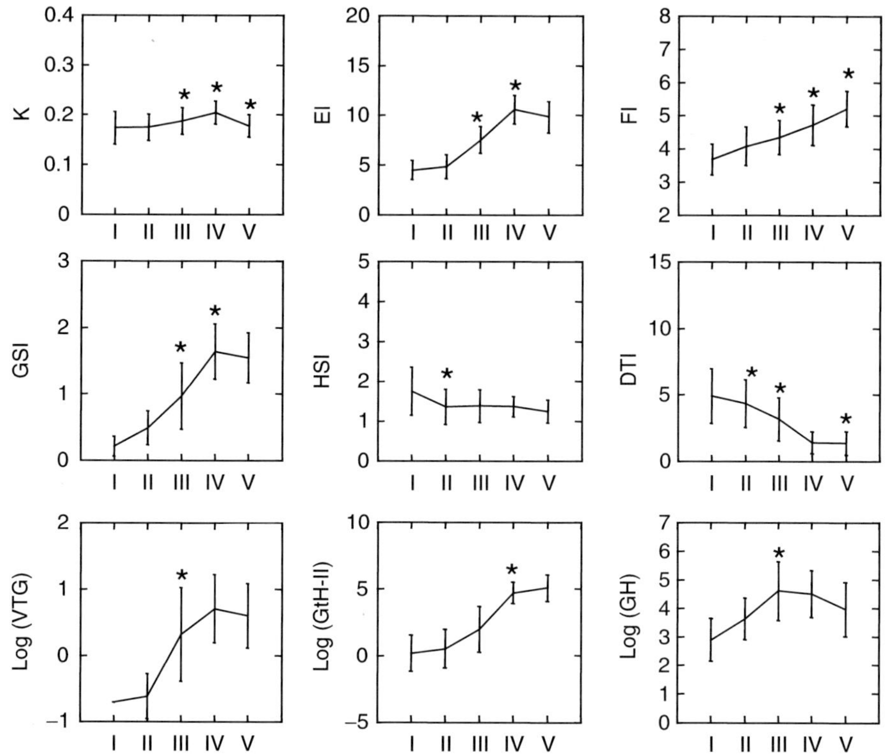

Fig. 2.5 Variations (mean ± SD) of certain morphological and physiological parameters during the silvering process, from resident eels (stage I) to female migratory eels (stage V). K: Fulton's condition factor, EI: Pankhurst eye index, FI: fin index, GSI: gonado-somatic index, HSI: hepato-somatic index, DTI: digestive tract index, vitellogenin, pituitary gonadotropin (gth-II), and GH: pituitary growth hormone. *: Significant difference from the value immediately before ($P < 0.05$), (see text for explanation of statistical tests)

(Table 2.2). Eels were sampled in the same manner as described above. Stages were assigned according to their physiological and morphological characteristics (Durif et al. 2008). Within the last silvering stage (V), we found small but significant differences in GSI according to the location (Fig. 2.6a). Landlocked eels of Lake Balaton in Hungary always displayed the lowest GSI, while eels caught in rivers during their downstream migration (Loire, Rhine and Sainte-Eulalie) showed the highest values. This suggests that high swimming activity (such as during downstream migration) further stimulates gonad development during silvering. Similarly, DTI values were lowest in actively migrating eels (Loire, Nive and Sainte-Eulalie).

Individuals obtained from the eel on-growing farm were all at stage III although they "looked" silver by their colour and eye size. Their mean GSI was 0.9% and

Table 2.2 Number of eels and origin

Location	Type of hydrosystem	Size of watershed (km^2)	Abbreviation used in graphs	Fishing gear	Number of eels
Lake Balaton, Hungary	Freshwater lake (no outlet to the sea)	5,775	Ba	Fyke net, electrofishing	240
Certes, France	Brackish water marsh	155	Bw-Fr	Fyke nets, eel pots	108
Grevelingen, Netherlands	Brackish water lake	140	Bw-Nl	Fyke nets	86
Helmond, Netherlands	Eel farm	–	Farm	none	12
Loire River, France	river	110,000	Lo	Stow net	390
Nive, France	river	1,000	Ni	Downstream trap	61
Rhine, France	Large river	224,000	Rh	Fyke net	529
Sainte-Eulalie, France	Coastal freshwater river	250	Ste	Downstream trap	61

none had undergone any regression of the digestive tract (DTI = 2.9%). During sampling it was found that most had food in their digestive tract. The only external feature that differentiated them from migratory eels at stage V was the length of the pectoral fin; fin index was 4 while it is approximately 5 in migratory eels (Fig. 2.6b).

2.3.3 Seasonal Changes

A detailed investigation of the seasonal dynamics of silvering was carried out on eels from the Grevelingen in the Netherlands (Van Ginneken et al. 2007a, b). The objective of this study was to obtain a finer seasonal description of the silvering process (start and duration) as well as a description of the physiological mechanisms involved (triggers and endocrine control).

Monthly samples of eels were collected between April and November in Lake Grevelingen. This is the largest brackish/saltwater lake of Western Europe with a total area of 14,000 ha. The lake is situated on the boundary between Zuid-Holland and Zeeland, the Netherlands and has a large standing stock of eels. Morphological, metabolic and endocrine parameters were measured.

Results showed a clear gradual increase in GSI and vitellogenin, but also in metabolites such as triglycerides, phospholipids, and cholesterol. Cortisol also showed a significant increase (one-way ANOVA $p < 0.05$) with highest values in September prior to downstream migration. From our observations in this study it became clear that a role for cortisol may be in mobilization of energy stores, especially in the European eel which has to cover a distance of 6,000 km to its spawning areas in the Sargasso Sea. Using large Blazka swim tunnels, (Van Ginneken and

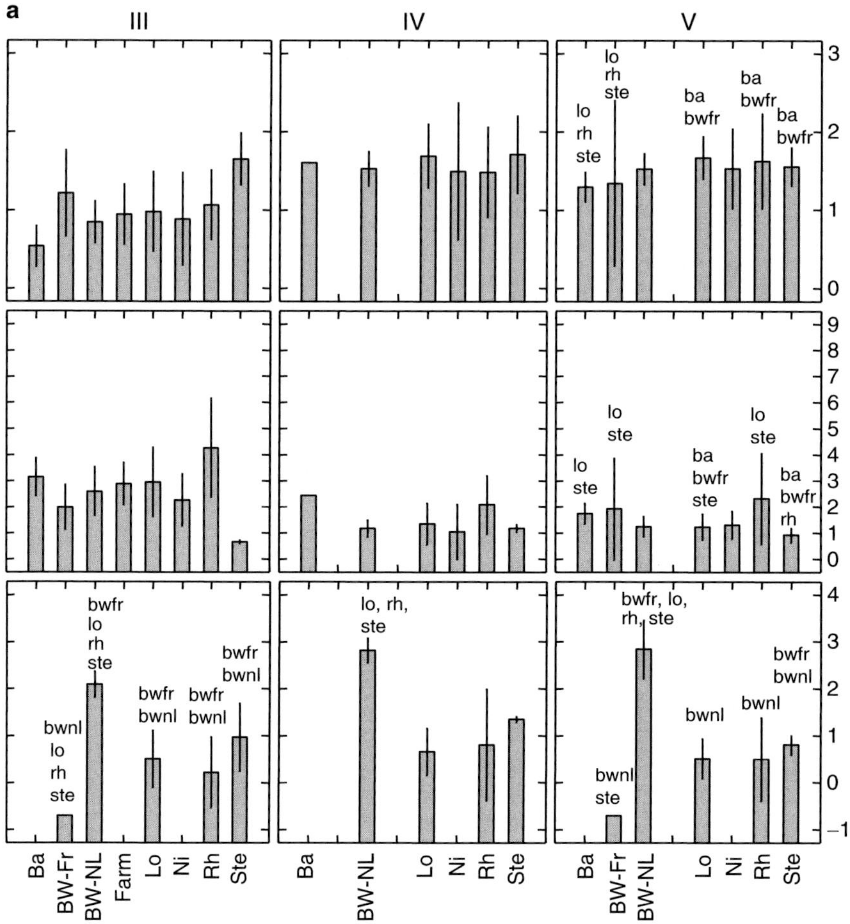

Fig. 2.6 **a** Variations (mean ± SD) in GSI, DTI and vitellogenin level in pre-migratory (stage III) and migratory eels (stages IV and V) from different locations. Explanations for abbreviations of locations are given in Table 2.2. Abbreviations on top of bars indicate significant differences ($P < 0.05$) with these locations

Van den Thillart 2000) demonstrated that for this tremendous swimming effort 40% of the energy reserves are needed while the remaining 60% can be used for gonad development. The observation that blood substrates like phospholipids and cholesterol are significantly (one-way ANOVA $p < 0.05$) increased in European silver eels corroborates the view that the major role of cortisol lays in the mobilization of the energy stores prior to and during migration.

August was clearly a cross-over month for silvering when stage IV and stage V eels appeared. A slight increase in T4 and T3 were found in eels caught in April. However it is not clear whether these changes were related to silvering, since such a cycle (minimum activity in summer and maximum activity in winter and spring) exists in many fish living at these latitudes (Swift 1960). Overall differences in T4

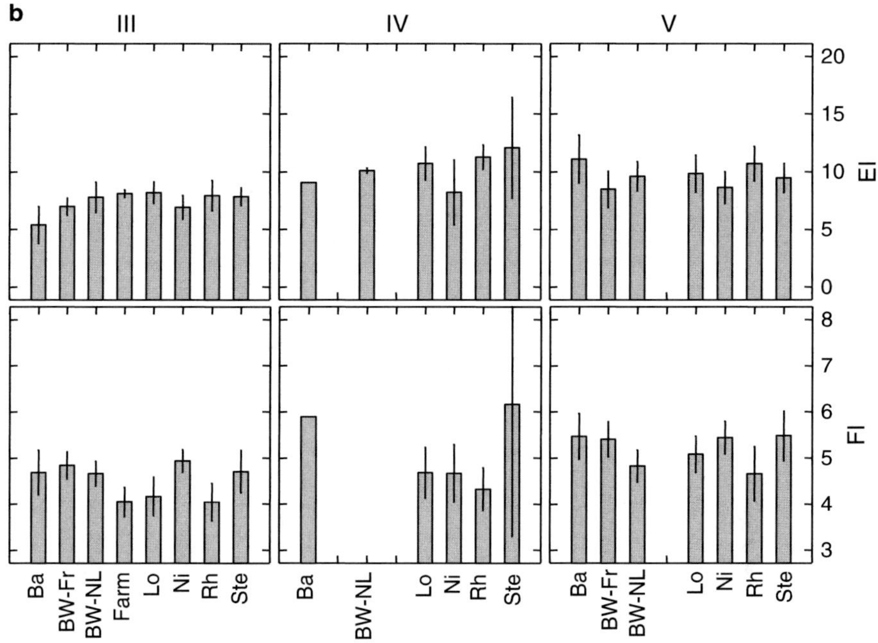

Fig. 2.6 (continued) **b** Variations (mean ± SD) in eye index (EI) and fin index (FI) in premigratory (stage III) and migratory eels (stages IV and V) from different locations. Explanations for abbreviations of locations are given in Table 2.2. No significant differences in these variables ($P < 0.05$) were detected between locations

and T3 were not significant between yellow and silver eels. A regular increase of 11-KT during silvering, with a maximum value in November during the migratory season was significant (one-way ANOVA, $p < 0.05$). It appears that androgens like testosterone play a more indirect role in reproduction in female fishes. They are produced in response to gonadotropin by the thecal layers of the oocytes and the levels gradually increase during oocyte growth and peaks during postvitellogenic growth (Nagahama 1994). We can conclude from the well-marked seasonal pattern in female testosterone, which lags behind but follows female estradiol (E2), that there is a relationship between the two ($r^2 = 0.31$). This apparent relationship between the two steroids supports the possibility that testosterone may act as a precursor for E2 synthesis during the vitellogenic season via aromatizing activity. The increased E2 profile in the period September-November suggests that in the period of gonad development the aromatizing enzymes are partially stimulated (Fig. 2.7).

In conclusion it appears clearly that silvering begins during the summer months (July–August). In the Grevelingen, changes coincided with a decrease in photoperiod and temperature. Analyses of commercial silver eel fishery data from the Loire River showed that the onset of downstream migration was linked to light level in terms of photoperiod and sunshine hours and migratory movements started earlier during years with low light level (Durif and Elie 2008).

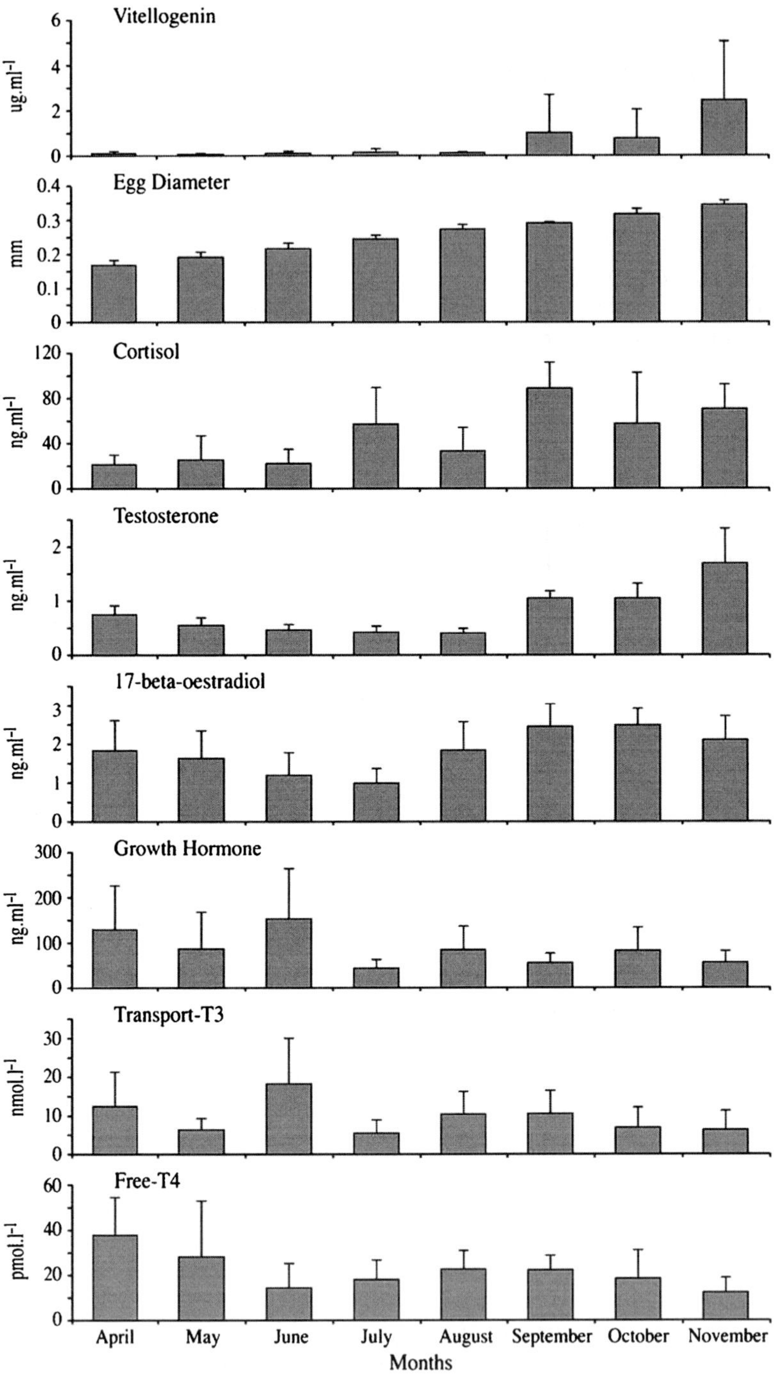

Fig. 2.7 Annual trends measured in eight female eels per month from Lake Grevelingen for maturation parameters and body hormones. April–July: yellow (sedentary stage) eels; August intermediate month; September–November: silver stage (migratory eels)

2.3.4 Age of Silver Eels

Age at silvering was obtained from most of the eels of the silvering stage study. The largest pair of otoliths (*sagittae*) was collected. Otoliths were later embedded in synthetic resin then polished to the nucleus with a polishing wheel. Etching was done using 10% EDTA. Yearly increments were revealed by staining with toluidine blue. Growth rings were then counted under a microscope. The age of individual eels was determined by the number of increments starting from the nucleus which was considered as year 1 of the eel's life.

Variability in age was extremely high. Resident eels (711 individuals) were on average 9.2 years old (min.: 3; max.: 21), pre-migrant eels (171 individuals) were 12 years old (min.: 6; max.: 22), and migrants (191 individuals) were 14.2 years old (min.: 5; max.: 24). The same variability existed within each location. Significant differences between locations were found within all stages (Fig. 2.8). Eels from the Loire were generally younger, especially resident eels with a mean age of 5.4, pre-migrant eels were 8.4, and migrant eels were 13.6. Landlocked eels of Lake Balaton were clearly on average the oldest eels obviously because they could not emigrate. Based on this, it could be hypothesised that eels would be older in large catchments where many obstacles were present. However this was not the case as migrating eels from smaller catchments such as the Nive and Ste Eulalie, were not younger on average then eels from the Rhine or the Loire (Fig. 2.8).

When body lengths of eels of the same age class were compared, migrant eels were always longer than resident eels indicating that migrating eels had benefited of a higher overall growth rate (Fig. 2.9). This suggests again that a high growth period precedes migration and therefore silvering.

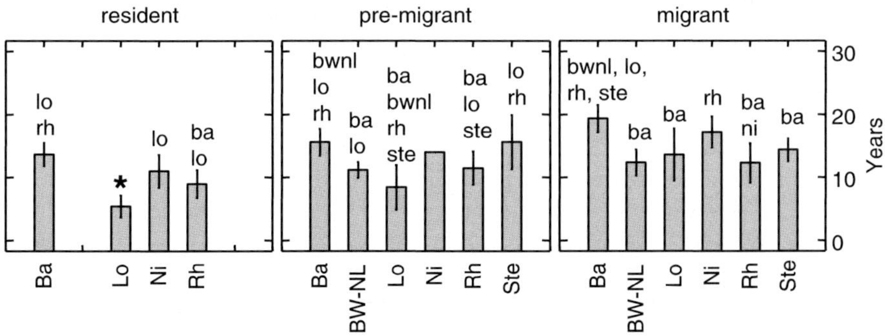

Fig. 2.8 Mean age (±SD) of resident (stages I and II), pre-migrant (stage III) and migrant eels (stages IV and V) from different locations: Lake Balaton (Ba), a brackish water lake in the Netherlands (BW-NL), Loire river (Lo), Nive river (Ni), Rhine river (Rh), Sainte-Eulalie river (Ste). Abbreviations (see Table 2.2 for explanation) on top of bars indicate significant differences (one-way ANOVA, $P < 0.05$) with these locations * indicates significant differences with all other locations (one-way ANOVA, $P < 0.05$)

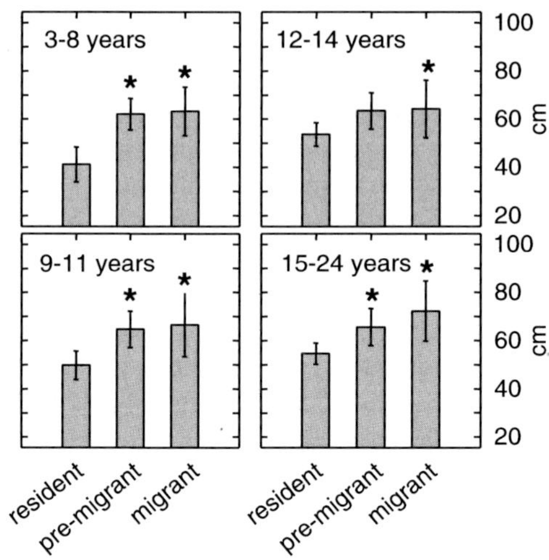

Fig. 2.9 Mean length (±SD) in mm of resident (stages I and II), pre-migrant (stage III) and migrant eels (stages IV and V) in different age groups. Age limits used to group eels were defined according to the 0.25, 0.50, and 0.75 quantiles of the age distribution of the sample. * significant difference with resident eels (one-way ANOVA, $P < 0.05$)

2.3.5 Changes in Skin Colour

Colour is the most obvious change during silvering. This criterion is most often used for stage determination (silver or yellow stage). Experienced fishermen or eel researchers can tell by its external features whether an eel is migratory or not. However, it is difficult for one to say which criteria are actually used in this rapid designation. Colour visualisation can be different from one person to another. The change in skin colour is more or less obvious. Countershading is not always present and tones are sometimes intermediate between yellow and a bronze color. We investigated to what degree these changes in colour were related to "maturity" of eels. A spectrophotometer (Minolta CM-508d) and CIELAB system were adjusted for reflectance, illuminant D65, and angle of 10°. Three measurements were taken on 194 individuals (comprising stages I, III, IV and V eels): on the back, on the belly, and just below the lateral line. Luminance (L^*) represents the degree of blackness (black = 0; white = 100). Red is represented by $+a^*$, green by $-a^*$, yellow by $+b^*$, and blue by $-b^*$.

Colour measurements on the back of eels could be directly linked to silvering. We found significant correlations (Pearson correlations, $P < 0.05$) between colour measurements on the back and GSI (gonado-somatic index), DTI (digestive tract index), and EI (eye index). $L_{back}{}^*$ was significantly correlated ($p < 0.05$) with GSI ($r^2 = 15$), DTI ($r^2 = 0.27$), and EI ($r^2 = 0.20$). Therefore, eels with a dark back (low L^*) had a high GSI, a low DTI, and a high EI. Significant correlations were also found with *b* values (yellow) on the back. Yellow decreased with GSI ($r^2 = 0.20$), and eye index

($r^2 = 0.24$), and increased with DTI ($r^2 = 0.36$). Values of a^* also showed significant correlations with DTI ($r^2 = 0.20$) and EI ($r^2 = 0.20$) indicating that regression of the digestive tract and eye size increased as "red" or bronze appeared on the back.

For each individual we evaluated the colour difference (ΔE) with a "yellow eel" using the Hunter-Scofield equation, which represents the distance in the *Lab* space between two colours: $\Delta E = ((\Delta L^*)^2 + (\Delta a^*)^2 + (\Delta b^*)^2)^{1/2}$. Control values for the "yellow eel" were obtained from averaging the $L^*a^*b^*$ values of stage I eels. Thus we obtained three values for each eel: ΔE_{back},$\Delta E_{lateral\ line}$ and ΔE_{belly}. Results showed that there was high individual variability within each stage. Only ΔE_{back} showed significant differences between stages (Kruskal-Wallis test, $p < 0.05$). The colour change was perceptible starting at stage IV since there was no significant difference between stage I and stage III eels. Stage IV showed the highest colour difference with "yellow eels" (Fig. 2.10).

2.3.6 Regression to a Yellow Stage

It is not known whether eels that are prevented from migrating go back to a "yellow" stage. Here, we investigated whether silver eels held captive beyond their normal migration, recovered "yellow" eel characteristics.

The first experimentation involved 104 female eels captured in November 2000 by a professional fisherman on the Loire River in France. Gonads were initially sampled on 29 individuals at the site of the commercial silver eel fishery. Mean GSI

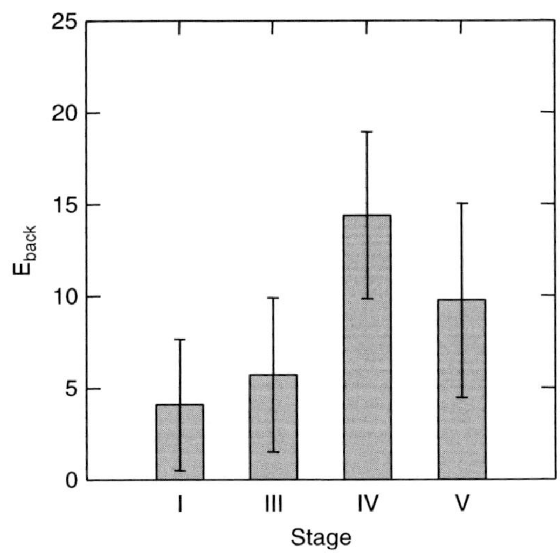

Fig. 2.10 Variations (mean ± SD) in colour differences measured on the back (δe_{back}) of different staged eels (resident: stage I, premigratory: stage III, and migratory stages IV and V)

of freshly caught migrating silver eels was equal to 1.8. The remaining 75 eels were placed in seawater tanks. After 2 months in the tank (January), GSI of eels showed a slight but significant decrease (Fig. 2.11). Colour values were still equivalent to measurements done at the fishery. In May (more than 6 months later), mean GSI was significantly lower (Mann and Whitney test $p < 0.05$), and it reached values corresponding to those found in yellow eels (lower than 1%). Skin colour also changed throughout the experiment. Values measured on the back of eels showed important variations at the very beginning of the experiment which were probably linked to handling of eels and to their change of environment: Luminance (L*)

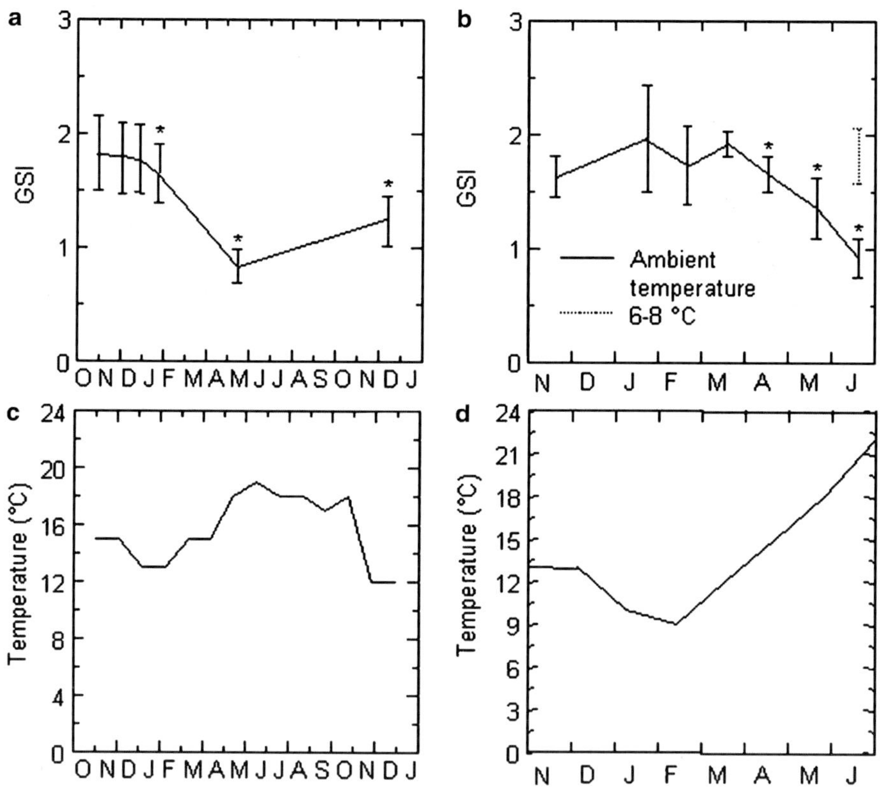

Fig. 2.11 Variations in GSI (±SD) of migratory eels which remained captive beyond their season of migration. Months are labeled on the x-axis. * indicates a significant difference from the value immediately before (Mann and Whitney test $P < 0.05$). (**a**) First set of experiments (November 2000–December 2001). Initial sampling was performed at the fishery (n = 29). Then eels were sacrificed at regular intervals (n = 46; n = 7; n = 7; n = 3; n = 13). (**b**) Second experimentation (November 2002–June 2003). Initial sampling was performed at the fishery (n = 22). Following samples respectively correspond to n = 5; n = 4; n = 5; n = 6; n = 6; n = 5 (ambient temperature). Eels kept at 6–8°C represent three individuals. (**c**) Mean monthly water temperature (2000–2001); (**d**) mean monthly water temperature (2002–2003)

and a* values both displayed significant changes (paired t-test, $p < 0.05$) when eels were placed in the tanks (Fig. 2.12a, b). Initial values were recovered 13 days later. In April, when GSI decreased, luminance (L^*) values started to increase, meaning that the colour of the back of eels was becoming lighter (Fig. 2.12a). The b^* value increased regularly until April, when eels appeared yellow, then remained constant until the last measurement at the end of September (Fig. 2.12c). The a^* value represents the bronze colour, typical of silver migrating eels. This parameter decreased significantly between April and May (paired t-test, $p < 0.05$). These events occurred when temperature was significantly higher (Fig. 2.11c) and equal to 18.3°C. The following December, eels displayed once again a significantly higher GSI, almost equivalent to that of silver migrants. No colour measurements were performed at that date.

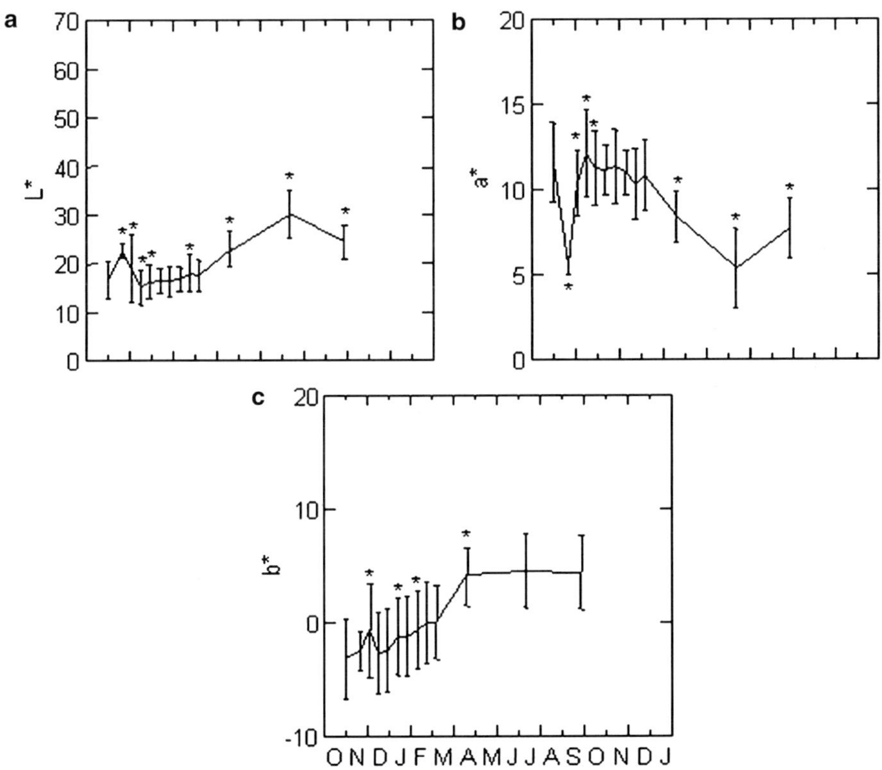

Fig. 2.12 Colour measurements performed on the back of eels during the first set of experiments (2000–2001). Values (mean ± SD) are given according to the CIE colour space. (**a**) L^* represents luminance and goes from 0 (black) to 100 (white); (**b**) a^* represents coordinates from green (negative values) to red (positive values); (**c**) b^* represents coordinates from blue (negative values) to yellow (positive values). *: Significant difference from the value immediately before (paired t-test, $P < 0.05$)

This experiment provides the first demonstration that silver eels prevented from migration may regress to yellow eels then "re-silver" the next year. Such silvering-regression-silvering cycles may likely occur in the wild when natural environmental conditions (such as climatic, hydrologic) or anthropogenic factors (migration barriers) may prevent migration.

In the second set of experiments (2002–2003), the mean GSI of freshly caught eels was equal to 1.6 (Fig. 2.11b). For eels kept at ambient temperature, this parameter started to decrease significantly (Mann and Whitney test $p < 0.05$) between March and April, thus approximately 5 months after their capture, and then again when temperature started to increase significantly. In June, eels had recovered yellow eel characteristics, as mean GSI was below 1%. At the same time, the GSI of eels maintained at 6–8°C had not significantly changed from its initial value (Mann and Whitney test $P < 0.05$) with a mean value was equal to 1.8%. However it was significantly different from eels kept at ambient temperature (Mann and Whitney test $P < 0.05$).

2.4 Conclusions

Silver eels come in all "shapes and sizes" and most often only the local fisherman will be able to tell whether an eel is truly migratory. The question here was whether these visible external differences really correspond to physiological differences. Do eels at the "migratory stage" all display the same morphological and physiological characteristics and can some of the variability be explained by giving a finer description of the silvering process?

Actively migrating eels display equivalent physiological characteristics regardless of their origin. The final stage of silvering (stage V) is reached while eels are actively swimming. Gonads of silver eels that are prevented from migrating (landlocked, migration barriers) are not as developed. Swimming also further induces regression of the digestive tract. Eels may "look" silver by their colour but still have low GSI (gonado-somatic index) and no regression of the digestive tract, such as farmed individuals which do not go beyond the pre-migrant stage III (probably due to continuous feeding, high water temperature, and low swimming activity).

There is now evidence that silver eels may revert to a yellow stage. GSI decreases and eels take on a yellow colour again. This occurs with increasing temperature, whereas eels kept at low temperature remain silver. Therefore, eels blocked or disturbed during their migration (by dams or river obstacles) may resume a sedentary life and wait for another year to pursue migration. In northern countries, where temperature remains low well into the spring, this may be a chance for a longer migration period and possible spring downstream runs. The role of temperature is probably also in the start of the silvering process. Decreasing temperature at the end of summer may be an environmental trigger for the silvering process. Two studies have shown that downstream migration starts earlier during cold summers (Vøllestad et al. 1986; Durif and Elie 2008).

The variability in age and size of female silver eels is tremendous. Female eels may start to silver when they have reached 50 cm in body length; however it is clearly more advantageous to attain a bigger size: energy stores are greater, swimming capacity is higher, and fecundity is increased. Because males successfully migrate at a small size (when they are 35–46 cm), it is obviously the fecundity issue that drives certain females to wait until they have reached a bigger size. The optimal size in terms of reproductive potential would be 70 cm (Durif et al. 2006). It is not clear still why the "decision" to silver is made at a certain time in the eel's life. The proportion of large eels is higher in the upstream parts of catchments, but they are also present nearer the sea (such as in the Grevelingen). The variability in morphological and physiological characteristics of silver eels decreases with the size of the catchment (Durif et al. 2005). Therefore it is probably related to diversity of habitat. There is now evidence that the timing of silvering within the eel's life cycle is closely related to somatic growth. In this chapter we have seen that all eels at the migratory stage showed a higher growth rate than resident eels regardless of their origin. Somatotropin was found to be highest in pre-migrant eels. It is well known that farmed animals, which have a high growth rate show early on external signs of silvering (eye size and colour). A higher growth period must be the signal for eels to silver. Depending on the habitat and growth conditions this may take more or less time; therefore eels in low productivity environments will take more time and consequently reach a bigger size. High productivity habitats will yield less fecund females but faster, while poorer environments will more slowly yield large fecund females. Eels appear to adopt both strategies. A reduction in the available habitat will greatly reduce possibilities for eels to choose between strategies and thus further decrease their already depleted population.

Acknowledgements This work has been supported by the European contract EELREP N°Q5RS-2001-01836.

References

Aroua S, Schmitz M, Baloche S, Vidal B, Rousseau K, Dufour S (2005) Endocrine evidence that silvering, a secondary metamorphosis in the eel, is a pubertal rather than a metamorphic event. Neuroendocrinology 82:221–232

Barni S, Bernocchi G, Gerzeli G (1985) Morphohistochemical changes in hepatocytes during the life cycle of the European eel. Tissue & Cell 17:97–109

Bergersen R, Klemetsen A (1988) Freshwater eel Anguilla anguilla L. from North Norway with emphasis on occurrence, food, age and downstream migration. Nordic Journal of Freshwater Research 64:54–66

Bertin L (1951) Les anguilles. Variation, croissance, euryhalinité, toxicité, hermaphrodisme juvénile et sexualité, migrations, métamorphoses. Payot, Paris

Beullens K, Eding EH, Gilson P, Ollevier F, Komen J, Richter CJJ (1997) Gonadal differentiation, intersexuality and sex ratios of European eel (Anguilla anguilla L.) maintained in captivity. Aquaculture 153:135–150

Bienarz K, Epler P, Malczewski B, Passakas T (1981) Development of European eel (Anguilla anguilla L.) gonads in artificial conditions. Aquaculture 22:53–66

Boucher-Firly S, Fontaine M (1933) Rapports existant entre quelques stades de développement de l'Anguille et le point de congélation de son sérum lors du passage en eau de mer. Compte rendu des seances de la Societe de Biologie, Paris 112:462

Burzawa-Gérard E, Baloche S, Leloup-Hatey J, Le Menn F, Messaouri H, Nunez-Rodriguez J, Peyon P, Roger C (1994) Ovogénèse chez l'anguille (Anguilla anguilla L.): ultrastructure de l'ovaire à différents stades de développement et implication des lipoprotéines au cours de la vitellogenèse. Bulletin Francais de la Pêche et de la Pisciculture 335:213–233

Callamand O, Fontaine M (1942) Activité thyroïdienne de l'Anguille au cours de son développement. Archives de Zoologie Expérimentale et Générale 82:129–136

Carlisle DB, Denton EJ (1959) On the metamorphosis of the visual pigments of Anguilla anguilla (L.). Journal of the Marine Biological Association of the United Kingdom 38:97–102

Colombo G, Grandi G, Rossi R (1984) Gonad differentiation and body growth in Anguilla anguilla L. Journal of Fish Biology 24:215–228

Dollerup J, Graver CM (1985) Repeated induction of testicular maturation and spermiation, alternating with periods of feeding and growth in silver eels, Anguilla anguilla (L.). Dana 4:19–39

Dufour S, Delerue-Le Belle N, Fontaine YA (1983a) Development of a heterologous radioimmunoassay for eel (Anguilla anguilla) gonadotropin. General and Comparative Endocrinology 49:403–413

Dufour S, Le Belle ND, Fontaine YA (1983b) Effects of steroid hormones on pituitary immunoreactive gonadrotropin in European freshwater eel, Anguilla anguilla L. General and Comparative Endocrinology 52:190–197

Dufour S, Lopez E, Le Belle N, Baloche S, Fontaine YA (1988) Stimulation of gonadotropin release and of ovarian development by the administration of a gonadoliberin agonist and of dopamine antagonists, in female silver eel pretreated with estradiol. General and Comparative Endocrinology 70:20–30

Dufour S, Bassompierre M, Montero M, Le Belle N, Baloche S, Fontaine YA (1991) Stimulation of pituitary gonadotropic function in female silver eel treated by a gonadoliberin agonist and dopamine antagonist. In: Scott AP, Sumpter JP, Kime DE, Rolfe MS (eds) Reproductive Physiology of Fish. FishSymp 91, Sheffield, pp. 54–60

Durif C, Dufour S, Elie P (2005) The silvering process of Anguilla anguilla: a new classification from the yellow resident to the silver migrating stage. Journal of Fish Biology 66:1025–1043

Durif C, Guibert A, Elie P (in press) Morphological discrimination of the silvering stages of the European eel. In: Casselman JM, Cairns DK (eds) Eels at the Edge: Science, Status, and Conservation Concerns. Proceedings of the 2003 International Eel Symposium. American Fisheries Society Symposium Publication, Bethesda, Maryland

Durif CMF, Elie P (2008) Prediction of downstream migration of silver eels in a large river catchment based on commercial fishery data. Fisheries Management and Ecology 15:127–137

Durif CMF, Dufour S, Elie P (2006) Impact of silvering stage, age, body size and condition on the reproductive potential of the European eel. Marine Ecology Progress Series 327:171–181

Dutil JD, Besner M, McCormick SD (1987) Osmoregulatory and ionoregulatory changes and associated mortalities during the transition of maturing American eels to a marine environment. American Fisheries Society Symposium 1:175–190

Epstein FH, Cynamon M, McKay W (1971) Endocrine control of Na-K-ATPase and seawater adaptation in Anguilla rostrata. General and Comparative Endocrinology 16:323–328

Es-Souni A, Ali MA (1986) Ultrastructure of the retinal pigmented epithelium of light and dark adapted young pigmented and mature silver eels Anguilla anguilla. Zoomorphology 106:179–184

Evans DH (ed) (1993) The Physiology of Fishes, Vol. CRC Press, Boca Raton, FL

Fontaine M (1975) Physiological mechanisms in the migration of marine and amphihaline fish. Advances in Marine Biology 13:241–355

Fontaine M (1985) Action de facteurs anormaux du milieu sur l'écophysiologie d'anticipation des poissons migrateurs amphihalins. Ichtyophysiologica acta 9:11–25

Han Y-S, Liao I-C, Tzeng W-N, Yu JY-L (2004) Cloning of the cDNA for thyroid stimulating hormone beta subunit and changes in activity of the pituitary- thyroid axis during silvering of the Japanese eel, Anguilla japonica. Journal of Molecular Endocrinology 32:179–194

Holmgren K, Wickström H, Clevestam P (1997) Sex-related growth of European eel, Anguilla anguilla, with focus on median silver eel age. Canadian Journal of Fisheries and Aquatic Sciences 54:2775–2781

Huang YS, Rousseau K, Le Belle N, Vidal B, Burzawa-Gérard E, Marchelidon J, Dufour S (1998) Insulin-like growth factor-I stimulates gonadotropin production from eel pituitary cells: a possible metabolic signal for induction of puberty. Journal of Endocrinology 159:43–52

Huang YS, Sbaihi M, Le Belle N, Dufour S (2001) Combination of testosterone with cortisol stimulates the silvering of female european eel (*Anguilla anguilla L.*). In: Yu JYL (ed) Proceedings of 4th Congress of the Asia and Oceania Society for Comparative Endocrinology. Institute of Zoology, Academia Sinica, Taipei, Taiwan, pp 215–222

Kleckner RC (1980) Swimbladder wall guanine enhancement related to migratory depth in silver phase Anguilla rostrata. Comparative Biochemistry and Physiology 65:351–354

Kleckner RC, Krueger WH (1981) Changes in swimbladder retial morphology in Anguilla rostrata during premigration metamorphosis. Journal of Fish Biology 18:569–577

Larsson P, Hamrin S, Okla A (1990) Fat content as a factor inducing migratory behaviour in the eel (Anguilla anguilla L.) to the Sargasso Sea. Naturwissenschaften 77:488–490

Le Belle N, Fontaine M (1987) Les anguilles mâles et femelles peuvent-elles frayer une seconde fois dans la nature? Ichtyophysiol Acta 11:2–5

Leloup-Hâtey J (1964) Fonctionnement de l'interrénal antérieur de deux Téléostéens: le Saumon atlantique et l'Anguille européenne. Annales de L'Institut Océanographique 42:224–337

Locket NA (1975) Some problems of deep-sea fish eyes. In: Ali MA (ed) Vision in Fishes. Plenum Press, New York/London, pp. 645–656

Lopez E, Fontaine YA (1990) Stimulation hormonale, in vivo, de l'ovaire d'anguille européenne au stade jaune. Reproduction Nutrition Development 30:577–582

Messaouri H, Baloche S, Hardy A, Leloup-Hatey J, Burzawa-Gérard E (1991) Hepatic 17β-estradiol receptors and vitellogenesis in Anguilla anguilla L. after treatment by 17β-estradiol alone or associated with bovine growth hormone. General and Comparative Endocrinology 82:238

Nagahama Y (1994) Endocrine regulation of gametogenesis in fish. International Journal of Developmental Biology 38:217–229

Pankhurst NW (1982a) Changes in body musculature with sexual maturation in the European eel. Journal of Fish Biology 21:417–428

Pankhurst NW (1982b) Relation of visual changes to the onset of sexual maturation in the European eel Anguilla anguilla L. Journal of Fish Biology 21:127–140

Pankhurst NW, Lythgoe JN (1983) Changes in vision and olfaction during sexual maturation in the European eel Anguilla anguilla (L.). Journal of Fish Biology 23:229–240

Pankhurst NW, Sorensen PW (1984) Degeneration of the alimentary tract in sexually maturing European *Anguilla anguilla* (L.) and American eels *Anguilla rostrata* (LeSueur). Canadian Journal of Zoology 62:1143–1148

Peyon P, Baloche S, Burzawa-Gerard E (1996) Potentiating effect of growth hormone on vitellogenin synthesis induced by 17 beta-estradiol in primary culture of female silver eel (Anguilla anguilla L.) hepatocytes. General and Comparative Endocrinology 102:263–273

Poole WR, Reynolds JD (1996) Growth rate and age at migration of Anguilla anguilla. Journal of Fish Biology 48:633–642

Rohr DH, Lokman PM, Davie PS, Young G (2001) 11-Ketotestosterone induces silvering-related changes in immature female short-finned eels, Anguilla australis. Comparative Biochemistry and Physiology A Comparative Physiology 130:701–714

Sorensen PW, Pankhurst NW (1988) Histological changes in the gonad, skin, intestine and olfactory epithelium of artificially-matured male American eels, Anguilla rostrata (LeSueur). Journal of Fish Biology 32:297–307

Stramke D (1972) Veränderungen am Auge des Europäischen Aales (*Anguilla anguilla*, L.) wärhrend des Gelb- und Blankaalphase. Institut für Hydrobiologie und Fischereiwissenschaft der Universität Hamburg 23:101–117

Svedäng H, Wickstrom H (1997) Low fat contents in female silver eels: Indications of insufficient energetic stores for migration and gonadal development. Journal of Fish Biology 50:475–486

Svedäng H, Neuman E, Wickström H (1996) Maturation patterns in female European eel: age and size at the silver eel stage. Journal of Fish Biology 48:342–351

Swift DR (1960) Cyclical activity of the thyroid gland of fish in relation to environmental changes. Symposia of the Zoological Society of London 2:17–27

Tesch FW (2003) The Eel. Fifth edition, Vol. Blackwell, Oxford

Utida S, Hirano T, Oide H, Ando M, Johnson DW, Bern HA (1972) Hormonal control of the intestine and urinary bladder in teleost osmoregulation. General and Comparative Endocrinology 3:317–327

Van Ginneken V, Durif C, Balm SP, Boot R, Verstegen MWA, Antonissen E, Van den Thillart G (2007a) Silvering of European eel (Anguilla anguilla L.): seasonal changes of morphological and metabolic parameters. Animal Biology 57:63–77

Van Ginneken V, Durif C, Dufour S, Sbaihi M, Boot R, Noorlander K, Doornbos J, Murk AJ, Van den Thillart G (2007b) Endocrine profiles during silvering of the European eel (Anguilla anguilla) living in saltwater. Animal Biology 57:453–465

Van Ginneken VJT, Van den Thillart GEEJM (2000) Eel fat stores are enough to reach the Sargasso. Nature 403:6766

Vøllestad LA (1992) Geographic variation in age and length at metamorphosis of maturing European eel: environmental effects and phenotypic plasticity. Journal of Animal Ecology 61:41–48

Vøllestad LA, Jonsson B (1986) Life-history characteristics of the European eel Anguilla anguilla in the Imsa River, Norway. Transactions of the American Fisheries Society 115:864–871

Vøllestad LA, Jonsson B, Hvidsten NA, Naesje TF, Haralstad O, Ruud-Hansen J (1986) Environmental factors regulating the seaward migration of European silver eels (Anguilla anguilla). Canadian Journal of Fisheries and Aquatic Sciences 43:1909–1916

Yamada Y, Zhang H, Okamura A, Tanaka S, Horie N, Mikawa N, Utoh H, Oka P (2001) Morphological and histological changes in the swim bladder during maturation of the Japanese eel. Journal of Fish Biology 58:804–814

Zacchei AM, Tavolaro P (1988) Lateral line system during the life cycle of Anguilla anguilla. Bollettino di Zoologia 3:145–153

Chapter 3
Silvering: Metamorphosis or Puberty?

Karine Rousseau, Salima Aroua, Monika Schmitz,
Pierre Elie, and Sylvie Dufour

Abbreviations 11-KT: 11-Ketotestosterone; CRH: Corticotropin releasing hormone; E2: Estradiol; FSH: Follicle-stimulating hormone; FSHβ: β subunit of FSH; GPα: Common subunit of gonadotropins; GH: Growth hormone; GRIF: Gonadotropin release-inhibiting factor; GSI: Gonadosomatic index; GTH: Gonadotropins; KClO4: Potassium perchlorate; LH: Luteinizing hormone; LHβ: β subunit of LH; PTU: Propythiouracil; SRIH: Somatostatin; T: Testosterone; T3: Triiodothyronine; T4: Thyroxine; TH(s): Thyroid hormone(s); TSH: Thyroid stimulating hormone; thyrotropin; Vg: Vitellogenin

3.1 Introduction

Eels have a complex migratory life cycle with the occurrence of two metamorphoses (for reviews see: Sinha and Jones 1975; Tesch 1977; Haro 2003). They represent a typical larval (first) metamorphosis, the leptocephalus larva being transformed into the glass eel. After this drastic transformation, the growth phase starts in the continental waters and glass eels become "yellow" eels. After many years in freshwater, the yellow eels transform into "silver" eels which stop growing and start their downstream migration towards the ocean and the area of reproduction. To allow this transition from sedentary life in freshwater to migrant life in seawater, eels undergo their second metamorphosis, known as silvering. Silvering not only preadapts the eel to deep-sea conditions (Sébert 2003), but also prepares for sexual maturation, which will only be completed during the oceanic migration towards the Sargasso Sea (Tesch 1982, 1989; Dufour and Fontaine 1985; Fontaine 1985; Dufour 1994). At the silver stage, eels are

K. Rousseau, S. Aroua, and S. Dufour
Museum National d'Histoire Naturelle, DMPA, UMR CNRS 5178 "Biology of Marine Organisms and Ecosystems", 7 rue Cuvier, CP 32, 75231 Paris cedex 05, France

M. Schmitz
Department of Biology, Karlstad University, S-651 88 Karlstad, Sweden

P. Elie
Unité Ressources Aquatiques Continentales, Cemagref-Bordeaux, 33612 Cestas Cedex, France

G. van den Thillart et al. (eds.), *Spawning Migration of the European Eel*,

blocked at a prepubertal stage (Dufour et al. 2003) but chronic administration of carp or salmon pituitary extract (gonadotropic treatment) is able to induce sexual maturation (Fontaine et al. 1964; Fontaine 1975). The lack of sexual maturation at the silver stage is due to a deficient production of pituitary gonadotropins (GtHs) (Dufour et al. 1983). Stimulation of synthesis and release of pituitary gonadotropin in the silver eel can be induced by combined treatment with a GnRH-agonist and a dopamine-antagonist, indicating that a double neuroendocrine blockade was responsible for the deficient pituitary gonadotropic function: a lack of endogenous stimulation by GnRH due to a deficient production of GnRH and a strong dopaminergic inhibition of GnRH action (Dufour et al. 1988, 1991; Vidal et al. 2004). At the yellow stage, in contrast, gonadotropic treatment has no significant effect on both gonad weight (Leloup-Hatey et al. 1989) and body color (Lopez and Fontaine 1990).

Eel silvering has always been considered as a metamorphosis. This is probably due to the great similarities which exist between the morphological changes observed during silvering and the ones occurring during smolting in salmonids. However, unlike smolting, silvering also includes some changes related to an onset of sexual maturation such as gonad development. These observations led to the hypothesis that silvering could correspond to a pubertal event. Indeed, puberty encompasses various morpho-physiological and behavioral changes, which, unlike metamorphic changes, are induced by sexual steroids (for review, see Romeo 2003). While metamorphosis is mostly triggered in vertebrates by the activation of the thyrotropic neuroendocrine axis, puberty is triggered by the activation of the gonadotropic axis. Our work aimed at investigating whether eel silvering should be considered as a "true" metamorphosis or as an onset of puberty, by comparing the profiles of pituitary and peripheral hormones during the transition from yellow to silver eels. This chapter reviews the morphological changes and describes the hormonal patterns occurring during eel silvering.

3.2 Definitions and Neuroendocrine Control of Metamorphosis and Puberty in Vertebrates

Metamorphosis and puberty are two major events of the postembryonic development in vertebrates. While they present common features regarding morphological and behavioural modifications, they are controlled by different neuroendocrine axes.

3.2.1 Metamorphosis

Metamorphosis allows the transition from one developmental stage in a specific environment (ecophase 1) to the next stage in a different environment (ecophase 2), and includes a migration between the two habitats. Metamorphosis corresponds to drastic changes in body shape, physiology and behaviour, and, unlike puberty, is encountered only in a few phyla/species.

In vertebrates, the most described metamorphosis is the transformation in amphibians of the aquatic larva (tadpole) into the terrestrial juvenile. Larval metamorphosis is also encountered in some other vertebrates, such as lampreys and some teleosts (elopomorphes and pleuronectiforms). In teleosts, different types of metamorphoses can be observed (Fig. 3.1). Youson (1988) proposed three pathways of ontogeny between the embryo and the adult: direct development from the embryo period leading to the juvenile and adult period (many examples such as goldfish); a typical indirect development involving a larval metamorphosis (first or "true" metamorphosis) leading to the juvenile period (example: flatfishes and eels); and a non-classical metamorphosis occurring during the juvenile period (example: the secondary metamorphosis, smolting, in salmons). Compared to larval metamorphosis, less drastic morphological, physiological and behavioural changes occur in juveniles of some migratory teleosts and are referred to as "secondary metamorphoses". This is the case of smolting in salmons and silvering in eels.

3.2.1.1 Larval Metamorphosis

Gudernatsch, in 1912, demonstrated for the first time the stimulatory role of the thyroid gland in the control of metamorphosis in amphibians. Indeed, when feeding tadpoles with thyroid gland extracts, he observed that their transformation to frogs was accelerated. Conversely, Allen (1916) was able to completely prevent metamorphosis by thyroidectomy. Since these experiments, the positive role of thyrotropic axis in the control of larval metamorphosis has been further studied in amphibians. The pair of thyroid glands is first detectable after embryogenesis when the tadpole

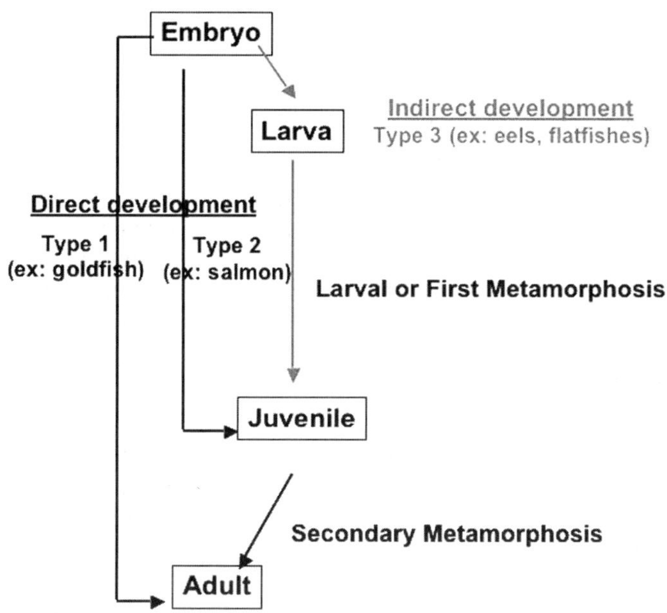

Fig. 3.1 Types of development in fish (adapted from Youson 1988)

begins to feed. Under the control of pituitary thyrotropin (TSH), these glands produce thyroid hormones (THs: thyroxine, T4 and triidothyronine, T3), which act on target organs via specific receptors. It was shown that as the thyrotropic axis was activated, a series of sequential morphological transformations occurred. An early change is the growth and differentiation of the limbs which in the absence of hormone still form, but will not progress beyond the bud stage. The final morphological change, tail resorption, occurs when the level of THs is highest at the climax of metamorphosis (for reviews: Kanamori and Brown 1996; Tata 2006).

Thyrotropin (TSH), a pituitary glycoprotein hormone, belonging to the same family as gonadotropins, controls the production of THs in amphibians (Manzon and Denver 2004), as classically shown in mammals (for review: McNabb 1992). Early studies of hypophysectomy and immunization demonstrated that TSH played a central role in amphibian metamorphosis (Dodd and Dodd 1976; Eddy and Lipner 1976). Indeed, Dodd and Dodd (1976) showed that the negative effect of hypophysectomy prior to metamorphosis could be reversed by treatment with mammalian TSH. Furthermore, passive immunization of tadpoles with an antiserum to bovine TSH prevented spontaneous metamorphosis (Eddy and Lipner 1976). Morphological and biochemical changes observed during metamorphosis, such as complete regression of tail and gills, *de novo* formation of bone, visual pigment transformation or functional differentiation of liver, can be induced by THs (for review: Tata 1998; Shi 1999). In contrast to the situation in mammals, in which the brain peptide discovered for its stimulatory control on TSH is TRH (for Thyrotropin Releasing Hormone) (Morley 1981), the brain neurohormone responsible for the activation of TSH production during amphibian metamorphosis is corticotropin-releasing hormone (CRH) and not TRH (for review: Denver 1999). In fact, in amphibians, the production and release of TSH by the pituitary appears to be regulated by different neuropeptides according to the life stage. Indeed, in premetamorphic amphibians, stimulation of the pituitary-thyroid axis is only responsive to CRH-like peptides and unresponsive to all other neurohormones tested, including TRH (Denver and Licht 1989; Denver 1993). Thus, CRH is the thyrotropin-releasing factor during the induction of metamorphosis in tadpoles, and the stimulatory action of TRH on TSH secretion develops after metamorphic climax (for review: Denver 1999; Tata 2006).

Although THs are the only obligatory signal for the initiation and completion of amphibian metamorphosis, other hormones can modulate the onset and progression of metamorphosis (Tata 2006). These include glucocorticoids and prolactin (PRL), which can accelerate and prevents TH-induced metamorphosis, respectively (for reviews: White and Nicoll 1981; Kikuyama et al. 1993).

In fishes, the involvement of the thyroid axis in larval metamorphosis (primary metamorphosis) was reported by Murr and Sklower (1928) who showed an increase in thyroid gland development of European eel leptocephalus larvae; this observation was further confirmed in another Anguilliform species, *Conger myriaster* (Kubota 1961; Yamano et al. 1991). This activation of the thyroid gland was also observed during larval flounder metamorphosis, *Paralichtys olivaceus* (Miwa and Inui 1987b). In addition, measurements of thyroid hormone body content indicated a significant increase of T3 and T4 in early stage of larval metamorphosis in flounder (Miwa et al. 1988; Tagawa et al. 1990) as well as in conger eel (Yamano et al. 1991). TH treatment induces flounder metamorphosis at both the whole body level (Inui and Miwa 1985; Miwa et al. 1988), and at the tissue level: erythrocytes (Miwa and Inui 1991), skeletal muscle (Yamano et al. 1991), gastric glands (Miwa et al. 1992; Huang et al. 1998; Soffientino and Specker 2001), chloride cells in the gill (Schreiber and Specker 2000) and bone remodelling for eye relocation (Solbakken et al. 1999; Okada et al. 2003). Experiments with TH deficiency led to the inhibition of pigmentation on the left side of the body and the inhibition of bone remodelling and right eye relocation (Inui and Miwa 1985; Miwa and Inui 1987a; Schreiber AM and Specker JT (1998); Okada et al. 2003). TH deficiency also inhibits the development of the anal fin pterygiophore and the body height reduction, the absorption of the dorsal fin ray, the formation of actinost and distal radials of the pectoral fin, and the pectoral fin shrinkage (Inui and Miwa 1985; Miwa and Inui 1987a; Okada et al. 2003). Both T4 and T3 directly stimulated, *in vitro*, the shortening of the cultured flounder fin rays isolated from prometamorphic larvae, and cortisol synergized with thyroid hormones (de Jesus et al. 1990). In contrary, both estradiol and testosterone attenuate the response of the fin rays to T3 *in vitro* (de Jesus et al. 1992). Similarly, when prometamorphic flounder larvae are immersed in solutions containing steroids, fin ray shortening is delayed, as well as eye migration and the development of benthic preference (de Jesus et al. 1992). Prolactin also antagonizes TH action *in vitro* and *in vivo* on fin ray shortening (de Jesus et al. 1993). Exogenous thyroid hormone has also been shown to stimulate metamorphoses in leptocephali of *Conger myriaster* (Kitajima et al. 1967) and of *Anguilla anguilla* (Vilter 1946), confirming the major role of THs in the induction of elopomorph larval metamorphosis. The role of THs would be crucial up to the final steps of glass eel metamorphosis. A central role for THs in the regulation of mechanisms leading to the colonization of continental habitats by *Anguilla anguilla* glass eels was first hypothesized by Fontaine and Callamand (1941) and demonstrated later by Edeline et al. (2004, 2005). Indeed, in European glass eels, river-colonizers exhibit an increased thyroid status compared to estuarine migrants (Edeline et al. 2004) and locomotor activity of glass eels was significantly increased by treatment with T4, while it was decreased by thiourea (Edeline et al. 2005). A correlation between decreasing T4 levels and the transition from pelagic toward benthic behaviour was shown in glass eels *A. anguilla* held in an aquarium (Jegstrup and Rosenkilde 2003). These data demonstrate the involvement of thyroid hormones as a major determinant of metamorphosis induction in eels, as in flatfish. Further studies are clearly needed to investigate the potential synergistic or antagonistic roles of other hormones as well as to determine the brain-pituitary control of thyroid function in teleosts.

3.2.1.2 Secondary Metamorphosis (Smolting)

Hyperactivity of the thyroid gland was also demonstrated by histological studies during the parr – smolt transition (secondary metamorphosis) of Atlantic salmon,

Salmo salar (Hoar 1939). The involvement of the thyroid axis in the control of smolting was later confirmed by Fontaine and Leloup (1960) by physico-chemical studies and more recently reviewed by Boeuf (1993). The administration of exogenous TH to juvenile parr – status salmonids results in morphological and physiological changes consistent with the parr – smolt transformation (for review: Boeuf 1993). For example, treatments with THs or with TSH have been able to induce the characteristic color change (silvering) of the skin, via an effect on purine and pteridine deposition (Chua and Eales 1971; Premdas and Eales 1976; Miwa and Inui 1985; Coughlin et al. 2001). THs are also able to induce the metabolic and muscular changes necessary for the oceanic migration. For instance, treatment of coho salmon, *Oncorhynchus kisutch*, with T3 in the diet accelerated the increase in concentration of adult forms of blood haemoglobin, while dietary PTU reduced this increase during smolting (for review: Hoar 1988). Thyroid hormones induce a shift towards slower isoforms of the muscle protein myosin heavy chain (Coughlin et al. 2001), and increased muscle contraction and maximum force (Katzman and Cech 2001). Moreover, when anti-thyroid drugs such as PTU were administered to smolting coho salmon, there was impairment of the body colour and retinal pigmentary changes (Sullivan et al. 1987; Alexander et al. 1998). Concerning olfaction, an early study showed that artificially increased TH levels induced parr to imprint to artificial odorants, while parr with unaltered hormone levels did not (Scholz 1980). More recently, using intraperitoneal implants of T3 for 16–20 days to mimic smolting, Lema and Nevitt (2004) demonstrated that T3 could induce olfactory cellular proliferation in juvenile coho salmon. Finally, the involvement of THs in stimulation of migratory behaviour (loss of positive rheotaxis and territoriality, and start of schooling) has been well documented (Baggerman 1963; Godin et al. 1974; Fontaine 1975; Youngson et al. 1985; Iwata et al. 1989; Boeuf and Le Bail 1990).

In conclusion, thyroid hormones are clearly involved in smolting. However, they are probably not sufficient to induce all the smolting-related changes. Other hormones such as GH and cortisol may be required for their synergizing action with THs in the control of osmoregulation and metabolism during the parr-smolt transformation. For example, administration of GH clearly improves hypo-osmoregulatory ability and seawater survival of parr (for review: Donaldson et al. 1979; Boeuf 1993), and induces other smolting-related changes, such as condition factor and skin pigmentation (for review: Donaldson et al. 1979).

A study in adult salmon suggests a role for CRH and/or TRH in the control of TSH, with variations possibly depending on species or physiological status (Larsen et al. 1998). Further studies are clearly needed to investigate which brain neuromediator is specifically implicated in the triggering of smolting.

3.2.2 Puberty

The term of puberty, first defined in humans, has been subsequently extended to the first acquisition of the capacity to reproduce in all mammalian and non-mammalian vertebrates as well as in invertebrates. By definition, puberty occurs in all species but only once in the life cycle.

In all vertebrates, puberty is triggered by the activation of the gonadotropic axis, constituted of a brain neuropeptide (gonadotropin-releasing hormone, GnRH), pituitary glycoprotein hormones (gonadotropins: luteinizing hormone, LH and follicle stimulating hormone, FSH) and gonadal steroids. Sex steroids induce the morpho-physiological and behavioural transformations characteristic of puberty.

The scheme (GnRH/LH-FSH/sex steroids) is largely conserved among vertebrates (Okuzawa 2002), even though additional controls may occur such as the dopaminergic inhibition of gonadotropin production in some teleosts (for reviews: Dufour et al. 2005; Pasqualini et al. 2004). Indeed, pioneer works from Peter and collaborators on goldfish, using hypothalamic lesions, demonstrated the existence of a GRIF (gonadotropin release-inhibiting factor) (Peter and Crim 1978; Peter et al. 1978; Peter and Paulencu 1980). Subsequent studies using agonists or antagonists *in vivo* (Chang and Peter 1983; Chang et al. 1984), primary culture of pituitary cells *in vitro* (Chang et al. 1990) and immunocytochemistry (Kah et al. 1984, 1986, 1987) provided evidence that GRIF was dopamine (DA). An inhibitory role of DA on the control of LH has been demonstrated in many adult teleosts at the time of ovulation and spermiation (catfish: De Leeuw et al. 1986; Van Asselt et al. 1988; coho salmon: Van der Kraak et al. 1986; rainbow trout: Linard et al. 1995; Saligaut et al. 1999; common carp: Lin et al. 1998; tilapia: Yaron et al. 2003). However, DA does not play an inhibitory role in all adult teleosts (Atlantic croaker: Copeland and Thomas 1989; gilthead seabream: Zohar et al. 1995). Concerning the early stages of gametogenesis and the control of puberty, the possible involvement of DA has only been studied, up to now, in bass, bream eel and mullet. In juvenile striped bass (Holland et al. 1998) and red seabream (Kumakura et al. 2003), data refuted a role for DA in the prepubertal control of gonadotropins, as GnRH alone was able to trigger precocious puberty. In contrast, in European eel, only a triple treatment with testosterone, GnRH agonist and pimozide (DA D2-type receptor antagonist) could induce increases in LH synthesis and release, indicating that removal of DA inhibition is required in the prepubertal eel for triggering GnRH-stimulated LH synthesis and release (Dufour et al. 1988; Vidal et al. 2004). A recent study in the grey mullet has demonstrated that D2 type receptor expressions in the brain and in the pituitary were high at the early and intermediate stages of puberty (Nocillado et al. 2007), when inhibition of the reproductive function by DA is particularly pronounced (Aizen et al. 2005).

3.3 "Metamorphic" and "Pubertal" Physiological Changes During Eel Silvering

During the silvering process, in addition to important behavioural changes such as the transition from a sedentary to a migrating behavior and the start of starvation, drastic morphological changes are observed. An extensive description of silvering morphometric changes is given in Chapter 2. Downstream migration is a common behavioural feature of silvering and smolting and is discussed in other chapters of this book. The capacity to adapt to seawater is also observed during these two "metamorphoses", but is less drastic during silvering compared to smolting, as yellow eels are already able to survive in salt water (Chapter 6). Modifications of swim bladder, which enable fish to live in deep water are reported and described in Chapter 5. Here, we briefly describe modifications, which concern organs (skin, eyes, gonads, metabolism and digestive tract) on which hormonal control has been studied.

3.3.1 Skin and Eyes

During the sedentary growth phase, the back of the eel is greenish-brown and the ventral surface is yellow, while at the time of the oceanic migration of reproduction, they are respectively blackish-brown and silvery-white. The silvering of the skin is thought to be the result of an increase of dorsal melanin and/or ventral purines (Bertin 1951; Pankhurst and Lythgoe 1982). This phenomenon is also observed in other migratory teleosts such as salmonids, where an increase in purine deposition in skin was observed during the parr-smolt transformation (Staley and Ewing 1992) and would correspond to an adaptation to the oceanic pelagic migration, limiting the visibility of the fish for their predators. In addition to this change in colour, the skin becomes thicker (Bertin 1951; Pankhurst 1982a).

The eye is probably one of the most modified sensory organs during silvering, as in the open ocean, conditions for vision are better, even considering that at great depths light intensity is low (Lythgoe 1979). During the transition from yellow to silver eel, increase in eye diameter and in retina surface area can be observed (Pankhurst 1982b). Moreover, structural modifications of the retina were reported, with an increase in the number of rods and a decrease in the number of cones (Pankhurst and Lythgoe 1983; Braekevelt 1985, 1988a). Finally, there is also change in visual pigment at the chromophore level with a predominance of rhodopsin in silver eels compared to porphyropsin in yellow eels (Carlisle and Denton 1959), which increases the blue light sensitivity of rods (Wood and Partridge 1993; Archer et al. 1995). All of these modifications of the visual system seem to be involved in vision in the deep sea and prepare the fish to its future reproductive migration. A similar shift in visual pigment dominance from porphyropsin to rhodopsin is observed during salmon smolting and has been proposed as an indicator of smolt status (Alexander et al. 1994).

These data show that similar morphological modifications of skin and eyes occur during both eel silvering and salmon smolting. These observations have traditionally made scientists consider silvering as a metamorphosis.

3.3.2 Gonads

The gonadosomatic index (GSI) increases from 0.3 in female yellow eels to ≥1.5 in silver eels with an increase of follicular diameter, thickening of the follicular wall and the appearance of many lipidic vesicles in the ooplasm (Fontaine et al. 1976; Lopez and Fontaine 1990). This increase in gonad size was shown to be a good criterion to estimate the state of advancement of the silvering process in the different eels (Marchelidon et al. 1999; Durif et al. 2005). Durif et al. (2005) described five stages with physiological and morphological validation. In this study, a growth phase (stages I and II), a pre-migrating stage (stage III) and a migrating phase (stages IV and V) were defined. Stages I and II correspond to the classical denomination "yellow" stage with a GSI < 0.4%; the gonads show small primary, non-vitellogenic oocytes, with a dense ooplasm and a dense nucleus with a large nucleolus (Aroua et al. 2005). Stage III corresponds to the classical denomination "intermediate" or "yellow/silver" stage with 0.4% ≤ GSI < 1.2%; oocytes are larger and a few lipidic vesicles are observed in the ooplasm, which indicates the initiation of the incorporation of lipid stores in the oocytes, also referred as "endogenous vitellogenesis" (Aroua et al. 2005). Stages IV and V correspond to the classical denomination "silver" stage with a GSI ≥ 1.2%; oocytes are further enlarged with a large nucleus and small nucleoli at a peripheral position and numerous lipid vesicles in the ooplasm, which is the oil-droplet stage of early vitellogenesis (Aroua et al. 2005). In the most advanced stage of silvering, vitellogenin can be observed in the ooplasm, as well as in the plasma, which corresponds to the start of exogenous vitellogenesis.

Considering the start of gonadal maturation, silvering should be considered as the initiation of puberty. As the development of gonads and sexual maturity are blocked at this stage and until the occurrence of oceanic reproductive migration, our group defined eel silvering as prepuberty. These data concerning gonadal maturation show that eel silvering is quite different from salmon smolting, which occurs before a growth phase and is not associated with changes related to reproduction.

3.3.3 Metabolism and Digestive Tract

At the silver stage, eels starve (Fontaine and Olivereau 1962) and this starvation is accompanied by a regression of the digestive tract (Han et al. 2003; Aroua et al. 2005; Durif et al. 2005). Eels need important metabolic changes to be able to accomplish their long oceanic migration, which is may be as much as 4,000 km for Japanese eels (*A. japonica*: Tsukamoto, 1992) and the American eel (*A. rostrata*: Tucker 1959; Mc Cleave et al. 1987) and 6,000 km for the European eel (*A. anguilla*: Schmidt 1923). To permit this long oceanic migration, animals have to accumulate energy stores at the yellow stage and then optimize the use of these stores, for both swimming and gonadal maturation. Van Ginneken and Van den Thillart (2000) demonstrated, using large swim-tunnels, that for their swim effort of 6,000 km, 40% of the European eels' energy reserves are needed while the remaining 60% of their energy stores can be used for gonad development. During silvering, changes in intermediary metabolism were observed (Boström and Johansson 1972; Lewander et al. 1974; Dave et al. 1974; Barni et al. 1985; Eggington 1986; Van Ginneken et al. 2007a). Dave et al. (1974) reported a slightly higher amount of unsaturated and longer fatty acids in muscle and a significantly lower level of 18:0 and 20:4n-6 in the liver of silver compared to yellow European eels. Comparable trends were more recently observed in the shortfin eel, *Anguilla australis* (De Silva et al. 2002). An increase in lipid (phospholipids, free fatty acids and cholesterol) content has also been reported recently at silvering in the blood of European eels (Van Ginneken et al. 2007a). According to Lewander et al. (1974), a redistribution of cholesterol occurs from other tissues to the gonads in silver eel. These metabolic variations would be amplified by environmental conditions encountered during the migration, such as depth and water temperature (Theron et al. 2000), as well as by locomotor activity (Van Ginneken et al. 2007b).

It is interesting to note that the switch to fasting that occurs during eel silvering appears rather definitive in comparison to other phases of fasting, observed at larval metamorphosis (in amphibians and glass eels), that are transitory and that correspond to a change of feeding behaviour.

3.4 Neuroendocrine Axes Involved in the Control of Silvering

As indicated previously, silvering consists of various morphological, physiological and behavioural changes. Among the modified organs, some are related to sensory organs, others to hydrostatic pressure or seawater adaptation, similarly to changes observed during smolting, which traditionally led to eel silvering being defined as a metamorphosis. However, unlike smolting, silvering also includes some changes related to an onset of sexual maturation such as gonad development, which led to the hypothesis that silvering corresponds to a pubertal event. Puberty, the major post-embryonic developmental event in the life cycle of all vertebrates, encompasses various morpho-physiological and behavioural changes, which unlike metamorphic changes are induced by sexual steroids (for review, see Romeo 2003). While metamorphosis is mostly triggered in vertebrates by the activation of the thyrotropic axis, puberty is triggered by the activation of the gonadotropic axis. In order to assess which neuroendocrine axis may be involved in the induction of silvering, we analyzed the profiles of pituitary and peripheral hormones during the transition from yellow to silver eels.

3.4.1 Thyrotropic Axis

Concerning eel silvering, Callamand and Fontaine (1942) observed hyperactivity of the thyroid gland before and during the downstream migration. It was then thought that the thyroid activation was responsible for the important morphological modifications observed during silvering. Especially, concerning eye changes, some cases of exophthalmy can be observed in humans having hyperthyroidian pathology (for reviews: Bradley 2001; Wiersinga and Bartalena 2002), supporting the former hypothesis of a role of thyroid hormones in eel eye changes. Moreover, Fontaine (1953) suggested that the thyroid activation during eel silvering at the moment of the downstream migration could be responsible for the increased locomotor activity of the animals, as is observed during smolting. In subadult American eels, elevated T4 plasma levels are correlated with increased locomotor activity under natural conditions (Castonguay et al. 1990).

Recently, our group and others have focused on the pituitary hormone controlling the synthesis of thyroid hormones, TSH (3.2). In contrast with the previous hypothesis, the study of the expression profiles of TSH showed a non-significant or a weak increase in TSH mRNA between yellow and silver eels (*Anguilla anguilla*: Aroua et al. 2005; *Anguilla japonica*: Han et al. 2004). Moreover, measurement of plasma levels of thyroid hormones in yellow and silver eels showed a moderate increase in thyroxine (T4) and no significant variations in triiodothyronine (T3) during silvering (*Anguilla anguilla*: Marchelidon et al. 1999; Aroua et al. 2005; *Anguilla japonica*: Han et al. 2004). These recent results suggest that the thyrotropic axis is poorly implicated in the neuroendocrine control of the silvering process. The weak variations observed in TSH mRNA and T4 plasma level could be involved in the increased activity of eels related to their migratory behavior. This role in motility may not be specific to the silver stage as thyroid hormones can also induce an increase in locomotor activity in yellow (Castonguay et al. 1990) and glass eels (Edeline et al. 2005).

Future studies should aim at investigating the brain control of pituitary TSH. Interestingly, preliminary data on eel pituitary cells suggest a role for CRH in the control of TSH, in addition to TRH (Pradet-Balade 1998).

3.4.2 Somatotropic Axis

Due to the implication of growth hormone (GH) in many physiological functions such as growth, metabolism, reproduction and immunity in various vertebrates, it is of interest to investigate the expression and synthesis of this hormone during silvering.

In a recent study, we focused on mRNA levels and pituitary GH content throughout the silvering process (Fig. 3.2). We showed no significant differences in mRNA levels between yellow and silver stage, and a significant decrease of GH pituitary content (Aroua et al. 2005). These data lead us to conclude that, unlike in salmonid smolting where GH has a strong role as a factor controlling osmoregulation and seawater adaptability (Björnsson 1997), in eel silvering the GH role is less critical, probably due to the fact that even at the yellow stage eels are able to pass into seawater.

Studies focusing on GH cell regulation have shown that basal release and synthesis of GH persist *in vitro*, in the absence of secretagogues or serum, using

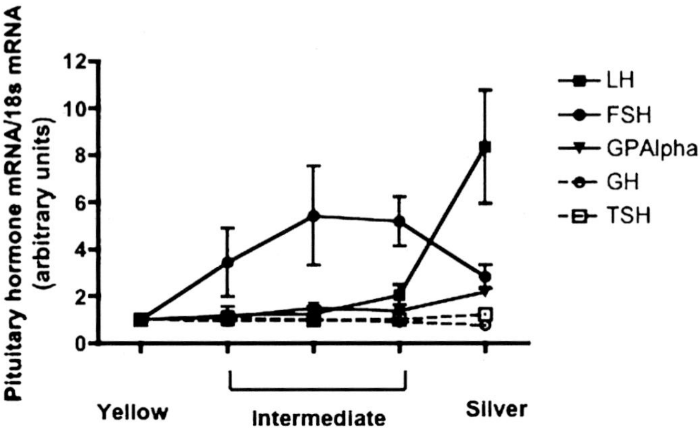

Fig. 3.2 Changes in pituitary hormone expression during eel silvering (after Aroua et al. 2005). FSH = follicle-stimulating hormone; LH = luteinizing hormone; Alpha = glycoprotein alpha subunit; GH = growth hormone; TSH = thyroid stimulating hormone

organ-cultured pituitaries (European eel: Baker and Ingleton 1975; Japanese eel: Suzuki et al. 1990) or primary cultures of primary cells (European eel: Rousseau et al. 1998, 1999; rainbow trout: Yada et al. 1991; turbot: Rousseau et al. 2001). All these observations lead to the suggestion that the major control of these cells *in vivo* is an inhibitory control (Rousseau and Dufour 2004). Our *in vitro* study in the European eel demonstrates that the brain inhibitory control of GH is exerted by somatostatin (SRIH) and that insulin-like growth factor 1 (IGF-1) exerts a negative feedback on GH. Thus, during silvering, the major regulation of the somatotroph cells, which are responsible of GH synthesis, would be an increase of their inhibitory control.

It is important to note that seasonal data suggest that a peak in GH and body growth may occur during summer (Durif et al. 2005). This suggests that while GH may not be involved in the control of the silvering process itself, the somatotropic axis may participate earlier in the initiation of the silvering. A similar growth surge is observed at puberty in mammals.

3.4.3 *Corticotropic Axis*

Only a few studies have focused on the corticotropic axis during the transition from yellow to silver stage. This is probably because of the difficulty of sampling blood in order to measure plasma cortisol levels without stressing animals. Recently, Van Ginneken et al. (2007a) demonstrated elevated cortisol levels in silver eels prior to migration. This is in agreement with the fact that during the downstream migration, eels are fasting and it is well know that the production of cortisol is induced in response to starvation. A role of cortisol may be to permit the mobilization of energy stores needed by the fish at this critical period.

Moreover, it was shown in salmonids that cortisol promotes seawater adaptability, acting in synergy with other pituitary hormones such as TSH and GH to stimulate gill (Na^+-K^+) ATPase activity and increase chloride cell number (for review: Boeuf 1993).

In addition to the effect of cortisol on energy mobilization and seawater adaptation, we previously demonstrated in the European eel that cortisol had also a strong positive effect on LH production *in vivo* as well as *in vitro* (Huang et al. 1999). This stimulation is stronger when eels are treated with a combination of cortisol and androgens, indicating synergistic action of these hormones on LH (Huang 1998; Sbaihi 2001). It is interesting to note that while in amphibians, cortisol has a synergistic effect with thyroid hormones on metamorphosis, a synergy between cortisol and sex steroids is observed in the control of eel silvering.

The various effects of cortisol demonstrated in the eel indicate that the corticotropic axis may play an important role throughout silvering by permitting energy mobilization and probably acting at the onset of puberty. Cortisol may therefore control the metabolic challenge occurring during both metamorphosis and puberty/reproduction in teleosts. It may also act in synergy with different other hormones to induce the various morphological changes observed: with sex steroids during silvering, with THs during larval metamorphosis and with GH and TH during smolting (Fig. 3.4).

3.4.4 Gonadotropic Axis

As highlighted by several authors, the silvering process is also characterized by gonadal modifications (Lopez and Fontaine 1990; Fontaine and Dufour 1991).

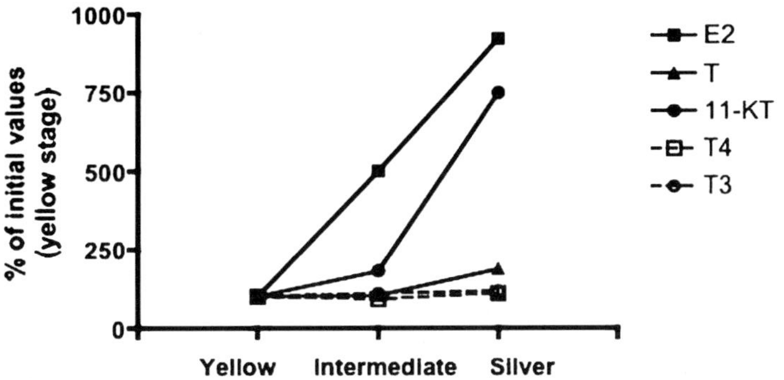

Fig. 3.3 Changes in peripheral hormone plasma levels during eel silvering (after Aroua et al. 2005). E2 = estradiol; T = testosterone; 11 KT = 11 ketotestosterone; T4 = thyroxine; T3 = triidothyronine

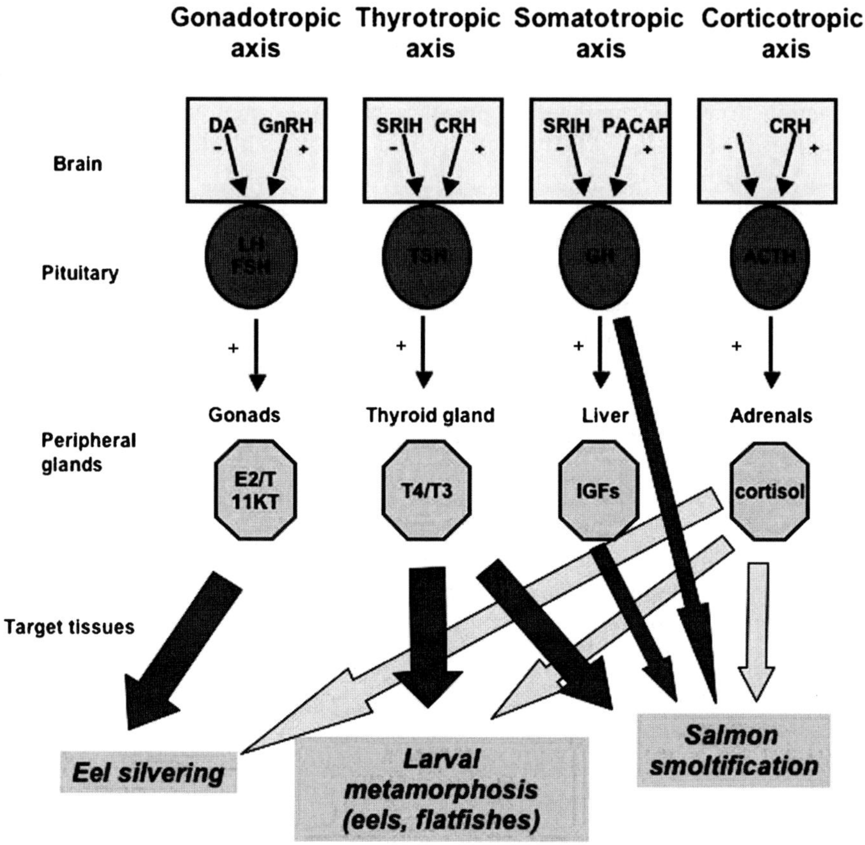

Fig. 3.4 Hormonal control of metamorphoses in teleost fish. Gonadotropic axis controls eel silvering, which should be considered as a "pubertal" event. Thyrotropic axis controls larval metamorphosis in fish, as in amphibians. It also controls salmon smoltification in synergy with somatotropic and corticotropic axes. Corticotropic axis acts in synergy with gonadotropic axis to control eel silvering, with thyrotropic axis to control larval metamorphosis and with both thyrotropic and somatotropic axes to control salmon smoltification. DA = dopamine; gnrh = gonadotropin-releasing hormone; FSH = follicle-stimulating hormone; LH = luteinizing hormone; TSH = thyroid stimulating hormone; SRIH = somatostatin; T4 = thyroxine; T3 = triiodothyronine; GH = growth hormone; igfs = insulin-like growth factors; E2 = estradiol; T = testosterone; 11 KT = 11 ketotestosterone; T4 = thyroxine; T3 = triidothyronine; CRH = corticotropin releasing hormone; ACTH = corticotropin. Target tissues include skin, eyes, digestive tract, muscles, cardiovascular system etc.

Recent field studies clearly demonstrated an increase in gonad weight and modifications of the oocyte structure throughout the silvering process (Marchelidon et al. 1999; Aroua et al. 2005; Durif et al. 2005). In addition, measurement of sexual steroids, estrogens (E2) and androgens (T and 11-KT), in the plasma showed an increase between yellow and silver stage (*A australis and A dieffenbachii*: Lokman et al. 1998; *A anguilla*: Sbaihi et al. 2001; Aroua et al. 2005; *A rostrata*: Cottrill et al.

2001; *A japonica*: Han et al. 2003) (Fig. 3.3). All these results suggest that the gonadotropic axis occupies an important position during the silvering process.

To complete the analysis of this axis, our recent studies have focused on the two pituitary hormones involved in the control of reproduction, the gonadotropins (luteinizing hormone, LH and follicle-stimulating hormone, FSH). In *A. japonica* (Han et al. 2003), as well as in *A. anguilla* (Aroua et al. 2005), variation on mRNA levels of the alpha and the beta subunits of the gonadotropins were observed throughout silvering. In *A. japonica*, authors observed a concomitant increase in mRNA of the different subunits, LHβ, FSHβ and the glycoprotein alpha (GPα) (Han et al. 2003). In *A. anguilla*, LH and FSH were shown to be differentially expressed during the silvering process, with an early increase in FSHβ expression and a late increase of LHβ expression (Aroua et al. 2005) (Fig. 3.2). These data suggest that FSH could play an early role in the activation of gonads, while LH may have an important role later in the silvering process. Indeed, a concomitance exists between the increase in FSH expression and the start of lipid incorporation in oocytes (also called "endogenous vitellogenesis"), which suggests that FSH could be responsible for the initiation of lipidic vitellogenesis. The early increase in FSH may also be responsible for the first increase in steroid production (E2), observed in intermediate eels. In contrast, the late increase in vitellogenin (Vg) plasma levels, concomitant with the late increase in LH expression, and the slight decrease in FSH, suggests that LH may participate in the induction of Vg production and initiation of the "exogenous vitellogenesis". Similarly, LH may also participate in the second increase in sex steroid levels, in silver eels.

In summary, among the different pituitary hormones that were studied, the gonadotropins, LH and FSH, showed the biggest variations during eel silvering.

3.4.5 Experimental Induction of Silvering Changes

All these results clearly indicate that the gonadotropic axis is activated during silvering. The interesting question is to understand if this activation of the gonadotropic axis is related to the "metamorphic" changes observed during the transition from yellow to silver eel.

Previous studies on the induction of eel sexual maturation showed an amplification of anatomic changes observed during silvering (Pankhurst 1982a, b, c). Thus, morphological changes observed during silvering such as increase of eye diameter (Boëtius and Larsen 1991), enhancement of silver-colour-body and decrease of gut weight (Pankhurst and Sorensen 1984) could be further induced by gonadotropic treatments (with human chorionic gonadotropin or carp pituitary extracts).

Experimental data using exogenous sex steroids are in agreement with this involvement of gonadotropic axis in the induction of silvering. Early studies showed that injections of male silver European eels with 17α-methyltestosterone resulted in enlarged eye diameter, increased skin thickness and darkened head and fins (Olivereau and Olivereau 1985). Similarly, implants of testosterone induced an increase of eye size in male silver eels (Boëtius and Larsen 1991). Moreover, a recent study in *Anguilla australis* showed that the external morphological changes such as the increase of eye diameter or the thicknening of the skin observed during silvering could be induced by 11KT (Rohr et al. 2001). Finally, in our recent studies, we showed that treatment with testosterone induced a decrease in the digestive tract-somatic index (Vidal et al. 2004; Aroua et al. 2005) and an increase in ocular index (Aroua et al. 2005), while E2 has no effect (Aroua et al. 2005). All these data suggest that the silvering changes observed during our experiments are androgen-dependent. In contrast, we recently showed that 3-month treatment of yellow eels with thyroid hormone (T3) did not induce any changes in ocular index and digestive tract-somatic index (Aroua et al. 2005). However, cortisol may have a synergistic action with steroids in this complex process of eel silvering, as we demonstrated that concomitant administration of E2, T and cortisol was most efficient in inducing the silvering of the skin in eels (Sbaihi 2001).

All these data demonstrate the involvement of androgens as crucial actors in the morpho-physiological changes of eel silvering. This is in contrast to smolting, in which a precocious sexual maturation and sex steroid administration are able to inhibit/prevent smolting and downstream migration (masu salmon: Ikuta et al. 1987; Munakata et al. 2001; amago salmon: Miwa and Inui 1986; Baltic salmon: Lundqvist et al. 1988; Atlantic salmon: Madsen et al. 2004).

3.5 Conclusions

All the external as well as internal modifications occurring during eel silvering have traditionally made scientists consider this event as a metamorphosis. However, as reviewed here, eel silvering may be primarily induced by the gonadotropic axis. There is an overall activation of this axis, with increases of gonadotropins and sex steroids. Moreover, exogenous sex steroids are able to induce peripheral morphological changes observed during this secondary metamorphosis. This let us regard silvering as a pubertal rather than a metamorphic event. The term "prepuberty" was first used by our group, as, during eel silvering, puberty is blocked and sexual maturation only occurs during the reproductive migration.

It is of great interest to note that while smolting and silvering share many similarities in term of morphological changes, the endocrinology of these two "secondary" metamorphoses drastically differs, with the major involvement of different neuroendocrine axes, the thyrotropic/somatotropic one for smolting and the gonadotropic one for silvering (Fig. 3.4). This suggests that "secondary" metamorphoses may have been acquired independently, via different endocrine mechanisms, during teleost evolution. The convergence between some morphological (skin silvering, eye size and pigments), metabolic and behavioural changes reflects that the control of the same peripheral target organs (skin, eye, muscle) and target genes is exerted by different hormonal receptors (thyroid hormone receptors in salmon versus androgen receptors in the eel). This discovery suggests an independent recruitment of different endocrine axes for the induction of secondary metamorphoses during teleost evolution.

Other endocrine axes may participate in the control of "metamorphoses". This is the case of the somatotropic and corticotropic axes acting in synergy with the thyrotropic axis (larval metamorphosis and smolting) or with the gonadotropic axis (silvering). Further studies should aim at investigating the neuroendocrine interactions controlling silvering, as well as the role of environmental and internal factors in the mechanisms leading to the activation of the gonadotropic axis.

Acknowledgements We thank S. Baloche, N. Le Belle, B. Vidal, J. Marchelidon and Drs. O. Kah, Y-S Huang, M. Montero, Y. Zohar, M. Sbaihi and F-A. Weltzien, for their contributions to the studies presented in this chapter. This work has been supported by the European contract EELREP N°Q5RS-2001-01836.

References

Aizen J, Meiri I, Tzchori I, Levavi-Sivan B, Rosenfeld H (2005) Enhancing spawning in the grey mullet (*Mugil cephalus*) by removal of dopaminergic inhibition. Gen Comp Endocrinol 142:212–221.

Alexander G, Sweeting R, McKeown B (1994) The shift in visual pigment dominance in the retinae of juvenile coho salmon (*Oncorhynchus kisutch*): an indicator of smolt status. J Exp Biol 195:185–197.

Alexander G, Sweeting R, McKeown B (1998) The effect of thyroid hormone and thyroid hormone blocker on visual pigment shifting in juvenile coho salmon (*Oncorhynchus kisutch*). Aquaculture 168:157–168.

Allen BM (1916) Extirpation experiments in *Rana pipiens* larva. Science 44:755–757.

Archer S, Hope A, Partridge JC (1995) The molecular basis for the green-blue sensitivity shift in the rod visual pigments of the European eel. Proc Biol Sci 262:289–295.

Aroua S, Schmitz M, Baloche S, Vidal B, Rousseau K, Dufour S (2005) Endocrine evidence that silvering, a secondary metamorphosis in the eel, is a pubertal rather a metamorphic event. Neuroendocrinology 82:221–232.

Baggerman B (1963) The effect of TSH and anti-thyroid substances on salinity preference and thyroid activity in juvenile Pacific salmon. Can J Zool 41:307–319.

Baker BI, Ingleton PM (1975) Secretion of prolactin and growth hormone by teleost pituitaries *in vitro*. II. Effect of salt concentration during long-term organs culture. J Comp Physiol 100:269–282.

Barni S, Bernocchi G, Gerzeli G (1985) Morphohistochemical changes in hepatocytes during the life cycle of the European eel. Tissue Cell 17:97–109.

Bertin L (1951) Les anguilles. Variation, Croissance, Euryhalinité, Toxicité, Hermaphrodisme Juvenile et Sexualité, Migrations, Métamorphoses. Paris: Payot.

Björnsson BT (1997) The biology of salmon growth hormone: from daylight to dominance. Fish Physiol Biochem 17:9–24.

Boëtius I, Larsen LO (1991) Effects of testosterone on eye size and spermiation in silver eels, *Anguilla anguilla*. Gen Comp Endocrinol 82:238.

Boeuf G (1993) Salmonid smolting: a pre-adaptation to the oceanic environment. In Fish Ecophysiology (Rankin JC and Jensen FB, eds). London: Chapman & Hall, pp. 105–135.

Boeuf G, Le Bail PY (1990) Growth hormone and thyroid hormones levels during smolting in different populations of Atlantic salmon. In Progress in Clinical Research (Epple A, Scanes CG and Steson MH, eds). New York: Wiley-Liss, Vol. 342, pp. 193–197.

Boström SL, Johansson RG (1972) Enzyme activity patterns in white and red muscle of the eel (*Anguilla anguilla*) at different developmental stages. Comp Biochem Physiol 42B:533–542.
Bradley EA (2001) Graves ophthalmopathy. Curr Opin Ophthalmol 12:347–351.
Braekevelt CR (1985) Retinal fine structure in the European eel *Anguilla anguilla*. IV. Photoreceptors of the yellow eel stage. Anat Anz 158:23–32.
Braekevelt CR (1988a) Retinal fine structure in the European eel *Anguilla anguilla*. VI. Photoreceptors of the sexually immature silver eel stage. Anat Anz 166:23–31.
Callamand O, Fontaine M (1942) L'activité thyroidienne de l'anguille au cours de son développement. Arch Zool Exp Gen 82:129–136.
Carlisle DB, Denton EJ (1959) On the metamorphosis of the visual pigments of *Anguilla anguilla* (L.). J Mar Biol Ass UK 38:97–102.
Castonguay M, Dutil JD, Audet C, Miller R (1990) Locomotor activity and concentration of thyroid hormones in migratory and sedentary juvenile American eels. Trans Am Fish Soc 119:946–956.
Chang JP, Peter RE (1983) Effects of dopamine on gonadotropin release in female goldfish, *Carassius auratus*. Neuroendocrinology 36:351–357.
Chang JP, Peter RE, Nahorniak CS, Sokolowska M (1984) Effects of catecholaminergic agonists and antagonists on serum gonadotropin concentrations and ovulation in goldfish: evidence for specificity of dopamine inhibition of gonadotropin secretion. Gen Comp Endocrinol 55:351–360.
Chang JP, Yu KL, Wong AO, Peter RE (1990) Differential actions of dopamine receptor subtypes on gonadotropin and growth hormone release *in vitro* in goldfish. Neuroendocrinology 51:664–674.
Chua D, Eales JG (1971) Thyroid function and dermal purines in brook trout, *Salvelinus fontinalis* (Mitchill). Can J Zool 49:1557–1561.
Copeland PA, Thomas P (1989) Control of gonadotropin release in the Atlantic croaker (*Micropogonias undulatus*): evidence for lack of dopaminergic inhibition. Gen Comp Endocrinol 74:474–483.
Cottrill RA, McKinley RS, Van der Kraak G, Dutil J-D, Reid KB, McGrath KJ (2001) Plasma non-esterified fatty acid profiles and 17b-oestradiol levels of juvenile immature and maturing adult American eels in the St. Lawrence River. J Fish Biol 59:364–379.
Coughlin DJ, Forry JA, McGlinchey SM, Mitchell J, Saporetti KA, Stauffer KA (2001) Thyroxine induces transitions in red muscle kinetics and steady swimming kinematics in rainbow trout (*Oncorhynchus mykiss*). J Exp Zool 290:115–124.
Dave G, Johansson M, Larsson A, Lewander K, Lidman U (1974) Metabolic and haematological studies on the yellow and silver shortfin phases of the European eel, *Anguilla anguilla* L. II. Fatty acid composition. Comp Biochem Physiol 47B:583–591.
de Jesus EG, Inui Y, Hirano T (1990) Cortisol enhances the stimulating action of thyroid hormones on dorsal fin-ray resorption of flounder larvae *in vitro*. Gen Comp Endocrinol 79:167–173.
de Jesus EG, Hirano T, Inui Y (1992) Gonadal steroids delay spontaneous flounder metamorphosis and inhibit T3-induced fin ray shortening *in vitro*. Zool Sci 9:633–638.
de Jesus EG, Hirano T, Inui Y (1993) The antimetamorphic effect of prolactin in the Japanese flounder. Gen Comp Endocrinol 93:44–50.
De Leeuw R, Goos HJ, Van Oordt PG (1986) The dopaminergic inhibition of the gonadotropin-releasing hormone-induced gonadotropin release: an *in vitro* study with fragments and cell suspensions from pituitaries of the African catfish, *Clarias gariepinus* (Burchell). Gen Comp Endocrinol 63:171–177.
De Silva SS, Gunasekera RM, Collins RO (2002) Some morphometric and biochemical features of ready-to-migrate silver and pre-migratory yellow stages of the shortfin eel of south-eastern Australian waters. J Fish Biol 61:915–928.
Denver RJ (1993) Acceleration of anuran amphibian metamorphosis by corticotropin releasing hormone-like peptides. Gen Comp Endocrinol 91:38–51.

Denver RJ (1999) Evolution of the corticotropin-releasing hormone signaling system and its role in stress-induced phenotypic plasticity. Ann NY Acad Sci 897:46–53.

Denver RJ, Licht P (1989) Neuropeptide stimulation of thyrotropin secretion in the larval bullfrog: evidence for a common regulator of thyroid and interrenal activity during metamorphosis. J Exp Zool 252:101–104.

Dodd MHI, Dodd JM (1976) The biology of metamorphosis. In Physiology of Amphibia (Lofts E, ed). New York: Academic Press, Vol. 3, pp. 467–599.

Donaldson E, Fagerlund U, Higgs D, McBride J (1979) Hormonal enhancement of growth. In Fish Physiology (Hoar WS, Randall DJ and Brett JR, eds). New York: Academic Press, Vol. VIII, pp. 456–597.

Dufour S (1994) Neuroendocrinologie de la reproduction de l'anguille: de la recherche fondamentale aux problèmes appliqués. Bull Fr Pêche Piscic 335:187–211.

Dufour S, Fontaine YA (1985) La migration de reproduction de l'anguille européenne (*Anguilla anguilla* L.): un rôle probable de la pression hydrostatique dans la stimulation de la fonction gonadotrope. Bull Soc Zool Fr 110:291–299.

Dufour S, Delerue-Le Belle N, Fontaine YA (1983) Development of a heterologous radioimmunoassay for eel (*Anguilla anguilla*) gonadotropin. Gen Comp Endocrinol 49:404–413.

Dufour S, Lopez E, Le Menn F, Le Belle N, Baloche S, Fontaine YA (1988) Stimulation of gonadotropin release and ovarian development, by the administration of a gonadoliberin agonist and of dopamine antagonists, in female silver eel pretreated with estradiol. Gen Comp Endocrinol 70:20–30.

Dufour S, Bassompierre M, Montero M, Le Belle N, Baloche S, Fontaine YA (1991) Stimulation of pituitary gonadotropic function in female silver eels treated by a gonadoliberin agonist and dopamine antagonist. In Proceedings of the Fourth International Symposium on Reproductive Physiology of Fish (Scott AP, Sumpter JP, Kime DE and Rolfe MS, eds). Sheffield: FishSymp 91, pp. 54–56.

Dufour S, Burzawa-Gérard E, Le Belle N, Sbaihi M, Vidal B (2003) Reproductive endocrinology of the European eel, *Anguilla anguilla*. In Eel Biology (Aida K, Tsukamoto K and Yamauchi K, eds). Tokyo: Springer, pp. 107–117.

Dufour S, Weltzien FA, Sébert ME, Le Belle N, Vidal B, Vernier P, Pasqualini C (2005) Dopaminergic inhibition of reproduction in teleost fishes: ecophysiological and evolutionary implications. Ann NY Acad Sci 1040:9–21.

Durif C, Dufour S, Elie P (2005) The silvering process of *Anguilla anguilla*: a new classification from the yellow resident to the silver migrating stage. J Fish Biol 66:1025–1043.

Eddy L, Lipner H (1976) Amphibian metamorphosis: role of thyrotropin-like hormone. Gen Comp Endocrinol 29:333–336.

Edeline E, Dufour S, Briand C, Fatin D, Elie P (2004) Thyroid status is related to migratory behavior in *Anguilla anguilla* glass eels. Mar Ecol Prog Ser 282:261–270.

Edeline E, Bardonnet A, Bolliet V, Dufour S, Elie P (2005) Endocrine control of *Anguilla anguilla* glass eel dispersal: effect of thyroid hormones on locomotor activity and rheotactic behavior. Horm Behav 48:53–63.

Eggington S (1986) Metamorphosis of the American eel, *Anguilla rostrata* LeSueur: II. Structural reorganisation of the locomotory musculature. J Exp Zool 238:297–309.

Fontaine M (1953) La fonction hypophyso-thyroidienne des poissons dans ses rapports avec leur morphologie et leur comportement. Journal du conseil international pour l'exploitation de la mer 19:23–38.

Fontaine M (1975) Physiological mechanisms in the migration of marine and amphihaline fish. Adv Mar Biol 13:241–255.

Fontaine M (1985) Action de facteurs anormaux du milieu sur l'écophysiologie d'anticipation des poissons migrateurs amphihalins. Ichtyophysiol Acta 9:11–25.

Fontaine M, Callamand O (1941) Sur l'hydrotropisme des civelles. Bull Inst Océanogr 811:1–6.

Fontaine M, Leloup J (1960) Etude du pouvoir de concentration des iodures par la thyroïde de deux salmonidés, la Truite arc-en-ciel (*Salmo gairdnerii* Rich) et le jeune saumon (*Salmo salar* L.). Arch Sci Physiol 14:15–24.

Fontaine M, Olivereau M (1962) Nutrition et sexualité chez les poissons. Ann Nutr Aliment 16:A125–A152.

Fontaine M, Bertrand E, Lopez E, Callamand O (1964) Sur la maturation des organes génitaux de l'anguille femelle (*Anguilla anguilla* L.) et l'émission spontanée des oeufs en aquarium. C R Acad Sci Paris 259: 2907–2910.

Fontaine YA, Dufour S (1991) The eels: from life cycle to reproductive endocrinology. Bull Inst Zool Acad Sinica 16:237–248.

Fontaine YA, Lopez E, Delerue-Le Belle N, Fontaine-Bertrand E, Lallier F, Salmon C (1976) Stimulation gonadotrope de l'ovaire chez l'anguille (*Anguilla anguilla* L.) hypophysectomisée. J Physiol, Paris 72:871–892.

Godin JG, Dill PA, Drury DE (1974) Effects of thyroid hormones on behaviour of yearling Atlantic salmon *Salmo salar* L. J Fish Res Board Can 31:1787–1790.

Gudernatsch JF (1912) Feeding experiments of tadpoles: the influence of specific organs given as food on growth and differentiation: a contribution to the knowledge of organs of internal secretion. Wilhelm Roux' Arch Entwi Mech Org 35:457–483.

Han Y-S, Liao I-C, Tzeng W-N, Huang Y-S, Yu JY-L (2003) Serum estradiol-17β and testosterone levels during silvering in wild Japanese eel *Anguilla japonica*. Comp Biochem Physiol 136B:913–920.

Han YS, Liao IC, Tzeng WN, Yu JYL (2004) Cloning of the cDNA for thyroid stimulating hormone ß subunit and changes in activity of the pituitary-thyroid axis during silvering of the Japanese eel, *Anguilla japonica*. J Mol Endocrinol 32:179–194.

Haro A (2003) Downstream migration of silver-phase anguillid eels. In Eel Biology (Aida K, Tsukamoto K and Yamauchi K, eds). Tokyo: Springer, pp. 215–222.

Hoar WS (1939) The thyroid gland of the Atlantic salmon. J Morphol 65:257–295.

Hoar WS (1988) The physiology of smolting salmonids. In Fish Physiology (Hoar WS and Randall DJ, eds). New York: Academic Press, Vol. 11B, pp. 275–343.

Holland MC, Hassin S, Zohar Y (1998) Effects of long-term testosterone, gonadotropin-releasing hormone agonist, and pimozide treatments on gonadotropin II levels and ovarian development in juvenile female striped bass (*Morone saxatilis*). Biol Reprod 59:1153–1162.

Huang L, Schreiber AM, Soffientino B, Bengtson DA, Specker JL (1998) Metamorphosis of summer flounder (*Paralichthys dentatus*): thyroid status and the timing of gastric gland formation. J Exp Zool 280:413–420.

Huang YS (1998) Rôle des steroids sexuels et des hormones métaboliques dans le contrôle direct hypophysaire de l'hormone gonadotrope (GtH-II) chez l'anguille européenne, *Anguilla anguilla*. Ph.D. thesis, University Paris VI, Paris, France.

Huang YS, Rousseau K, Sbaihi M, Le Belle N, Schmitz M, Dufour S (1999) Cortisol selectively stimulates pituitary gonadotropin beta-subunit in primitive teleost, *Anguilla anguilla*. Endocrinology 140:1228–1235.

Ikuta K, Aida K, Okumoto N, Hanyu I (1987) Effects of sex steroids on the smoltification of masu salmon, *Oncorhynchus masou*. Gen Comp Endocrinol 65:99–110.

Inui Y, Miwa S (1985) Thyroid hormone induces metamorphosis of flounder larvae. Gen Comp Endocrinol 60:450–454.

Iwata M, Yamanome T, Tagawa M, Ida H, Hirano T (1989) Effects of thyroid hormones on phototaxis of chum and coho salmon juveniles. Aquaculture 82:329–338.

Jegstrup IM, Rosenkilde P (2003) Regulation of post-larval development in the European eel: thyroid hormone level, progress of pigmentation and changes in behaviour. J Fish Biol 63:168–175.

Kah O, Chambolle P, Thibault J, Geffard M (1984) Existence of dopaminergic neurons in the preopric region of the goldfish. Neurosci Lett 48:293–298.

Kah O, Dubourg P, Onteniente B, Geffard M, Calas A (1986) The dopaminergic innervation of the goldfish pituitary. An immunocytochemical study at the electron-microscope level using antibodies against dopamine. Cell Tissue Res 244:577–582.

Kah O, Dulka JG, Dubourg P, Thibault J, Peter RE (1987) Neuroanatomical substrate for the inhibition of gonadotrophin secretion in goldfish: existence of a dopaminergic preoptico-hypophyseal pathway. Neuroendocrinology 45:451–458.

Kanamori A, Brown DD (1996) The analysis of complex developmental programmes: amphibian metamorphosis. Genes Cells 1:429–435.

Katzman S, Cech JJ Jr (2001) Juvenile coho salmon locomotion and mosaic muscle are modified by 3′,3′,5′-tri-iodo-L-thyronine (T3). J Exp Biol 204:1711–1717.

Kikuyama S, Kawamura K, Tanaka S, Yamamoto K (1993) Aspects of amphibian metamorphosis: hormonal control. Int Rev Cytol 145:105–148.

Kitajima C, Sato T, Kawanishi M (1967) On the effect of thyroxine to promote the metamorphosis of a conger eel-preliminary report. Bull Jpn Soc Sci Fish 33:919–922.

Kubota SS (1961) Studies on the ecology, growth and metamorphosis in conger eel, *Conger myriaster* (Brevoort). J Fac Fish Mie Prefectural Univ 5:190–370.

Kumakura N, Okuzawa K, Gen K, Kagawa H (2003) Effects of gonadotropin-releasing hormone agonist and dopamine antagonist on hypothalamus-pituitary-gonadal axis of prepubertal female red seabream (*Pagrus major*). Gen Comp Endocrinol 131:264–273.

Larsen DA, Swanson P, Dickey JT, Rivier J, Dickhoff WW (1998) In vitro thyrotropin-releasing activity of corticotropin-releasing hormone-family peptides in coho salmon, Oncorhynchus mykiss. Gen Comp Endocrinol 109:276–285.

Leatherland JE, Hilliard RW, Macey DJ, Potter IC (1990) Changes in serum thyroxine and triidothyronine concentrations during metamorphosis of the Southern Hemisphere lamprey *Geotria australis*, and the effect of propylthiouracil, triiodothyronine and environmental temperature on serum thyroid hormone concentrations of ammocoetes. Fish Physiol Biochem 8:167–177.

Leloup-Hatey J, Hardy A, Quérat B (1989) Responses of gonado-somatic indices and plasma levels of sex steroids to gonadotrophic stimulation in freshwater European eel at yellow and silver stages. Gen Comp Endocrinol 74:251.

Lema SC, Nevitt GA (2004) Evidence that thyroid hormone induces olfactory cellular proliferation in salmon during a sensitive period for imprinting. J Exp Biol 207:3317–3327.

Lewander K, Dave G, Johansson ML, Larsson A, Lidman U (1974) Metabolic and hematological studies on the yellow and silver phases of the European eel, *Anguilla anguilla* L. I. Carbohydrate, lipid, protein and inorganic ion metabolism. Comp Biochem Physiol B47:571–581.

Lin HR, Van der Kraak G, Zhou XJ, Liang JY, Peter RE, Rivier JE, Vale WW (1998) Effects of [D-Arg6, Trp7, Leu8, Pro9NEt]-luteinizing hormone-releasing hormone (sGnRH-a) and [D-Ala6, Pro9NEt]- luteinizing hormone-releasing hormone (LHRH-a), in combination with pimozide or domperidone, on gonadotropin release and ovulation in the Chinese loach and common carp. Gen Comp Endocrinol 69:31–40.

Linard B, Bennami S, Saligaut C (1995) Involvement of estradiol in a catecholamine inhibitory tone of gonadotropin release in rainbow trout (*Oncorhynchus mykiss*). Gen Comp Endocrinol 99:192–196.

Lokman PM, Vermeulen GJ, Lambert JGD, Young G (1998) Gonad histology and plasma steroid profiles in wild New Zealand freshwater eels (*Anguilla dieffenbachii* and *A. australis*) before and at the onset of the natural spawning migration. I. Females. Fish Physiol Biochem 19:325–338.

Lopez E, Fontaine YA (1990) Stimulation hormonale *in vivo* de l'ovaire d'anguille européenne au stade jaune. Reprod Nutr Dev 30:577–582.

Lundqvist H, Clarke WC, Johansson H (1988) The influence of precocious sexual maturation on survival to adulthood of river stocked Baltic salmon, *Salmo salar*, smolts. Ecography 11:60–69.

Lythgoe JN (1979) The Ecology of Vision. Oxford: Clarendon Press.

Madsen SS, Skovbølling S, Nielsen C, Korsgaard B (2004) 17-β estradiol and 4-nonylphenol delay smolt development and downstream migration in Atlantic salmon, *Salmo salar*. Aquat Toxicol 68:109–120.

Manzon RG, Denver RJ (2004) Regulation of pituitary thyrotropin gene expression during xenopus metamorphosis: negative feedback is functional throughout metamorphosis. J Endocrinol 182:273–285.

Marchelidon J, Le Belle N, Hardy A, Vidal B, Sbaihi M, Burzawa-Gérard E, Schmitz M, Dufour S (1999) Etude des variations de paramètres anatomiques et endocriniens chez l'anguille européenne (*Anguilla anguilla*) femelle, sédentaire et d'avalaison: application à la caractérisation du stade argenté. Bull Fr Pêche Piscicult 355:349–368.

Mc Cleave JD, Kleckner RC, Castonguay M (1987) Reproductive sympatry of American and European eels and implications for migration and taxonomy. Proc R Soc Lond 1:286–297.

McNabb FMA (1992) Thyroid Hormones. Englewood Cliffs, NJ: Prentice Hall, pp. 283.

Miwa S, Inui Y (1985) Effects of L-thyroxine and ovine growth hormone on smoltification of amago salmon (*Oncorhynchus rhodurus*). Gen Comp Endocrinol 58:436–442.

Miwa S, Inui Y (1986) Inhibitory effect of 17αmethyltestosterone and estradiol 17β on smoltification of sterilized amago salmon *Oncorhynchus rhodurus*. Aquaculture 53:21–39.

Miwa S, Inui Y (1987a) Effects of various doses of thyroxine and triidothyronine on the metamorphosis of flounder (*Paralichtys olivaceus*). Gen Comp Endocrinol 67:356–363.

Miwa S, Inui Y (1987b) Histological changes in the pituitary-thyroid axis during spontaneous and artificially-induced metamorphosis of larvae of the flounder *Paralichtys olivaceus*. Cell Tissue Res 249:117–123.

Miwa S, Inui Y (1991) Thyroid hormone stimulates the shift of erythrocyte populations during metamorphosis of the flounder. J Exp Zool 259:222–228.

Miwa S, Tagawa M, Inui Y, Hirano T (1988) Thyroxine surge in metamorphosing flounder larvae. Gen Comp Endocrinol 70:158–163.

Miwa S, Yamano K, Inui Y (1992) Thyroid hormone stimulates gastric development in flounder larvae during metamorphosis. J Exp Zool 261:424–430.

Morley JE (1981) Neuroendocrine control of thyrotropin secretion. Endocrine Rev 2:396–436.

Munakata A, Amano M, Ikuta K, Kitamura S, Aida K (2001) The involvement of sex steroid hormones in downstream and upstream migratory behavior of masu salmon. Comp Biochem Physiol 129B:661–669.

Murr E, Sklower A (1928) Untersuchungen über die inkretorischen Organe der Fische. I. Das verhalten der Schilddrüse in der Metamorphose des Aales. Zeitschr Vergleichende Physiol 7:279–288.

Nocillado JN, Levavi-Sivan B, Carrick F, Elizur A (2007) Temporal expression of G-protein-coupled receptor 54 (GPR54), gonadotropin-releasing hormones (GnRH), and dopamine receptor D2 (drd2) in pubertal female grey mullet, *Mugil cephalus*. Gen Comp Endocrinol 150:278–287.

Okada N, Tanaka M, Tagawa M (2003) Bone development during metamorphosis of the Japanese flounder (*Paralichthys olivaceus*): differential responses to thyroid hormone. In The Big Fish Bang. Proceedings of the 26th Annual Larval Fish Conference (Browman HI and Skiftesvik AB, eds). Bergen: Institute of Marine Research, 11 pp.

Okuzawa K (2002) Puberty in teleosts. Fish Physiol Biochem 26:31–41.

Olivereau M, Olivereau J (1985) Effects of 17 α-methyltestosterone on the skin and gonads of freshwater male silver eels. Gen Comp Endocrinol 57:64–71.

Pankhurst NW (1982a) Changes in the skin-scale complex with sexual maturation in the European eel, *Anguilla anguilla* (L). J Fish Biol 21:549–561.

Pankhurst NW (1982b) Relation of visual changes to the onset of sexual maturation in the European eel, *Anguilla anguilla* (L). J Fish Biol 21:127–140.

Pankhurst NW (1982c) Changes in body musculature with sexual maturation in the European eel, *Anguilla anguilla* (L). J Fish Biol 21:417–428.

Pankhurst NW, Lythgoe JN (1982) Structure and color of the tegument of the European eel *Anguilla anguilla* (L.). J Fish Biol 21:279–296.

Pankhurst NW, Lythgoe JN (1983) Changes in vision and olfaction during sexual maturation in the European eel *Anguilla anguilla* (L.). J Fish Biol 23:229–240.

Pankhurst NW, Sorensen PW (1984) Degeneration of the alimentary tract in sexually maturing European *Anguilla anguilla* (LeSueur). Can J Zool 62:1143–1149.

Pasqualini C, Vidal B, Le Belle N, Sbaihi M, Weltzien F-A, Vernier P, Zohar Y, Dufour S (2004) Un contre-pouvoir au contrôle de la reproduction par la GnRH chez les poisons Téléostéens:

l'inhibition dopaminergique. Rôle ancestral et conservation différentielle chez les Vertébrés? J Soc Biol 198:61–67.
Peter RE, Crim LW (1978) Hypothalamic lesions of goldfish: effects on gonadal recrudescence and gonadotropin secretion. Ann Biol Anim Bioch Biophys 18:819–823.
Peter RE, Paulencu CR (1980) Involvement of the preoptic region in the gonadotropin release-inhibition in the goldfish. Neuroendocrinology 31:133–141.
Peter RE, Crim LW, Goos HJTh, Crim JW (1978) Lesioning studies on the gravid female goldfish: neuroendocrine regulation of ovulation. Gen Comp Endocrinol 35:391–401.
Premdas FH, Eales JG (1976) The influence of TSH and ACTH on purine and pteridine deposition in the skin of rainbow trout (*Salmo gairdneri*). Can J Zool 54:576–581.
Pradet-Balade B (1998) Evolution de la regulation de la fonction thyréotrope: étude chez les Téléostéens. Ph.D. thesis, University Paris XI, Paris, France.
Rohr DH, Lokman PM, Davie PS, Young G (2001) 11-ketotestosterone induces silvering-related changes in immature female short-finned eels, *Anguilla australis*. Comp Biochem Physiol 130A:701–714.
Romeo RD (2003) Puberty: a period of both organizational and activational effects of steroid hormones on neurobehavioural development. J Neuroendocrinol 15:1185–1192.
Rousseau K, Dufour S (2004) Phylogenetic evolution of the neuroendocrine control of growth hormone: contribution from teleosts. Cybium 28:181–198.
Rousseau K, Huang YS, Le Belle N, Vidal B, Marchelidon J, Epelbaum J, Dufour S (1998) Long-term inhibitory effects of somatostatin and insulin-like growth factor 1 on growth hormone release by serum-free primary culture of pituitary cells from European eel (*Anguilla anguilla*). Neuroendocrinology 67:301–309.
Rousseau K, Le Belle N, Marchelidon J, Dufour S (1999) Evidence that corticotropin-releasing hormone acts as a growth hormone-releasing factor in a primitive teleost, the European eel (*Anguilla anguilla*). J Neuroendocrinol 11:385–392.
Rousseau K, Le Belle N, Pichavant K, Marchelidon J, Chow BKC, Boeuf G, Dufour S (2001) Pituitary growth hormone secretion in turbot, a phylogenetically recent teleost, is regulated by a species-specific pattern neuropeptides. Neuroendocrinology 74:375–385.
Saligaut C, Linard B, Breton B, Anglade I, Bailhache T, Kah O, Jego P (1999) Brain aminergic systems in salmonids and other teleosts in relation to steroid feedback and gonadotropin release. Aquaculture 177:13–20.
Sbaihi M (2001) Interaction des stéroïdes sexuels et du cortisol dans le contrôle de la reproduction et du métabolisme calcique chez un téléostéen migrateur, l'anguille (*Anguilla anguilla*). Ph.D. thesis, University Paris VI, Paris, France.
Sbaihi M, Fouchereau-Peron M, Meunier F, Elie P, Mayer I, Burzawa-Gérard E, Vidal B, Dufour S (2001) Reproductive biology of the conger eel from the south coast of Brittany, France and comparison with the European eel. J Fish Biol 59:302–318.
Schmidt J (1923) Breeding places and migration of the eel. Nature 111:51–54.
Scholz AT (1980) Hormonal regulation of smolt transformation and olfactory imprinting in coho salmon. Ph.D. thesis, University of Wisconsin, Madison, WI.
Schreiber AM, Specker JL (1998) Metamorphosis in the summer flounder, *Paralichthys dentatus*: stage-specific developmental response to altered thyroid status. Gen Comp Endocrinol 111:156–166.
Schreiber AM, Specker JL (2000) Metamorphosis in the summer flounder, *Paralichthys dentatus*: thyroidal status influences gill mitochondria-rich cells. Gen Comp Endocrinol 117:238–250.
Sébert P (2003) Fish adaptations to pressure. In Fish adaptation (Val AL and Kapoor BG, eds). Enfield: Science Publishers, pp. 73–95.
Shi Y-B (1999) Amphibian Metamorphosis. From Morphology to Molecular Biology. New York: Wiley.
Sinha VPR, Jones JW (1975) The European Freshwater Eel. Liverpool: Liverpool University Press.
Soffientino B, Specker JL (2001) Metamorphosis of summer flounder, *Paralichthys dentatus*: cell proliferation and differentiation of the gastric mucosa and developmental effects of altered thyroidal status. J Exp Zool 290:31–40.

Solbakken JS, Norberg B, Watanabe K, Pittman K (1999) Thyroxine as a mediator of metamorphosis of Atlantic halibut, *Hippoglossus hippoglossus*. Environ Biol Fish 1–2:53–65.
Staley KB, Ewing RD (1992) Purine levels in the skin of juvenile coho salmon (*Oncorhynchus kisutch*) during parr-smolt transformation and adaptation to seawater. Comp Biochem Physiol 101B:447–452.
Sterba G, Schneider J (1961) Zur Wirkung von Kalium-perchlorat (KClO4) auf Ammocoeten. Naturwissenschaften 48:485–486.
Sullivan CV, Darling DS, Dickhoff WW (1987) Effects of triiodothyronine and propylthiouracil on thyroid function and smoltification of coho salmon (*Oncorhynhcus kisutch*). Fish Physiol Biochem 4:121–135.
Suzuki S (1986) Induction of metamorphosis and thyroid function in the larval lamprey. In Frontiers in Thyroidology (Mederios-Neto G and Gaitan E, eds). New York: Plenum Press, Vol. 1, pp. 667–670.
Suzuki R, Kishida M, Hirano T (1990) Growth hormone secretion during longterm incubation of the pituitary of the Japanese eel, *Anguilla japonica*. Fish Physiol Biochem 8:159–165.
Tagawa M, Miwa S, Inui Y, de Jesus EG, Hirano T (1990) Changes in thyroid hormone concentrations during early development and metamorphosis of the flounder, *Paralichthys olivaceus*. Zool Sci 7:93–96.
Tata JR (1998) Hormonal Signalling and Postembryonic Development. Berlin: Springer.
Tata JR (2006) Amphibian metamorphosis as a model for the developmental actions of thyroid hormone. Mol Cell Endocrinol 246:10–20.
Tesch FW (1977) The Eel: Biology and Management of Anguillid Eels. London: Chapman & Hall.
Tesch FW (1982) The Sargasso sea eel expedition 1979. Helgoländer Meeresun 35:263–277.
Tesch FW (1989) Changes in swimming depth and direction of silver eels (*Anguilla anguilla* L.) from the continental shelf to the deep sea. Aquat Living Resour 2:9–20.
Theron M, Guerrero F, Sébert P (2000) Improvement in the efficiency of oxidative phosphorylation in the freshwater eel acclimated to 10.1 MPa hydrostatic pressure. J Exp Biol 203:3019–3023.
Thomson AJ, Sargent JR (1977) Changes in the levels of chloride cells and (Na^+ + K^+)-dependent ATPase in the gills of yellow and silver eels adapting to seawater. J Exp Zool 200:33–40.
Tsukamoto K (1992) Discovery of the spawning area for Japanese eel. Nature 356:789–791.
Tucker DW (1959) A new solution to the Atlantic eel problem. Nature 183:495–501.
Van Asselt LA, Goos HJ, Smit-Van Dijk W, Speetjens PAM, Van Oordt PGWJ (1988) Evidence for the involvement of D2 receptors in the dopaminergic inhibition of gonadotropin release in the African catfish, *Clarias gariepinus*. Aquaculture 72:369–378.
Van der Kraak G, Donaldson EM, Chang JP (1986) Dopamine involvement in the regulation of gonadotropin release in coho salmon. Can J Zool 64:1245–1248.
Van Ginneken V, Van den Thillart G (2000) Eel fat stores are enough to reach the Sargasso. Nature 403:156–157.
Van Ginneken V, Durif C, Balm SP, Boot R, Verstegen MWA, Antonissen E, Van den Thillart G (2007a) Silvering of European eel (*Anguilla anguilla* L.): seasonal changes of morphological and metabolic parameters. Anim Biol 57:63–77.
Van Ginneken V, Dufour S, Sbaihi M, Balm P, Noorlander K, de Bakker M, Doornbos J, Palstra A, Antonissen E, Mayer I, Van den Thillart G (2007b) Does a 5500-km swim trial stimulate early sexual maturation in the European eel (*Anguilla anguilla* L.)? Comp Biochem Physiol 147A:1095–1103.
Vidal B, Pasqualini C, Le Belle N, Holland MC, Sbaihi M, Vernier P, Zohar Y, Dufour S (2004) Dopamine inhibits luteinizing hormone synthesis and release in the juvenile European eel: a neuroendocrine lock for the onset of puberty. Biol Reprod 71:1491–1500.
Vilter V (1946) Action de la thyroxine sur la metamorphose larvaire de l'Anguille. CR Seances Soc Biol Ses Fil 140:783–785.
White BA, Nicoll CS (1981) Hormonal control of amphibian metamorphosis. In Metamorphosis (Gilvert LI and Frieden E, eds). New York: Plenum Press, pp. 363–396.
Wiersinga WM, Bartalena L (2002) Epidemiology and prevention of Graves' ophthalmopathy. Thyroid 12:855–860.

Wood P, Partridge JG (1993) Opsin substitution induced in retinal rods of the eel, *Anguilla anguilla* (L.): a model for G-protein-linked receptors. Proc R Soc Lond B 254:227–232.
Yada T, Urano A, Hirano T (1991) Growth hormone and prolactin gene expression and release in the pituitary of rainbow trout in serum-free culture. Endocrinology 129:1183–1192.
Yamano K, Tagawa M, de Jesus EG, Hirano T, Miwa S, Inui Y (1991) Changes in whole body concentrations of thyroid hormones and cortisol in metamorphosing conger eel. J Comp Physiol B 161:371–375.
Yaron Z, Gur G, Melamed P, Rosenfeld H, Elizur A, Levavi-Sivan B (2003) Regulation of fish gonadotropins. Int Rev Cytol 225:131–185.
Youngson AF, Scott DCB, Johnstone R, Pretswell D (1985) The thyroid system's role in the downstream migration of Atlantic salmon (*Salmo salar* L.) smolts. Aquaculture 45:392–393.
Youson JH (1988) First metamorphosis. In Fish Physiology (Hoar WS and Randall DJ, eds). San Diego, CA: Academic Press, Vol. XI, Part B, pp. 135–196.
Youson JH (1997) Is lamprey metamorphosis regulated by thyroid hormones? Amer Zool 37:441–460.
Zohar Y, Harel M, Hassin S, Tandler A (1995) Broodstock management and manipulation of spawning in the gilthead seabream, *Sparus aurata*. In Broodstock management and Egg and Larval Quality (Bromage N and Roberts RJ, eds). London: Blackwell, pp. 94–117.

Chapter 4
Silver Eel Migration and Behaviour

Maarten C.M. Bruijs and Caroline M.F. Durif

4.1 Introduction

After their first transatlantic migration, larvae metamorphose into 'glass eels' and swim into estuarine areas of river deltas. The glass eels develop into the so-called elvers (~65 mm) which swim up the rivers. They enter inland waters of northern Africa and Europe during different periods; Morocco: September–October; Spain, Portugal, South France: November–December; North France: January–March; British Islands and the Netherlands: February–April and Scandinavia: April-May. After reaching a length of 30 cm, elvers become known as yellow eels. These specimens have moved inland to coastal seawater or inland freshwater and continue to grow for some 8–15 years (males) up to 10–18 years (females), but they even may become much older.

After their period of growth, in preparation of their return trip to the spawning grounds, the eels transform into silver eel. Once the eel has undergone pre-puberty or "silvering", which is accompanied by marked changes in morphology, body constitution and other features (see Chapter 2), it is ready to start its spawning migration. Its lifestyle changes radically. As we have seen in Chapter 2, silver eels stop feeding, they have already acquired salinity tolerance while still living in freshwater and finally they have begun their puberty. As European eels, *Anguilla anguilla*, spawn in the Sargasso sea, the silver eels must migrate from inland or coastal waters over a distance of about 6,000 km to reach this area. Somehow silver eels are triggered by environmental factors to start their migration.

M.C.M. Bruijs
KEMA Technical and Operational Services, P.O. Box 9035, 6800 ET Arnhem, The Netherlands

C.M.F. Durif
Center for Ecological and Evolutionary Synthesis, University of Oslo, P.O. Box 1066, Blindern, 0316 Oslo. Institute of Marine Research, 5392 Storebø, Norway

G. van den Thillart et al. (eds.), *Spawning Migration of the European Eel,*

Downstream migration has been subject to investigations on physiological and ecological topics for many years, not only to resolve the mysterious life-cycle of this catadromous teleost species, but first and foremost to understand the reason for its global decline. During their downstream migration in river systems, silver eel have to pass many types of barriers such as large barrages, flood-control dams, flood gates, weirs, hydropower stations, sluices, pumping stations and fisheries. Such barriers are abundant in the many regulated European river systems and inland waters. These barriers are clear obstructions for downstream movement and cause a risk for the survival of the silver eel. They also have a clear impact on river flows with much diversity in flow patterns. For better understanding of the impact by barriers on downstream migration of silver eels and the development and implementation of technical and management measures, improved detailed knowledge of silver eel behaviour at barriers is necessary.

As with many diadromous fish species, eels are threatened by the presence of dams and hydroelectric facilities in rivers. Turbine entrainment as well as impingement of silver-phase eels on the screens causes massive mortality of eels (Travade and Larinier 1992; Hadderingh and Bakker 1998). Up to 100% of the eels entering the intakes may be injured (Larinier and Travade 1998), depending on the type of turbine and flow conditions. The situation has become critical for eels and management solutions on a European level are urgently needed. With respect to eel behaviour, information can be used to develop techniques to protect silver eels at turbine intakes of hydropower facilities. Knowledge of eel behaviour around barriers and timing of migration events may lead to successful protective measures (Richkus and Dixon 2003; Durif et al. 2003; Bruijs et al. 2003) (Durif and Elie 2008).

In 1996, the Committee of Ministers of the Benelux Economic Union decided to guarantee free fish migration in the hydrographic river catchments of the Benelux countries by 2010. This has already resulted in the construction of fish passages at weirs and hydropower stations in the Belgian and Dutch parts of the river Meuse. The European Water Framework Directive (EC 2000) requires undisturbed migration for fish in European river systems, and the European Commission has mandated the preparation of a European eel action plan (EC 2005, 2007). These developments will increase pressure for measures that protect eels and other species during their downstream migration. The need to manage the European eel fisheries has been recently addressed under the Common Fisheries Policy. A major stumbling block is the inadequate knowledge on the basic biology of the eel, crippling the attempt to set up rational management strategies. Furthermore it is well known that substantial mortality appears with downstream migrating eels passing the turbines of hydropower stations. Hydropower stations are widespread in many European rivers and might have detrimental effects on the population level of the European eel.

The objective of this chapter is to bring together existing and new information on the behaviour of downstream migrating silver eels related to environmental migration triggers, migration events in rivers, obstructions in migration routes and behaviour when silver eels encounter mechanical and behavioural barriers designed to guide them towards safe bypasses at hydropower plants.

4.2 Migratory Behaviour of the Silver Eel

Eels migrate in large groups during narrow specific periods: they gather from streams and small river systems in the catchment area of large European river systems like the rivers Rhine, Meuse, and Loire. As Lowe (1952) wrote, downstream runs of silver eels "will depend on three things: the number of silver eels available in the lake or river, the external conditions, and the responsiveness of the eels to these external conditions". In many biological phenomena it is important that physical events (i.e. environmental factors) coincide with biological ones. Once the eel has undergone its "pre-pubertal" metamorphosis (i.e. silvering) it is ready to start its spawning migration. Its behaviour changes and it becomes receptive to certain environmental factors which will trigger the downstream movements. The urge to migrate is apparently so strong that they will even leave the water to escape if necessary (Tesch 2003). Runs of silver eels typically occur at night and during heavy rainfall (Bertin 1951). This has long been described by fishermen. Consequently migratory movements have been correlated with environmental factors that result in increased discharge (rainfall, flood events, dam openings) and low light conditions (wind, increased turbidity, atmospheric depression, moon phases) (Otamura 2002). Migration also coincides with the decrease of temperature in the autumn. In this section, we will review existing literature on the effect of environmental factors on downstream migration. Habitats of eels are extremely variable. They are found in freshwater and saltwater, lakes, ponds, marshes, rivers and estuaries, and yet they must rely on the same cues for the onset of migration. The relative influence of environmental factors is seemingly different according to the type of habitat. Their effects on eel are not clearly understood.

4.2.1 Downstream Migration Period and Timing

Most of the knowledge on the timing and dynamics of downstream migration comes from the commercial fishing industry. Downstream runs of European silver eels typically start in the autumn and may last until early spring. Authorized fishing periods centre on the time of major runs. In France, fishing for silver eels is allowed from 1 October to February 15. Although little is known about possible runs in September, the majority of silver eels in the Loire River migrate in November (Durif and Elie 2008). The first migratory movements obviously depend on the timing of the silvering process. Silver migratory individuals (at stage V, see Chapter 2 for details of classification) first appear in August and September.

There are reports of downstream runs occurring in the spring. These often occur in areas regulated by dams such as in the study by Feunteun et al. (2000) and Acou et al. 2008 an or in the Dutch canals (in Deelder 1954). Spring migrations of eels, in addition to the autumn runs, have also been reported in Lake Fardume in Sweden, but these silver eels often had food items in their gut (Westin 2003). The downstream migration of silver eels has been extensively investigated by means of telemetry in the Dutch part of river Meuse by Bruijs et al. (2003) and Winter et al. (2006). The downstream migration occurred predominantly during October to November 2002 and January to February 2003. However, some eels were found to start their downstream migration a long period after their initial release. These 'slow migrators' were detected during March 2003 and even up to November 2005. Eels migrating beyond winter are probably latecomers for the autumn migration. In southern Norway (river Imsa) migration can start as early as July but the main peak is in September and October, and most individuals have migrated by November. (Hvidsten 1985; Haraldstad et al. 1985;Vøllestad et al. 1986). Timing of migration is similar in Ireland with an early start in July–August and a maximum in September and October (Poole et al. 1990). From these studies, it appears that eels start migrating earlier at northern latitudes, such as in the river Imsa in Norway (Vøllestad et al. 1986). Bergersen and Klemetsen (1988) studied eels from the northern limit of its distribution area, along the north Norwegian coast. Silver eels mostly migrated in August starting in July. By leaving earlier, these eels, which have more distance to cover, will likely reach the spawning grounds at the same time as other subpopulations.

Downstream migration generally occurs within narrow timeframes. It was found at the Lith-Alphen fisheries on the River Meuse that during most years about 20 days yielded 50% of the total silver eel catches, and in 1 year just 10 days yielded more than 60% of the total catch. Because this fishery does not operate when discharge is too high, it is likely that >50% of the total number of eels passing this location occurs in less than 20 days, i.e. the migrating silver eel population in one season passes during <20 downstream migration events (Bruijs et al. 2003; Winter et al. 2006).

Investigations described by Oberwahrenbrock (1999) in the River Mosel at the Fankel hydropower station (Germany), showed that during 3 months of nightly samplings (by means of one anchored stow net (10×5 m) in the tailrace of the hydropower station), a migration peak was observed during 1 night (during waning of the moon + increase of river discharge) in which 67% of all migrating eels passed and 90% of all eels passed during the 10 days of largest catches (i.e. 10 days of highest migration activity).

Monitoring of silver eel migration was carried out between 1999 and 2001 on the River Nive in France (Durif 2004; Durif et al. 2003; Gosset et al. 2005). In 1999 migration occurred over a period of 19 days out of 60 sampling nights. 75% of the total number of eels caught in the trap was captured during 8 consecutive nights. In 2000, migration was monitored for 75 days and eels migrated during 20 of those days. Thirty six percent of the silver eels migrated during 2 consecutive nights. Finally in 2001, monitoring took place during 90 days and eels were caught during 22 days with 40% of the eels being captured over 4 consecutive days.

4.2.2 Water Temperature

The effect of temperature is not clear and the range of temperatures during which migration takes place is extremely variable according to different studies. Overall, downstream runs occur when the temperature decreases. In Norway, on the River Imsa, eels migrate between 4°C and 18°C (Vøllestad et al. 1986). Maximum migration was observed at a water temperature of 9°C and migration decreased at both higher and lower temperatures. Hvidsten (1985) gives a maximum threshold of 14°C for silver eel movements. In Spain, the range is between 10°C and 16°C (Lobon-Cervia and Carrascal 1992). Vøllestad et al. (1986) found that eels migrated earlier when summer temperatures were lower.

On the River Nive, eels migrated between 7°C and 17°C (Fig. 4.1). Data obtained from a commercial silver eel fishery on the Loire River in France showed that eels migrated between −3°C and 21°C; most migration occurred however between 6°C and 15°C. Runs at both sites occurred during both increases and decreases in temperature.

These data show that the temperature range at which eels migrate is fairly wide and it is not possible (and maybe not relevant) to set a threshold. It is probable that temperature acts more on the physiology of eels (i.e. silvering), than on migratory movements. In any case, what the individual eel perceives is probably more a temperature difference rather than a definite value.

4.2.3 Photoperiod

Vøllestad et al. (1994) showed that photoperiod affected downstream migration using tagged silver eels. They observed that eels would migrate faster as the daylight decreased. They suggested that photoperiod had an effect on silvering. Durif et al. (2005) also hypothesized that photoperiod (or the decrease in temperature) activates the last stages of the silvering process in a similar way that these environmental factors affect smolting in Salmonids. The obvious advantage would be to synchronize the onset of migration via physiological preparation, so that the future spawners would be physiologically ready for migration at the same time.

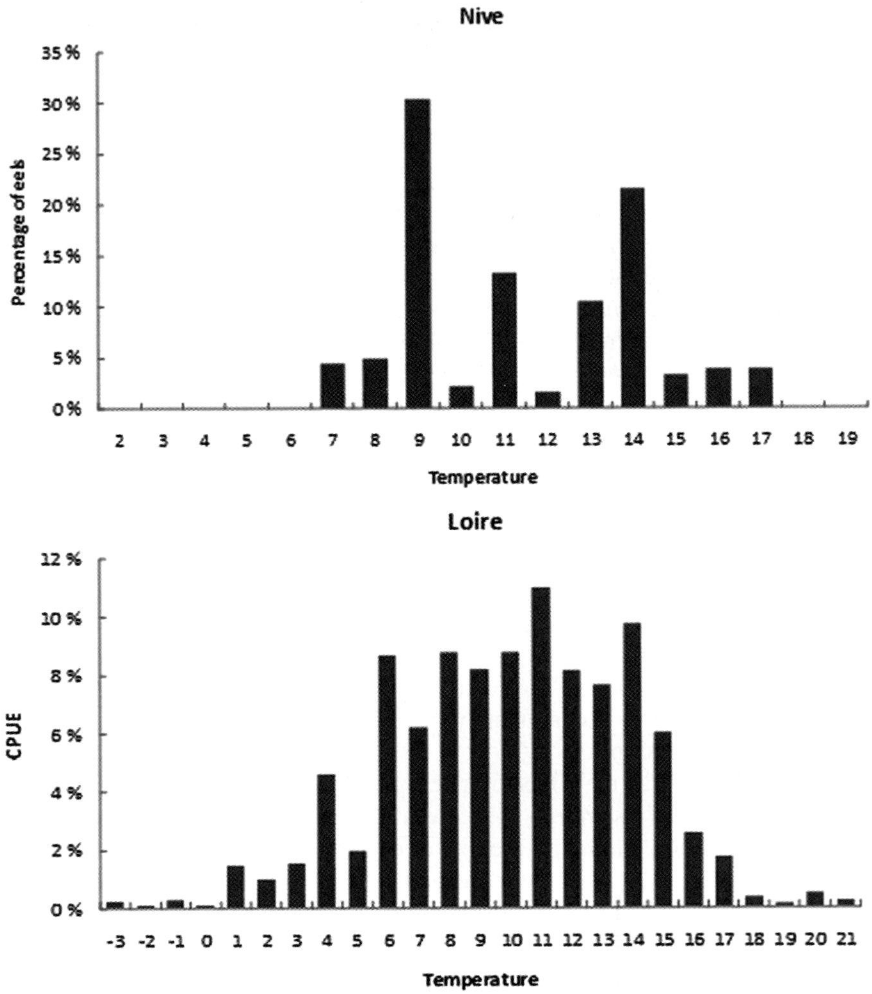

Fig. 4.1 Percentage of silver eels caught in the downstream trap on the river Nive and CPUE of the silver eel commercial fishery on the Loire vs. water temperature (pooled data for the periods 1999–2001 for the Nive and 1990–2001 for the Loire)

4.2.4 Discharge

River discharge has an influence on the migration of the eel. During high discharge the catch is larger than during low discharge (Tesch 1977). Jens (1952–1953) assumed that it is not the discharge, but a high water level that stimulates eel migration. Vøllestad et al. (1986), Jonsson (1991), Deelder (1954) and Lowe (1952) did not agree with this. The water level in the Dutch polders is kept at a constant level.

When the water level rises at another location, the water level within the polders remains the same, only the water current increases. When this happens, the migration of silver eel also increases (Deelder 1954; Durif et al. 2003).

Figures 4.2a–c show the results of the Nive River study in France (Gosset et al. 2005). During all three migratory seasons, close links between discharge and runs of eels were observed. Eels were caught in the downstream trap either during the flood or the day after. Data from the commercial silver eel fishery, located downstream of the much larger catchment, the River Loire (117,000 km^2), were also analysed for comparisons with the Nive catchment (1,030 km^2). The relationship between CPUEs (calculated as number of eels/the hours spent fishing) and discharge was not as clear every year, probably resulting from different perspectives due to the size of the catchments, i.e. longer distances between the start off point of the eel and the observer (fishery), (Figs. 4.2a–c). CPUEs were generally not proportional to discharge, and during

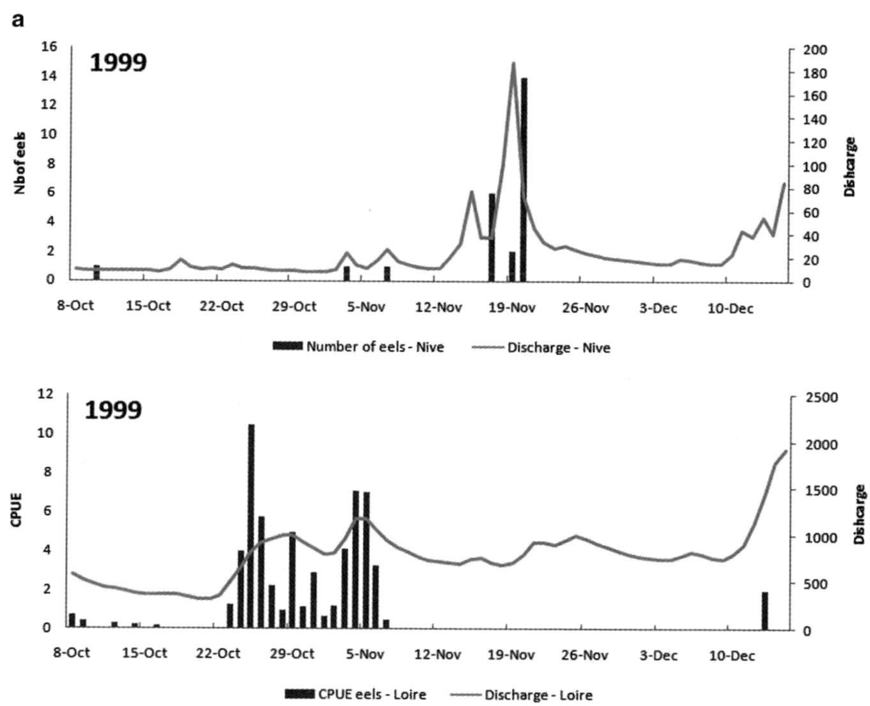

Fig. 4.2 (**a**) Comparison of downstream runs of silver eels and discharge on the Loire and Nive rivers in 1999 (Durif et al. 2003; Durif and Elie 2008; Gosset et al. 2005). Runs are expressed as number of eels caught in the trap on the Nive and as CPUEs on the Loire. Discharge is in m^3 s^{-1}. (**b**) Comparison of downstream runs of silver eels and discharge on the Loire and Nive rivers in 2000 (Durif et al. 2003; Durif and Elie 2008; Gosset et al. 2005). Runs are expressed as number of eels caught in the trap on the Nive and as CPUEs on the Loire. Discharge is in m^3 s^{-1}. (**c**) Comparison of downstream runs of silver eels and discharge on the Loire and Nive rivers in 2001 (Durif et al. 2003; Durif and Elie 2008; Gosset et al. 2005). Runs are expressed as number of eels caught in the trap on the Nive and as CPUEs on the Loire. Discharge is in m^3 s^{-1}

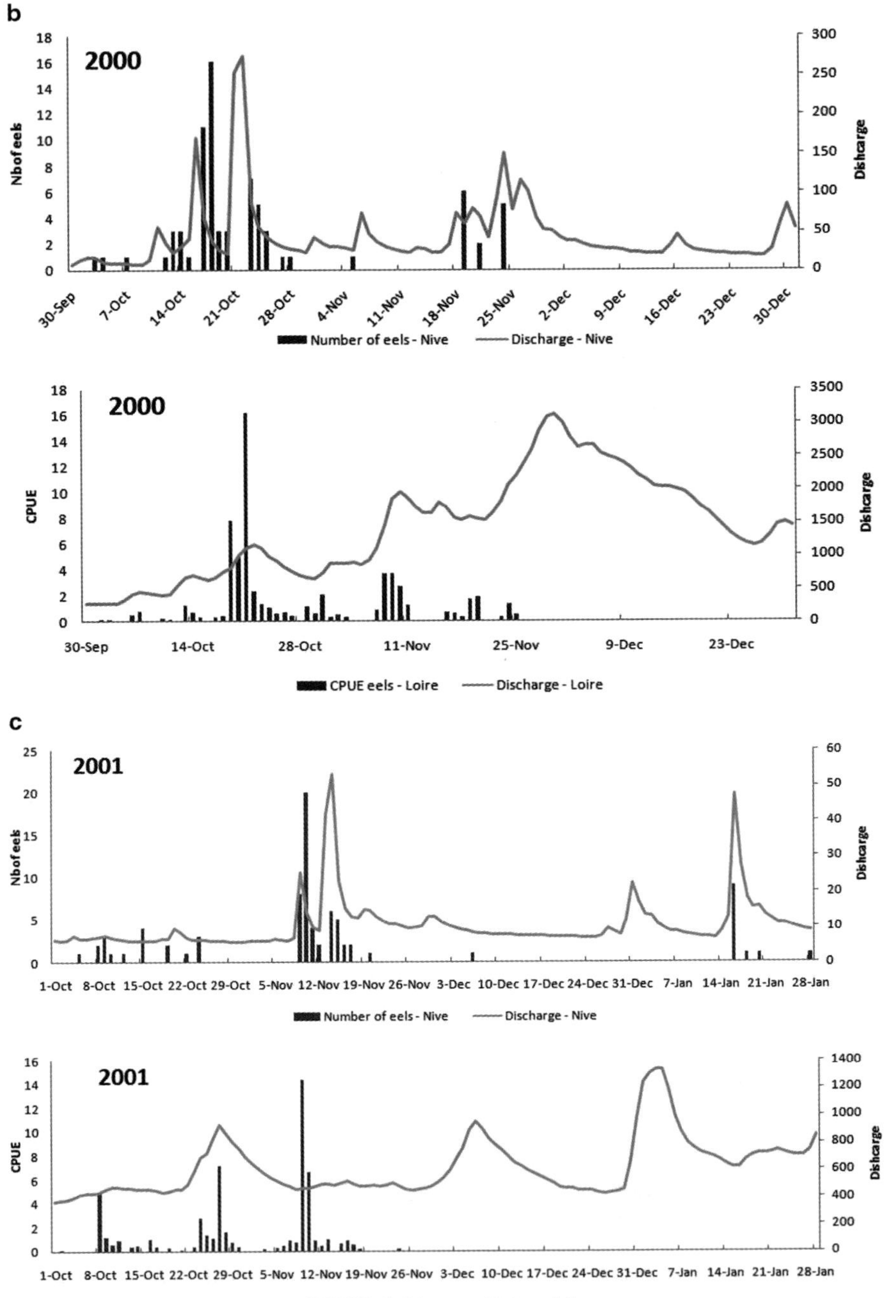

Fig. 4.2 (continued)

the 12 years of data few migration events were actually associated with high discharge (Durif and Elie 2008). In 2001, three main runs occurred and only the second run could be associated with a flood event. Data from 1996 is also shown (Fig. 4.3) when a major run (28% of the total catch of the season in 1 night) occurred without any variation in discharge at the fishery or in the Loire tributaries. The only event that occurred prior to migration was heavy rainfall, but which did not have any effect on discharge.

Migration of silver eels in the River Meuse, as shown by the cumulative passage of eels at detection stations (Fig. 4.4), showed that most of the eel migration events took place at distinct moments during a couple of weeks in autumn when the river discharge started to increase (Bruijs et al. 2003; Winter et al. 2006). The main event occurred between 25 and 28 October 2002 during the first increase of river discharge. This migration event started upstream and reached the lower part of the river about 3 days later. Further analysis of the results showed that river discharge affects the migration route of silver eel in the downstream area of the River Meuse, where eels follow the route of highest discharge (Jansen et al. 2007). This is in accordance with the results found for distribution of eels over the alternative ways near a hydropower complex. Both results provide a clear indication of equal distribution of eels over the different migration routes in accordance with the river flow.

Migrating during high flow is clearly advantageous as it takes less energy. The dynamics of downstream migration are obviously linked to discharge since the swimming speed of the eel will necessarily depend on flow velocity. However, whether discharge constitutes an actual trigger is questionable. In any case, it is not necessary for the simple reason that many areas present no perceptible current and therefore this factor is probably not the essential trigger.

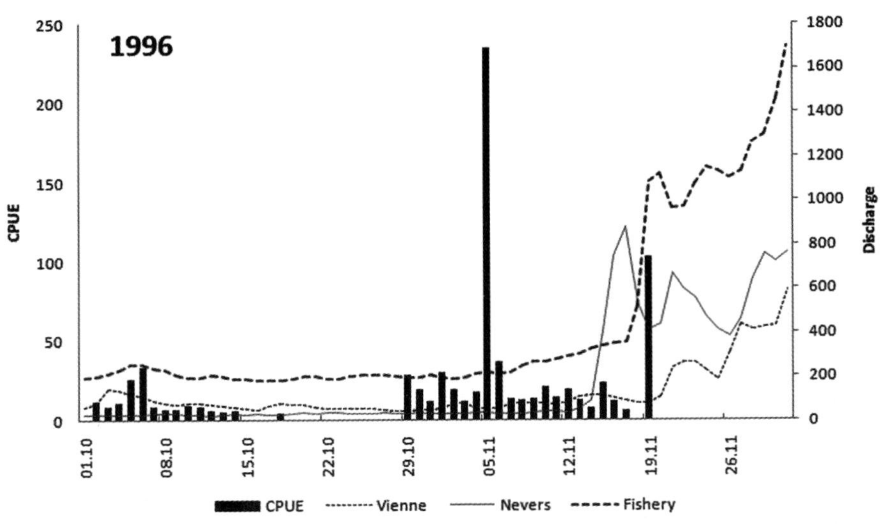

Fig. 4.3 Comparison of downstream runs of silver eels on the Loire River and discharge ($m^3 s^{-1}$) of several tributaries in 1996

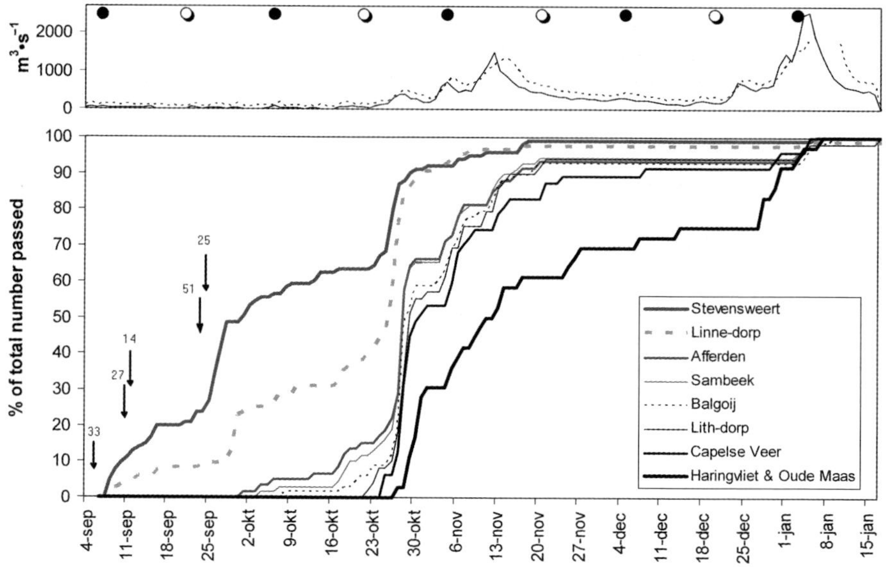

Fig. 4.4 Cumulative percentage of total passages of tagged silver eel per detection station (bottom panel). Arrows indicate the timing and number of eels released. River discharge ($m^3\ s^{-1}$) at Linne (—) and Alphen (- - -) and lunar phase (○ full moon; • new moon) are shown in the top panel (Bruijs et al. 2003)

4.2.5 *Moonlight and Light*

The effect of the moon appears to be dual. It is well known that eels are strongly photophobic, and a full moon will inhibit migration. There seems to be indeed an internal rhythm, but which can be obscured by hydrological or meteorological factors (Cullen and McCarthy 2003). From research and from experience of professional fisherman, it appears that silver eel are most active around the last quarter (Tesch 2003). Increased catches are observed in a period ranging from about 4 days prior to the last quarter, up to 2 days after the last quarter. Boetius (1967) and Jens (1952–1953) state that it is not the moonlight, but the moon phase that is the influencing factor. This is because eels in a closed tank also showed increased activity at the same time that eels in the river were active (Boetius, 1967) and because during cloudy nights the moon also had an influence. Lowe (1952), Deelder (1954), Hain (1975), Haraldstad et al. (1985) and Vøllestad et al. (1986) disagree with this. During a study by Haraldstad et al. (1985) it was observed that the migration was highest during the first quarter and stopped completely during full moon. The conclusion was drawn that migration is not influenced by phase, but depends on light. Deelder (1954) argues that the eel is exposed to the moon all of its life and it cannot be assumed that its preference is not continued when it is cloudy. Also, a night with new moon is always darker than a night with full moon, even when it is cloudy. Korringa (1947) suggests that moonlight triggers an internal rhythm that is present whether there is light or not, such as is also observed for other animals. In that case the moonlight is the influencing factor. Around the last quarter the night is darkest. At new moon the sun and moon go down and set at the same time. A few days later the moon goes down later, so that during the first hours there is light. At full moon it is light during almost the full night, but from a few days after full moon the moon comes up a while after sunset and then the first hours are dark (Hain 1975). Independent statistical studies by Deelder (1954) and Jens (1952–1953) show that the migration of silver eel is largely influenced by the moon. That effect is however not the same everywhere. In the upper part of the Rhine, migration occurs before the last quarter, in the Baltic Sea and in Dutch inland waters after the last quarter.

Commercial fishery data from the Loire River (Durif and Elie 2008) showed that runs of eels were higher during the last quarter and the new moon (respectively 36% and 29%) compared to the full moon (17%) and the first quarter (18%). Although these differences were statistically significant (χ^2 test, $p < 0.001$), Analysis of Variance showed that the moon phase factor only explained 3% of the variability in CPUEs. No significant differences were found on the Nive River; eel runs occurred equally during full moon.

It is difficult to come to a conclusion on the effect of the lunar cycle. Endogenous activity has been shown (Jens 1952–1953; Boetius 1967), but it is clearly obscured by other factors (i.e. water turbidity, cloudy skies) that suggest that luminosity is the factor mostly affecting the runs.

4.2.6 Atmospheric Pressure

Professional fisherman and researchers have observed that the migration of silver eel often occurs during storms (i.e. atmospheric depression). In a similar way to the effect of the lunar phase, atmospheric pressure has often been cited as having an effect on the eel's behavior when they are in captivity, in a lake or in a closed water body (where no flow is perceptible). Eels become more active (active swimming and escapement behaviour) during thunderstorms (Lowe 1952). Deelder (1954) found a strong correlation between runs and the occurrence of microseismic oscillations (due to passage of depressions). Hvidsten (1985) also indicated that atmospheric pressure is statistically correlated with migration peaks. According to Okamura et al. (2002), atmospheric depression would trigger eel seaward migration (i.e. estuarine migration) where no rainfall or variation in flow can be detected.

4.2.7 Turbidity/Darkness

As mentioned above, eels are strongly photophobic and the inhibition caused by light surpasses the stimulatory effect of flow according to Hadderingh et al. (1999). They avoid bright lights. This strong behaviour is used to deflect eels from hydroelectric turbines (see below). From the commercial fishery data on the Loire, it was observed that low light levels resulted in early migration, as did decreasing temperatures. All the events that tend to decrease the light level can be considered as favourable for downstream migration. During the silver eel study on the river Nive, downstream runs were mostly correlated with turbidity ($R^2 = 0.67$, Durif et al. 2003). Many authors indicate the positive effect of wind, which also increases turbidity. Eels usually stop migrating during daylight hours: commercial silver eel fishing always takes place at night. Eels sometimes pause for several weeks if there is an obstacle or when the light level is too high (Lowe 1952; Vøllestad et al. 1994; Durif et al. 2003; Watene et al. 2003). In the Middle Ages, silver eels were caught by lighting fires on the riverbanks, thus stopping the movements of downstream migrants (Bertin 1951). Bertin (1951) also mentions the experiments of Pedersen in 1906, who observed that eels stopped swimming when a beam of bright light was shone on the river.

4.2.8 Relevance of Migration Parameters

The migration of silver eel mainly occurs during the period after full moon and increases of river flow, often due to rainy periods in autumn. The main downstream migration events occur after sunset and before midnight. The most important parameters that induce the migration of silver eel therefore seem to be river flow and darkness/turbidity. These triggers are found to be significantly related to the start of migration in most studies. However, although silver eels commonly descend rivers when discharge rises (Euston et al. 1997; Bruijs et al. 2003; Durif et al. 2003), not all discharge events are accompanied by migration events. Also, darkness seems not always to be a prerequisite. Behrmann-Godel and Eckmann (2003) found during a study at a hydropower dam that the onset of migration coincided with the first flood event that followed the full moon, but was independent of daytime, because migration and turbine passage occurred during both day and night. These factors are thought to play a role in the efficiency of energy use by the animals and in protection from predators during the downstream migration.

4.3 Migration Barriers and Their Effects

4.3.1 Migration Barriers

During their downstream migration, silver eels meet many, principally manmade, obstacles that need to be passed before they reach the mouth of the river and the sea. Different types of barriers for downstream migration of silver eels exist all over Europe, presenting problems through obstruction of downstream movement as well as risk for survival of silver eels. Such barriers are, for example, large barrages, flood-control dams, flood gates, weirs, hydropower stations, sluices, pumping stations and fisheries, which are abundant in many European river systems and inland waters. Most are built to enable water management, for example to control the water discharge and water level in rivers to ensure shipping, water conservation for irrigation etc. Such structures have an impact on river flows with much diversity in flow patterns. River systems like the Rhine, Meuse, Loire and Gironde are especially highly regulated. But the mouths of rivers into the sea are also often closed by dams and sluices, such as in the Netherlands, for protection against floods. Estuaries provide a gradual transition zone in salinity and temperature. Before entering the sea, the silver eel has the time for the necessary physiological adaptation to marine conditions. However, flood control sluices provide a sharp division between fresh and marine water. There is no brackish water zone for the fish to adapt to changing osmotic conditions. Silver eels and other fish are directly confronted with marine conditions which may have physiological consequences.

All these barriers significantly hinder the migration of silver eels and other migratory fish species. On a diurnal scale, timing of migration activities shows a higher number of detections at night, especially during the first half of the night. It was found by Bruijs et al. (2003) that the diurnal differences were stronger at the hydropower stations where 63% of all detections took place between 19 and 24 h, than at the river stations with 35% between 19 and 24 h. This is a clear indication that eel show hesitation to pass hydroelectric facilities and wait until flow conditions are more favourable. Moreover, pumping- and hydropower stations cause mortality among downstream migrating silver eels that pass through the pumps and turbines. When silver eels migrate all barriers present in the main river need to be passed before reaching the sea. Silver eels that live far upstream obviously need to pass more obstacles. Also, when an eel is living at distance from the main river course, such as in river branches, small streams and polders, it may need to pass extra barriers such as weirs and pumping stations. For better understanding of the problems of barriers for downstream migration and technical measures, improved detailed knowledge of silver eel behaviour at barriers is necessary (Kroes et al. 2006).

During downstream migration in rivers, eels appear to use the direction of stream flow as an orienting stimulus (Carton 2001; Jansen et al. 2007). This behavioural response to water currents is known as rheotaxis. Barriers in river systems are specifically built to alter the water flow: the main river flow is stopped and any excessive water is spilled in a controlled manner over the weir or spillway. If enough head is present at such a location, a hydropower station is often present. In that case the main river flow is directed through the turbines in order to produce energy. It is clear that migrating eels, following the direction of the main river flow, are hampered in their migration and are directed mainly through a turbine or alternatively over the adjacent weir.

It is not only through causing difficulty in finding a proper route to pass that barriers, hydropower stations, pumping stations and fisheries cause mortality among migrating eels which is potentially of great significance to overall eel stock. Fisheries and modern pumping stations are though to be most lethal for eels as in principle all specimens that enter the fishing nets are landed and modern pumps are 100% lethal. Passage through the turbines of hydropower plants is lethal to a lesser extent, however, up to 50% of the turbine-passing eels may be killed depending on the type of turbine and when a series of stations is present in a river system, cumulative mortality may rise up to 100% (Dönni et al. 2001; Dumont et al. 2005; Dumont 2006).

The problem of downstream migration has been known for more that 200 years (Gerhardt 1893). However, only recent publications summarise the worldwide knowledge and technology standard for fish protection and downstream bypasses, i.e. to protect fish from entering intakes and to provide a safe bypass (Larinier and Travade 2002; Richkus and Dixon 2003; ATV-DVWK 2004; Dumont et al. 2005). Facilities to protect fish from entrainment and to enable a safe downstream migration have been installed at about 50 European hydropower stations (Gosset and Travade 1999; Hadderingh et al. 1999; Adam et al. 1999; Adam 2000). While there are cases that provide ample experience on the implementation and efficiency of fish protection and bypass facilities, the transfer of this knowledge to other water systems is yet to be done.

For a long time, many barriers have been impassable for any migratory fish species, both up and downstream. In recent years, fish passages are planned for many weirs and hydropower facilities in European rivers, at least to facilitate upstream migration. Even more protective measures for fishes can be expected as a result of the EC Water Framework Directive (EC 2000) which requires an undisturbed migration for fishes in river systems. Also, the EU has published guidelines for the protection of eels (EC 2005, 2006). For downstream passage, eels may use a fish passage, or in some cases specific downstream bypasses. Although upstream migration facilities are being installed at many barriers, only at a very limited number of hydropower plants are functional downstream migration facilities in operation.

4.3.2 *Effects of Barriers on Downstream Migrating Silver Eel*

In large regulated river systems, weirs and hydropower facilities are the most important barriers as they not only hinder migration, they also cause mortality among downstream passing fish populations. Next to the danger of hydropower plants, silver eels may also be affected by large drops over weirs or spill ways (Kroes et al. 2006) and by impingement and entrainment at pumping stations and industrial cooling water intakes. Increased flow velocities in front of intakes may act as an attraction flow for downstream migrating fish, or when the intake flow is higher than the maximum swimming capacity, fish may not be able to escape from the intake flow velocity. But especially pumping stations, which are used for water management in lowland polder areas in the Netherlands, Germany and Flanders, are thought to cause major damage to fish when these try to leave the polder area. Mortality due to passage through pumping stations is thought to be much higher than with hydropower stations. In most cases no alternative routes, i.e. fish passages or weirs, are available, thus the pumping station is the only route. Also, the pump types that are in use cause 100% mortality. In summary, depending on the type of barrier, different types of damage to fish can occur (Adam and Bruijs 2006):

- In dams with heads over 10 m and/or an impact of hard structures in the tail water region, fish may suffer lethal injuries.
- Screens and trash racks in front of water intakes cause damage when fish are impinged and pressed against the racks by the approach velocity.
- Turbines and pumps cause mortality by direct contact with turbine blades, shear stress, cavitation and pressure differences.
- Fish that are entrained through the screens at cooling or potable water intakes die in the following zone.
- Fish that have passed and survived hydraulic structures are easy prey for predatory fish and fish-eating birds in the tailwater.
- A fast change of osmotic conditions at flood control sluices in coastal areas may provoke high mortality rate among the fish, especially when freshwater fish species are discharged into marine water.

However, a major cause of direct mortality among downstream migrating silver eels are hydropower stations, of which many have been built in European rivers in combination with a weir. These stations cause problems for downstream migrating fishes, in particular for the diadromous species which maintain a two-phase life-history involving extensive migrations between sea and freshwater. In many European rivers like the Rhine and Meuse numerous hydropower stations have been installed. The types of turbines used in European rivers are Kaplan, Straflo or Francis turbines. In the larger rivers these stations can have a capacity from 5 up to circa 100 MWe, like Iffezheim in the River Rhine.

The operation of hydropower stations in rivers produces potential injury to downstream migrating fishes if they become entrained in turbine-intake flow and pass along fast moving turbine blades. Early reports of fish mortality at European hydropower stations were published by Raben Von (1955, 1957) and by Berg (1986). In these studies, the average mortality rates for the European eel, which is very susceptible to injury of turbine blades, ranged between 15% and 38%. However, eels are not only entrained through the turbines; impingement on the trash racks may also cause severe mortality (Adam and Bruijs 2006). Trash racks prevent the intake of debris. When the approach velocities towards the trash rack or screens are higher than the swimming capacity, fish are impinged on the screen (Fig. 4.5).

In the period 1990–2002, several investigations on downstream fish migration at hydropower stations have been carried out or are underway all over the world. On the first International Catadromous Eel Symposium held in St. Louis (USA) in 2000, several presentations dealt with problems of eel passage at hydroelectric power stations in Canada, the United States, New Zealand and France (Dixon 2003). In Germany, Holzner (2000) studied the impact of fish passage through the turbines of Dettelbach hydroelectric power station on the river Main, a tributary

Fig. 4.5 Eel collected from the trash rack at the Wahnhausen power station at the river Fulda in Hessen, Germany (photo by Adam B)

of the river Rhine. In the Netherlands fish passage mortality was studied at Linne hydroelectric power station on the river Meuse in 1990/1991 (Hadderingh and Bakker 1998) and in 1999, as well as in 2001 and 2002 (Bruijs et al. 2003).

Injuries and mortality are established directly after collecting the fish; this mortality can be defined as direct mortality. Delayed mortality, for example mortality after 24 h, cannot be excluded but in many cases was not investigated. The most frequent type of injury of eel are bisections (chopped), broken backbones and internal haemorrhages. Other fish species show considerably lower mortalities. This difference is probably caused by the relatively great length of the eel resulting in a higher strike probability of the turbine blades during the passage through the turbines.

In Europe, most applied turbine types are Kaplan turbines. Different eel mortalities are found: at Linne, the highest lethal injuries, up to 30% of the fish that pass the turbines, have been found for eels, 38% mortality at Neckarzimmern Germany, 22% mortality at Dettelbach Germany, 20% at Obernau Germany, 24% at Beauharnois Canada, and 37% at Raymondville USA (Hadderingh and Bruijs 2002). Of special importance for silver eels is the cumulative mortality due to passage through a series of hydroelectric power stations. Typically, 15% to 25% of silver eels passing through the turbines of a hydroelectric power station in a large river, such as the Meuse or the Mosel, are fatally injured. Therefore, the average survival rate amounts to 80% or 0.8 per facility. The passage through a second facility leads to an average rate of 0.8*0.8 = 0.64. The total survival rate of downstream migrating silver eels after the passage of five hydroelectric power stations drops to approximately 30%. A higher number of obstacles virtually prevent all silver eels from reaching the sea. Dönni et al. (2001) made a calculation of the cumulative mortality of downstream migrating silver eels for the River Rhine between Lake Constance and Basel. The cumulative mortality of downstream migrating silver eels, passing the maximal number of 11 hydroelectric power stations in this stretch of river, amounted 93%.

Bruijs et al. (2003) investigated the cumulative mortality of silver eels in the River Meuse by means of telemetry. In September 2002, 150 silver eels were surgically implanted with Nedap-transponders and released at the catch site, upstream in the River Meuse. Of these, 121 started to migrate downstream of which 37% successfully reached the North Sea (Winter et al. 2006). Hydropower mortality was at least 9% and assessed to be 16–26%. Fisheries mortality was at least 16% (reported recaptures) and estimated to be 22–26%. Also a difference was found in diurnal pattern; 63% of the eels that passed through the turbines did so during the first 5 h of the night after sunset, whereas for the stations on free-flowing sections this was only 35%.

4.4 Eel Behaviour at Barriers

4.4.1 Hesitation at Trash Racks

In order to investigate the technical possibilities and efficiency of mechanical systems in front of the intakes of hydropower stations, a number of laboratory studies have been performed focusing on eel behaviour, including Adam (1999), Amaral (2003), Adam et al. (2002). Only a few studies have investigated the specific behaviour of eels during their natural downstream migration and passage of hydropower stations (Haro et al. 2000; Durif et al. 2003; Bruijs et al. 2003; Behermann-Godel and Eckmann 2003).

It was found during investigations in model flumes, that eels react differently to mechanical barriers than do other fish species (Adam and Schwevers 2001; Amaral 2003). When eels approach the intake area of hydropower stations they show hesitation when confronted with the trash racks. Although the main river flow that is used by silver eels to migrate downstream in general runs through the turbines, the majority stop and try to seek alternative routes. Some eels even return in an upstream direction and come back only weeks or months later (Bruijs et al. 2003; Winter et al. 2006; Jansen et al. 2007). When no alternative routes are available or when these are difficult to find, the eel eventually will pass through the trash racks. The eel may swim close up to the trash rack, without passing, and stay there for a while, swimming up and down.

It can be concluded that eels show a clear hesitation to pass the trash racks at hydropower station as well as an upstream-orientated escaping movement in front of trash racks. The upstream-orientated escaping movements of eels in front trash racks were clearly observed in flume tank experiments and described by Adam (1999, 2000) and Adam et al. (1999) (Fig. 4.6).

The typical behaviour of the silver eel in front of trash racks has also been observed the field. For example, during fishing efforts in the headwater with fyke

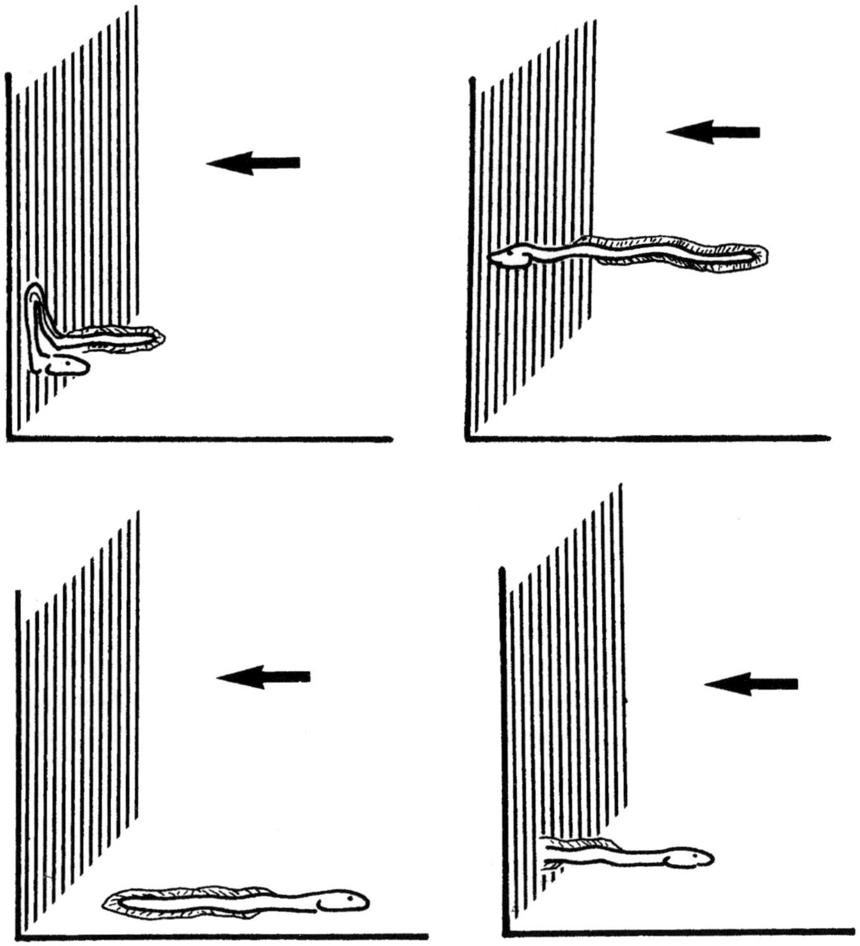

Fig. 4.6 Different phases of the reaction of eel (return behaviour) in front of a mechanical barrier (Adam et al. 1999)

nets placed with the opening in a downstream direction, eels that have returned from the trash rack entered the fyke nets in an upstream direction (RH Hadderingh, 2000, personal communication). The observations of the upstream-orientated escaping movements of eels in front of trash racks are furthermore confirmed by investigations of Holzner (2000) at the hydropower station of Dettelbach on the river Main in Germany and by Haro et al. (2000) at the Cabot hydropower station on the Connecticut River (USA). Figure 4.7 (Brown 2005; Brown et al. in press) provides a clear example of the movement of eels in the fore bay. The eels were found to move around at a range of depths, spending much time at or near the bottom. The eels are also found in the main flow in the fore bay for most of the time.

Bruijs et al. (2003) found clear differences between tagged eels in passage behaviour at hydropower stations, i.e. eels showing hesitation in front of the intakes; 40%

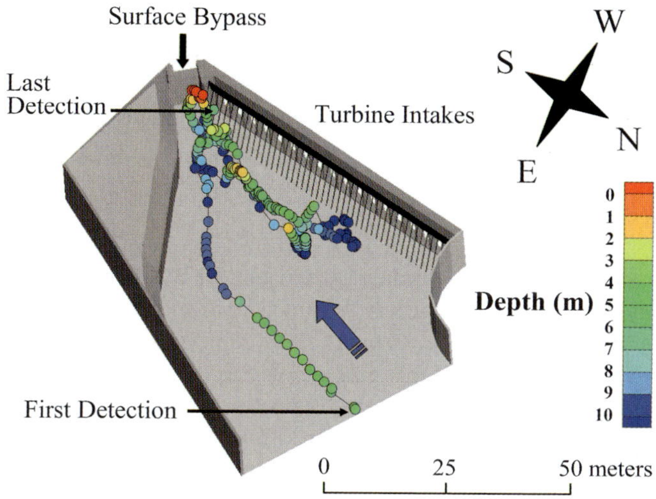

Fig. 4.7 Track of depth-telemetered eel in Cabot Station fore bay, Connecticut River, Massachusetts. The depth of each detected position is colour coded; dark blue is at or near the bottom, green is in the mid-water column, and red is at or near the surface (see legend) (Brown 2005; Brown et al. 2008)

of the eels showed recurrence, in contrast to the river stations where this hardly occurred, indicating a hesitation to pass the trash rack (Winter et al. 2006; Jansen et al. 2007). Eel passage of the river stations was characterized by usually only one or a series of detections with 2 min intervals, whereas the passage of the detection stations at the intakes of hydropower stations showed a different pattern (Winter et al. 2006). Apart from eels that were detected once, one group showed recurrence with intervals above 2 min, varying from several hours to several weeks. Another group showed stationary behaviour indicated by a series of detections with 2 min intervals. Eels showing odd behaviour might seek alternative routes to pass the hydropower station, for example by migration through the fish passage or over the weir.

Behrmann-Godel and Eckmann (2003) found that when migrating eels arrived at the hydropower station, they either passed through the turbines immediately or stayed upstream of the powerhouse for up to 8 days, showing a characteristic circling behaviour. Circling eels repeatedly approached the trash rack, sprinted upstream, and finally passed through the turbines with the next high water discharge.

4.4.2 Making Use of Silver Eel Behaviour to Bypass Hydropower Stations

The damages caused to silver eels occur during the passage of turbines, where they suffer mechanical injuries as a result of being hit by the runners and turbine blades or from pressure fluctuations and turbulences (shear stress) due to the flow conditions. In addition, in the long term, fish-friendly turbines are not expected to replace the old ones at the many existing hydroelectric power stations. In order to prevent damage of eel by turbine passage, several types of barriers, which are basically divided into mechanical and behavioural screening techniques, can be applied.

4.4.2.1 Mechanical Barriers

Laboratory and field research studies carried out all over the world have proved the suitability of mechanical barriers mounted in front of turbines to physically prevent eels from entering the turbines. Laboratory experiments (Adam et al. 1999; Schwevers 1998) showed that eels are able to escape at velocities $<0.5\,m\,s^{-1}$. Eels that come in direct physical contact with the trash rack react by turning and swimming upstream. In addition, Adam (2000) concluded, based on laboratory observations, that eels show an initial period of hesitation to pass the trash rack and searching behaviour but at a later stage tried to pass the bars. Eels of 70 cm length have been found to pass trash racks with a bar width of 2 cm, and 40–50 cm eels passed through 18 mm racks (Jens 1987).

Under certain conditions, the silver eel is thus very capable of maintaining itself in front of trash racks at water intakes for a certain period of time. Field observations of this behaviour have been made by Haro et al. (1999) at the Cabot hydropower station on the Connecticut River. Eels of up to 90 cm showed a strong hesitation to pass racks of 102 mm bar width, but in the end did pass the trash rack. This behaviour was also observed by professional fisherman in the river Mosel (R Eckmann, 1999, personal communication) and at investigations in the river Main (Holzner 2000) as well as by Bruijs et al. (2003) in the River Meuse.

Gosset et al. (2005) investigated the behaviour of eels in the forebay by means of radiotelemetry. Almost half of the eels returned upstream of the headrace after their release, and most eventually migrated downstream over the adjacent dam during appropriate environmental conditions. Upon arrival at the power plant, eels displayed foraging behaviour in the forebay with frequent displacements interrupted by long resting periods in zones with low current. The repulsive effect of the trash rack located in front of the turbine intake increased with increasing turbine discharge. The study indicated that a trash rack with a smaller bar-spacing (around 20 mm), associated with an appropriate bypass, could deflect a large proportion of the female eels from the turbines. However, the risk of mortality due to impingement on the trash rack was not investigated. This behaviour can be used as a way to divert silver eel from turbines, which must be combined with altered turbine management and the availability of appropriate bypasses or alternative route over an adjacent weir.

Results obtained in laboratory and field studies show that the following values for mechanical barriers should not be exceeded (DWA 2005):

- Internal width of mechanical barrier: $d <= 15\,mm$
- Approach velocity before the barrier: $v <= 0.5\,m\,s^{-1}$

Middle-sized and large hydroelectric power stations usually employ barriers with d = 50 to 150 mm and the approach velocity may reach up to 1.0 m s^{-1}. In most hydroelectric power stations, the construction of a 15 mm barrier would create serious technical and economical difficulties. Furthermore, there exist to date no appropriate devices for the removal of the accumulated debris at such fine barriers. Today the employment of the described barriers is limited to hydroelectric power stations with a maximum discharge rate of 20 m^3 s^{-1}, whereas normal facilities, such as those in the rivers Meuse, Rhine and the Mosel reach levels of up to 400 m^3 s^{-1}. Mortality among downstream migrating silver eels can therefore only be prevented if passage through turbines is avoided.

4.4.2.2 Behavioural Barriers

Fish protection at water intakes has traditionally been achieved by fine physical screens. However, such systems are costly to purchase, operate and maintain. They may become blocked easily by debris, restricting water flow resulting in loss of effectiveness. Behavioural deterrent systems are an alternative, used where mechanical fish screening is impracticable. Fish have a number of well-developed senses and are able to detect and react to light, sound, temperature, pressure change and many other stimuli. The relative sensitivity and capacity to react to any of these stimuli varies with species and life stage. To be effective, the stimulus must be strong enough to repel fish at a range where they are not at risk of being involuntarily drawn into turbines by the strength of the water current. For eels, light is an effective stimulus. In general, several reactions to light can be expected including general behaviour towards natural light, such as the diurnal rhythm of fish in relation to the surrounding illumination level where the activity of the fish changes with the illumination level. Reactions of fish as a result of observing artificial light sources absent in their natural environment can be exploited with fish deflection systems.

With respect to the development of a suitable deflection method at water intakes, two behavioural characteristics of eels are very important. Firstly, eels are negatively phototactic. During the day eels hide in the bottom. They are nocturnal and the downstream migration of most silver eels occurs at night (Haraldstad et al. 1985; Tesch 1977). This behaviour is demonstrated by the catch of eels in the cooling water sieves of Bergum thermal power station (situated at a lake) where eels almost exclusively impinge during the dark period (RH Hadderingh, 2001 personal communication). The downstream migration of silver eels takes place almost completely during the night. Van Drimmelen (1951), Lowe (1952), Bräutigam (1961, 1962) and Hölke (1964) used the light-avoidance reaction of eels for commercial fishery purposes and increased their catch by directing eels to fishing nets by means of underwater lamps. Secondly, eels are strongly attracted to water currents; the highest commercial catches are achieved in the main stream of rivers (Tesch 1977). The preference of migrating eels for areas with the highest current might be explained by the saving of energy as the need for active swimming is reduced (Hansen and

Jonsson 1985) and a shorter migration period is required to reach the sea (Thorpe et al. 1981). The application of light barriers has been extensively investigated by Hadderingh (1982) and Hadderingh et al. (1992, 1999). The results (i.e. the effectiveness) of a light system are varying and strongly depend on different factors such as: the angle between the barrier and the flow direction, the water velocity, turbidity and the availability of an effective bypass.

4.4.2.3 Bypasses

Most of the described observations on eel behaviour are discussed with respect to the design of appropriate downstream passage facilities. The eels' behaviour clearly indicates that it is possible to provide a safe passage facility near the trash racks. As the eels are not willing to pass the trash rack they will seek for an alternative route, provided the conditions are appropriate. Such an alternative passage or bypass must be found by the eel through a clear attractive flow. Investigations on designs for the effectiveness of such bypass facilities will increase in the near future as it may provide cost-effective solution for eel passage at hydropower plants.

To achieve good results in deflecting fish, both the deflecting part and the bypass part of the system has to be successful. Till recently little attention was paid to the design of bypasses. A properly functioning bypass has to fulfil a number of conditions such as: dimensions of the opening, water velocity and illumination level. Intensification of laboratory and practical research on bypasses is therefore recommended. To be successful the bypass must comply with a number of requirements. In France, much research has been done on this aspect by Larinier et al. (1996) and harinier (1998) but these did not concern eels.

Gosset et al. (2005) tested the efficiencies of two potential bypasses for downstream migration. These consisted of a surface and a bottom sluice installed on the spillway of a small hydroelectric power plant in France. Total efficiency of both bypasses, evaluated on the basis of downstream movement of radiotagged eels, ranged from 56% to 64%. However, preferred passage through the bottom bypass for both tagged and untagged eels was confirmed by telemetry, as three to four times as many eels transited through the bottom bypass compared to the surface one.

After entering the bypass system, the fish must be transported through a tube or drain system. To prevent damage, sharp angles, rough walls and shocks must be prevented. Also, the maximum current must be 12 m s^{-1} and the diameter of the tubes/drains must be large enough to prevent clogging. In order to prevent damage at the outlet, the outlet must be located in a horizontal plane. Also, the maximum height must be 1–3 m.

At non-lethal barriers, such as weirs and dams, a spillway or small bypass with sufficient attraction flow will do. The eels search for possibilities to pass the barriers and will make use of the flow through such facilities, as these are the only routes. At hydropower stations, where the main river flow runs through the turbines, directing eel towards the bypass is more difficult. The findings described above regarding the hesitation reaction of eel in front of a hydropower facility, combined with the observations of their searching behaviour, indicate that there are good possibilities to divert eels from trash racks and inlet channels of hydropower facilities. The simplest way is through an existing fish pass. However, in many cases the entrances are not optimal: they are often located too far away from the area in which the eels search and the attraction flow is too small to attract them. A bypass needs to be located close to the power station inlet. The entrances could or should be located at the bottom, mid-depth and at the surface.

The behavioural pattern of eels in front of passable and impassable trash racks up to an approach velocity of 0.5 m s^{-1} is characteristic for eels. It has led to the development of alternative bypass systems, such as the so-called Bottom Gallery® (by Floecksmühle and IFÖ, Germany). It consists of a bottom-oriented bypass system crossing the inlet of a hydropower station. However, it has not been tested yet under real conditions in front of a hydropower station. Also other bypass constructions that enhance the passage of silver eel through alternative routes have been designed, but so far no real implementations have been made.

4.4.2.4 Prediction of Downstream Migration

As an alternative to mechanical and behavioural fish protection systems, protection can be achieved by means of turbine management that takes into account the migration behaviour and migration timing of silver eels. To operate an effective turbine management system, information is needed on the timing of silver eel migration, i.e. a prediction tool is required. Attempts at modeling downstream migration have been made (Hvidsten 1985; Vøllestad et al. 1986; Vøllestad et al. 1994; Euston et al. 1997; Haro et al. 2003), but results in terms of prediction of number of migrating eels are low because models have to take into account not only environmental factors, but also the number of eels physiologically ready to migrate (i.e. silver-phase eels) in a given year as well as the type of hydrosystem (river, lake, marsh) and the presence of obstacles (dams and hydropower stations). Another approach relies in using the innate locomotor activity of eels at the time of downstream migration to predict the runs (Durif et al. 2008). Such a tool may be modelled information or an early warning system.

A tool to assess the distinct periods with silver eel peak migration is the Migromat system (Adam 2000). The Migromat predicts migration events in free-living eels by monitoring activity of eels held in tanks close to the river. The system contains two tanks, each with five connected compartments, which are continuously supplied with river water (Fig. 4.8). Each tank is stocked with 60 silver eels. Displacements of PIT-tagged eels between the compartments are detected by frame antennae around the openings between compartments. Increases in activity level, i.e. eel passages between compartments, indicate pre-migratory restlessness and predict migration timing. For its application, the Migromat systems should be installed on the riverbank close to the hydropower stations. The warnings provided by the system allow for turbine management, i.e. the turbines can be closed down during migrating peak periods of silver eel to offer them a save passage over the weirs and bypasses.

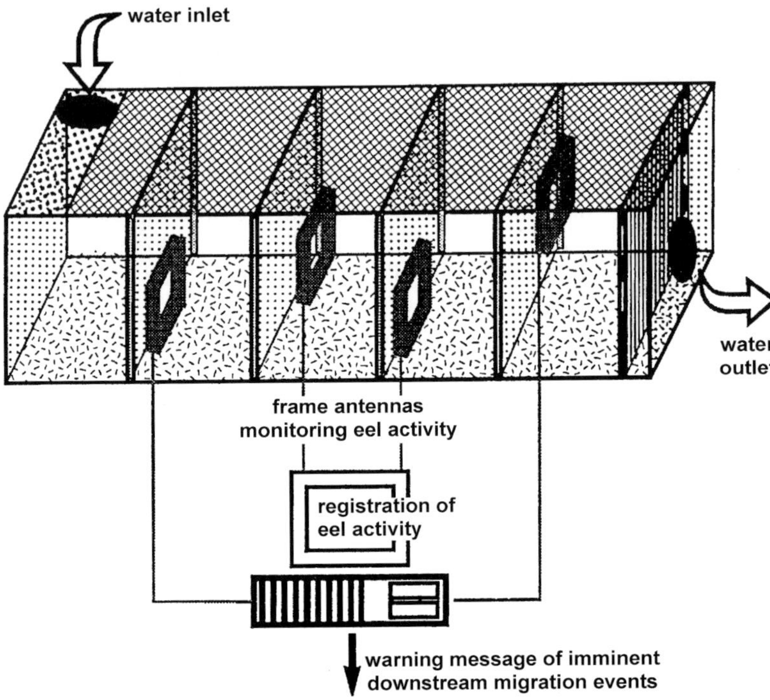

Fig. 4.8 Schematic overview of the Migromat system. Only one of the two tanks is shown

Bruijs et al. (2003) investigated the application of the Migromat at Linne hydropower station in the River Meuse. The correspondence of migration events in the river found by the different monitoring experiments with the warnings provided by the Migromat verify that the system accurately registers the pre-migratory restlessness of eels, thereby predicting the downstream migration events of silver eels with high precision (Adam and Bruijs 2006; Bruijs et al. in press). An example of this prediction is shown in Fig. 4.9. The increase of activity within the Migromat corresponds with an increase of turbine passages as monitored by means of telemetry. This increased turbine passage occurs only several hours after the warning is sent out by the system, leaving sufficient time for the hydropower station operator to adjust the turbine operations.

The prediction of migration events by this early warning system enables an eel-friendly turbine operating management of hydropower plants, by which a high percentage of downstream migrating eels can be saved. Application of the Migromat throughout the migration season of 2002, would have reduced the total mortality by hydropower in the Dutch section of the River Meuse by a maximum of 69.4%, assuming that all eels pass over the weir or through the fishpass without hesitation (Bruijs et al. in press). Closing down the turbines for longer migration periods e.g. a number of months during the autumn, being the main migration period for silver eel, means a substantial loss of electricity production for the electricity companies. Closing down the turbines during short periods with peak migration will be a better option.

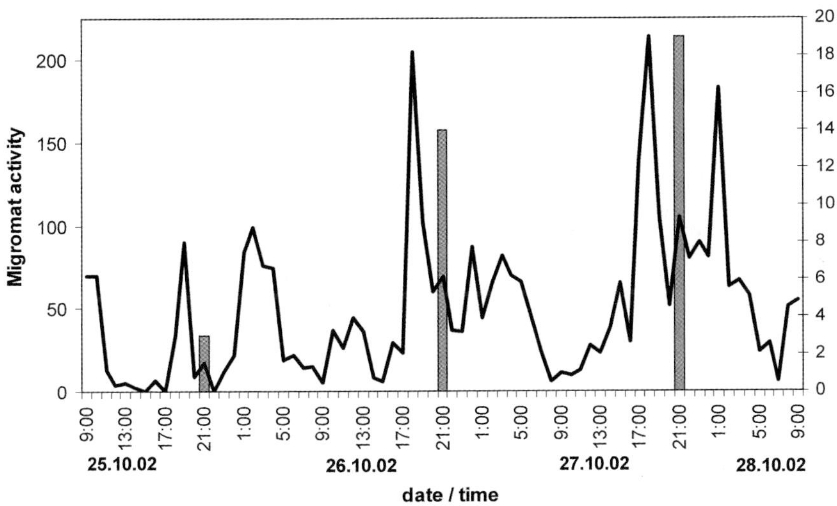

Fig. 4.9 Number of eels passing the Linne hydropower station as determined by the telemetry system (bars) and eel activity index in the Migromat system (line), 25–28 October 2002

The Migromat system is applied as a protective measure at the Wannhausen hydropower station in the river Fulda in Germany. Due to the limited bar width of 2 cm and the high approach velocity of ~1 m s^{-1} silver eels impinged on the trash rack, leading to damage of vertebrae and organs (Schwevers 1998), see also Fig. 4.3. Based on the warnings by the Migromat, the flow of the turbines of the Wannhausen hydropower station is halved in order to maintain an approach velocity of 0.5 m s^{-1}. This enables the eels to escape from the trash rack. By implementing this measure, the damage was decreased significantly (Adam B and Dumont U, 2007, personal communication).

Similar experiments were carried out during the silver eel monitoring study on the River Nive (Durif et al. 2008). PIT-tagged eels were placed in tanks fitted with a flat board antenna (Fig. 4.10). Eels showing escaping behaviours and bursts of activity were automatically recorded by the antenna. Counts of activity (i.e. number of detections) were compared with the number of wild migrating eels caught in the trap as well as changes in environmental factors. Results showed that activity of eels was always high during the first 3–4 days after they were placed into the tank. After this acclimation period, they only showed peaks of activity during specific periods, and sheltered under the clay tiles for the rest of the time. Video observations showed that during active periods, eels would either swim in circles along the walls of the tanks at moderate speed or try to escape by jumping above surface level towards the water inflow.

Although correlations between the number of eels caught in the trap and the activity counts were not statistically significant, both were closely related (Fig. 4.11, top panel). Most migratory eels in 2000 were caught between 13/10 and 25/10 when captive eels were most active. During this period, the four peaks of activity

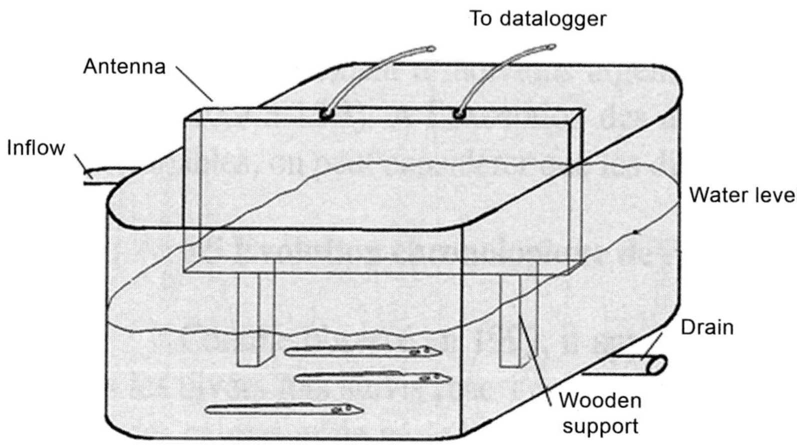

Fig. 4.10 Schematic of the eel activity detection tank

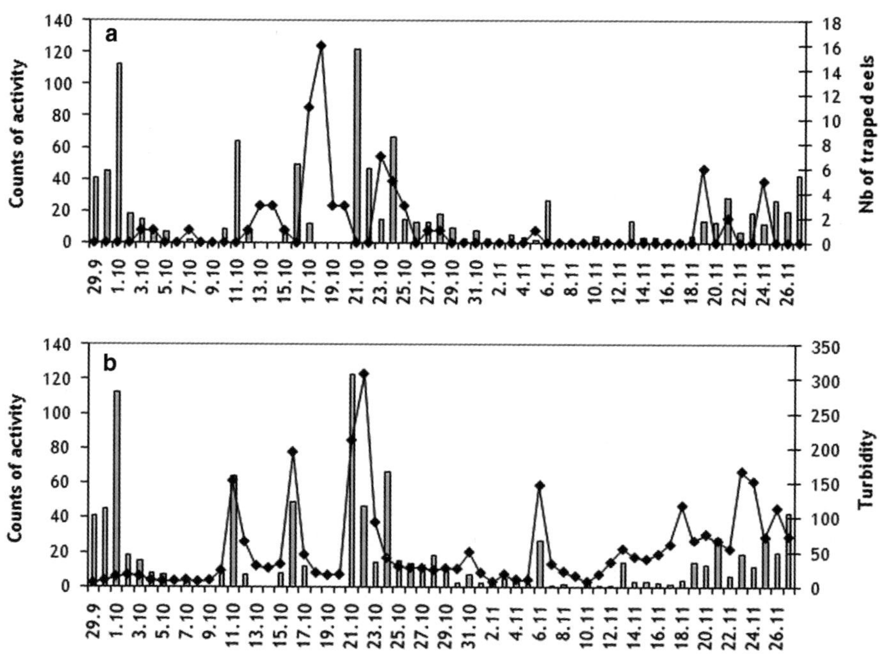

Fig. 4.11 Results of the study on locomotor activity of silver eels on the river Nive in 2000. Counts of activity (bars) are compared with the number of wild silver eel caught in a trap

were closely followed, approximately 24 h later, by captures of migrants in the trap. Activity in the tank was then reduced until 18/11, when eels became agitated again concurrently with a second wave of downstream migration. These activity peaks were also closely related to increases in turbidity (Fig. 4.11, bottom panel) and both were highly correlated ($R^2 = 0.45$ when the acclimation period was removed).

4.5 Conclusions

Worldwide, populations of the eel, *Anguilla anguilla*, are undergoing a severe decline. The causes for this are not clear because a large part of the eel's life cycle remains unknown (i.e. transatlantic migration and spawning). However, it is quite clear that inland, anthropogenic impacts have a disastrous effect. Indeed, the result of human activities, such as obstruction of downstream migration causes great mortality. Hydropower has a clear and significant effect on the total survival and escapement of adult silver eel. To implement efficient mitigation measures, we need to understand what triggers the eel to migrate.

Recent studies on the silver phase eels have shown that it is important to differentiate the physiological modifications that lead to the silver stage (i.e. silvering, see Chapter 2) from the actual migratory movements. Eels are physiologically ready to migrate at the end of summer and this will probably determine for the most part the onset of migratory movements. Temperature and photoperiod variations appear to trigger silvering in fish that are physiological ready (length, age, fat content). However, migratory movements seem to depend on other factors. Habitats of eels are extremely variable. They are found in freshwater and saltwater, lakes, ponds, marshes, rivers and estuaries, and yet they must rely on the same cues for the onset of downstream movements. River flow is usually well correlated with runs of eels, but it is probably not a necessary trigger since many water bodies do not present any variation in flow. Knowledge obtained from the Migromat early-warning system as well as the study of locomotor activity of silver eels on the river Nive, also support this, since eels in the tanks could not perceive the flow (Durif et al. 2008). A decrease in light level is essential for eel migration. Increased turbidity causes greater activity in silver eels. More research on sensory mechanisms of eels is needed to understand what causes the increased locomotor activity.

The European Commission has published a regulation to establish a framework for the protection and sustainable use of the stock of the European eel (EC 2007), which, among other measures, includes the target that management actions must allow an escapement of 40% of the silver eel population from each European river basin. These developments will increase pressure for measures that protect eels and other species during their downstream migrations.

Obstructions in rivers may cause significant delay and mortality among downstream-migrating silver eel. To investigate the relative role of, in particular, hydropower stations, research on eel behaviour in relation to dam and turbine passage has been conducted, but much work is needed still. The knowledge obtained in laboratory studies and field observation on the eel's specific behaviour can be used to develop methods to guide silver eels along obstructions. As they migrate in groups and collectively react to environmental cues, a combination of protective measures based on their behaviour is thought to be optimal.

References

Acou A, Laffaille P, Legault A, Feunteun E (2008) Migration pattern of silver eel (*Anguilla anguilla*, L.) in an obstructed river system. Ecology of Freshwater Fish 17: 432–442.

Adam B (1999) Aalabwanderung – ergebinsse von Versuchen in Modellgerinnen. Arbeiten des Deutschen Fischereiverbandes 70: 37–68.

Adam B (2000) MIGROMAT® – ein Frühwarnsystem zur Erkennung der Aalabwanderung. Wasser and Boden 52(4): 16–19.

Adam B and Bruijs MCM (2006) General requirements of fish protection and downstream migration facilities. Investigations on the river Meuse system. In Free Passage for Aquatic Fauna in Rivers and Other Water Bodies. Proceedings of the International DWA Symposium on Water Resources Management. April 3–7, 2006, Berlin, Germany.

Adam B and Schwevers U (2001) Planungshulfen fur den Bau funktionsfahiger Fischaufstiegsanlagen. Bibliothek Natur and Wissenschaft, Band 17: 64S.

Adam B, Schwevers U and Dumont U (1999) Beiträge zum Schutz abwandernder Fische – Verhaltensbeobachtungen in einem Modellgerinne – Solingen (Verlag Natur and Wissenschaft), Bibliothek Natur und Wissenschaft 16: 63S.

Adam B, Schwevers U and Dumont U (2002) Rechen- und Bypassanordnungen zum Schutz abwandernder Aale. Wasserwirtschaft 92(4–5): 43–46.

Amaral SV, Winchell FC, McMahon BJ and Dixon DA (2003) Evaluation of angled bar racks and louvers for guiding silver phase American eels. American Fisheries Society, Symposium 33, pp. 367–376.

ATV-DVWK (2004) ATV-DVWK-Themen WW-8.1: Fischschutz- und Fischabstiegsanlagen – Bemessung, Gestaltung, Funktionskontrolle. Hennef (ATV-DVWK – Deutsche Vereinigung für Wasserwirtschaft, Abwasser und Abfall e.V.) 256S.

Behrmann-Godel J and Eckmann R (2003) A preliminary telemetry study of the migration of silver European eel (*Anguilla anguilla* L.) in the River Mosel, Germany. Ecology of Freshwater Fish 12: 196–202.

Berg R (1986) Fish passage through Kaplan turbines at a power plant on the River Neckar and subsequent eel injuries. Vie Milieu 36: 307–310.

Bergersen R, Klemetsen A (1988) Freshwater eel *Anguilla anguilla* L. from North Norway with emphasis on occurrence, food, age and downstream migration. Nordic Journal of Freshwater Research 64: 54–66.

Bertin L (1951) Les anguilles. Variation, croissance, euryhalinité, toxicité, hermaphrodisme juvénile et sexualité, migrations, métamorphoses, Vol. Paris: Payot.

Boetius J (1967) Experimental indication of lunar activity in European silver eels, *Anguilla anguilla*. Medelelser fra Danmarks Fiskeri og Havundersøgelser 6(1): 1–6.

Bräutigam R (1961) Über Versuche zur Intensivierung des Blankaalfangs durch die Kombination von Lichtsperren und Großreusen und ihre grundsätzlichen Bedingungen. Fischereiforschung; 4; (1/2); S. 19–25.

Bräutigam R (1962) Intensivierung des Blankaalfangs mit Hilfe von Lichtsperrketten und Bedingungen der gewerblichen Anwendung. Fischereiforschung: 5: 6: S. 8–15.

Bruijs MCM, Polman HJG, Van Aerssen GHFM, Hadderingh RH, Winter HV, Deerenberg C, Jansen HM, Schwevers U, Adam B, Dumont U and Kessels N (2003) Management of silver eel: Human impact on downstream migrating eel in the river Meuse. EU-Report Contract Q5RS-2000-31141.

Bruijs MCM Hadderingh RH, Schwevers U, Adam B, Dumont U, Winter HV (In press) Managing human impact on downstream migrating European eel (*Anguilla anguilla*) in the River Meuse. In Eels at the Edge science, status and conservation concerns. Casselman JM and Cairns D (eds). Bethesda, MD: American Fisheries Society, Symposium 58, Bethesda, Maryland.

Brown LS (2005) Downstream passage of silver phase American eels at a small hydroelectric facility. M.Sc. thesis, University of Massachusetts, Amherst, 122 pp.

Brown LS, Haro A and Castro-Santos T (2008) Three-dimensional movement of silver-phase American eels (*Anguilla rostrata*) in the forebay of a small hydroelectric facility. In Casselman JM and Cairns D (eds) Eels at the Edge. Bethesda, MD: American Fisheries Society, Symposium 58 (in press).

Brown LS, Haro A and Castro-Santos T (in press) Three-dimensional movement of silverphase American eels (*Anguilla rostrata*) in the forebay of a small hydroelectric facility. In Eels at the edge: science, status, and conservation concerns. Casselman JM and Cairns DK (eds.) American Fisheries Society, Symposium 58, Bethesda, Maryland.

Carton G (2001) Do fish have feelings? Discovering how fish detect water flows. Water & Atmosphere 9(3): 10–11.

Cullen P and McCarthy TK (2003) Hydrometric and meteorological factors affecting the seaward migration of silver eels (*Anguilla anguilla*, L.) in the lower River Shannon. Environmental Biology of Fishes 67: 349–357.

Deelder CL (1954) Factors affecting the migration of the silver eel in Dutch inland waters. Journal du Conseil Permanent International Pour I'Exploration de la Mer, 20: 177–185.

Dixon DA (ed) (2003) Biology, Management, and Protection of Catadromous Eels, Vol. Bethesda, MD: American Fisheries Society, Symposium 33.

Dönni W, Maier KJ and Vicentini H (2001) Bestandsentwicklung des Aals (*Anguilla anguilla*) im Hochrhein. Mitteilungen zur Fischerei Nr. 69. Bundesamt für Umwelt, Wald und Landschaft (BUWAL), Bern.

Dumont U, Danderer P and Schwevers U (2005) Handbuch Querbauwerke. Düsseldorf (Ministerium für Umwelt und Naturschutz, Landwirtschaft und ländlichen Raum NRW), 212S.

Dumont U (2006) Report on the restoration of the longitudinal connectivity of the river Sieg. Ingenieubüro Floecksmühle. March, 2006. pp 15.

Durif C (2004) Thesis abstract: The downstream migration of the European eel *Anguilla anguilla*: Characterisation of migrating silver eels, migration phenomenon, and obstacle avoidance. Cybium 28: 26.

Durif C, Elie P, Gosset C, Rives J, and Travade F (2003) Behavioral study of downstream migrating eels by radio-telemetry at a small hydroelectric power plant. In Dixon DA (ed) Biology, Management, and Protection of Catadromous Eels. Bethesda, MD: American Fisheries Society, Symposium 33, pp. 343–356.

Durif C, Dufour S, and Elie P (2005) The silvering process of *Anguilla anguilla*: A new classification from the yellow resident to the silver migrating stage. Journal of Fish Biology 66: 1025–1043.

Durif CMF and Elie P (2008) Prediction of downstream migration of silver eels in a large river catchment based on commercial fishery data. Fisheries Management and Ecology 15: 127–137.

Durif CMF, Travade F, Rives J, Elie P and Gosset C (2008) Relationship between locomotor activity, environmental factors, and timing of the spawing migration in the European eel, *Anguilla anguilla*. Aquatic Living Resources 21: 163–217.

DWA, 2005. DWA Topics: Fish protection technologies and downstream fishways. Dimensioning, Design, Effectiveness Inspecion. July 2005. Deutsche Vereinigun fur Wasserwirtschaft, Abwasser und Abfall e.V. pp 226.

EC (2000) Water Framework Directive, 2000/60/EC.

EC (2005) Proposal for the recovery of the European eel stock, COM(2005) 472.

EC (2006) Proposal for a Council Regulations establishing measures for the recovery of the stock of the European eel. 9247/06. May 11, 2006.

EC (2007) Proposal for a Council Regulation establishing measures for the recovery of the stock of European eel. 13139/05 PECHE 203 – COM(2005) 472 final. June 5, 2007.

Euston ET, Royer DD, and Simons CL (1997) Relationship of emigration of silver American eels (*Anguilla rostrata*) to environmental variables at a low head hydro station. Proceedings of the International Conference on Hydropower 1997: 549–558.

Feunteun E, Acou A, Lafaille P, and Legault A (2000) The European Eel: Prediction of spawner escapement from continental population parameters. Canadian Journal of Fisheries and Aquatic Sciences 57: 1627–1635.

Gerhardt (1893) Über Aalleitern und Aalpässe. Zeitschrift für Fischerei 1 S. 194–199.
Gosset C and Travade F (1999) Etude de dispositifs d'aide a la migration de dévalaison des salmonidae: Barrières comportementales. Cybium 23(suppl 1): 45–66.
Gosset C, Travade F, Durif C, Rives J, Elie P (2005) Tests of two types of bypass for downstream migration of eels at a small hydroelectric power plant. River Research and Applications 21: 1095–1105.
Hadderingh RH (1982) Experimental reduction of fish impingement by artificial illumination at Bergum power station. Int. Revue ges. Hydrobiol. 67: 887–900.
Hadderingh RH, Van der Stope JW and Habraken JWPM (1992) Deflecting eels from water inlets of power stations with light. Irish Fisheries Investigations Series A (Freshwater) 36, 80–89.
Hadderingh RH and Bakker HD (1998) Fish mortality due to passage through hydroelectric power stations on the Meuse and Vecht rivers. In Jungwirth M, Schmutz S and Weiss S (eds) Fish Migration and Fish Bypasses. Oxford: Fishing News Books, pp. 315–328.
Hadderingh RH and Bruijs MCM (2002) Hydroelectric power stations and fish migration. Tribune de l'eau 55(619–620/5–6): 79–87.
Hadderingh RH, Beijer RFLJ De, Van Aerssen GHFM and Van Der Velde G (1999) Reaction of silver eels on underwater light sources and water currents: An experimental study. Regulated Rivers: Research and Management 15: 356–371.
Hain JHW (1975) The behaviour of migratory eels, Anguilla anguilla, in response to current, salinity and lunar period. Helgolander Wissenschaftliche Meeresuntersuchungen. 27: 211–233.
Hansen LP and Jonsson B (1985) Downstream migration of hatchery-reared smolts of Altantic salmon (*Salmo salar*) in the river Imsa, Norway. Aquaculture 45: 237–248.
Haraldstad O, Vøllestad LA and Jonsson B (1985) Descent of European silver eels, *Anguilla anguilla*, in a Norwegian watercourse. Journal of Fish Biology 26: 37–41.
Haro A, Castro-Santos T and Boubée J (2000) Behavior, and passage of silver-phase Ameican eels, *Anguilla rostrata* (LeSueur) at a small hydroelectric facility. Dana 12:33–42.
Haro A, Castro-Santos T, Whalen K, Wippelhauser G and McLaughlin L (2003) Simulated effects of hydroelectric project regulation on mortality of American eels. In: D.A. Dixon (ed.) Biology, Management, and Protection of Catadromous Eels. Bethesda, MD: American Fisheries Society, pp. 357–365.
Hölke H (1964) Aalfang mit Hilfe von elektrischen Licht in der Fluss-Fischerei. Deutsche Fischerei Zeitung 11: 82–88.
Holzner M (2000) Neue Versuche zur Schaden minimierung bei der Aalabwanderung. SVK Fischereitagung, Künzell, March 1–2, 2000, pp. 8.
Hvidsten NA (1985) Yield of silver eel and factors effecting downstream migration in the stream Imsa, Norway. Institute of Freshwater Research Drottningholm Report 62: 75–85.
Jansen HM, Winter HV, Bruijs MCM and Polman HJG (2007) Just go with the flow? Route selection and mortality during downstream migration of silver eels in relation to river discharge. ICES Journal of Marine Science 64: 1437–1443.
Jens G (1952–1953) Über den lunaren Rhythmus der Blankaalwanderung. Arch. für Fischereiwissenschaft. 4: 94–110.
Jonsson N (1991) Influence of water flow, water temperature and light on fish migration in rivers. Nordic Journal of Freshwater Research 66: 20–35.
Korringa P (1947) Relations between the moon and periodicity in the breeding of marine animals. Ecological Monographs 17: 347–381.
Kroes MJ, Gough P, Schollema PP and Wanningen H (2006) From sea to source; Practical guidance for restoration of fish migration in European rivers.
Larinier M (1998) Upstream and downstream fish passage experience in France. In: Jungwirth, M., Schmutz, S. & Weiss, S. (eds.) Fish migration and fish bypasses. Fishing News Books, Oxford. 127–145.
Larinier M and Travade F (1998) Petits aménagements hydroélectriques et libre circulation des poissons migrateurs. Houille Blanche 8: 46–51.
Larinier M and Travade F (2002) Downstream migration: Problems and facilities. Bulletin Français de la Pêche et de la Pisciculture 634: 181–207.

Larinier M, Porcher JP, Travade F and Gosset C (1996) Passes à Poissons. Expertise, conception des ouvrages de franchissement. Conseil Supérieur de la Pêche, Paris, pp. 336.

Lobon-Cervia J and Carrascal M (1992) Seasonal timing of silver eels in a Cantabrian stream. Archiv fur Hydrobiologie 125: 121–126.

Lowe RH (1952) The influence of light and other factors on the seaward migration of the silver eel, *Anguilla anguilla*. Journal of Animal Ecology 21(2): 275–309.

Oberwahrenbrock K (1999) Aalschutzinitiative Rheinland-Pfalz/RWE-Energie AG., Projektfortshrittsbericht 1, Stand der Arbeiten an den Projektzielen a, b und c zum Januar 1999. pp 78.

Okamura A, Yamanda Y, Mikawa N, Tanaka S and Oka HP (2002) Exotic silver eels *Anguilla anguilla* in Japanese waters: seaward migration and environmental factors. Aquatic Living Resources 15, 335–341.

Poole WR, Reynolds JD, Moriarty C (1990) Observations on the silver eel migrations of the Burrishoole river system, Ireland, 1959 to 1988. Internationale Revue Der Gesamten Hydrobiologie 75: 807–815.

Raben Von K (1955) Kaplanturbinen und Fische. Wasserwirtschaft 45: 196–200.

Raben Von K (1957) Über Turbinen und ihre schädliche Wirkung auf Fische. Zeitschrift für Fischerei und deren Hilfwissenschaften, Band VI N.F. Heft 1–8: 171–182.

Richkus WA and Dixon DA (2003) Review of research and technologies on passage and protection of downstream migrating catadromous eels at hydroelectric facilities. In Dixon DA (ed) Biology, Management, and Protection of Catadromous Eels. Bethedsa, MD: American Fisheries Society, Symposium 33, pp. 377–388.

Schwevers U (1998) Die Biologie der Fischwanderung, Bibliothek Natur and Wissenschaft Band 11, VNW Verlag Natur and Wissenschaft Solingen, pp. 84. Tesch F-W (1977) The Eel. London: Chapman & Hall.

Tesch FW (2003) The Eel. Fifth edition. Oxford: Blackwell.

Thorpe JE, Ross LG, Struthers G and Watts W (1981) Tracking Atlantic salmon (*Salmo salar* L.) through Loch Voil, Scotland. Journal of Fish Biology 19: 519–537.

Todd PR (1981) Timing and periodicity of migrating New Zealand freshwater eels (*Anguilla* spp.). New Zealand Journal of Marine and Freshwater Research 15: 225–235.

Travade F and Larinier M (1992) La migration de dévalaison: Problèmes et dispositifs. Bulletin Francais de la Pêche et de la Pisciculture 326/327: 165–176.

Van Drimmelen DE (1951) Vangst van schieraal met behulp van licht. Visserij Nieuws 3: 137–140

Vøllestad LA, Jonsson B, Hvidsten NA, Næsje TF, Haraldstad O and Ruud-Hansen J (1986) Environmental factors regulating the seaward migration of European silver eels. Canadian Journal of Fisheries and Aquatic Sciences 43: 1909–1916.

Vøllestad LA, Jonsson B, Hvidsten NA and Naesje TF (1994) Experimental test of environmental factors influencing the seaward migration of European silver eels. Journal of Fish Biology 45: 641–651.

Watene EM, Boubée JAT and Haro A (2003) Downstream movement of mature eels in a hydroelectric reservoir in New Zealand. In Dixon DA (ed) Biology, Management, and Protection of Catadromous Eels. Bethesda, MD: American Fisheries Society, Symposium Series 3, pp. 295–305.

Westin L (2003) Migration failure in stocked eels *Anguilla anguilla*. Marine Ecology Progress Series 254: 307–311.

Winter HV, Jansen HM and Bruijs MCM (2006) Assessing the impact of hydropower and fisheries on downstream migrating silver eel, *Anguilla anguilla*, by telemetry in the River Meuse. Ecology of Freshwater Fish 15: 221–228.

Part III
Performance

Chapter 5
High Pressure Resistance and Adaptation of European Eels

Philippe Sébert, Aurélie Vettier, Aline Amérand, and Christine Moisan

Abbreviations BL: Body length; COX: Cytochrome c oxidase; d: Density; EC: Energy charge; FPT: Final preferred temperature; HP: Hydrostatic pressure; PRI: Pressure resistance index; Ptr: Pressure threshold; PUFA: Polyunsaturated fatty acids; ROS: Reactive oxygen species; SB: Swim bladder

5.1 Introduction

The fish is often used as a model for experiments under pressure because it has three main advantages (Barthélémy 1985): (a) It is a vertebrate which has an organisation not very different from a mammal; for some authors, due to the high number of species, the fish is even the typical vertebrate (Bone et al. 1995); (b) The ectothermic quality of fish enables study of pressure/temperature interactions; (c) Because they breathe water, fish can be used to study separately the effects of hydrostatic pressure and/or the effects of gas pressure, which is useful in understanding mammalian physiology (see Sébert 1997).

The adverse effects of high pressure on fishes have been known since the 19th century (Bert 1878; Regnard 1885) and are well reviewed in the literature (Gordon 1970; Sébert and Macdonald 1993; Sébert 2003). However, the great majority of these studies are concerned with the biological effects of high hydrostatic pressure (HP) *per se* considered, like temperature, as a thermodynamic factor (see Somero 1991; Sébert 2003). In other words, the effects of pressure on fishes have been studied more from a fundamental than from an ecophysiological point of view. In fact, in regard to pressure, we can consider three types of fishes: those which never encounter variations in pressure, and are unable to adapt to pressure effects (Sébert 2003); those which always live at great depth and have a poor resistance to low (atmospheric) pressure (Siebenaller and Somero 1989; Somero 1991); and finally those which, as the eel, *Anguilla*, live a part of their life under pressure and thus must adapt to its adverse effects. In terms of

P. Sébert, A. Vettier, A. Amérand, and C. Moisan
Laboratoire ORPHY, UFR Sciences, 6, Avenue Le Gorgeu, CS 93837, 29238 Brest Cedex 3 – France

G. van den Thillart et al. (eds.), *Spawning Migration of the European Eel,*

environmental adaptation, the eel is a fascinating animal, and an excellent model to examine almost any environmental effects in the field of fish biology (Owen 2001).

5.2 Eel and Hydrostatic Pressure

Eel biology and the eel life cycle are largely described in other chapters of this book and are not reviewed here. Most of the eel life cycle is in the yellow non-migratory stage, at atmospheric pressure in lakes, rivers or estuaries: this is mainly a period (the longest) of development and growth. However, the most important periods are under pressure at sea: the hatching and migration of larvae from the Sargasso Sea to the rivers, and many years later, the migration of silver eels from the rivers back to the Sargasso Sea. The time under pressure is relatively short, but requires many adaptations in order to cope with the high energy demand in a new environment (adaptation to temperature, salinity, pressure, swimming, etc.).

The pressures to which the eels are exposed and how they dive are not known. The possible range is from 200 to 700 m (2 to 7 MPa; Fricke and Kaese 1995; Aoyama et al. 1999; McCleave and Arnold 1999; Jellyman and Tsukamoto 2002; Tesch 2003; Van Ginneken and Maes 2005), although one eel has been photographed at 2,000 m (Robins et al. 1979). Moreover it seems that during its migration under pressure, eels perform daily vertical migrations (Tesch 1978). Most of the results reported in this chapter concerns experiments performed at 10.1 MPa hydrostatic pressure (10.1 MPa # 100 atm # 100 bars # 1,000 m H_2O) in fresh water.

5.3 Is the Yellow Non-migrating Eel Able To Cope with Pressure Effects?

Before discussing the ability of silver eels to migrate at great depths, i.e. to explore the ecophysiological aspects of the migration, it is necessary to report what is known about the effects of high pressure on fishes. In fact, migration towards the Sargasso Sea is a big challenge for the eel which must acclimate to changes in salinity and temperature, to starvation, absence of solar light and, above all, to continuous swimming at high pressure. In other words, migration represents a high energy demand which must be satisfied in order to ensure the survival of the species. In this chapter, we will focus on the energetic aspects knowing that pressure has evidently other specific, sometimes adverse effects (see for review; Macdonald 1984; Sébert and Macdonald 1993; Sébert 1997).

It is clear that the silvering process (the metamorphosis from yellow non-migrating eel to silver migrating one) is a preparation for migration i.e. to all the concomitant changes in environmental factors. In this sense, its importance in the pressure acclimatisation processes can be only deduced and elucidated by comparing pressure effects on silver and yellow eels. Numerous physiological data have been obtained in yellow eels before and after pressure exposure (Tables 5.1 to 5.4). When a fish is exposed to high hydrostatic pressure, at constant temperature and normoxia, it exhibits a strong motor activity concomitant with an increase in oxygen consumption. The pressure at which these symptoms appear depends on species, size, rate of compression, water temperature, salinity (Ponat 1967; Flügel 1972; Vettier et al. 2005) and the metabolic rate before compression (see Sébert and Macdonald 1993; Table 5.5). In order to compare different conditions and /or species,

Table 5.1 Physiological data (mean ± SEM) in heart, brain and plasma from yellow freshwater eels, *Anguilla anguilla*, under normal and high pressure

		10.1 MPa		
	0.1 MPa	1–3 days	30 days	References
Heart				
Isometric twitch tension, mN	3.2	3.7 (SW)	–	Gennser et al. 1990
Fc, bpm	28 ± 1.2	50 ± 2.0*	–	Sébert and Barthélémy 1985b
Oxygen consumption, mmol h^{-1} kg^{-1}	1.0 ± 0.03	6 (end of compression)	0.67 ± 0.05*	Simon et al. 1989
Respiratory frequency min^{-1}	35	50	–	Belaud 1975
Q10	2.3 ± 0.3	–	1.2 ± 0.3	Sébert et al. 1995
Plasma				
Ht,%	28.8 ± 0.9	–	27.1 ± 1.1	Sébert et al. 1991
Proteins, g L^{-1}	32.5 ± 1.5	–	31.4 ± 0.8	Sébert et al. 1991
Lactates, mM	1.65 ± 0.42	–	0.71 ± 0.12	Sébert et al. 1991
Na^+, mEq L^{-1}	144 ± 6	–	151 ± 5	Sébert et al. 1991
Cl^-, mEq L^{-1}	59.5 ± 2.9	–	73.3 ± 3.1*	Sébert et al. 1991
K^+, mEq L^{-1}	8.6 ± 0.6	–	7.1 ± 0.5	Sébert et al. 1991
Ca^{++}, mEq L^{-1}	7.6 ± 0.3	–	7.3 ± 0.4	Sébert et al. 1991
Mg^{++}, mEq L^{-1}	3.6 ± 0.1	–	4.6 ± 0.2	Sébert et al. 1991
Osmolarity, mosm L^{-1}	333 ± 17	–	365 ± 11	Sébert et al. 1991
Glucose, g L^{-1}	1.1 ± 0.2	1.5 ± 0.2	–	Simon 1990
Total FFA, nmol L^{-1}	501 ± 22	265 ± 61*	–	Simon 1990
NE, nmol L^{-1}	5.7 ± 1.7	14.7 ± 1.7*	–	Sébert et al. 1987
E, nmol L^{-1}	11.1 ± 1.6	20.2 ± 2.1*	–	Sébert et al. 1987
Brain				
NE, nmol g^{-1}	1.40 ± 0.072	1.42 ± 0.085	–	Sébert et al. 1986
E, nmol g^{-1}	0.24 ± 0.010	0.28 ± 0.009	–	Sébert et al. 1986
DA, nmol g^{-1}	0.83 ± 0.041	1.01 ± 0.032	–	Sébert et al. 1986
DOPAC, nmol g^{-1}	0.27 ± 0.016	0.30 ± 0.022	–	Sébert et al. 1986
5-HT, nmol g^{-1}	0.89 ± 0.095	1.06 ± 0.136	–	Sébert and Barthélémy 1985a
5-HIAA, nmol g^{-1}	0.21 ± 0.011	0.29 ± 0.028*	–	Sébert and Barthélémy 1985a
Gly, nmol g^{-1}	1,394 ± 59	1,390 ± 76	–	Sébert and Barthélémy 1985b
Gln, nmol g^{-1}	14,377±439	15,834 ± 674	–	Sébert and Barthélémy 1985b
GABA, nmol g^{-1}	2,560 ± 99	2,530 ± 76	–	Barthélémy et al. 1991
MDA, nmol g^{-1}	6.1 ± 0.5	24.2 ± 6.5*	–	Sébert and Barthélémy 1985a

Experimental pressure 10.1 MPa (1,000 m depth); rate of compression 1.0 MPa min^{-1}.
TW = 15–17°C. SW = seawater.
*Statistically different ($p < 0.05$ or better) from controls at 0.1MPa. Values are mean ± SEM

Table 5.2 Physiological data (mean ± SEM) in gills and liver from yellow freshwater eels, *Anguilla anguilla*, under normal and high pressure

		10.1 MPa		
	0.1 MPa	1–3 days	30 days	References
Gill				
Water content, %	75.0 ± 0.5	–	76.0 ± 0.5	Sébert et al. 1991
Na^+, μEq g_{ww}^{-1}	52.2 ± 3.8	–	59.5 ± 1.7	Sébert et al. 1991
Cl-, μEq g_{ww}^{-1}	24.0 ± 1.5	–	31.3 ± 1.5*	Sébert et al. 1991
K^+, μEq g_{ww}^{-1}	64.3 ± 0.8	–	62.9 ± 2.0	Sébert et al. 1991
Na^+,K^+ATPase, μmol_{Pi} mg_{Prot}^{-1} h^{-1}	9.8 ± 0.6	–	4.5 ± 0.04*	Sébert et al. 1991
Mg^{++}ATPase, μmol_{Pi} mg_{Prot}^{-1} h^{-1}	7.51 ± 0.8	–	3.0 ± 0.3*	Sébert et al. 1991
Anisotropy	0.220± 0.001	–	0.215±0.001*	Sébert P et al., 1989 unpublished data
Unsaturation index	191 ± 8	–	197 ± 7	Sébert P et al., 1989 unpublished data
Saturation ratio	0.48 ± 0.03	–	0.37 ± 0.05*	Sébert P et al., 1989 unpublished data
Number Cl^-, cells mm^{-2}	798 ± 174	–	3,095 ± 403*	Dunel-Erb et al. 1996
Fractional area, μm^2 mm^{-2}	6,804 ± 1,316	–	46,194±4,470*	Dunel-Erb et al. 1996
Number of mucus cells, aff	17.8 ± 1.1	–	4.4 ± 0.6	Dunel-Erb et al. 1996
Liver				
Proteins, mg 100 mg_{ww}^{-1}	10.0 ± 0.5	9.5 ± 0.4	10.1 ± 0.3	Simon 1990
Lactate, μmol g_{ww}^{-1}	0.81 ± 0.32	0.84 ± 0.28	0.95 ± 0.43	Simon 1990
Glycogen, μg 100 mg_{ww}^{-1}	0.90 ± 0.17	0.67 ± 0.10	0.65 ± 0.11	Simon 1990
Total FA, μmol g_{ww}^{-1}	2.2 ± 0.6	5.0 ± 0.5*	4.2 ± 0.5*	Simon 1990
IDH, μmol_{subst} min^{-1} kg^{-1}	10.1 ± 0.99	11.3 ± 2.5	15.1 ± 0.75	Simon 1990
MDH, μmol_{subst} min^{-1} kg^{-1}	608 ± 34.4	671 ± 45	782 ± 50.9	Simon 1990
CS, μmolsubst min^{-1} kg^{-1}	6.19 ± 0.735	9.06 ± 1.22	6.47 ± 0.819	Simon 1990
COX, μmol_{subst} min^{-1} kg^{-1}	1.38 ± 0.178	0.713 ± 0.71[a]	1.58 ± 0.118*	Simon 1990
GOT, μmol_{subst} min^{-1} kg^{-1}	92.2 ± 10.23	134 ± 24.2	142 ± 21.97	Simon 1990
GPT, μmol_{subst} min^{-1} kg^{-1}	33.7 ± 3.25	33.9 ± 3.8	37.1 ± 1.7	Simon 1990
PFK, μmol_{subst} min^{-1} kg^{-1}	0.12 ± 0.019	0.15 ± 0.01	0.16 ± 0.023	Simon 1990
PK, μmol_{subst} min^{-1} kg^{-1}	5.23 ± 0.559	4.53 ± 0.43	5.02 ± 0.372	Simon 1990
LDH, μmol_{subst} min^{-1} kg^{-1}	10.28±1.289	13.2 ±1.1	9.78 ±0.896	Simon 1990
GPD, μmol_{subst} min^{-1} kg^{-1}	1.43 ± 0.422	1.68 ± 0.31	2.22 ± 0.416	Simon 1990
CPK, μmol_{subst} min^{-1} kg^{-1}	10.7± 1.7	7.29± 0.87	7.75 ± 0.715	Simon 1990
ATP, nmol g_{ww}^{-1}	834 ± 119	–	749 ± 109	Simon 1990
ADP, nmol g_{ww}^{-1}	589 ± 36	–	730 ± 108	Simon 1990
AMP, nmol g_{ww}^{-1}	321 ± 34	–	449 ± 80	Simon 1990
IMP, nmol g_{ww}^{-1}	1,108±775	–	490 ±122*	Simon 1990
EC	0.64 ± 0.03	–	0.57 ± 0.04	Simon 1990

(continued)

Table 5.2 (continued)

	0.1 MPa	10.1 MPa 1–3 days	10.1 MPa 30 days	References
Membrane extracts				
PC, % total	59.2 ±2.27	–	50.5 ± 1.2*	Sébert et al. 1994
PE, % total	11.6 ±1.4	–	19.8 ±1.5*	Sébert et al. 1994
Cholesterol, nmol mg_{Prot}^{-1}	11.5 ± 2.3	–	20 ± 3.6*	Sébert et al. 1994

The experimental pressure was 10.1 MPa (1,000 m depth) and the rate of compression was 1.0 Mpa min^{-1}. TW = 15–17°C.
*Statistically different ($p < 0.05$ or better) from controls at 0.1MPa. PC: phosphatidylcholine; PE: phosphotidylethanolamine. Values are mean ±SEM.

Table 5.3 Physiological data (mean ± SEM) in red (R) and white (W) muscles from yellow and silver freshwater eels, *Anguilla anguilla*, under normal and high pressure

		0.1 MPa	10.1 MPa 1–3 days	10.1 MPa 30 days	References
Red muscle fibres respiration					
ADP/O	R	2.52 ± 0.04	–	2.87 ± 0.05*	Théron et al. 2000
MO2, μmol min^{-1} g^{-1}	W	0.10 ± 0.01	–	0.11 ± 0.01	Vettier and Sébert 2004
	R	0.29 ± 0.05	–	0.28 ± 0.04	Vettier and Sébert 2004
$OH^{\bullet}$ ng.g_{ww}^{-1} min^{-1}	R	6.5 ± 0.9	–	5.5 ± 1.2	Amérand et al. 2005
Membranes					
Anisotropy,	R	0.247± 0.002	–	–	Vettier et al. 2006
Unsaturation, index	R	336 ± 3	–	–	Vettier et al. 2006
Saturation ratio	R	045 ± 0.01	–	–	Vettier et al. 2006
Cholesterol, nmol $nmol^{-1}$ Phospholipids	R	0.109 ± 0.005	–	–	Vettier et al. 2006
Glycolysis					
Glycolytic flux, $nmol_{G6P}$ min^{-1} mg_{Prot}^{-1}	W	8.11 ± 0.9	–	14.8 ± 1.3*	Sébert et al. 1998
Transition time, s		0.6		0.4*	Sébert et al. 1998
Muscle composition					
Mean fibre area, μm^2	W	1,747 ± 55	–	1,458 ± 50	Simon et al. 1991
	R	291 ± 7	–	285 ± 71	Simon et al. 1991
Water content, mg $100 mg^{-1}$	W	70 ± 1	–	71 ± 1	Simon et al. 1991
Protein content, mg $100 mg^{-1}$	W	3.7 ± 0.3	–	2.5 ± 0.2	Simon et al. 1991
	R	2.2 ± 0.2	2.2 ± 0.3	2.3 ± 0.2	Simon et al. 1991
Lactate, μmol g_{ww}^{-1}	W	49.4 ± 4.5	43.2 ± 1.6	42.2 ± 6.7	Simon 1990
	R	3.1 ± 4.2	22.3 ± 6.1	23.4 ± 3.0	Simon 1990
Glycogen, μg g_{ww}^{-1}	W	0.4 ± 0.04	0.1 ± 1.6*	0.3 ± 0.0	Simon 1990
	R	2.1 ± 0.4	1.6 ± 0.2	1.8 ± 0.3	Simon 1990
Fatty acids, μmol g_{ww}^{-1}	W	1.7 ± 0.3	2.1 ± 0.4	1.9 ± 0.3	Simon 1990
	R	2.7 ± 0.5	0.2 ± 0.1*	2.8 ± 0.6	Simon 1990

The experimental pressure is 10.1 MPa (1,000 m depth) and the rate of compression is 1.0 MPa min^{-1}. TW = 15–17°C.
*Statistically different ($p < 0.05$ or better) from controls at 0.1MPa. the data from Simon (1990) and Simon et al. (1991) concern yellow eels. Values are mean ± SEM

Table 5.4 Enzymes activities (µmolsubst min^{-1} kg^{-1}) and energetic nucleotides (nmol gww^{-1}) in red and white muscles from yellow freshwater eels, Anguilla anguilla, under pressure (mean ± SEM)

	White muscle			Red muscle		
	0.1 MPa	ST	LT	0.1 MPa	ST	LT
IDH	0.74 ± 0.06	0.79 ± 0.08	1.25 ± 0.23	0.68 ± 0.30	2.33 ± 0.30*	1.47 ± 0.31
MDH	43.6 ± 6	59 ± 58	51.3 ± 2.5	39.5 ± 7.4	180 ± 23*	80 ± 15
CS	1.29 ± 0.16	1.22 ± 0.1	1.38 ± 0.11	1.16 ± 0.17	5.09 ± 1.53	2.28 ± 0.59
COX	0.025 ± 0.007	0.012 ± 0.0045	0.05 ± 0.01*	0.060 ±0.019	0.065 ± 0.01	0.088 ± 0.027
GOT	5.11 ± 0.09	7.0 ± 0.8	7.38 ± 0.37*	6.02 ± 0.66	20.36 ± 3.31*	11.03 ± 1.90
GPT	0.34 ± 0.02	0.28 ± 0.02	0.48 ± 0.04*	0.26 ± 0.06	0.36 ± 0.03	0.26 ± 0.03
PFK	3.78 ± 1.97	2.53 ± 1.22	6.85 ± 1.59	0.77 ± 0.441	0.34 ± 0.15	1.40 ± 0.29
PK	130 ± 6	111.4 ± 12.8	150 ± 18	95.2 ± 22.6	93.9 ± 12.6	97.3 ± 13.0
LDH	376 ± 43	529 ± 25	406 ± 33	226 ± 22	370 ± 50*	262 ± 36
GPD	0.057 ± 0.007	0.092 ± 0.008	0.068 ± 0.01	0.081 ± 0.015	0.160 ± 0.008*	0.132 ± 0.013*
CPK	891 ± 38	756 ± 36	946 ± 68	559 ± 83	578 ± 115	504 ± 87
ATP	2,937 ± 267	–	2,457 ± 218	1,536 ± 212	–	2,130 ± 340
ADP	900 ± 46	–	870 ± 105	346 ± 59	–	522 ± 81
AMP	112 ± 28	–	143 ± 42	39 ± 9	–	45 ± 5
IMP	915 ± 321	–	759 ± 246	220 ± 78	–	277 ± 66
EC	0.85 ± 0.02	–	0.85 ± 0.03	0.89 ± 0.01	–	0.88 ± 0.01

The experimental pressure was 10.1 MPa (1,000 m depth) and the rate of compression was 1.0 MPa min^{-1}. TW = 15–17°C.
*Statistically different ($p < 0.05$ or better) from controls at 0.1MPa; ST: short-term pressure exposure; LT: long-term pressure exposure. From Simon (1990). Values are mean ± SEM

Table 5.5 Oxygen consumption (mmol.h^{-1} kg^{-1}) of eels at different stages

	0.1 MPa	End of compression	Ratio
Glass	4.2	24	5.7
Yellow	1.2 ± 0.04	3.7 ± 0.1	3.1
Silver	1.6 ± 0.2	2.5 ± 0.5	1.6

The fish were compressed at a rate of 1.0 MPa min^{-1} up to 10.1MPa. data from Vettier and Sébert (2004). Note the lower pressure sensitivity of silver eels. Values are mean ± SEM (N = 5 in each group of adults). Results for glass eels have been obtained from one group of 120 individuals (Sébert, P 1992, unpublished data)

it is useful to consider some parameters such as the pressure threshold (Ptr) which is the pressure level corresponding to the appearance of the above symptoms and /or the maximal oxygen consumption generally observed at the end of the compression period (Table 5.5). It has been recently suggested that at least a part of the strong motor activity observed during compression is due to disturbance of the hydrodynamical regulation of buoyancy resulting in a deflated swim bladder (see Section 5.4.2; Speers-Roesch et al. 2004).

At the end of the compression period, performed at 0.2 MPa min^{-1} or 1 MPa min^{-1} depending on the experiment, the oxygen consumption progressively decreases despite the fact that a strong motor activity is maintained (Fig. 5.1). This pattern appears to be common to several fish species (see, Fontaine 1929; Naroska 1968). Measurements performed after 3 h at 10.1 MPa hydrostatic pressure show a decrease in muscle ATP content and energy charge together with a decrease in cytochrome oxidase activity. As oxygen transport is not altered with pressure (the circulatory convection is generally increased: Belaud et al. 1976; Gennser et al. 1990; Pennec and LeBras 1988; together with catecholamine release: Sébert et al. 1986), it has been proposed that hydrostatic pressure induces a state resembling histotoxic hypoxia, by decreasing membrane fluidity (Sébert et al. 1987; Sébert 1993). The first 48 h under pressure appear critical. In fact, if a fish has an efficient anaerobic metabolism to compensate for the deleterious pressure effects on aerobic processes, it can survive: this is the case for eel and goldfish, *Carassius auratus*, which can produce sufficient energy from the anaerobic pathway but not for trout, *Oncorhynchus mykiss*. After this critical period, the animal is generally able to acclimate to the effects of high hydrostatic pressure (Johnstone et al. 1989; Simon et al. 1989). The term "generally" is chosen because our experience shows that since about 10–15 years ago, the resistance of yellow eels to compression seems to decrease concomitantly with the observed decline in population. Now, when yellow eels are exposed to 10.1 MPa hydrostatic pressure for a long period (about 1 month), there exist numerous acclimatization processes which concern energy production. A complete description of these acclimatization processes has been given earlier (Sébert 2003), and will be summarised below. From the maximal value reached at the end of the compression period, oxygen consumption decreases progressively (over 6–8 days) to a level which represents about 65% of the oxygen consumption before compression (Fig. 5.1). Note that such a decrease (−35%) observed in pressure acclimatised fishes corresponds to the effects of a 3–4°C decrease in temperature, considering $Q_{10} = 2$: this is in the range of what is generally observed for the relationship temperature/pressure for several physiological processes i.e. about −2°C to −5°C for 10.0 MPa (Brauer et al. 1985). This acclimatization process includes restoration of muscle nucleotides (ATP, ADP, AMP) and energy charge and, although the aerobic pathway appears restored to the level observed at atmospheric pressure (0.1 MPa), also of the anaerobic capacity (Sébert et al. 1998). The restoration of the aerobic pathway during pressure acclimatization is accompanied by an improvement of oxidative phosphorylation efficiency together with readjustment of relative activities of the respiratory chain complexes (Théron et al. 2000). This seems to lead to a decrease in electron leaks and consequently to a decrease in reactive oxygen species, ROS, production (Amerand et al. 2005, 2006). If the eel is recompressed during the week after decompression, the acclimatization process takes place within some hours, at least as judged from the steady state in oxygen consumption (Simon et al. 1989). This indicates that the mechanisms which have been involved in the acclimatization process have produced effects which are still present. Measurements performed after 1 month under pressure show that the acclimatization and thus the improvement in oxidative phosphorylation is due, at least in great part, to the restora-

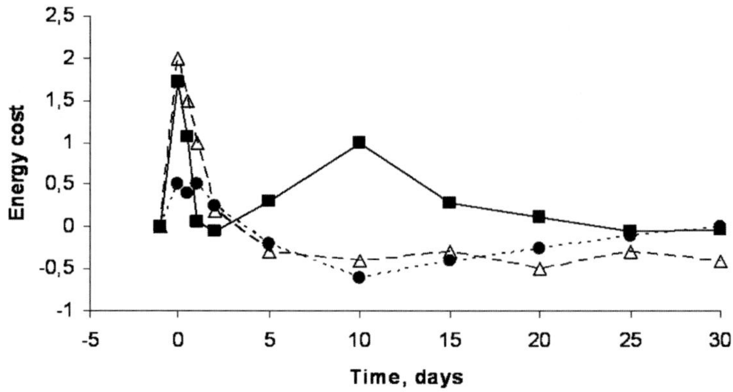

Fig. 5.1 Energy cost of pressure exposure (mmol h^{-1} kg^{-1}). Eels are compressed at 0.2 MPa min^{-1} to 10.1 MPa (yellow eels: triangles and silver eels: squares) or to 6.1 MPa (circles-silver eels) at 1.0 MPA.min^{-1} (for 12 h) then decompressed (for 12 h)

tion of membrane fluidity by increasing unsaturated fatty acids proportions, at the expense of saturated ones, in the phospholipid bilayer (Section 5.4.3.7). To explain all these modifications, which take place if the yellow eel is exposed to high pressure for a long time, we have proposed that at atmospheric pressure yellow eels have a suprafunctioning of the mitochondrial apparatus (despite the resulting energy cost, but they are still feeding). When pressure is applied, its effects are to restore a normal mitochondrial functioning allowing the eel to cope with pressure without a deleterious decrease in energy production (Sébert and Théron 2001).

It is interesting to point out that at least two morphological changes are observed after 1 month under pressure. The first is at the white muscle level, where there is an increase in small diameter fibres at the expense of large fibres, leading to an overall 16% decrease in mean fibre area (Table 5.3) together with a 32% decrease in protein content (Simon et al. 1991). The second is that, in studies of the gill epithelium of freshwater yellow eels acclimatised to HP, Dunel-Erb et al. (1996) found a significant decrease in the number of mucus cells and a large increase in density and in fractional area of chloride cells on the apical surface (Table 5.2). This is thought to compensate, at least partly, for the impairment of ATPase by high pressure (Sébert et al. 1991). It is surprising that the yellow eel under pressure increases its ability to excrete salts although it stays in fresh water.

Concluding this section, yellow freshwater non-migrating eels are able to acclimate to the hydrostatic pressure effects. The acclimatization process mainly concerns the aerobic pathway (improvement of oxidative phosphorylation efficiency) by way of homeoviscous adaptation. The question could be raised about the ability of yellow eels to migrate. In this stage, the eel is able to cope with pressure effects, changes in salinity and /or temperature and is known to have the capacity to swim for a long distance (Van Ginneken et al. 2005a). But the yellow eel is unable to reproduce. Thus we must consider that what is observed in yellow eels reflects potential ability to migrate but they do not migrate because they are not sexually ready! The metamorphosis, i.e. the silvering process, appears strictly necessary.

5.4 Silver Eels Under Pressure

5.4.1 Effects of Metamorphosis (Silvering Process) on Energy Metabolism

The morphological changes which appear during the different eel life stages are reviewed by Tesch (2003) and Durif et al. (2005). It is known that during the silvering process, red muscle volume increases (Pankhurst 1982) as is observed in pelagic fishes which can swim at depth (Altringham and Ellerby 1999). The volume increase is probably due more to an increase in fat and mitochondria than to an increase in number of muscle fibres (Lewander et al. 1974; Pankhurst 1982). Moreover, activities of enzymes involved in the aerobic pathway are higher in silver than yellow eels

Table 5.6 Effects of long term exposure of male silver eels to high pressure

	Muscle	0.1 MPa	10.1 MPa 3 days	10.1 MPa 25 days	10.1 MPa 50 days
Muscle fibre respiration, μmol min^{-1} g^{-1}	R	0.48 ± 0.0	0.59 ± 0.05	0.57 ± 0.04	0.53 ± 0.03
	W	0.14 ± 0.01	0.21 ± 0.02	0.20 ± 0.02	0.14 ± 0.01
Water muscle content, %	W	61.9 ± 1.6	63.7 ± 1.3	61.2 ± 1.1	60.2 ± 1.8
Protein content, mg g^{-1}	R	20.7 ± 1.0	18.7 ± 0.7	20.3 ± 0.6	26.7 ± 1.4
	W	23.0 ± 0.6	23.4 ± 1.3	24.1 ± 0.8	–
ATP, μmol g^{-1}	R	1.7 ± 0.1	1.14 ± 0.24	1.19 ± 0.18	1.6 ± 0.7
	W	2.3 ± 0.3	2.4 ± 0.2	2.5 ± 0.3	2.4 ± 0.4
AS, μmol g^{-1}	R	1.95 ± 0.1	1.38 ± 0.25	1.58 ± 0.24	2.1 ± 0.12
	W	2.8 ± 0.3	2.8 ± 0.2	3.1 ± 0.3	3.2 ± 0.52
EC	R	0.93 ± 0.01	0.90 ± 0.01	0.87 ± 0.01	0.89 ± 0.01
	W	0.92 ± 0.01	0.92 ± 0.01	0.98 ± 0.01	0.97 ± 0.01
Gill water content, %	–	81 ± 0.3	81 ± 0.4	82 ± 0.2	80 ± 0.3
COX activity, μmol min^{-1} g^{-1}	R	79.0 ± 2.0	71.0 ± 5.0	82.2 ± 2.0	120.6 ± 19.3
	W	9.3 ± 1.6	7.6 ± 1.3	6.1 ± 0.4	8.4 ± 2.2
Myoglobin, mg g^{-1}	R	1.7 ± 0.13	–	1.7 ± 0.24	–

Values post decompression: Vettier and Sébert (2004) and unpublished data. The experimental pressure was 10.1 MPa (1,000 m depth) and the rate of compression was 1 MPa min^{-1}. R is for red muscle and W for white muscle. Values are mean ± SEM

(Boström and Johansson 1972; Egginton 1986) together with a change in the main energy stores from glycogen in the yellow stage to fat in migrating fish (Lewander et al. 1974; Barni et al. 1985; Zara et al. 2000). It is probable that the hormonal changes observed during metamorphosis (Marchelidon et al. 1999; Chapter 3) help in the use of fat as the main substrate for aerobic energy production. Concomitantly, an increase in myoglobin content is observed (Egginton 1986). Clearly, silver eels have higher aerobic capacities than yellow eels and this is in agreement with the observation that the mass-specific power output of silver phase slow muscle is greater than that of yellow phase slow muscle (Ellerby et al. 2001). Note that twitch tension of swimming muscle of seawater adapted eels increases with pressure and is maximal at 20 MPa (Wardle et al. 1987), together with absence of pressure sensitivity for cardiac contraction (Gennser et al. 1990). In terms of muscle mechanics, it cannot be excluded that fast white muscle is recruited to help in obtaining optimal swimming performance (Ellerby et al. 2001); the fact that aerobic capacities increase in white muscle during silvering (Boström and Johanson 1972) and/or pressure exposure argues in this sense (Tables 5.4 and 5.6).

Thus, metamorphosis from yellow to silver stages prepares the eel to cope with many of the special conditions linked to migration at depth: energy metabolism (see above), morphological changes (Tesch 2003; Durif et al. 2005), ability to swim (Ellerby et al. 2001; Durif et al. 2005; Thillart in this book), salinity (Fontaine 1975; Lecomte-Finiger and Yahyaoui 1990; Bertin 1951; Thomson and Sargent 1977), obscurity (Carlisle and Denton 1959; Pankhurst and Lythgoe 1983; Tesch 2003), floatability (Kleckner 1980; Pelster 1997), absence of feeding activity (Zara et al. 2000; Durif et al. 2005), and evidently reproduction by way of sexual maturation. But what about pressure?

5.4.2 Pressure Effects and Pressure Resistance

Unfortunately, adult migrating silver eels have rarely been caught (Van Ginneken and Maes 2005; with a camera: Robins et al. 1979). Consequently, it is impossible to describe the physiological adjustments which occur when the eel is really adapted to a deep environment and not to only one or several factors. However, the experimental data obtained and the comparison of silver and yellow eels helps in the understanding of the processes involved.

As for yellow eels, effects of pressure on silver eels have been studied only using a hyperbaric chamber (Johnstone et al. 1989; Nilsson et al. 1981) except for Fontaine et al. (1985) who used caging at different depths. They observed significant ovarian development and a strong increase in GTH level which was not due to decreased light (Dufour and Fontaine 1985). Moreover, recent data from our team (Vettier 2005) show a significant decrease in hepato-somatic index and a trend to increase in gonado-somatic index after 1 month under pressure, variations which are typical of the silvering process (Durif et al. 2005). These results suggest that hydrostatic pressure is involved in the triggering of sexual development and plays a positive role in eel reproduction (Sébert et al. 2007), even if a certain level of energy release (by swimming) seems a prerequisite to induce sexual maturation (Nilsson et al. 1981). We cannot exclude the hypothesis that high pressure can modify gene expression (Bartlett et al. 1989, 1993, 1995; Simon et al. 1994) and thus protein synthesis involved in all life processes. How do silver eels cope with pressure effects (which require energy) at the same time as they must swim for 6,000 km to reach the Sargasso Sea?

As we have seen previously, specific pressure effects can depend on several factors, including rate of compression, temperature, pressure used, species, and developmental stage. Thus, different protocols can be used to expose silver eels to high pressure: Fig. 5.1 shows that the diving pattern is not very important, at least at the level of animal metabolism evaluated from oxygen consumption. It must be pointed out that nobody knows exactly how silver eels reach great depths: is it abruptly starting at the continental plateau or progressively? This latter pattern is the more evoked in relation to some works showing vertical migrations, at a weekly time scale, but for released fish (Tesch 1978; Fricke and Kaese 1995; McCleave and Arnold 1999). Indeed, information about the depth of spawning has been extrapolated from data on the release of hormone-treated females tagged with transmitters, and on larval catches. Again, the number of telemetry studies and the number of radio transmitter-tagged animals is low. Releasing female adults in the Sargasso Sea demonstrated a preference for the upper zone of the ocean at depths of 250–270 m and at temperature around 19°C (Fricke and Kaese 1995). However, in the study of Tesch (1989) the maximum swimming depth of hormone treated silver female eels in the Sargasso Sea was nearly 700 m. Hormone treated female Japanese silver eels, *Anguilla japonica*, tagged with ultrasonic transmitters were released at their supposed spawning grounds in the western Pacific Ocean. These eels preferred relatively shallow water, swimming at a depth ranging from 81–172 m and at relatively high temperatures of 18–28°C (Aoyama et al. 1999). Interestingly, the catch of *Anguilla* larvae of less than 5 mm length confirmed these observations. The smallest (probably just hatched) larvae were found at depths between 50 and 300 m with temperatures of 24°C to 18°C respectively (Castonguay and McCleave 1987). Those temperatures are close to the final preferred temperature (FPT) of sexually mature *Anguilla rostrata* (17.5°C), so spawning probably takes plays in the upper 200 m of the ocean at temperatures close to FPT (Haro 1991). Although the preceding authors suggest depths between 20 and 700 m for spawning, in reality it is perhaps higher during migration. Indeed, the increase in swim bladder function in silver when compared to yellow eels extends the maximal depth at which silver eels can maintain swim bladder volume (Kleckner 1980). Sébert (2007) has recently calculated that the migration depth is potentially around 2,000 m. This calculation is based on the measured pressure threshold values (Ptr, the pressure at which the animal exhibits tremors and/or convulsions indicating muscle dysfunction and consequently inability for swimming) in migrating silver eels compared to those observed in deep living fishes. Ptr is estimated to be about 240 atm at surface temperature which means that at depth temperature, Ptr is probably near to 400 atm. As Ptr generally corresponds to twice the depth of living, the latter can be estimated to be about 200 atm (2,000 m depth). However, the effects of hydrostatic pressure on energy metabolism of silver eels appear of little importance (Vettier 2005 and Table 5.6). Submitting red permeabilized muscle fibres to 10.1 MPa HP induces a small decrease (from 0% to 25% depending on the eel origin) in oxygen consumption (Vettier and Sébert 2004). However, when the intact eel is exposed to 10.1 MPa for 3, 30 or 50 days, the observed changes are small and the eels are able to acclimate. An interesting point is that pressure exposure induces an improvement of aerobic capacities in red but also in white muscle. In the white muscle, this improvement is realised by way of a cytochrome c oxidase (COX) activity increase and a trend to increase ATP content (Vettier and Sébert 2004). COX seems to play an important role in pressure acclimatization (Theron et al. 2000; Sébert and Theron 2001), which is not surprising considering that it is the last enzyme complex of the respiratory chain and is responsible for oxygen consumption.

Interestingly, the COX activity appears to increase with the maturation stage and /or the pressure exposure (Fig. 5.2). The improvement in aerobic capacities, whatever the muscle type, is evidently in relationship with the high energy demand due to swimming activity during the migration. Clearly, silver eels have higher aerobic capacities than yellow eels at atmospheric pressure (Egginton 1986). Pressure exposure can improve these capacities, helping the silver eels to resist its effects. However, several parameters are able to modulate the pressure resistance and /or acclimatization. In order to estimate the role of different parameters in the acclimatization process, it was necessary to have a viable index. In this aim, knowing that pressure resistance and thus success of the acclimatisation, depends mainly on events during the first hours under pressure, a Pressure Resistance Index (PRI) has been proposed (Vettier 2005). The higher the PRI, the higher the fish fitness. The PRI is calculated from normalised values of Pressure threshold (the pressure at which the fish exhibits tremors) and maximal oxygen consumption observed at the end of compression period. These two parameters, which allow a good estimation of pressure sensitivity (Sébert and Macdonald 1993), are combined with the energy charge values, EC, observed after decompression because EC reflects the possibility for the fish to restore a "normal" energetic state i.e. to acclimate (Simon et al. 1992). To illustrate this, experiments performed in our laboratory have shown that trout, yellow eel and silver eel have PRI values of about –22, 1 and 16 respectively.

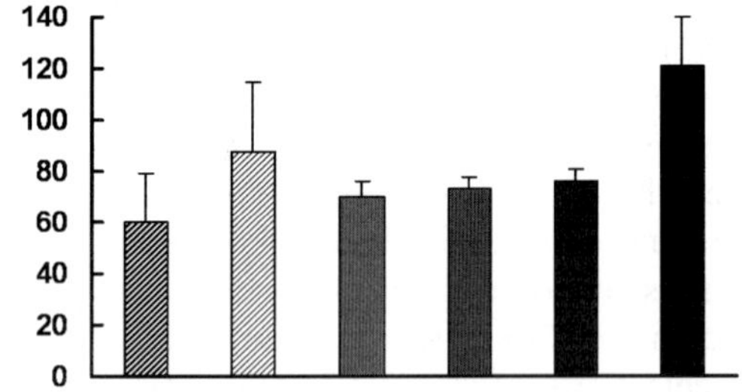

Fig. 5.2 COX activity (mU.g^{-1}) in eel red muscle. Y: yellow stage; S: silver stage; the number besides the stage is the duration (days) of exposure to 10.1 MPa hydrostatic pressure

5.4.3 *Factors Modulating Pressure Resistance*

5.4.3.1 Freshwater or Not

Yellow and silver eels can live in different aquatic environments, for example freshwater or marsh with different salinities. Knowing that these eels will migrate in seawater (silver eels are very well pre-adapted to sea water; Thomson and Sargent 1977; Lecomte-Finiger and Yahyaoui 1990), is there an advantage for them to live in brackish water? We have compared silver males from these two origins. If the two populations are similar in terms of size, some physiological characteristics at 0.1 MPa are very different: density and muscle water content are higher in saltwater but COX activity, energy charge, protein contents of red muscle and haematocrit are lower (Vettier and Sébert 2004). Clearly, red muscle from marsh eels has less good muscle quality (more water, less proteins) which induces apparent lower aerobic capacities at 0.1 MPa. However, it appears that after some days under pressure (10.1 MPa), many of these parameters are improved, in contrast to what is observed in fishes living in freshwater. Although the two groups were tested in freshwater (Vettier and Sébert 2004; Vettier 2005), it appears that growing in salt water (even if the salt concentration is low, as in the marsh) has disadvantages at 0.1 MPa but allows of a faster acclimatisation to pressure effects: about 3 days are needed in contrast to several weeks for freshwater eels. However, we cannot confirm that this result is exclusively due to differences in salinity because the two populations are different in origin and we do not know exactly what the physico-chemical properties of the two aquatic environments and the food quality and availability are.

5.4.3.2 Parasitism

Most eel populations are known to be highly infected by *Anguillicola crassus*, the development of which appears to depend on the temperature (Knopf et al. 1998). This nematode damages the swim bladder (Molńar et al. 1994; Würtz and Taraschewski 2000; Chapter 9). Damaging the gas gland epithelium could impair the production of salts, e.g. lactate, that play a central role in gas secretion (Kobayashi et al. 1989) which is necessary to compensate for the high density of muscle tissue and skeletal elements (Pelster 2004). If such damages, which seem to depend on the fish size (Lefebvre et al. 2002), appear not too deleterious for non-migrating eels, they can become severe in migrating silver eels where gas secretion rates are about five times greater in order to maintain swim bladder volume and thus extend the maximal depth at which they can migrate (Kleckner 1980). Thus it has been shown that these damages could impede the supposed vertical displacements that eels perform during their migration to the Sargasso Sea, although the effects on swimming capacities have also been discussed (Sprengel and Luchtenberg 1991; Kirk et al. 2000; Nimeth et al. 2000). Our experiments have shown that parasitism, whatever its level, does not modify pressure resistance

(Vettier et al. 2003), bearing in mind that the latter is evaluated from short term pressure exposure. However, parasitism does not seem to have drastic effects (Kelly et al. 2000; Roche et al. 2003) on the ability of the eels to acclimatise to sea water and hydrostatic pressure (caging experiments, Fontaine et al. 1990), although some of our results suggest that infected eels do not have the respiratory and metabolic capacities to migrate and to cope with all the changes the migratory activity entails prior to reproduction (Vettier and Sébert 2004).

5.4.3.3 Nutritional State

Silver eels stop feeding even before migration (Zara et al. 2000; Durif et al. 2005); this goes together with a progressive reduction of the digestive tract (Tesch 2003). It is known that pressure acclimatisation is achieved mainly by a restoration of membrane fluidity by increasing the recruitment of polyunsaturated fatty acids, PUFA (see Sébert 2003 and Section 5.2). Synthesis of PUFA is more difficult in fishes living in seawater (Watanabe 1982; Bell et al. 1986; Green and Selivonchick 1987) which is the condition during migration. Thus, eels must probably have enough PUFA stored before starting towards the ocean. In this sense, it is probable that availability and quality of food in the aquatic environment prior to migration plays an important role (Vettier and Sébert 2004), dietary history being very important for metabolic machinery (Moon 1983). Now, the question is about the usefulness of fasting during migration. Several studies have shown that food deprivation for 6 months exerts little metabolic effect (Love 1970; Moon 1983; Cornish and Moon 1985) but can compromise muscle contractile proteins (Moon and Latham 1984). As it seems that eels have sufficient fat stores to migrate (Van Ginneken and Van Den Thillart 2000), it can be supposed that feeding is not necessary and that proteins from the digestive tract could be used to build up gonads. Another interesting aspect is that muscle fibres from long-term fasting eels are more resistant to high pressure (in terms of aerobic capacities) than fibres from eels having fasted for a short time (see Fig. 5.3). Perhaps this can be explained by the fact that, at least in ectotherms, dietary restriction allows locomotor superiority

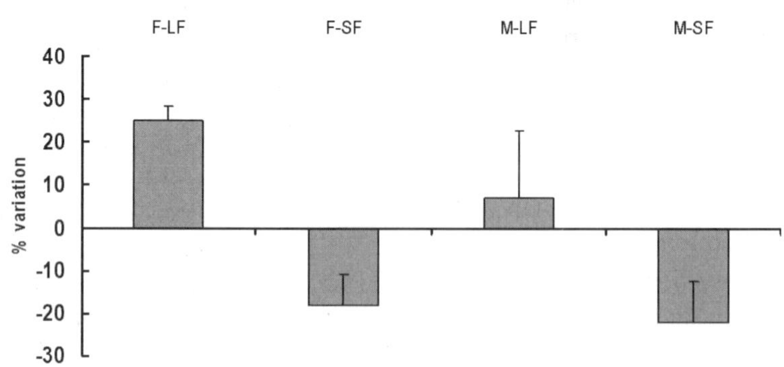

Fig. 5.3 Muscle fibre respiration changes at 10.1 MPa (in % of variation by comparison with values at 0.1 MPa). F: females; M: males; LF: long fasting; SF: short fasting

(Le Galliard et al. 2004), and reduces the cost of protein assimilation (Owen 2001), although some results seem to show that digestion can increase maximal oxygen capacity (Hicks and Bennet 2004). Perhaps it can be concluded that fasting helps migration by enhancing pressure resistance. But fasting could have also "mechanical" effects. Indeed, as they use only fat stores, the density of the fish is expected to progressively increase (Vettier 2005) as it spends longer in seawater (McCleave 1977). Indeed, although the relative composition of the body does not change during swimming (Van Ginneken et al. 2005b; Chapter 8), the volume of fat decreases together with the body mass. Due to lower lipid density, the eel loses more volume than mass and consequently animal density is expected to increase, which could help it to swim deeper and deeper and/or to counteract the higher density of coldwater encountered at depth without extra effort to maintain neutral floatability.

5.4.3.4 Water Temperature

Water temperature is probably not a problem for the migrating eel because, at least in the yellow stage, it is known to tolerate a large temperature range. However, we do not know at what depth and thus temperature the eel migrates: some authors suggest that with the optimal temperature for sexual development being a little over 20°C, the spawning depths could be about 200m (Nilsson et al. 1981). If low temperature is known to increase aerobic capacity of fish swimming muscle (Guderley 1990), it has negative effects on membrane fluidity, at least for short term exposure (Hazel 1995) and interacts with pressure effects at the mitochondrial level (Sébert et al. 2004). Moreover, it is well established that acute temperature changes have detrimental effects on both central nervous system processes and contractile performance in muscle of fishes (see Block 1991). Thus we can assume that eels migrate at a depth where temperature is not too low, in the 13–22°C range, indicating their great thermal tolerance during oceanic migration (Tesch 2003). However, it is not clear if the eels swim towards a rising temperature gradient or avoid colder water (Ekman 1932; Westin and Nyman 1979). The vertical migrations (up to 300m, deeper during the day than at night) can induce large temperature variations: do they have a metabolic regulation role (Block 1991) or is it to avoid predators? We have no answer.

The conjunction of low temperature and high pressure is not good in terms of thermodynamics, but it has been shown that, in eels acclimated to high pressure, which is the case for migrating silver eels, temperature has little effect on metabolism (Sébert et al. 1995).

5.4.3.5 Buoyancy

Due to a density (d = 1.05–1.20) higher than seawater (d = 1.026) fishes must use different mechanisms to maintain neutral buoyancy and thus to stay in the water column (Pelster 1997). If this is not the case, they sink. The best known apparatus is a gas-filled cavity, the swim bladder (SB). Active and passive mechanisms also exist. The main active process consists in producing hydrodynamic lift, due to water flow over wing-shaped fins and due to hovering, swimming at higher speed and/or at the optimal tilt angle by combining pectoral fin lift, body lift and lift from the tilted thrust vector. These active processes are independent of depth but represent an energy cost (Strand et al. 2005). The passive mechanisms consist mainly, in fishes living at depth, accumulating lipids (even in the SB) and/or water in different tissues; decreasing content of heavy ions, in order to decrease the fish density (Pelster 1997).

The swim bladder raises interesting problems in the migrating silver eels. If physoclistous fishes (closed SB) are placed under increased hydrostatic pressure, they secrete gases into their SB and if the pressure is reduced, they reabsorb gases. Physostomatous fishes have a SB with a pneumatic duct and sometimes also a gas gland and a rete: this is the case for the eel. Fishes in seawater, such as the migratory eel, need a SB occupying about 5% of the body volume (Bone et al. 1995). The very efficient counter-current system of the *rete mirabile* allows accumulation of gas in the SB with pressure gradients which may be up to several hundred atmospheres (Kobayashi et al. 1989; Pelster and Scheid 1992; Pelster 1997 for reviews) but the energy cost to pressurise gases is very low (Strand et al. 2005). The diffusive gas loss is reduced by an increase in the crystalline guanine content of the SB wall during the yellow-silver eel metamorphosis (Kleckner 1980; Bone et al. 1995). It is known that nitrogen, CO_2 and even argon can accumulate depending on the depth but the main gas is oxygen (Schölander and Van Dam 1953; Pelster and Scheid 1992). This has two consequences. The first is illustrated by the following calculation. Considering a migrating silver eel with a body volume of 1 L (body mass around 1.1 kg); it has a SB volume of about 50 mL. If they dive from the surface to 1,000 m depth (10.1 MPa), maintenance of the SB volume requires a gas secretion of 5,000 mL which represents about 220 mmol O_2! The secretion rate values are in the range of 1 to 3 mL h^{-1} (Tytler and Blaxter 1973; Kleckner 1980; Goolish 1992): it increases during the yellow-silver stages transition in the eel but decreases with the ambient pressure (Kleckner 1980). So, even using a high rate value of 2 mL h^{-1} (0.1 mmol O_2), it takes 2,500 h (more than 3 months) to restore the SB volume, knowing that gas density increases with pressure and thus is less efficient in terms of buoyancy. This suggests that when starting their migration, silver eels must dive slowly and progressively in order to limit excessive oxygen consumption. A second consequence is that at high pressure oxygen becomes toxic for tissues by way of reactive oxygen species, ROS. D'Aoust (1969) has shown that the high oxygen pressures which can be measured inside the SB can induce rigidity of the fins, alterations of breathing and gill function and then tetanic contraction or muscle paralysis. These symptoms are evidently not compatible with swimming activity. However in the eel SB epithelium, a high activity of glutathione reductase, which degrades ROS, has been measured whilst the enzyme was not detectable in muscle tissue: this protects the tissues from the O_2 damage (Pelster and Scheid 1992).

Now, at what depth does the eel migrate and does it undergo vertical migrations as suggested by different authors (Tesch 1978; McCleave and Arnold 1999; see also the review by Van Ginneken and Maes 2005)? Kleckner (1980) has calculated from experiments on silver eels that they can maintain their SB volume up to 150 m depth, which is not very deep when compared to the observations of Robins et al. (1979) or Bailey et al. (2005). As a starting hypothesis, an average depth of 1,000 m can be considered and Tesch (1978) has observed vertical daily migrations up to 400 m during eel tracking experiments. Considering the eel model used above (1 kg body mass with a SB volume of 50 mL) that means that every day, if the volume of SB is maintained for neutral buoyancy, it must alternatively secrete and reabsorb about 30 mL of gas. Gas reabsorption is more rapid than secretion (Tytler and Blaxter 1973; Kleckner 1980) which probably needs more than 12 h (see above) to be completed and thus cannot be realized in the frame of daily vertical migrations. All the above observations suggest that SB is probably insufficient to ensure neutral buoyancy and *a fortiori* vertical migrations as suggested by Speers-Roesch et al. (2004) for *Goregonus hoyi*. This is particularly relevant if we consider that migrating eels have their SB more and more infested with *Anguillicola crassus*. Clearly, it seems reasonable to think that, as in other deep-living fishes (Bone et al. 1995; Pelster 1997), migrating eels progressively abandon gas as a source of static lift because there are difficulties in regulating buoyancy with gases over a wide range of pressure (and thus vertical migrations). The extensive review of buoyancy at depth by Pelster (1997) shows that tissue composition in terms of lipids, water, proteins and ions can help in the process but all the mechanisms involved require energy which becomes unavailable for swimming and reproduction. Some authors have suggested that the lipid content of eggs can help with buoyancy (see Pelster 1997) but this strategy, or that concerning the increase in lipid tissue content, are without relevance in the eel because it does not eat during migration: lipids must come only from fat stores which progressively decrease, probably inducing an increase in density. We raise the hypothesis that maintaining buoyancy using the swim bladder is not really feasible at low energy cost in the migrating eel: consequently, it probably progressively sinks from Europe to the Sargasso Sea. However, this phenomenon is probably somewhat limited, firstly by the fact that temperature decreases with depth making water denser, secondly by using the hydrodynamic lift induced by the swimming activity and helped by the enlarged pectoral fins. Results from possible changes of membrane fluidity and COX activity (Vettier 2005) during the migration suggest that this particular fish progressively improves its pressure resistance and thus can dive deeper and deeper. This agrees with the hypothesis and can explain why adult eels have never been found in the Sargasso Sea.

5.4.3.6 Sex

Studying silver eels from different origins has shown that some differences exist due to the sex of the animal. This is true in terms of aerobic capacities or pressure resistance (Vettier and Sébert 2004) although the sex expression is not always synchronized with metamorphosis (Beullens et al. 1997). In fact, maximal oxygen consumption of muscle fibres has been measured after 1 month under pressure. Then, considering that red muscle represents 30% of total body mass and white muscle

6% of total muscle mass (Goolish 1991) with an increase in red muscle proportion (about +60%) with the silvering process (Eggington 1986) we have estimated the maximal aerobic capacity of males and females under pressure. It is clear that such estimations are only indicative (but realistic, see Kimberly et al. 1997) because composition is probably somewhat different in the two sexes as indicated by the difference in densities. The swimming cost has been estimated to about 2.2 mmol O_2 h^{-1} kg^{-1} by Van Ginneken and Van den Thillart (2000) which represents, under pressure, more than 40% of the maximal abilities for females but less than 30% for males. Thus, males have a lower (but good) pressure resistance (Pressure Resistance Index: about 12 and 20 for males and females respectively) with a better aerobic capacity (about 7.5 mmol O_2 h^{-1} kg^{-1} and 4.5 mmol O_2 h^{-1} kg^{-1} for males and females respectively) with the same characteristics for contraction of isolated fibres (Rossignol et al. 2006). The swimming cost has been estimated on a basis of a 0.5 Body Length s^{-1} speed (BL s^{-1}). That means that, if the travel represents 6,000 km, females need about 230 days (about 7 months) and males about 350 days (about 12 months) to reach the spawning area. This is evidently impossible to ensure reproduction! However, it is known that males generally start their migration sooner in the season, i.e. about 50 days before the females: thus they can migrate for 280 days (230 + 50) to meet the females at the same time in the Sargasso Sea. That means they must swim at a speed of about 21.5 km day^{-1} or 0.25 m s^{-1} which represents, in our example, about 0.65 BL s^{-1}. Considering that the cost of swimming increases with resistance i.e. with the square of speed, which means that the relative cost of swimming under pressure for the males would be about 40% to 50% of the total aerobic capacities, as in females, which is acceptable as is the estimated speed of 0.6–0.7 BL s^{-1} (Vettier et al. 2006). This simple calculation suggests that males meet the females by starting sooner and swimming faster, which is possible due to their higher aerobic capacities under pressure. However, several questions stay without clear answers. It is known that a linear relationship exists between swimming speed *vs* tail beat frequency (McCleave 1977) which depends on water temperature (Ellerby et al. 2001). Is it the same in males and females? Moreover, we have no clear data to confirm that the males have sufficient fat stores to ensure such an energy requirement, their high density being probably a sign of lower fat content. Indubitably, physiology of migrating males must be studied more because the majority of the available data are from females.

5.4.3.7 Silvering

Effects of silvering on eel physiology have been comprehensively described and many data are available in this book. However, as we have previously shown that yellow eels are able to acclimatize to high hydrostatic pressure, our question is "how the silvering process can help and /or participate in the acclimatization processes?" Considering that the ocular index can give an estimation of the silvering stage, it appears that pressure resistance, evaluated with the PRI, is higher overall when the silvering process is advanced. However, whatever the silver stage, all the silver eels are able to acclimatize to high pressure (Nilsson et al. 1981; Fontaine et al. 1985; Johnstone et al. 1989; Vettier and Sébert 2004). Moreover, results reported above show that long term exposure to high pressure (10.1 MPa) does not generally change the energetic metabolism features of the silver eel significantly, leading to the hypothesis (Vettier et al. 2005) that the silvering process induces, in the yellow eel, several effects similar to those observed after a long term pressure exposure: increases in COX activity, in gill chloride cell number, decreases in pressure sensitivity, muscle protein contents and reactive oxygen species.

Indeed, the increase in COX activity agrees with the improvement of aerobic metabolism by pressure or silvering (Boström and Johanson 1972; Lewander et al. 1974; Egginton 1986). Improvement of aerobic metabolism corresponds to a better efficiency of the respiratory chain and coupled oxidative phosphorylation (Theron et al. 2000; Sébert et Theron 2001) by way of a lower electron leak, explaining the observed decrease in reactive oxygen species as hydroxy/radical OH$^{\bullet}$ (Amerand et al. 2005, 2006). At the same time, it appears that pressure sensitivity, estimated from the pressure threshold (Ptr) at which eels exhibit a strong motor activity during the compression, is the same in pressure-naïve silver eels at 0.1 MPa as in yellow eels after 1 month under pressure. Such an increase in Ptr is known as an index of pressure adaptation (Sébert and Macdonald 1993). Thus silver eels appear prepared for pressure effects and when they encounter pressure for the first time they have the same behaviour as pressure-acclimated yellow eels. It is a well known fact that most of the pressure acclimatisation process is performed by the way of restoring a normal membrane fluidity (homeoviscous theory) (White and Somero 1982; Cossins and Macdonald 1984; Macdonald and Cossins 1985), the cell membrane appearing as a metabolic pacemaker (Else and Hulbert 2003) and the major target for the HP effects. Evidently, other features of membrane organisation can also influence membrane function (Lee 1991; Hazel 1995) but membrane fluidity, determined by anisotropy measurements, is a good index of the membrane state. Consequently, if the main effects of hydrostatic pressure concern cell membranes and that we consider silver eels as pre-adapted to high pressure we must observe a higher membrane fluidity in silver eels (lower anisotropy) than in yellow eels after 1 month under pressure. This hypothesis has been verified and confirmed by measuring, at atmospheric pressure, red muscle membrane fluidity and composition in silver and yellow eels. The higher fluidity observed in silver eels is ensured by increasing the recruitment of polyunsaturated fatty acids in the membrane phospholipids (Vettier et al. 2006), which is the common way to ensure homeoviscosity with the help of cholesterol as a modulator (Dave et al. 1975).We can thus consider that one of the aims of the silvering process is to perform the main adaptations before migration because they have a non-negligible energy cost. Thus, on entering the migratory process the silver eel can devote all its available energy to swimming activity and gonad development, which would be probably problematic if the fishes must ensure them together with adaptation to their new environment.

5.5 Potential Pressure Effects on Reproduction, Eggs and Larval Development

As seen in the preceding sections, high hydrostatic pressure (10.1 MPa) has potential effects on fish energy metabolism and consequently on migration for reproduction. As seen previously, we estimate the maximum migration depth at about 2,000 m. Such a depth and the corresponding pressure of 20.1 MPa probably has no greater effects than the pressure tested (10.1 MPa) on yellow and silver eels' muscle energy metabolism as such. However, nothing allows us to reject the possibility that, when the fish has to swim (using its muscles) and to resist high pressure (that is to say when the fish is in the natural conditions of migration), things could be different. This needs further experiments which are scheduled in our lab.

Most of the studies actually performed on eels have the aim of explaining the decline of the population and to find a solution to stop this phenomenon. One of the often explored possibilities is to control reproduction by artificially obtaining eggs, then larvae able to survive and to develop into glass eels. Although high pressure is known to induce significant increases in pituitary LH content and gonadosomatic index (Fontaine et al. 1985; Vettier et al. 2005) experiments conducted until now to obtain larvae have not been very successful. We are convinced that one of the main reasons is that the pressure factor has been neglected.

Spawning seems to be a surface-related event. Eggs were found to rise in the water column (Balon 1975) up to the surface with maximum speeds of 2.24 m per hour. The main hatching times for European eels are between 47 and 60 h after fertilization (see Van Ginneken et al. 2005). During these times eggs will rise 105–134 m. Assuming that hatching should take place in the food-rich upper water layers, these shallow depths represent spawning depths. However, for the Japanese eel it was observed that high pressure delays embryonic development and hatching times (Hiroi et al. 2003). However, egg ascent in the water column must be accompanied by a volume increase. This latter has to be compensated by some material; it can be supposed that water entry will be facilitated. Indeed, Beney et al. (1997) have pressurized vesicles of egg yolk phosphatidylcholine in pure water and their volumes were measured by optical microscopy. It appears that the volume of vesicles decreases at high pressure (250 MPa) by 16% which is more than can be accounted for by bulk compression of water (10%). This arises from the lateral compression of the bilayer reducing the area, and hence the volume, of the vesicle. Water moves out to accommodate the change. We can reasonably suppose that reversal of this physical effect occurs when pressure decreases, which can modify buoyancy of eggs and then embryos. The buoyancy of eggs has often been used as an indicator in the assessment of egg quality, especially in studies of marine teleosts that spawn pelagic eggs, such as red sea bream *Pagrus major* (Watanabe et al. 1984a, b), sea bass *Dicentrarchus labrax* (Carrillo et al. 1989), yellowtail *Seriola quiqueradiata* (Mushiake et al. 1994) and the Japanese eel (Seoka et al. 2003). In these species, the ratio of buoyant eggs to total eggs spawned, measured several hours after fertilization, correlates positively with egg hatchability, which is based on the fact that most of the unfertilized eggs and the fertilized eggs that have ceased developing do not remain buoyant for very long.

Till the 1960s, pressure shocks were used to induce triploidy (Dasgupta 1962) in amphibians or in fishes (Rottmann et al. 1991; Gillet et al. 2001) or even tetraploidy in fish embryos (Yamazaki and Goodier 1993). Marsland (1938, 1950, 1970) had shown that high pressures cause inhibition of embryonic development in shallow-water marine invertebrates. It was shown on sea urchin embryos that pressure inhibits the formation of the furrow, normally formed when cells enter into cellular division (Begg et al. 1983). The effects of pressure are different in deep-sea organisms, which silver eels can be considered to be after several months under pressure. Studies done on high hydrostatic pressure effects on embryonic development have also been performed on echinoderms living at depth. It was demonstrated that these organisms nonetheless did not present normal development at too high temperatures but they also need pressure. Experiments have demonstrated that the best pressure is the pressure at which they normally live (Young and Tyler 1993; Tyler and Young 1998). In *Echinus affinis*, the embryos are truly barophilic and could not develop under pressures lower than 1,000 m depth. Moreover, in *Echinus acutus*, which lives at about 1,000 m depth, development at atmospheric pressure is possible but the number of abnormalities decreases with the depth. In the same manner, the asteroid *Plutonaster bifrons* has the highest percentage of normal development near the peak of the species distribution (2,000 m), developmental rate being retarded at pressures higher and lower than that at this depth (Young et al. 1996). Similar conclusions concerning the requirement of high pressure for embryos to develop have been drawn from the vent worms *Alvinella pompejana* and *Riftia pachyptila* (Pradillon et al. 2001). Although they are not well known, pressure effects on egg buoyancy, on larval development and viability are certainly of importance and need further experiments.

5.6 Concluding Remarks

Hydrostatic pressure has potential effects on living organisms from the cell to the organismic levels. The eel is thus concerned mainly during its migration for reproduction but eggs and larvae are also potential targets for mechanical effects of pressure. Most of the pressure effects on the eel have been studied on the non-migratory yellow stage. When hydrostatic pressure is applied for a short term period, aerobic metabolism is altered by the way of a decrease in membrane fluidity which can induce inhibition of oxidative phosphorylation. However, the yellow eel is able to acclimatize to the pressure effects by modifying membrane phospholipid composition (which has an energy cost) in order to maintain homeoviscosity i.e. optimal fluidity. Indeed, it appears clearly that pressure acclimatization induces a state where yellow eels resemble silver eels at atmospheric pressure. Probably, such an acclimatized yellow eel has the metabolic ability to migrate but the metamorphosis to the silver eel stage (silvering process) is required for sexual maturation and reproduction, the aim of the migration. The results obtained from silver eels show that they have low sensitivity to pressure effects (they are pressure resistant). Several observations suggest that silver eels are really ready to cope with the new environment they will encounter during the migration. The set-up of adaptative mechanisms before migration allows them to save energy during it and consequently to optimize the energy budget which is restrained to fat stores. For example, the silvering process allows the eels to modify their membrane composition in order to counterbalance the potential damaging pressure effects without energy loss. However, pressure resistance of silver eels appears modulated by several factors such as sex, origin, nutritional state, temperature. The combination of factors such as, for example, pressure exposure and swimming activity remains to be studied.

Acknowledgements This work has been supported by the European contract EELREP N°Q5RS-2001-01836.

References

Altringham JD, Ellerby DJ (1999) Fish swimming: patterns in muscle function. J. Exp. Biol. 202(23):3397–3403

Amerand A, Vettier A, Sébert P, Cann-Moisan C (2005) In vitro effect of hydrostatic pressure exposure on hydroxyl radical production in fish red muscle. Redox Rep. 10:25–28

Amerand A, Vettier A, Sébert P, Moisan C (2006) Does hydrostatic pressure has an effect on reactive oxygen species in the eels. Undersea Hyperb. Med. 33:157–160

Aoyama J, Hissmann K, Yoshinaga T, Sasai S, Uto T, Ueda H (1999) Swimming depth of migrating silver eels *Anguilla japonica* released at seamounts of the West Mariana Ridge their estimated spawning sites. Mar. Ecol. Prog. Ser. 186:265–269

Bailey DM, Genard B, Collins MA, Rees JF, Unsworth SK, Battle EJV, Bagley PM, Jamieson AJ, Priede IG (2005) High swimming and metabolic activity in the deep-sea eel *Synaphobranchus kaupii* revealed by integrated *in situ* and *in vitro* measurements. Physiol. Biochem. Zool. 78:335–346

Balon EK (1975) Reproductive guilds of fishes: a proposal and definition. J. Fish. Res. Board Can. 32:821–864

Barni S, Bernocchi G, Gerzeli G (1985) Morphohistochemical changes in hepatocytes during the life cycle of the European eel. Tissue Cell 17:97–109

Barthélémy L (1985) Le poisson modèle scientifique en hyperbarie. Bull. Inst. Oceanogr. 4:9–31

Barthélémy L, Cann-Moisan C, Simon B, Caroff J, Sébert P (1991) Concentrations encéphaliques d'acide gamma-aminobutyrique (Gaba) et de glutamine chez un poisson (*Anguilla anguilla* L.) soumis à une pression hydrostatique de101 ATA. Med. Sub. Hyp. 1:35–47

Bartlett D, Wright M, Yayanos AA, Silverman M (1989) Isolation of a gene regulated by hydrostatic pressure in a deep-sea bacterium. Nature 342:572–574

Bartlett DH, Chi E, Wright ME (1993) Sequence of the Omph gene from the deep-sea bacterium *photobacterium* Ss9. Gene 131:125–128

Bartlett DH, Kato C, Horikoshi K (1995) High pressure influences on gene and protein expression. Res. Microbiol. 146:697–706

Belaud A, Barthélémy L, Lesaint J, Peyraud C (1976) Trying to explain an effect of *per se* hydrostatic pressure on heart rate in fish. Aviat. Space Environ. Med. 47:252–257

Begg DA, Salmon ED, Hyatt HA (1983) The changes in structural organization of actin in the sea urchin egg cortex in response to hydrostatic pressure. J. Cell. Biol. 97:1795–1805

Belaud A (1975) Contribution à l'étude de quelques réactions physiologiques de l'anguille (*Anguilla anguilla* L.) soumise à diverses conditions hyperbares. Ph.D. thesis, UBO, Brest, pp. 271

Bell MV, Henderson RJ, Sargent JR (1986) The role of polyunsaturated fatty acids in fish. Comp. Biochem. Physiol. 83B:711–719

Beney L, Perrier-Cornet JM, Wayert M, Gervais P (1997) Modification of phospholipids vesicles induced by high pressure: influence of bilayer compressibility. Biophys. J. 72:1258–1263

Bert P (1878) La pression barométrique. Masson, Paris: Reprinted in 1979 by the Editions du Centre National de la Recherche Scientifique, Paris

Bertin L (1951) les anguilles variation croissance euryhalinité toxicité hermaphrodisme juvénile et sexualité migrations métamorphoses. Payot, Paris

Beullens K, Eding EH, Ollevier F, Komen J, Richter CJJ (1997) Sex differentiation changes in length weight and eye size before and after metamorphosis of European eel (*Anguilla anguilla* L.) maintained in captivity. Aquaculture 153:151–162

Block BA (1991) Endothermy in fish: thermogenesis ecology and evolution. In: Hochachka PW and Mommsen TP (eds) Phylogenetic and Biochemical Perspectives. Elsevier, Amsterdam, pp. 269–311

Bone Q, Marshall NB, Blaxter JHS (1995) Biology of Fishes. Blackie Academy and Professional, London

Bostrom SL, Johansson RG (1972) Enzyme activity patterns in white and red muscle of the eel (*Anguilla anguilla*) at different developmental stages. Comp. Biochem. Physiol. 42B:533–542

Brauer RTW, Jordan MR, Miller CG, Johnson ED, Dutcher JA, Sheeman ME (1985) Interaction of temperature and pressure in intact animals. In: Péqueux AJR, Gilles R (eds) High Pressure Effects on Selected Biological Systems. Springer, Berlin, pp. 3–28

Carrillo M, Bromage N, Zanuy S, Serrano R, Prat F (1989) The effect of modifications in photoperiod on spawning time ovarian development and egg quality in the sea bass (*Dicentrarchus labrax* L.) Aquaculture 81:351–365

Carlisle DB, Denton EJ (1959) On the metamorphosis of the visual pigments of *Anguilla anguilla* (L.). J. Mar. Biol. Ass. UK 38:97–102

Castonguay M, McCleave JD (1987) Vertical distributions diel and ontogenic vertical migrations and net avoidance of leptocephali of *Anguilla* and other common species in the Sargasso Sea. J. Plankton. Res. 9:195–214

Cornish I, Moon TW (1985) Glucose and lactate kinetics in American eel *Anguilla rostrata*. Am. J. Physiol. 249:R67–R72

Cossins AR, Macdonald AG (1984) Homeoviscous theory under pressure: the molecular order of membrane from deep-sea fish. Biochim. Biophys. Acta 776:144–150

D'Aoust BG (1969) Hyperbaric oxygen: toxicity to fish at pressures present in their swimbladders. Science 163:576–578

Dasgupta S (1962) Induction of triploidy by hydrostatic pressure in the leopard frog *Rana pipiens*. J. Exp. Zool. 151:105–121

Dave G, Johansson-Sjobeck ML, Larsson A, Lewander K, Lidman U (1975) Metabolic and hematological effects of starvation in the European eel *Anguilla anguilla* L. I. Carbohydrate lipid protein and inorganic ion metabolism. Comp. Biochem. Physiol. 52A:423–430

Dufour S, Fontaine YA (1985) La migration de reproduction de l'anguille européenne (*Anguilla anguilla* L.): rôle probable de la pression hydrostatique dans la stimulation de la fonction gonadotrope. Bull. Soc. Zool. Fr. 110:291–299

Dunel-Erb S, Sébert P, Chevalier C, Simon B, Barthélémy L (1996) Morphological changes induced by acclimatation to high pressure in the gill epithelium of the freshwater yellow eel. J. Fish. Biol. 48:118–122

Durif C, Dufour S, Elie P (2005) The silvering process of *Anguilla anguilla*: a new classification from the yellow resident to the silver migrating stage. J. Fish Biol. 66:1025–1043

Egginton S (1986) Metamorphosis of the American eel *Anguilla rostrata* Leseur: changes in metabolism of skeletal muscle. J. Exp. Biol. 237:173–184

Ekman S (1932) Prinzipielles uber die wanderungen und die tiergeographische stellung des europaischen aales *Anguilla anguilla* (L.) Zoogeographica 1:85–106

Ellerby DJ, Spierts IL, Altringham JD (2001) Slow muscle power output of yellow- and silver-phase European eels (*Anguilla anguilla* L.): changes in muscle performance prior to migration. J. Exp. Biol. 204:1369–1379

Else PL, Hulbert AJ (2003) Membranes as metabolic pacemakers. Clin. Exp. Pharmacol. Physiol. 30:559–564

Flügel H (1972) Adaptation and acclimatization to high pressure environments. In: Brauer RW (ed) Barobiology and the Experimental Biology of the Deep Sea. University of North Cardina Chapel Hill, NC, pp. 69–88

Fontaine M (1929) De l'augmentation de la consommation d'O des animaux marins sous l'influence des fortes pressions ses variations en fonction de l'intensité de la compression. C. R. Acad. Sci. 188:460–461

Fontaine M (1975) Physiological mechanisms in the migration of marine and amphihaline fish. Adv. Mar. Biol. 13:241–355

Fontaine YA, Dufour S, Alinat J, Fontaine M (1985) L'immersion prolongée en profondeur stimule la fonction hypophysaire gonadotrope de l'anguille européenne (*Anguilla anguilla* L.) femelle. C. R. Acad. Sci. Paris 300:83–87

Fontaine YA, Le Belle N, Lopez EBQ, Alinat J, Vidal B, Barthélémy L, Sébert P, Alinat J, Petter AJ (1990) Infestation de populations françaises d'anguilles (*Anguilla anguilla* L.) par des nématodes (*Anguillicola crassus*): essais thérapeuthiques évaluation de risques potentiels liés à l'écophysiologie de l'hôte. Ann. Parasitol. Hum. Comp. 65:64–68

Fricke H, Kaese R (1995) Tracking of artificially matured eels (*Anguilla anguilla*) in the Sargasso Sea and the problem of the eel's spawning site. Naturwissenschaften 85:290–291

Gennser M, Karpe F, Ornhagen HC (1990) Effects of hyperbaric pressure and temperature on atria from ectotherm animals (*Rana pipiens* and *Anguilla anguilla*). Comp. Biochem. Physiol. 95A:219–228

Gillet C, Vauchez C, Haffray P (2001) Triploidy induced by pressure shock in Arctic charr (*Salvelinus alpinus*): growth survival and maturation until the third year. Aquat. Living Resour. 14:327–334

Goolish EM (1991) Aerobic and anaerobic scaling in fish. Biol. Rev. 66:33–56

Goolish EM (1992) Swimbladder function and buoyancy regulation in the killifish *Fundulus heteroclitus*. J. Exp. Biol. 166:61–81

Gordon MS (1970) Hydrostatic pressure. In: Hoar WS, Randall DJ (eds) Fish Physiology. Academic Press, New York/London, pp. 445–464

Greene DHS, Selivonchick DP (1987) Lipid metabolism in fish. Prog. Lipid Res. 26:53–85

Guderley H (1990) Functional significance of metabolic responses to thermal acclimation in fish muscle. Am. J. Physiol. 259:R245–R252

Haro AJ (1991) Thermal *preferenda* and behaviour of Atlantic eels (genus *Anguilla*) in relation to their spawning migration. Environ. Biol. Fish 31:171–184

Hazel JR (1995) Thermal adaptation in biological membranes: is homeoviscous adaptation the explanation? Annu. Rev. Physiol. 57:19–42

Hicks JW, Bennett AF (2004) Eat and run: prioritization of oxygen delivery during elevated metabolic states. Respir. Physiol. Neurobiol. 144:215–224

Hiroi J, Yasumasu S, Kawazu K, Kaneko T (2003) Hatching enzymes in the Japanese eel. In: Aida K, Tsukamoto K, Yamauchi K (eds) Eel Biology. Springer, Tokyo, pp. 445–456

Jellyman D, Tsukamoto K (2002) First use of archival transmitters to track migrating freshwater eels *Anguilla dieffenbachii* at sea. Mar. Ecol. Prog. Ser. 233:207–215

Johnstone DF, Macdonald AG, Mojsiewicz WR, Wardle CS (1989) Preliminary experiments in the adaptation of the European eel (*Anguilla anguilla*) to hydrostatic pressure. J. Physiol. 417:87

Kelly CE, Kennedy CR, Brown JA (2000) Physiological status of wild European eels (*Anguilla anguilla*) infected with the parasitic nematode *Anguillicola crassus*. Parasitology 120(2):195–202

Kimberly A, Hammond KA, Diamond J (1997) Maximal sustained energy budgets in human and animals. Nature 386:457–462

Kirk RS, Lewis JW, Kennedy CR (2000) Survival and transmission of *Anguillicola crassus* Kuwahara Niimi & Itagaki, 1974 (Nematoda) in seawater eels. Parasitology 120(3):289–295

Kleckner RC (1980) Swim bladder volume maintenance related to initial oceanic migratory depth in silver-phase *Anguilla rostrata*. Science 208:1481–1482

Knopf K, Wurtz J, Sures B, Taraschewski H (1998) Impact of low water temperature on the development of *Anguillicola crassus* in the final host *Anguilla anguilla*. Dis. Aquat. Organ. 33:143–149

Kobayashi H, Pelster B, Scheid P (1989) Water and lactate movement in the swimbladder of the eel *Anguilla anguilla*. Respir. Physiol. 78:45–57

Lecomte-Finiger R, Yahyaoui A (1990) La microstrucutre de l'otholite au service de la connaissance du dévelopment larvaire de l'anguille européenne. C. R. Acad. Sci. Paris 308:1–7

Lee AG (1991) Lipids and their effects on membrane proteins: evidence against a role for fluidity. Prog. Lipid Res. 30:323–348

Lefebvre F, Contournet P, Crivelli AJ (2002) The health state of the eel swimbladder as a measure of parasite pressure by *Anguillicola crassus*. Parasitology 124(4):457–463

Le Galliard JF, Clobert J, Ferrière R (2004) Physical performance and Darwinian fitness in lizards. Nature 432:502–505

Lewander K, Dave G, Johansson ML, Larsson A, Lidman U (1974) Metabolic and hematological studies on the yellow and silver phases of the European eel *Anguilla anguilla* L. I. Carbohydrate lipid protein and inorganic ion metabolism. Comp. Biochem. Physiol. 47B:571–581

Love RM (1970) The Chemical Biology of Fishes. Academic Press, New York

Macdonald AG (1984) The effects of pressure on the molecular structure and physiological functions of cell membrane. Phil. Trans. R. Soc. Lond. 304B:47–68

Macdonald AG, Cossins AR (1985) The theory of homeoviscous adaptation of membranes applied to deep-sea animals. Symp. Soc. Exp. Biol. 39:301–322

Marchelidon J, Le Belle N, Hardy A, Vidal B, Sbaihi M, Burzawa-Gerard E, Schmitz M, Dufour S (1999) Study of variations of anatomical and endocrine parameters in sedentary and downstream migrating female European eels (*Anguilla anguilla*): application to the characterization of the silver stage. Bull. Fr. Pêche Pisc. 355:349–368

Marsland DA (1970) Pressure-temperature studies on the mechanisms of cell division. In: Zimmerman AM (ed) High Pressure Effects on Cellular Processes. Academic Press, New York, pp. 260–312

Marsland DJ (1938) The effects of high hydrostatic pressure upon cell division in Arbacia eggs. J. Cell. Comp. Physiol. 12:57–70

Marsland DJ (1950) The mechanisms of cell division: temperature-pressure experiments on the cleaving eggs of *Arbacia punctulata*. J. Cell. Comp. Physiol. 3:205–227

McCleave JD (1977) Migratory mechanisms in larval and adult American and European eels (*Anguilla rostrata* and *A anguilla*). Nat. Geog. Soc. 18:517–528

McCleave JD, Arnold GP (1999) Movements of yellow- and silver-phase European eels (*Anguilla anguilla* L) tracked in the western north sea. Ices J. Mar. Sci. 56:510–536

Molnar K, Szekely C, Perenyi M (1994) Dynamics of *Anguillicola crassus* (Nematoda: Dracunculoidea) infection in eels of Lake Balaton Hungary. Folia Parasitol. (Praha) 41:193–202

Moon TW (1983) Metabolic reserves and enzyme activities with food deprivation in immature American eels *Anguilla rostrata* (Lesueur). Can. J. Zool. 61:802–811

Moon TW, Latham R (1984) Myofibrillar ATPase activities in red and white myotomal muscles of the American eel *Anguilla rostrata* (Lesueur). Comp. Biochem. Physiol. 79A:151–153

Mushiake K, Kawano K, Sakamoto W, Hasegawa I (1994) Effects of extended daylength on ovarian maturation and hCG induced spawning in yellowtail fed moist pellets. Fish Sci. 60:647–651

Naroska V (1968) Vergleichende untersuchugen über den einfluss des hydrostatischen druckes auf überlebensfähigkeit und stoffwechsel intensität mariner evertebraten und teleosteer Kieler. Meeresforschung 24:95–123

Nilsson L, Nyman L, Westin L, Ornhagen H (1981) Simulation of the reproductive migration of European eels (*Anguilla anguilla* L.) through manipulation of some environmental factors under hydrostatic compression. Spec. Sci. Technol. 4:475–484

Nimeth K, Zwerger P, Wurtz J, Salvenmoser W, Pelster B (2000) Infection of the glass-Eel swimbladder with the nematode *Anguillicola crassus*. Parasitology 121(1):75–83

Owen SF (2001) Meeting energy budgets by modulation of behaviour and physiology in the eel (*Anguilla anguilla* L.). Comp. Biochem. Physiol. 128A:631–644

Pankhurst NW (1982) Changes in body musculature with sexual maturation in the European eel *Anguilla anguilla* (L.). J. Fish Biol. 21:417–428

Pankhurst NW, Lythgoe JN (1983) Changes in vision and olfaction during sexual maturation in the European eel *Anguilla anguilla* (L.) and American eel *Anguila rostrata* (Lesueur). J. Fish Biol. 23:229–240

Pelster B (1997) Buoyancy at depth. In: Randall DJ, Farell AP (eds) Deep Sea Fishes Fish Physiology. Academic press, San Diego, CA, pp. 195–238

Pelster B (2004) pH regulation and swimbladder function in fish. Respir. Physiol. Neurobiol. 144:179–190

Pelster B, Scheid P (1992) Countercurrent concentration and gas secretion in the fish swim bladder. Physiol. Zool. 65:1–16

Pennec JP, Le Bras YM (1988) Dial and seasonal rhythms of the heart rate in the common eel (*Anguilla anguilla* L.): role of cardiac innervation. Exp. Biol. 47:155–160

Ponat A (1967) Untersuchungen zur zellulären druckresistenz verschiedener evertebrates der nord-und ostsee Kieler. Meeresforschung 23:21–47

Pradillon F, Shillito B, Young CM, Gaill F (2001) Developmental arrest in vent worm embryos. Nature 413:698–699

Regnard P (1885) Phénomènes objectifs que l'on peut observer sur les animaux soumis aux hautes pressions. C. R. Sci. Soc. Biol. 37:510–515

Robins CR, Cohen DM, Robins C (1979) The Eels *Anguilla* and *Histiobranchus* photographed on the floor of the deep Atlantic in the Bahamas. Bull. Mar. Sci. 29:401–405

Roche H, Buet A, Ramade F (2003) Caractéristiques écophysiologiques d'une population d'anguilles de camargue exposée à une pollution clandestine par des polluants organiques persistants. Rev. Ecol. 58:103–126

Rossignol O, Sébert P, Simon B (2006) Effects of pressure acclimatization on silver eel (*Anguilla anguilla* L.) slow muscle contraction. Comp. Biochem. Physiol. 143A:234–238

Rottmann RW, Shireman JV, Chapman FA (1991) Induction and Verification of Triploidy in Fish. Southern Regional Aquaculture Center, pp. 421

Scholander PF, Van Dam L (1953) Composition of the swimbladder gas in deep sea fishes. Biol. Bull. Mar. Biol. Lab. 104:75–86

Sébert ME, Amérand A, Vettier A, Weltzien FA, Pasqualini C, Sébert P, Dufour S (2007) Effects of high hydrostatic pressure on the pituitary-gonad axis in the European eel *Anguilla anguilla* (L.). Gen. Comp. Endocrinol. 153:289–298

Sébert P (1993) Energy metabolism of fish under hydrostatic pressure: a review. Trends Comp. Biochem. Physiol. 1:289–317

Sébert P (1997) Pressure effects on shallow-water fishes. In: Randall DJ, Farell AP (eds) Deep Sea Fishes Fish Physiology. Academic Press, San Diego, CA, pp. 279–324

Sébert P (2003) Fish adaptations to pressure. In: Val AL, Kapoor BG (eds) Fish Adaptations. Science Publisher, Enfield, pp. 73–95

Sébert P (2007) Fish muscle function and pressure. In: Sébert P, Onyango D, Kapoor BG (eds) Fish Life in Special Environments. Science Publisher, Enfield (in preparation)

Sébert P, Barthélémy L (1985a) Effects of high hydrostatic pressure per se 101 atm on eel metabolism. Respir. Physiol. 62:349–357

Sébert P, Barthélémy L (1985b) Hydrostatic pressure and adrenergic drugs (agonists and antagonists): effects and interactions in fish. Comp. Biochem. Physiol. 82C:207–212

Sébert P, Macdonald AG (1993) Fish. In: Macdonald AG (ed) Effects of High Pressure on Biological Systems. Springer, Berlin, pp. 147–196
Sébert P, Theron M (2001) Why can the eel unlike the trout migrate under pressure? Mitochondrion 1:79–85
Sébert P, Barthélémy L, Caroff J (1986) Catecholamine content (as measured by the HPLC method) in brain and blood plasma of the eel: effects of 101 ATA hydrostatic pressure. Comp. Biochem. Physiol. 84C:155–157
Sébert P, Barthélémy L, Caroff J, Hourmant A (1987) Effects of hydrostatic pressure *per se* (101 ATA) on energetic processes in fish. Comp. Biochem. Physiol. 86A:491–495
Sébert P, Péqueux A, Simon B, Barthélémy L (1991) Effects of long term exposure to 101 ATA HP on blood gill and muscle composition and of some enzyme activities of the fw eel (*Anguilla anguilla* L.). Comp. Biochem. Physiol. 98B:573–577
Sébert P, Meskar A, Simon B, Barthélémy L (1994) Pressure acclimation of the eel and liver membrane composition. Experientia 50:121–123
Sébert P, Simon B, Barthélémy L (1995) Effects of a temperature increase on oxygen consumption of yellow fresh water eel exposed to high hydrostatic pressure. Exp. Physiol. 80:1039–1046
Sébert P, Peragon J, Barroso JB, Simon B, Melendez-Hevia E (1998) High hydrostatic pressure (101 ATA) changes the metabolic design of yellow freshwater eel muscle. Comp. Biochem. Physiol. 121B:195–200
Sébert P, Theron M, Vettier A (2004) Pressure and temperature interactions on cellular respiration: a review. Cell. Mol. Biol. 50:491–500
Seoka M, Yamada S, Iwata Y, Yanagisawa T, Nakagawa T, Kumai H (2003) Differences in the biochemical content of buoyant and non-buoyant eggs of the Japanese eel *Anguilla japonica*. Aquaculture 216:355–362
Siebenaller JF, Somero GN (1989) Biochemical adaptation to the deep sea. CRC Crit. Rev. Aquat. Sci. 1:1–25
Simon B (1990) Métabolisme énergétique de l'anguille (*Anguilla anguilla* L.): effets d'expositions de courte durée (3 h) et de longue durée (1 mois) à 101 ATA de pression hydrostatique. Ph.D. thesis, UBO, Brest pp. 105
Simon B, Sébert P, Barthélémy L (1989) Effects of long-term exposure to hydrostatic pressure *per se* (101 ATA) on eel metabolism. Can. J. Physiol. Pharmacol. 67:1247–1251
Simon B, Sébert P, Barthélémy L (1991) Eel *Anguilla anguilla* (L.) muscle modifications induced by long-term exposure to 101 ATA hydrostatic pressure. J. Fish Biol. 38:89–94
Simon B, Sébert P, Cann-Moisan C, Barthélémy L (1992) Muscle energetics in yellow fresh water eel (*Anguilla anguilla* L.) exposed to high hydrostatic pressure (101 ATA) for 30 days. Comp. Biochem. Physiol. 102B:205–208
Simon B, Barthélémy L, Sébert P (1994) Comparison of some isoenzymes on the yellow and silver phases of the European eel (*Anguilla anguilla* L.). Aquaculture 25:937–942
Somero GN (1991) Hydrostatic pressure and adaptation to the deep sea. In: Prosser CL (ed) Environmental and Metabolic Animal Physiology. Wiley-Liss, New York, pp. 167–204
Speers-Roesch B, Lingwood D, Stevens ED (2004) Effects of temperature and hydrostatic pressure on routine oxygen uptake of the bloater (*Coregonus hoyi*). J. Great Lakes Res. 30:70–81
Sprengel G, Luchtenberg H (1991) Infection by endoparasites reduces maximum swimming speed of European smelt *Osmerus eperlanus* and European eel *Anguilla anguilla*. Dis. Aquat. Organ. 11:31–35
Strand E, Jorgensen C, Huse G (2005) Modelling buoyancy regulation in fishes with swimbladders: bioenergetics and behaviour. Ecol. Model. 185:309–327
Tesch FW (1978) Telemetric observations on the spawning migration of the eel (*Anguilla anguilla*) west of the European continental shelf. Environ. Biol. Fish 3:203–209
Tesch FW (1989) Changes in swimming depth and direction of silver eels (*Anguilla anguilla* L.) from the continental shelf to the deep sea. Aquat. Living Resour. 2:9–20
Tesch FW (2003) The Eel. Fifth edition. Thorpe JE, Oxford

Theron M, Guerrero F, Sébert P (2000) Improvement in the efficiency of oxidative phosphorylation in the freshwater eel acclimated to 10.1 MPa hydrostatic pressure. J. Exp. Biol. 203(19):3019–3023

Thomson AJ, Sargent JR (1977) Changes in the levels of chloride cells and ($Na^+ + K^+$)-dependent ATPase in the gills of yellow and silver eels adapting to seawater. J. Exp. Zool. 200:33–40

Tyler PA, Young CM (1998) Temperature and pressure tolerances in dispersal stages of the genus Echinus (*Echinodermata: Echinoidea*): prerequisites for deep-sea invasion and speciation. Deep-Sea Res. II 45:253–277

Tytler P, Blaxter JHS (1973) Adaptation by cod and saithe to pressure changes. Neth. J. Sea Res. 7:31–45

Van Ginneken V, Van Den Thillart G (2000) Eel fat stores are enough to reach the Sargasso. Nature 403:156–157

Van Ginneken V, Maes G (2005) The European eel (*Anguilla anguilla* Linnaeus) its lifecycle evolution and reproduction: a literature review. Rev. Fish Biol. Fish. 15:367–398

Van Ginneken V, Antonissen E, Müller UK, Booms R, Eding E, Verreth J, Van Den Thillart G (2005a) Eel migration to the Sargasso: remarkably high swimming efficiency and low energy costs. J. Exp. Biol. 208:1329–1335

Van Ginneken V, Vianen G, Muusze B, Palstra A, Verschoor L, Lugten O, Onderwater M, Van Schie S, Niemantsverdriet P, Van Heeswijk R, Eding E, Van Den Thillart G (2005b) Gonad development and spawning behavior of artificially-matured European eel (*Anguilla anguilla* L.). Anim. Biol. 55:203–218

Vettier A (2005) La migration de reproduction de l'anguille européenne (*Anguilla anguilla* L): effets de la pression hydrostatique et de la métamorphose. Ph.D. thesis, UBO, Brest, pp. 90

Vettier A, Sébert P (2004) Pressure resistance of aerobic metabolism in eels from different water environments. Mitochondrion 3:347–354

Vettier A, Szekely C, Sébert P (2003) Are yellow eels from Lake Balaton able to cope with high pressure encountered during migration to the Sargasso Sea? The case of energy metabolism. Anim. Biol. 53:329–338

Vettier A, Amérand A, Cann-Moisan C, Sébert P (2005) Is the silvering process similar to the effects of pressure acclimatization on yellow eels? Respir. Physiol. Neurobiol. 145:243–250

Vettier A, Labbé C, Amérand A, Da Costa G, Le Rumeur E, Moisan C, Sébert P (2006) Hydrostatic pressure effects on eel mitochondrial functioning and membrane fluidity. Undersea Hyperb. Med. 33:149–156

Wardle CS, Tetteh-Lartey N, Macdonald AG, Harper AA, Pennec JP (1987) The effect of pressure on the lateral swimming muscle of the European eel *Anguilla anguilla* and the deep sea eel *Histiobranchus bathybius*; results of Challenger cruise 6b/85. Comp. Biochem. Physiol. 88A:595–598

Watanabe T (1982) Lipid nutrition in fish. Comp. Biochem. Physiol. 73B:3–15

Watanabe T, Itoh A, Kitajima C, Fujita S (1984a) Effect of dietary protein levels on reproduction of red sea bream. Bull. Jpn. Soc. Sci. Fish. 50:1015–1022

Watanabe T, Itoh A, Murakami A, Tsukashima Y, Kitajima C, Fujita S (1984b) Effect of nutritional quality of diets given to broodstock on the verge of spawning on reproduction of red sea bream. Bull. Jpn. Soc. Sci. Fish. 50:1023–1028

Westin L, Nyman L (1979) Activity orientation and migration of Baltic eel (*Anguilla anguilla* (L.)) Rapp. Cons. Explor. Mer. 174:631–644

White FN, Somero GN (1982) Acid base regulation and phospholipid adaptation to temperature time courses and physiological significance of modifying the milieu for protein function. Physiol. Rev. 62:40–90

Würtz J, Taraschewski H (2000) Histopathological changes in the swimbladder wall of the European eel *Anguilla anguilla* due to infections with *Anguillicola crassus*. Dis. Aquat. Organ. 39:121–134

Yamazaki F, Goodier J (1993) Cytogenetic effects of hydrostatic pressure treatment to suppress the first cleavage of salmon embryos. Aquaculture 110:51–59

Young CM, Tyler PA (1993) Embryos of the deep-sea echinoid *Echinus affinis* require high pressure for development. Limnol. Oceanogr. 38:178–181

Young CM, Tyler PA, Gage JD (1996) Vertical distribution correlates with pressure tolerances of early embryos in the deep-sea asteroid *Plutonaster bifrons*. J. Mar. Biol. Ass. UK 76:749–757

Zara V, Palmieri L, Giudetti A, Ferramosca A, Capobianco L, Gnoni GV (2000) The mitochondrial tricarboxylate carrier: unexpected increased activity in starved silver eels. Biochem. Biophys. Res. Commun. 276:893–898

Chapter 6
Acclimation to Seawater in the European Eel *Anguilla anguilla*: Effects of Silvering

J. Cliff Rankin

6.1 Introduction

6.1.1 Osmoregulation

Teleost fishes regulate the osmotic concentration of their body fluids at about 30–40% of the level of oceanic sea water, as do most vertebrates (including ourselves!). This process is known as osmoregulation, which will be briefly summarized following the outline given in Rankin and Davenport (1981), which uses the eel as an example. The most primitive chordates, the hagfish, Family *Myxinidae*, have blood isoosmotic with sea water and are confined to the marine environment, where the phylum Chordata (which includes the vertebrates) must have originated. Later vertebrates are thought to have evolved from ancestors who lived in freshwater, which they were only able to colonize by reducing their blood osmotic concentration to minimize two problems; osmotic entry of water and diffusional loss of salts, which are serious problems as fish gills must have a large surface area and thin epithelium to facilitate oxygen uptake from the water. In marine teleost fishes blood composition is similar to that of fishes in freshwater, resulting in the opposite problems; osmotic entry of water and diffusional loss of salts.

In the face of these dissipative forces homeostasis is maintained by active processes in the gills, kidneys and guts. Most teleost species only possess the mechanisms to osmoregulate (and therefore survive) in either sea water or freshwater; they are said to be stenohaline. A small minority are euryhaline, being able to move between fresh and salt water either at certain stages in their life cycle (e.g. salmon) or at any time (e.g. eels, flounder). The basic features of eel osmoregulation have long been known but in recent years there has been a resurgence of interest in its endocrine control mechanisms. The functions of the osmoregulatory organs will be briefly described, followed an overview of the hormones implicated and finally a more detailed discussion of their role in the eel life cycle.

J.C. Rankin
Department of Chemical and Biological Science, University of Huddersfield, Queensgate, HD1 3DH UK

G. van den Thillart et al. (eds.), *Spawning Migration of the European Eel,*

6.1.1.1 Gills

A comprehensive review of the several functions of fish gills has been published recently (Evans et al. 2005). Gills of marine fishes excrete excess Na^+ (sodium ions) and Cl^- (chloride ions) that enter across the gills and alimentary canal from the higher ambient concentrations. Mitochondria-rich cells in eel gills were suggested as the site for ion extrusion and named "chloride-secreting cells" by Keys and Willmer (1932). Many studies have been carried out on gill histology in relation to salinity. In *Anguilla japonica* two types of chloride cell (A and B) were recognized which were both more numerous 3 days after FW > SW (freshwater to seawater) transfer, after which type A cells continued to increase in number until 2 weeks, whereas type B cells decreased and were rarely seen at 2 weeks post-transfer. This, and the structure of the cells and the way it changed suggested that type B was a transitional stage in the development of the typical (type A) marine chloride cell (Shirai and Utida 1970). In *A. anguilla* it was suggested that new chloride cells differentiated in the hours following FW > SW transfer, with a second wave of mitotic divisions between days 2 and 5–7 post transfer, peaking at days 3 or 4 in addition to hypertrophy of existing cells (Olivereau 1970). In *A. japonica* the activity of the enzyme Na^+, K^+-ATPase (sodium + potassium-activated adenosine triphosphatase) in the gills increased following FW > SW transfer with the greatest increase between 5 and 7 days (Kamiya and Utida 1968) in proportion to an increase in the number of chloride cells, where the enzyme was localized (Utida et al. 1971).

A model for the ion transport mechanisms in marine chloride cells proposed by Silva et al. (1977) has been confirmed by biochemical and molecular biological techniques. Chloride cells are filled with a branching tubular system formed by invaginations of the basolateral membrane, rich in the enzyme Na^+, K^+-ATPase (see Evans 2005 et al. for references). This uses the energy from conversion of one ATP molecule to ADP to pump three Na^+ out of the cell and two K^+ in, thus maintaining a very low cytoplasmic Na^+ concentration. Na^+ can enter the cytoplasm down its electrochemical gradient via the NKCC (sodium, potassium, 2 chloride) cotransporter (Cutler and Cramb 2002). The potential energy of the sodium gradient enables the K^+ and Cl^- ions to enter the cell against electrochemical gradients that favor the exit of these ions. The K^+ does not remain in the cell but exits through K_{ir} (hypertonicity-inducible inward rectifier) potassium channels found in eel chloride cells by Suzuki et al. (1999). Cl^- ions exit through CFTR (cystic fibrosis transmembrane regulator) channels in the apical membrane (Marshall and Singer 2002). These mechanisms result in a net transfer of Cl^- ions from the fish extracellular fluids to seawater.

In freshwater, the gills actively take up Na^+ and Ca^{2+} ions and in most species, but not apparently eels, Cl^- ions. The mechanisms are not as clear as in seawater gills; freshwater chloride cells and/or pavement epithelial cells may be involved and Na^+ uptake appears to involve a V-type H^+ ATPase which extrudes H^+ across the apical membrane, creating an electrical potential gradient favoring uptake of Na^+ ions which are pumped across the basolateral membranes into the extracellular fluid (Evans 2005). This chapter is not about the molecular mechanisms of osmoregulation; the details above are given because the transporters/exchangers present in gill cells are very useful to determine the extent of seawater or freshwater adaptation. NKCC is characteristic of seawater osmoregulation; V-type ATPase of freshwater osmoregulation. Na^+, K^+-ATPase is involved in both, but its activity is higher in seawater fish gills. Study of different isoforms can give more detailed information. Genomic studies should give much more useful information but the genomes of only two teleosts have been completely sequenced, the zebrafish, *Danio rerio*, and the fugu, *Fugu rubrides*, are both stenohaline and of limited value in relation to the enormous potential contribution of fish studies to elucidating osmoregulatory mechanisms common to all vertebrates. However, the very recent application of microarray technology to gene profiling in relation to salinity acclimation in *Anguilla anguilla* (Kalujnaia et al. 2007) promises to throw much light on the proteins involved in osmoregulation and its endocrine control. If the European eel should become extinct it will be a tragedy for physiological research as well as all the other devastating consequences. In fact a great deal of the research carried out on teleost osmoregulation has used members of the genus *Anguilla* including, amongst others, the closely related Japanese eel, *A. japonica* and American eel, *A. rostrata*.

6.1.1.2 The Alimentary Canal

Marine fishes lose water osmotically. Their survival depends on replacing it by drinking. If the ingested water is drained from an oesophageal fistula plasma sodium and haematocrit rise rapidly resulting in death in a few days; reintroducing the drained water into the stomach results in a situation similar to that in intact eels (Takei et al. 1998). As the seawater is hypertonic to the body fluids there will be an osmotic gradient from body fluids to gut lumen opposing water uptake. Drinking rates can be estimated by adding a non-absorbable marker to the water, allowing the fish to drink for a measured time interval and then comparing the gut contents of the marker with the water concentration (Maetz and Skadhauge 1968). In the first hour after FW > SW transfer *A. anguilla* drinking rate increased about threefold ($p < 0.01$) over the FW value, and then declined to about half the FW value over the next few hours. A slow increase then ensued peaking at 1 week, followed by a slow reduction to values not significantly different to those in FW after 3 to 6 weeks (Kirsch and Mayer-Gostan 1973).

Imbibed water first passes along the oesophagus which is impermeable to Na^+ and Cl^- ions and to water in freshwater eels but becomes highly permeable to the two ions, which diffuse into the blood (to be removed by the gills) after FW > SW transfer. The "desalted" water enters the stomach which, in contrast to the oesophagus, is permeable to water which passes from the body fluids into the lumen (Hirano and Mayer-Gostan 1976). The eel has still not gained any water; this happens in the intestine.

In cultured *A. japonica* body weight decreased to a minimum 2 days after FW > SW transfer and thereafter partially recovered. Water and salt uptake across isolated intestinal sacs bathed on both sides by Ringer solution (i.e. in the absence of an osmotic gradient) increased following FW > SW transfer, both reaching a peak after 5 days. The absorbed fluid was approximately isoosmotic (Oide and Utida 1967). Skadhauge (1969, 1974) showed that the in situ perfused intestine of *A. anguilla* was capable of fluid absorption against an osmotic gradient, the maximum hypertonicity of the absorbate being proportional to the salinity to which the eels were acclimated.

Water uptake is solute-linked, dependent on active NaCl uptake. Eel intestinal cells have an absorptive form of cotransporter 2, NKCC2 (Cutler and Cramb 1996) located in the apical membrane, with the K^+ recycling through basal channels and the Na^+ and Cl^- exiting through the basolateral membranes into the extracellular fluid via Na^+, K^+-ATPase and chloride channels respectively (see review by Lionetto and Schettino 2006). Cell membranes have a very low permeability to water unless water channel molecules are present. Martinez et al. (2005) found that *A. anguilla* intestine expressed homologues of the aquaporin AQP1 and the aquaglyceroporin AQPe, the mRNA of the former showing large increases in abundance following seawater acclimation. The role of the alimentary canal in maintaining water balance in seawater-acclimated eels has recently been reviewed by Ando et al. (2003).

6.1.1.3 Kidneys

Freshwater fishes gain water osmotically. It is eliminated renally. In *A. anguilla* at 12°C this amounted to 26.4 ml (kg body weight)$^{-1}$ (24 h)$^{-1}$ of dilute urine ([Na^+] 13.1 mM compared to 150 mM in plasma). After 7 days in seawater it had decreased to 6.06 ml (kg body weight)$^{-1}$ (24 h)$^{-1}$, $p < 0.001$ (Chester Jones et al. 1969). This 77% decrease in urine flow could be accounted for by a 71% decrease in GFR (glomerular filtration rate). GFR and urine flow rate decreased within a few hours of FW > SW transfer, accompanied by a decrease in blood pressure (both ventral and dorsal aortic). Kirsch (1972) found similar results at 13.5°C with a urine flow rate of 28.6 ml (kg body weight)$^{-1}$ (24 h)$^{-1}$ in FW reducing by 86.2% 24 h after FW > SW transfer. In teleosts GFR is dependent on renal perfusion pressure (Brown et al. 1993). In the rainbow trout, *Oncorhynchus mykiss* (formerly known as *Salmo gairdneri*), changes in GFR result in changes in the proportion of individual glomeruli that are filtering (Brown et al. 1978) with seawater acclimated fishes having a large proportion of glomeruli either perfused but non-filtering or non-perfused. Marine teleosts are losing water so need to reduce urinary losses to a minimum. The role of the kidney seems to be to eliminate excess divalent ions, mainly Mg^{2+} and Ca^{2+} (Oikari and Rankin 1985).

The large increase in haematocrit in the first day following FW > SW transfer in *A. anguilla* (Ball et al. 1971) shows that considerable haemoconcentration accompanies the loss in body weight (reflecting water loss) in *A. anguilla* (Ball et al. 1971). This reduction in blood volume could account for the hypotension leading to antidiuresis (Chester Jones et al. 1969) but in fact renal function is under complex endocrine control.

6.1.2 Endocrine Control of Osmoregulation

The adrenocortical steroid hormone aldosterone regulates sodium balance in higher vertebrates. Removal of the interrenal, the adrenocortical homologue in the eel is possible but drastic surgery is required. It resulted in impairment of osmoregulation, reduced serum Na^+ concentrations in freshwater-acclimated and increased concentrations in seawater-acclimated European eels (Chan et al. 1967). The mineralocorticoid hormone in teleosts is cortisol. Partial hypophysectomy (removal of the pars distalis or anterior pituitary) of *A. japonica* prevented the increase in water absorption in isolated intestinal sacs which normally follows FW > SW transfer ($p < 0.02$). Cortisol injection increased water absorption ($p < 0.05$), restoring it to control levels (Hirano 1967), suggesting that the ACTH (adrenocorticotropic hormone)–interrenal system plays an important role in salt adaptation of eel. (Plasma cortisol fell to extremely low levels following hypophysectomy in *A. anguilla*: Ball et al. 1971). In intestinal sacs taken from adrenalectomized *A. anguilla* water transport was reduced ($p < 0.001$); it was restored to the same value as in sham-adrenalectomized eels by cortisol injections (Gaitskell and Chester Jones 1970).

Some hypophysectomized teleosts die in FW unless injected with the anterior pituitary hormone prolactin, but eels can survive (Fontaine et al. 1949). Plasma sodium levels fall; this can be slowed, but not reversed, if prolactin treatment is started immediately after hypophysectomy (Olivier and Chartier-Baraduc 1966). The hyponatremia is due to an increased sodium outflux but sodium loss is small, explaining the eel's survival. Prolactin reduces the elevated sodium outflux ($p < 0.001$) (Maetz et al. 1967). Many subsequent observations led to the idea that cortisol is the seawater-adapting hormone and prolactin the freshwater-adapting hormone in teleosts. A more recent consensus is that cortisol is also involved in promoting ion uptake from freshwater, where it interacts with prolactin, whilst in seawater the GH/IGF2 (growth hormone/insulin-like growth factor) axis plays an important role in addition to cortisol, although studies of this system in eels appear to be lacking (McCormick 2001).

Other endocrine systems are involved in control of drinking in eels: angiotensin II (ANGII) stimulates drinking and atrial natriuretic peptide (ANP) and bradykinin inhibit it (reviewed by Takei 2002). ANP inhibits intestinal Na^+ absorption as well as drinking in SW *A. japonica* (Tsukada et al. 2005) These combined actions reduced plasma $[Na^+]$ whereas immunoneutralization of circulating ANP increased it (Tsukada and Takei 2006). All three members of the eel natriuretic peptide system; ANP, VNP (ventricular natriuretic peptide) and CNP appear to be involved in the control of osmoregulatory mechanisms (Takei and Hirose 2002). CNP may have an osmoregulatory function in freshwater *Anguilla japonica*; infusion at physiological rates increased influx of Na^+ from the environment (Rankin and Takei, unpublished observations cited in Takei and Hirose 2002).

Intracranial injection of eel Angiotensin II stimulated drinking in the Japanese eel whilst several hormones, including eel ANP, inhibited it. However, the most potent inhibitor of drinking was ghrelin (Kozaka et al. 2003). Ghrelin is a hormone found in mammalian stomach which stimulates growth hormone secretion (Kojima and Kangawa 2005). It has also been found in Japanese eel stomach. Plasma levels increase transiently after transfer to sea water and it has been suggested that it participates in osmoregulation (Kaiya et al. 2006).

ANGII, as well as ACTH, stimulates cortisol secretion in the rainbow trout, *Oncorhynchus mykiss* (Arnold-Reed and Balment 1994). Calcitonin gene-related peptide (CGRP) may be involved in the endocrine control of the eel gill (Lafont et al. 2006). Further discussion this and other aspects of the endocrine control of osmoregulation can be found in Chapter 5.

6.1.2.1 Hormones Affecting the Kidney

The posterior pituitary hormone AVT (arginine vasotocin) is diuretic at high (probably pharmacological, see Warne 2002) doses but antidiuretic at low doses in *A. anguilla* (Henderson and Wales 1974; Babiker and Rankin 1978). The pressor effects of high doses would be expected to be diuretic but when pressor effects were eliminated in a perfused eel kidney preparation AVT was antidiuretic (Rankin et al. 1972). This was also the case in rainbow trout, where AVT significantly ($p < 0.001$) reduced the population of filtering glomeruli (Amer and Brown 1995). The renal and vascular effects of AVT in teleosts are of unclear physiological significance in teleosts, where there are no consistent differences in plasma concentrations between stenohaline freshwater and marine species or between euryhaline species acclimated to FW or SW (Brown and Balment 1997); in *A. anguilla* FW plasma AVT was $0.9 \pm 0.1\,\text{fmol ml}^{-1}$ and SW plasma it was $1.0 \pm 0.1\,\text{fmol ml}^{-1}$. Angiotensin II is another vasoactive peptide in which pressor actions might result in diuresis but in an *in situ* perfused trout kidney preparation concentrations in the physiological range were antidiuretic (Dunne and Rankin 1972).The trout appears to have a local renin-angiotensin system which may be involved in the control of urine flow rate (Brown and Balment 1997).

Adrenalectomy led to a reduction in GFR and urine flow in both yellow and silver freshwater-acclimated eels. These effects were prevented by daily injections of cortisol, the mineralocorticoid hormone of teleosts (Chan et al. 1969). Atrial Natriuretic Peptide (ANP) and Ventricular Natriuretic Peptide (VNP) levels increase in *A. japonica* plasma 6h after FW > SW transfer (p, 0.05 in both cases, gradually returning to FW levels thereafter (Kaiya and Takei 1996). ANP infusion at physiological rates into intact SW-acclimated eels reduced urine flow and increased urinary Na^{+} concentration, although excretion rate did not change (Takei and Kaiya 1998).

6.2 Does the Process of Silvering Affect Osmoregulatory Capacity?

Early work on European eels by Chan et al. (1967) included measurements of blood serum ionic composition of yellow and silver eels in freshwater and after 4 to 5 weeks acclimation to sea water. Experiments were performed on a total of 99 eels in 1963 and 1965. We did not have an osmometer at the time but serum sodium ion concentrations are a good reflection of osmoregulatory capability. There was no evidence that the silver eels had improved capacity to maintain serum sodium concentrations in sea water: in fact the group of sea water yellow eels had a lower serum sodium concentration than the two groups of silver eels (Fig. 6.1). There was clear evidence for endocrine control of osmoregulation. Adrenalectomy and/or removal of the Corpuscles of Stannius resulted in reduced serum sodium concentrations in freshwater- and increased concentrations in seawater-acclimated eels. There were large changes in body water content: body weight increased by 20.9% in freshwater- and decreased by 9.6% in seawater-acclimated eels following adrenalectomy. It is likely that the adrenalectomy was not complete. Subsequent experiments where a more drastic procedure was followed to ensure removal of all adrenocortical cells from yellow eels showed that they were unable to survive in sea water for more than 48 h without cortisol replacement therapy (Mayer et al. 1967), large changes being observed in sodium fluxes across the gills.

The question of acclimation of eels to sea water was re-investigated with experiments being carried out in four localities; the laboratories of the Museum National d'Histoire Naturelle in Concarneau and Paris, France; the Cemagref research station in St. Seurin, France and the Institute of Biology, Leiden University, The Netherlands. Running sea water was available at Concarneau and re-circulating sea water systems at the other locations.

6.2.1 Sea Water Challenge Test

Yellow eels can clearly acclimate well to sea water, even to artificial double-strength sea water (Maetz and Skadhauge 1968; Skadhauge 1969) and they are found living in coastal waters. Although Chan et al. (1967) found that yellow eels seemed to acclimate better to sea water than silver eels it was not possible to determine the

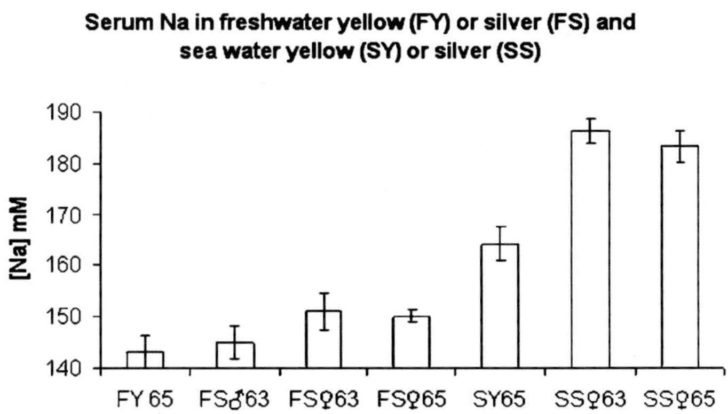

Fig. 6.1 Serum sodium concentrations (mm) in long-term seawater-acclimated eels. FY = yellow eels in freshwater, FS = silver eels in freshwater, SY = yellow eels in sea water, SS = silver eels in sea water. ♀ = female, ♂ = male. The numbers refer to the year of the experiment, e.g. 65 = 1965 (Data from Chan et al. 1967)

origin of the eels (bought from Billingsgate Fish Market) so the yellow and silver groups (a distinction based solely on skin coloration) might not have been strictly comparable. As part of the EELREP study, measurements of plasma osmolality were made on freshwater-acclimated yellow eels from the river Loire kept at Concarneau; long term seawater-acclimated silver eels from the Loire kept at St. Seurin and long term seawater-acclimated eels from the Grevelingen kept at Leiden. The osmolalities were almost identical: Loire yellow eels in freshwater 294 ± 5.3 mOsm kg^{-1}; Loire silver eels in sea water 299.6 ± 2.8 mOsm kg^{-1}; Grevelingen silver eels in sea water 299.9 ± 4.3 mOsm kg^{-1} (all values mean ± standard error of mean). Silver eels can obviously osmoregulate very well in sea water.

In order to investigate the possibility of small differences in osmoregulatory capacity between yellow and silver eels the short-term response to abrupt transfer from freshwater to seawater was investigated. Salmonids also undergo a process of morphological and physiological changes prior to migration from fresh- to seawater called smolting (Boeuf 1993) that resembles silvering in some ways (Fontaine et al. 1995). Study of smolting has benefited greatly from the adoption of a standardized test, the 24 (or 48) h seawater challenge test (Clarke 1982). Fish are transferred directly to sea water and after 24 (or 48) h one of several key indicators of osmoregulatory capacity is measured. For example, the activity or expression (Seidelin et al. 2001) of the enzyme Na^+, K^+-activated ATPase, which plays a key role in sodium excretion across the gills in seawater fishes. Activity increases and remains high when smolts, which are preadapted to enter sea water, are transferred, but not in the case of parr or post-smolts. A number of studies on eels have shown that Na^+, K^+-activated ATPase activity is greater in seawater- than in freshwater-acclimated eels (e.g. Sargent and Thomson 1974) and that it increases following transfer of eels from freshwater to seawater (e.g. Utida et al. 1971).

A sea water challenge test for eels was developed using yellow eels (mean mass = 102 g) from the River Loire. Experiments were carried out in the Marine Laboratory at Concarneau in collaboration with Drs. M. Fouchereau-Peron and A-G. Lafont. Eels were transferred from freshwater to 1,000 mOsm l^{-1} seawater and at intervals were anaesthetized in either MS222 (Sigma) or clove oil. Syringes were rinsed with either ammonium or lithium heparin solution and air-dried. A blood sample was taken rapidly by puncture of the caudal vessels into a heparinized syringe and immediately centrifuged. Samples were taken over a period of 24 days post-transfer. Osmolality was measured on 10 μl aliquots using a vapor pressure osmometer (Wescor model 5520, USA). Further 10 μl plasma aliquots were diluted with 1,000 μl distilled water (101 × dilutions) and analyzed for inorganic cations and anions using an ion chromatograph (Dionex series 4500i, USA). In some instances the plasma was deep frozen for subsequent measurement of osmolality and dilution for ionic analysis; in other cases these procedures were carried out immediately. As a test for possible inaccuracies in osmolality readings due to protein breakdown with freezing and thawing, one set of measurements was made on the same samples both directly and after freezing. The increase in osmolality after freezing was just over 1% (the limit of accuracy of the instrument) and not statistically significant (paired t test: before freezing 287.7 ± 3.2; after freezing and thawing 290.9 ± 4.8 mean ± SEM n = 11).

Gill samples (ca. 400 mg) were cut from the distal half of filaments in the centre of several gill arches, placed in ice-cold SEIDM buffer (300 mM sucrose, 20 mM Na_2-EDTA, 50 mM imidazole, 0.1% sodium deoxycholate, 10 mM β-mercaptoethanol, pH 7.3) in 1.5 ml Eppendorf tubes and frozen in liquid nitrogen. They were returned to the University of Southern Denmark in liquid nitrogen and then assayed for Na^+, K^+-activated ATPase activity according to McCormick (1993) using a microplate reader (SPECTRAmax PLUS, Molecular Devices, Sunnyvale, CA). Protein content was measured by a micro-assay based on the method of Lowry et al. (1951) and enzymatic activity was normalized to protein concentration and expressed as μmol ADP [mg protein]$^{-1}$ h^{-1}.

6.2.1.1 Evaluation of Sea Water Challenge Test

Plasma osmolality increased following transfer to sea water and reached a plateau between 24 and 48 h after transfer (Fig. 6.2). Measurements of plasma sodium ion concentration showed a similar pattern but were more variable. Regression analysis of the results over the first 24 h showed that plasma osmolality was just over double the sodium concentration as expected, but that the correlation was not particularly good:

$$\text{Osmolality} = 2.09\,[Na^+] + 37.5,\ r^2 = 0.868\ (n = 16)$$

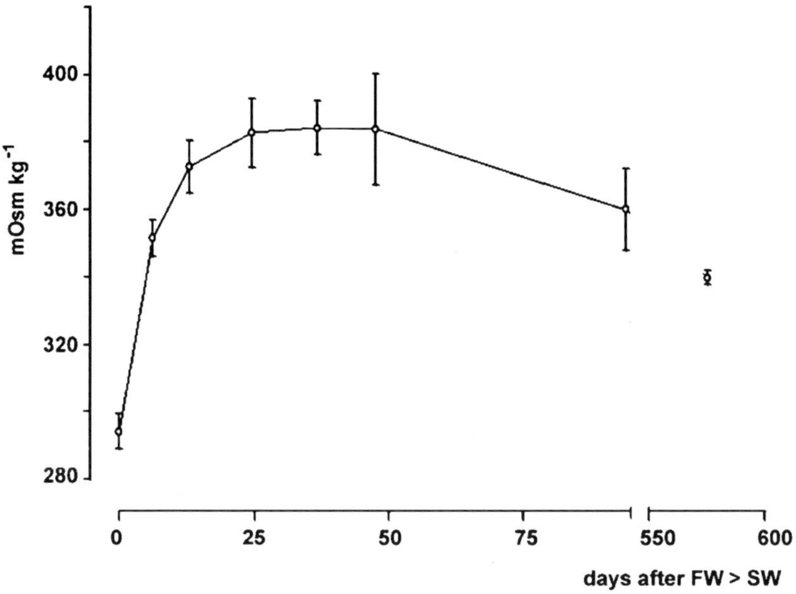

Fig. 6.2 Plasma osmolality (mOsm kg^{-1}) of yellow eels transferred from freshwater to sea water at time zero

Gill Na^+, K^+-ATPase activity did not increase over this period; in fact ANOVA showed that there were no significant changes with time over the whole 24 days (Fig. 6.3). It was concluded that plasma osmolality was the best measure of ability to withstand a sea water challenge and that 36 h after transfer was the most appropriate time at which to measure it.

6.2.1.2 Results of Sea Water Challenge Test

Eels were obtained from Lake Grevelingen and a commercial hatchery, Royaal BV, in the Netherlands and from the River Loire, France, as part of the EELREP project. The hatchery eels were not classified as yellow or silver but both yellow and silver eels were obtained from the lake and the river. They were subjected to 36 h sea water challenge tests in experiments carried out in Concarneau, Paris and Leiden. The results are shown in Fig. 6.4. The plasma osmolality 36 h after transfer to sea water was lowest in the silver eels from the Loire. This value was significantly different from those for the Loire yellow and hatchery groups ($P < 0.05$) and from the Greveling yellow group ($P < 0.001$). There were no other significant differences between the groups (ANOVA followed by Tukey test).

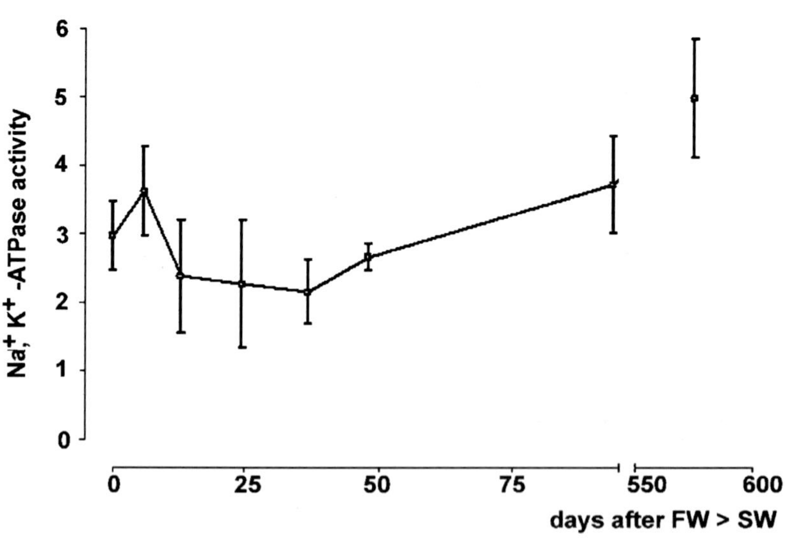

Fig. 6.3 Plasma Na^+, K^+-ATPase activity (μmol ADP [mg protein]$^{-1}$ h^{-1}) of yellow eels transferred from freshwater to sea water at time zero

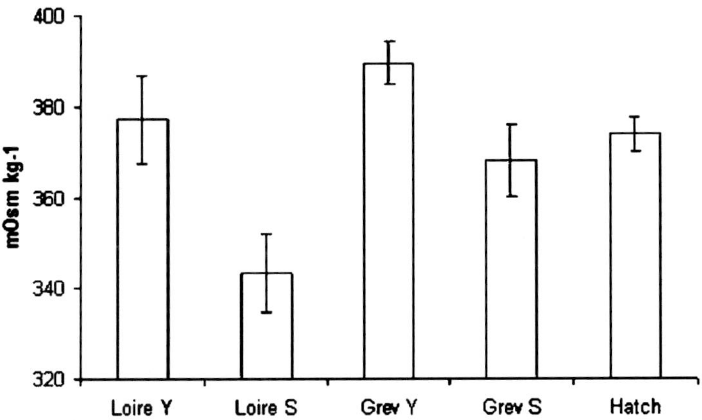

Fig. 6.4 Plasma osmolalities (mOsm kg^{-1}) at 36 h after freshwater to sea water transfer in sea water challenge tests. Loire = eels from the river Loire, Grev = eels from lake Grevelingen, Hatch = hatchery eels from Royaal BV, Y = yellow eels, S = silver eels

6.3 Does Maturation Affect Osmoregulatory Capacity?

6.3.1 Intestinal Function

The question of whether maturation affects osmoregulation arises, particularly in view of the apparent degeneration of the intestine, an essential organ in marine osmoregulation. Closer examination of intestines of partially matured eels at St. Seurin showed that the reduction in size of the intestine appeared to be due to atrophy of the muscle layers, not the epithelium. The ability of isolated epithelia to absorb fluid in the absence of an osmotic gradient by solute-linked water transport was studied by filling sealed intestinal sacs with eel Ringer solution (Rankin and Maetz 1971), incubating them in identical solution and weighing them at regular intervals. Intestines from three hormone-injected eels proved capable of absorbing between 12% and 35% of the luminal fluid over a 2 h period, in fact more than was observed with intestines from control eels. There is no reason to suppose that the intestines of maturing eels do not function normally from the osmoregulatory point of view.

6.3.2 Gill Na^+, K^+-ATPase Activity

During an attempt to artificially induce maturation by hormone injection at St. Seurin over a 16 week period gill samples were frozen from control and injected males and females. No differences were found between control and injected eels. Gonadosomatic index (GSI) was calculated as a measure of relative maturity. Gill

Na+, K+-ATPase activity remained constant over a wide range of GSIs in both females and males (Fig. 6.5).

6.3.3 Plasma Osmolality and Ionic Concentrations

In a successful maturation experiment in Leiden, blood samples were taken from female Grevelingen eels before hormone treatment started and after maturation was complete. There were no significant changes in plasma osmolality, potassium or magnesium concentrations. There was a significant 24 mEq l^{-1} *decrease* in sodium concentration, possible related to the highly significant 16 mEq l^{-1} increase in calcium concentration, although some of the calcium might have been bound to vitellogenin (Table 6.1).

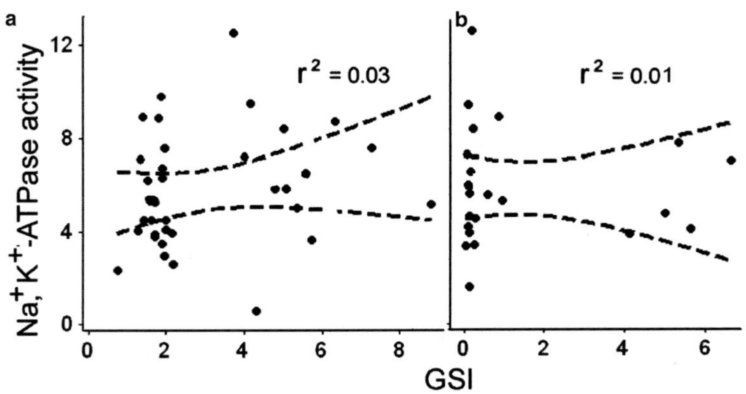

Fig. 6.5 a. Plasma ATPase activity (µmol ADP [mg protein]$^{-1}$ h^{-1}) compared to gonadosomatic index (GSI) in female silver eels at St. Seurin. Values include controls and hormone-treated eels. **b**. Plasma ATPase activity (µmol ADP [mg protein]$^{-1}$ h^{-1}) compared to gonadosomatic index (GSI) in male silver eels at St. Seurin. Values include controls and hormone-treated eels

Table 6.1 Plasma osmolality and ionic concentrations in control and mature (hormonally-induced) Grevelingen eels. Values are given as mean ± standard error of mean (n). P values calculated using Dunnett's multiple comparison test following ANOVA

	Control eels	Mature eels	P
Osmolality mOsm kg^{-1}	299.9 ± 4.30 (10)	315.4 ± 8.05 (13)	nsd
[Na^+] mM	157.9 ± 6.65 (10)	134.0 ± 3.36 (15)	<0.05
[K^+] mM	2.61 ± 0.27 (10)	4.17 ± 0.56 (15)	nsd
[Mg^{2+}] mM	1.44 ± 0.19 (9)	3.16 ± 0.71 (14)	nsd
[Ca^{2+}] mM	1.83 ± 0.10 (9)	9.96 ± 2.16 (14)	<0.001

6.4 Conclusions

The results of the sea water challenge test showed that both yellow and silver eels acclimate well in the short term to abrupt transfer from freshwater to sea water, as had previously been shown in the long term (Chan et al. 1967). Silver eels from the Loire showed a small but significant reduction in plasma osmolality 36 h after transfer compared to the other groups but there was no significant difference between yellow and silver Grevelingen eels. Are there therefore any signs of pre-adaptation as far as regulation of plasma osmolality is concerned? It is perhaps of interest that the rise in Na^+, K^+-ATPase activity following transfer of yellow Loire eels to sea water was very slow and non-significant. Forrest et al. (1973) found that its activity rose very slowly after transfer of yellow *Anguilla rostrata* to sea water. If the eels were treated with cortisol before transfer the enzyme activity increased to levels found in seawater-acclimated eels and did not increase further following transfer to sea water. Injection of yellow American eels with cortisol, in addition to increasing Na^+, K^+-ATPase activity, causes their ventral surface to turn silver (Epstein et al. 1971).

Thomson and Sargent (1997) scraped epithelial cells from European eel gill arches. They counted samples of cells to estimate the number of chloride cells and assayed Na^+, K^+-ATPase activity. When yellow eels were transferred from freshwater to sea water the percentage of chloride cells increased from about 2% to about 5% after 6 days and to 6% from 13 to 21 days. For silver eels the chloride cells formed about 6% of the total in freshwater and remained about the same after transfer to sea water. Na^+, K^+-ATPase activity had increased 2.5 times 2 weeks after yellow eels were transferred to sea water but there was only a slight increase in silver eels. Freshwater chloride cells are found on the gill lamellae in addition to the filaments, where seawater chloride cells are found, and are also involved in freshwater osmoregulation and calcium uptake in salmonids (Perry and Wood 1985; Perry and Laurent 1989). In the Japanese eel the ratio lamellar to filamentary chloride cells was much lower in yellow eels than in silver eels; silver eels caught in brackish water during downstream migration had very few chloride cells on the lamellae (Kaneko et al. 2003). These and other studies suggest that, although the yellow eel acclimated to freshwater has no difficulty coping with abrupt transfer to sea water, the silver eel does show pre-adaptations to marine life as far as gill chloride cells are concerned. There may be an advantage in having the machinery for marine osmoregulation in place before feeding stops preparatory to the spawning migration, during which all the energy reserves of the body will be required for swimming and maturation in addition to maintaining homeostasis. This energy required for the latter includes the cost of osmoregulation in a hyperosmotic environment which the eel continues to carry out efficiently up to the time of maturation. The situation resembles that in pressure acclimation where the yellow eel can acclimatize to high pressure but the silver eel is already prepared (Vettier et al. 2005, see also Chapter 5).

Acknowledgements I would like to thank the following who helped greatly with the practical work. Dr. Guido van den Thillart, Dr. A. Palstra and the staff at the Institute of Biology, University of Leiden. Dr. M. Fouchereau-Peron, Dr. A-G Lafont, Dr. S. Dufour and the staff of the Museum

National d'Histoire Naturelle in Concarneau and Paris. Dr. S. S. Madsen and Ms. J. D. Hansen at the University of Southern Denmark, Odense. Dr. P. Elie and the staff of the Cemagref field station at St. Seurin. The work was supported by the EELREP grant Q5RS-2001-01836 from the European Union.

References

Amer S, Brown JA (1995) Glomerular actions of arginine vasotocin in the *in situ* perfused trout kidney. Am. J. Physiol. 269: R775–R780

Ando M, Mukuda T, Kozaka T (2003) Water metabolism in the eel acclimated to sea water: from mouth to intestine. Comp. Biochem. Physiol. 136B:621–633

Arnold-Reed DE, Balment RJ (1994) Peptide hormones influence in vitro interrenal secretion of cortisol in the trout, *Oncorhynchus mykiss*. Gen. Comp. Endocrinol. 96:85–91

Babiker MM, Rankin JC (1978) Neurohypophysial hormonal control of kidney function in the European eel (*Anguilla anguilla* L.) adapted to sea-water or freshwater. J. Endocrinol. 76:347–358

Boeuf G (1993) Salmonid smolting: a preadaptation to the oceanic environment. In "Fish Ecophysiology" eds. Rankin JC, Jensen FB, pp. 105–135. Chapman & Hall, London

Ball JN, Chester Jones I, Forster ME, Hargreaves G, Hawkins EF, Milne KP (1971) Measurement of plasma cortisol levels in the eel *Anguilla anguilla* in relation to osmotic adjustments. J. Endocrinol. 50:75–96

Brown JA, Balment RJ (1997) Teleost renal function: regulation by arginine vasotocin and by angiotensins. In "Ionic Regulation in Animals" eds. Hazon N, Eddy FB, Flik G, pp. 150–164. Springer, Berlin

Brown JA, Jackson BA, Oliver JA, Henderson IW (1978) Single nephron filtration rates (SNGFR) in the trout, *Salmo gairdneri*. Pflügers Arch. 377:101–108

Brown JA, Rankin JC, Yokota SD (1993) Glomerular haemodynamics and filtration in single nephrons of non-mammalian vertebrates. In "New Insights in Vertebrate Kidney Function" eds. Brown JA, Balment RJ, Rankin JC, pp. 1–44. Cambridge University Press, Cambridge

Chan DKO, Chester Jones I, Henderson IW, Rankin JC (1967) Studies on the experimental alteration of water and electrolyte composition of the eel (*Anguilla anguilla* L.) J. Endocrinol. 37:297–317

Chan DKO, Rankin JC, Chester Jones I (1969) Influences of the adrenal cortex and the corpuscles of Stannius on osmoregulation in the European eel (*Anguilla anguilla* L.) adapted to freshwater. Gen. Comp. Endocrinol. 2: Suppl. 2:342–353

Chester Jones I, Chan DKO, Rankin JC (1969) Renal function in the European eel (*Anguilla anguilla* L.): Changes in blood pressure and renal function of the freshwater eel transferred to sea-water. J. Endocrinol. 43:9–19

Clarke WC (1982) Evaluation of the sea-water challenge test as an index of marine survival. Aquaculture 28:177–183

Cutler CP, Cramb G (1996) Two isoforms of the $Na^+/K^+/2Cl^-$ cotransporter are expressed in the European eel (*Anguilla anguilla*). Biochem. Biophys. Acta 1566:92–103

Cutler CP, Cramb G (2002) Two isoforms of the Na/K/2Cl cotransporter are expressed in the European eel (*Anguilla anguilla*). Biochim. Biophys. Acta 1566:92–103

Dunne JB, Rankin JC (1972) Effects of atrial natriuretic peptide and Angiotensin II on salt and water excretion by the perfused rainbow trout kidney. J. Physiol. 446:92P

Epstein FH, Cynamon M, McKay W (1971) Endocrine control of Na-K-ATPase and seawater adaptation in *Anguilla rostrata*. Gen. Comp. Endocrinol. 16:323–328

Evans D, Piermarini PM, Choe KP (2005) The multifunctional fish gill: Dominant site of gas exchange, osmoregulation, acid-base regulation, and excretion of nitrogenous waste. Physiol. Rev. 85:97–177

Fontaine M, Callamand O, Olivereau M. (1949) Hypophyse et euryhalinité chez l'anguille. Compt. Rend. Hebd. Acad. Sci. 228, 513–514
Fontaine Y-A, Pisam M, Lemoal C, Rambourg A (1995) Silvering and gill mitochondria-rich cells in the eel, *Anguilla anguilla*. Cell. Tissue Res. 281:465–471
Forrest JN, Cohen AD, Schon DA, Epstein FP (1973) Na transport and Na-K-ATPase in gills during adaptation to seawater: effects of cortisol. Am. J. Physiol. 224:709–713
Gaitskell RE, Chester Jones I (1970) Effects of adrenalectomy and cortisol injection on the *in vitro* movement of water by the intestine of the freshwater European eel (*Anguilla anguilla* L.). Gen. Comp. Endocrinol. 15:491–493
Henderson IW, Wales NAM (1974) Renal diuresis and antidiuresis after injections of arginine-vasotocin in the freshwater eel (*Anguilla anguilla* L.). J. Endocrinol. 61:487–500
Hirano T (1967) Effect of hypophysectomy on water transport in isolated intestine of the eel *Anguilla japonica*. Proc. Jpn. Acad. 43:793–796
Hirano T, Mayer-Gostan N (1976) Eel esophagus as an osmoregulatory organ. Proc. Nat. Acad. Sci. 73:1348–1350
Kaiya H, Takei H (1996) Changes in plasma atrial and ventricular natriuretic peptide concentrations after transfer of eels from freshwater to seawater or vice versa. Gen. Comp. Endocrinol. 104:337–345
Kaiya H, Tsukada T, Yuge S, Mondo H, Kangawa K, Takei Y (2006) Identification of eel ghrelin in plasma and stomach by radioimmunoassay and histochemistry. Gen. Comp. Endocrinol. 148:375–382
Kalujnaia S, McWilliam IS, Zaguinaiko VA, Feilen AL, Nicholson J, Hazon N, Cutler CP, Balment RJ, Cossins AR, Hughes M, Cramb G (2007) Salinity adaptation and gene profiling analysis in the European eel (*Anguilla anguilla*) using microarray technology. Gen. Comp. Endocrinol. 152:274–280
Kamiya M, Utida S (1968) Changes in activity of sodium-potassium-activated adenosinetriphosphatase in gills during adaptation of the Japanese eel to sea water. Comp. Biochem. Physiol. 26:675–685
Kaneko T, Hasegawa S, Sasai S (2003) Chloride cells in the Japanese eel during their early life stages and downstream migration. In "Eel Biology" eds. Aida K, Tsukamoto K, Yamauchi K, pp. 457–468. Springer, Tokyo
Keys AB, Willmer EN (1932) "Chloride-secreting cells" in the gills of fishes with special reference to the common eel. J. Physiol. 76:368–378
Kirsch R (1972) The kinetics of peripheral exchanges of water and electrolytes in the silver eel (*Anguilla anguilla* L.) in fresh water and in sea water. J. Exp. Biol. 57:489–512
Kirsch R, Mayer-Gostan N (1973) Kinetics of water and chloride exchanges during adaptation of the European eel to sea water. J. Exp. Biol. 58:105–121
Kojima M, Kangawa K (2005) Ghrelin: structure and function. Physiol. Rev. 85:495–522
Kozaka T, Fujii Y, Ando M (2003) Central effects of various ligands on drinking behavior in eels acclimated to seawater. J. Exp. Biol. 206:687–692
Lafont A-G, Fitzpatrick T, Rankin JC, Dufour S, Fouchereau-Peron M (2006) Possible role of calcitonin-gene related peptide in osmoregulation via the endocrine control of the gill in a teleost, the eel, *Anguilla anguilla*. Peptides 27:812–819
Lionetto MG, Schettino T (2006) The Na^+-K^+-$2Cl^-$ cotransporter and the osmotic stress response in a model salt transport epithelium. Acta Physiol. 187:115–124
Lowry OH, Rosebrough NJ, Farr AL, Randall RJ (1951) J. Biol. Chem. 193, 265–275.
Maetz J, Skadhauge, E (1968) Drinking rates and gill ionic turnover in relation to external salinities in the eel. Nature 217:371–373
Maetz J, Mayer N, Chartier-Baraduc MM (1967) La balance minérale du sodium chez *Anguilla anguilla* en eau de mer, en eau douce et au cours de transfert d'un milieu à l'autre: Effets de l'hypophysectomie et de la prolactine. Gen. Comp. Endocrinol. 8:177–188
Marshall WS, Singer TD (2002) Cystic fibrosis transmembrane conductance regulator in teleost fish. Biochem. Biophys. Acta 1566:16–27

Martinez A-S, Cutler CP, Wilson GD, Phillips C, Hazon N, Cramb G (2005) Regulation of expression of two aquaporin homologs in the intestine of the European eel: effects of seawater acclimation and cortisol treatment. Am. J. Physiol. 288:R1733–R1743

Mayer N, Maetz J, Chan DKO, Forster M, Chester Jones I (1967) Cortisol, a sodium excreting factor in the eel (*Anguilla anguilla* L.) adapted to sea water. Nature 214:1119–1120

McCormick SD (1993) Methods for nonlethal gill biopsy and measurement of Na^+, K^+-ATPase activity. Can. J. Fish. Aquatic. Sci. 50, 656-658

McCormick SD (2001) Endocrine control of osmoregulation in teleost fish. Am. Zool. 41:781–794

Oide M, Utida S (1967) Changes in water and ion transport in isolated intestines of the eel during salt adaptation and migration. Mar. Biol. 1:102–106

Oikari AOJ, Rankin JC (1985) Renal excretion of magnesium in a freshwater teleost, *Salmo gairdneri*. J. Exp. Biol. 117:319–333

Olivereau M (1970) Réaction des cellules à chlorures de la branchie après passage en eau de mer chez l'Anguille européene. Compt. Rend. Soc. Biol. 164:1951–1955

Olivier M, Chartier-Baraduc MM (1966) Action de la prolactine chez l'anguille intact et hypophysectomisée II Effet sur les électrolytes plasmatiques (Na K Ca). Gen. Comp. Endocrinol. 7:27–36

Perry SF, Wood CM (1985) Kinetics of branchial calcium uptake in the rainbow trout: Effects of acclimation to various external calcium levels. J. Exp. Biol. 116:411–433.

Perry SF, Laurent P (1989) Adaptational responses of rainbow trout to lowered external NaCl concentration: Contribution of the branchial chloride cell. J. Exp. Biol. 147:147–168.

Rankin JC, Maetz J (1971) A perfused teleostean gill preparation: Vascular effects of neurohypophysial hormones and catecholamines. J. Endocr. 51:621–635.

Rankin JC, Davenport D (1981) Animal osmoregulation. Blackie, Glasgow

Rankin JC, Moss NG, Babiker MM (1972) Study of urine production by a perfused kidney preparation of the eel (*Anguilla anguilla* L.) with special reference to hormone actions. Gen. Comp. Endocrinol. 18:618

Sargent JR, Thomson AJ (1974) Nature and properties of inducible sodium-plus-potassium ion-dependent adenosine-triphosphatase in gills of eels (*Anguilla anguilla*) adapted to freshwater and sea-water. Biochem. J. 144:69–75

Seidelin M, Madsen SS, Cutler CP, Cramb, G (2001) Expression of gill vacuolar-type H^+-ATPase B subunit, and Na^+, K^+-ATPase α_1 and β_1 subunit messenger RNAs in smolting *Salmo salar*. Zool. Sci. 18:315–324

Shirai N, Utida S (1970) Development and degeneration of the chloride cell during seawater and freshwater adaptation of the Japanese eel, *Anguilla japonica*. Z. Zellforsch. 103:247–264

Silva P, Solomon R, Spokes K, Epstein FH (1977) Ouabain inhibition of gill Na-K-ATPase: Relationship to active chloride transport. J. Exp. Zool. 199:419–426

Skadhauge E (1969) The mechanism of salt and water absorption in the intestine of the eel (*Anguilla anguilla*) adapted to waters of various salinities. J. Physiol. 204:135–158

Skadhauge E (1974) Coupling of transmural flows of NaCl and water in the intestine of the eel (*Anguilla anguilla*). J. Exp. Biol. 60:535–546

Suzuki Y, Itakura M, Kashiwagi M, Nakamura N, Matsuki T, Naito N, Takano K, Fujita T, Hirose S (1999) Identification by differential display of a hypertonicity-inducible inward rectifier potassium channel highly expressed in chloride cells. J. Biol. Chem. 274:11376–11382

Takei Y (2002) Hormonal control of drinking in the eel: an evolutionary approach. In "Osmoregulation and Drinking in Vertebrates" eds. Hazon N, Flik G, pp. 61–82. BIOS Scientific Publishers, Oxford

Takei Y, Hirose S (2002) The natriuretic peptide system in eels: a key endocrine system for euryhalinity? Am. J. Physiol. 282:R940–R951

Takei Y, Kaiya H (1998) Antidiuretic effect of eel ANP infused at physiological doses in conscious, seawater-adapted eels, *Anguilla japonica*. Zool. Sci. 15:399–404

Takei Y, Tsuchida T, Tanakadate A (1998) Evaluation of water intake in seawater adaptation in eels using a synchronized drop counter and pulse injector system. Zool. Sci. 15:677–682

Thomson AJ, Sargent JR (1977) Changes in the levels of chloride cells and (Na^+ + K^+ ATPase in the gills of yellow and silver eels adapting to sea water. J. Exp. Zool. 200:33–40

Tsukada T, Takei Y (2006) Integrative approach to osmoregulatory action of atrial natriuretic peptide in seawater eels. Gen. Comp. Endocrinol. 147:31–38

Tsukada T, Rankin JC, Takei Y (2005) Involvement of drinking and intestinal sodium absorption in hyponatremic effect of atrial natriuretic peptide in seawater eels. Zool. Sci. 22:77–85

Utida S, Kamiya M, Shirai N (1971) Relationship between the activity of the Na^+ -K^+ -activated adenosinetriphosphatase and the number of chloride cells in eel gills with special reference to sea-water adaptation. Comp. Biochem. Physiol. 38A:443–447

Vettier A, Amérand A, Cann-Moisan C, Sébert P (2005) Is the silvering process similar to the effects of pressure acclimatization on yellow eels? Resp. Physiol. Neurobiol. 145:243–250

Warne JM (2002) The role of arginine vasotocin in teleost fish osmoregulation. In “Osmoregulation and Drinking in Vertebrates” eds. Hazon N, Flik G., pp 83–95. BIOS Scientific Publishers, Oxford

Chapter 7
Calcitonin Gene-Related Peptide (CGRP), a New Endocrine Factor Involved in the Complex Multi-hormonal Control of the Gill Osmoregulation: Evidence from Studies in the European Eel, *Anguilla anguilla*

Anne-Gaëlle Lafont, Sylvie Dufour, and Martine Fouchereau-Peron

7.1 Introduction

7.1.1 The CT/CGRP Peptide Family in Vertebrates

7.1.1.1 Situation in Mammals

In mammals, alternative tissue-specific processing of initial mRNA from the calcitonin gene generates two distinct peptides: calcitonin (CT) in the thyroid C cells and calcitonin gene-related peptide (CGRP) in the peripheral and central nerves (Copp et al. 1962; Amara et al. 1982). In this group, CT plays an endocrine role in the control of calcium metabolism, as a hypocalcaemic and hypophosphataemic hormone (Azria 1989). At high doses, CGRP has the same calciotropic actions as CT, which likely result from interaction with CT receptor (Raynaud et al. 1994; Cornish et al. 2001). In mammals, the major localization of CGRP is in the central and peripheral nervous system (Amara et al. 1985; Lee et al. 1985; Skofitsch and Jacobowitz 1985). Its wide distribution in the brain suggests a neuromediator or neuromodulator function and an involvement in various brain functions (Van Rossum et al. 1997; Wimalawansa 1996, 1997). CGRP is also involved in the control of cardiovascular functions by acting locally as a hypotensive factor on heart and blood vessels (Gennari and Fischer 1985; Tippins 1986) and in the local control of digestive functions by inhibiting gastric acid secretion (Gennari and Fischer 1985; Taché et al. 1984).

Muséum National d'Histoire Naturelle, Département des Milieux et Peuplements Aquatiques, USM 0401, UMR CNRS 5178, Biologie des Organismes Marins et Ecosystèmes, 75231 Paris Cedex 05 and Station de Biologie Marine 29900 Concarneau

7.1.1.2 Situation in Non-mammalian Vertebrates

In non-mammalian vertebrates, alternative splicing was also demonstrated generating two distinct transcripts for CT and CGRP such as in chicken, salmon and fugu (Minvielle et al. 1987; Jansz and Zandberg 1992; Clark et al. 2002), indicating an ancient origin of both CGRP and CT in vertebrates. CT, synthesized in the ultimobranchial body (UBB) cells, seems to be involved in the regulation of calcium homeostasis in birds, reptiles and anuran amphibians (Copp et al. 1967; Pearse and Cavalheira 1967; Azria 1989; Suzuki et al. 1997; Ogawa et al. 2003; Yaoi et al. 2003). In teleosts the current data remain contradictory concerning the implication of this peptide in calcium regulation and the role of CT has still to be determined in several species (Yamauchi et al. 1978; Hirano et al. 1981; Chakrabarti and Mukerjee 1993; Mukherjee et al. 2004). Concerning CGRP, this neuropeptide could play a role in the central and peripheral nervous systems of sauropsids and anuran amphibians, as in mammals, and would represent an ubiquitous distribution in some teleost species with a possible involvement in osmoregulation (Kline et al. 1988, 1989; Fouchereau-Peron et al. 1990; Fouchereau-Peron 1998; Lanuza et al. 2000; Martinez-Garcia et al. 2002; Suzuki et al. 2002).

7.1.2 Endocrine Control of Gill Osmoregulation in Teleosts

7.1.2.1 Osmoregulation in Teleosts

During their life cycle, eels will perform two major oceanic migrations. Leptocephali larva will metamorphose into glass eels that migrate from seawater (SW) into estuaries using passive tidal transport. After adjusting to freshwater (FW), they will begin a long sedentary phase of growth, feeding and sexual differentiation. During this phase they will metamorphose into yellow eels and then, after numerous behavioural, morphological and physiological changes, transform into silver eels. At this stage, eels will stop feeding, transfer again into SW and perform their second oceanic migration to rejoin their site of reproduction. All these stages are under the control of environmental and physiological stimuli that will trigger the different transformations and adaptations. Among these modifications, we will further detail in this introduction the osmoregulatory adaptation to the different environments.

During their migration, teleosts are subjected to changes in their external environment resulting in a potential high osmotic stress (Fig. 7.1). In fact, in FW fishes need to compensate for osmotic water uptake, whereas in SW, fishes lose water by osmosis. Teleosts are able to maintain their plasma osmolality, which is hyper-osmotic to FW and hypo-osmotic to SW. Thus, FW teleosts must counteract the passive gain of water and the diffuse loss of ions by producing a large amount of dilute urine from kidneys and actively taking up salt ions across the gills. In contrast, SW teleosts need to compensate for the passive gain of ions and loss of water. This is performed by drinking SW, taking up salt ions and water along the digestive

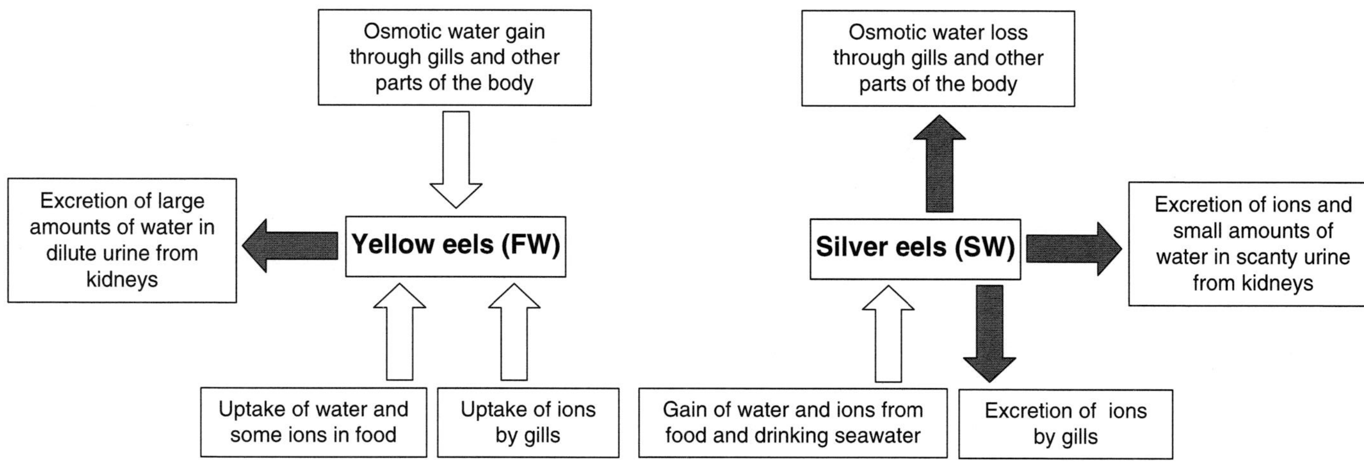

Fig. 7.1 Regulation of the ionic and osmotic concentrations in teleosts in relation to the environmental conditions (freshwater, seawater). In this figure the freshwater teleosts are symbolised by the yellow eels and the seawater teleosts by the silver eels. In FW, yellow eels must counteract the passive gain of water and the diffusive loss of ions by producing a large amount of dilute urine from kidneys and actively taking up salt ions across the gills. In contrast, in SW, silver eels need to compensate for the passive gain of ions and loss of water by drinking SW, taking up salt ions and water along the digestive tract, actively excreting ions through the gills and producing a limited amount of isotonic urine from the kidneys

tract, actively excreting ions through the gills and releasing a limited amount of isotonic urine from the kidneys.

Various organs are involved in osmoregulatory mechanisms such as gills, kidney, intestine and skin, and are under the control of a multi-hormonal complex. The regulation of ion balance in these organs has been the subject of several reviews (Utida et al. 1972; Henderson et al. 1985; Takei 1993; McCormick 2001; Evans et al. 2005; Sakamoto and McCormick 2006). In this introduction we chose to focus on the gills in the context of new data concerning a new endocrine actor in osmoregulation, CGRP (Calcitonin Gene-Related Peptide).

Gills represent an original osmoregulatory organ for aquatic animals, when compared to the situation in tetrapods. The branchial epithelium is an important site of ion and water exchange in fishes. Chloride cells, pavement cells, mucus cells and neuroepithelial cells constitute the four distinct cell types present in the gill epithelium (Perry 1997). The basolateral membrane of the chloride cells contains the Na^+, K^+-ATPase. This enzyme is essential in SW for eliminating excess ions, whereas in FW it constitutes one of the enzymes involved in active ion uptake. The chloride cells change morphologically and functionally during adaptation to different salinities (Chang et al. 2002; Hirose et al. 2003; Laiz-Carrion et al. 2005; Lee et al. 2003; Marshall 2002). For instance in FW, Pisam et al., described two subtypes of chloride cells, α and β subtypes (Pisam et al. 1987; Pisam and Rambourg 1991). The α cells are thought to be precursors of the characteristic chloride cells found in the gills of SW-adapted fishes whereas β cells are found only in FW fishes (Manzon 2002; Perry 1997; Shikano and Fujio 1998). The morphological and functional differences between FW and SW chloride cells seem to be in part related to the Na^+, K^+-ATPase activity which is under complex hormonal control. In fact, many paracrine, autocrine and endocrine hormones interact on the organs involved in osmoregulatory mechanisms in fishes depending on environmental salinity. In this introduction we focus on the main endocrine hormones involved at least directly in the control of gill osmoregulation.

7.1.2.2 Prolactin

Prolactin (PRL) is a pituitary hormone that plays a major endocrine role in the FW adaptation of many teleost species by regulating the hydromineral balance. In the 1950s, studies on killifish, *Fundulus heteroclitus*, demonstrated that hypophysectomised fish were able to survive in FW only when treated with PRL (Burden 1956; Pickford and Phillips 1959). In contrast, Pickford et al. later on demonstrated that injection of PRL inhibited gill Na^+, K^+-ATPase activity in hypophysectomised killifish in FW (Pickford et al. 1970). Many studies have since reached contradictory results concerning the effect of PRL on the activity of this enzyme in the teleost gill, probably due to the use of heterologous hormones (Manzon 2002; McCormick 2001). For instance, ovine PRL seems to be able to bind to growth hormone receptors as well as to PRL receptors in teleosts (Prunet et al. 1994). However, it is generally accepted that PRL is the FW-adapting hormone in most teleost species, mainly by decreasing water uptake and increasing Na^+ and Cl^- retention in the osmoregulatory organs, such as the gills (Manzon 2002). PRL injection of hypophysectomised channel catfish, *Ictalurus punctatus*, in FW restored the plasma osmolality of the FW control fish. The plasma osmolality of hypophysectomised catfish acclimated to 30% SW remained at the same level as that of the environment, whereas the plasma osmolality of the control fish increased significantly to levels higher than that of the environment (Eckert et al. 2001). These results suggest a stimulation by PRL of the ion uptake by the gills with an increase in plasma ionic concentrations and osmolality, both in FW and dilute SW (Eckert et al. 2001). In the same way, PRL treatment of SW fishes results in an inhibition of gill ion excretion and an increase in plasma ion concentration and osmolality (Brown and Brown 1987; Hirano 1986; Madsen and Bern 1992; McCormick 2001).

This positive effect of PRL on ion uptake by the gills appears to be related to the action of this hormone on the morphology, distribution and number of chloride cells in this osmoregulatory organ. For instance, in the SW-adapted tilapia, *Oreochromis niloticus*, PRL injection resulted in a decrease of α cell size in the gills, a dedifferentiation to the FW α cell morphology and a concomitant appearance of FW β cells (Pisam et al. 1993). These modifications seem to affect the gill Na^+, K^+-ATPase activity and therefore the ion balance. In a similar way, in two species of sea bream, *Sparus sarba* and *Sparus aurata*, PRL treatment induced a decrease in the gill Na^+, K^+-ATPase activity and an increase in plasma osmolality and ion concentration (Kelly et al. 1999; Mancera et al. 2002). PRL may be also involved in the regulation of water balance by acting on the osmotic permeability of the gills and by increasing the mucus secretion of the gill mucus cells (Manzon 2002).

PRL receptors have been characterized in many teleosts and the highest expression levels were observed in the major osmoregulatory organs (kidney, gill and intestine) (Auperin et al. 1995; Higashimoto et al. 2001; Le Rouzic et al. 2001; Prunet et al. 2000; Sandra et al. 1995; Tse et al. 2000). In tilapia, *Oreochromis niloticus*, and sea bream, *Sparus aurata*, PRL receptors are present in the chloride cells of the gills (Sandra et al. 2000; Santos et al. 2001). The expression of these receptors has been shown to be dependent on external salinity. For instance, in tilapia, *Oreochromis mossambicus*, an increase in the external salinity is followed by a decrease in the expression of the gill PRL receptors (Shiraishi et al. 1999). All these results suggest an important role of PRL in the adaptation to a hypoosmotic environment and the maintenance of ion balance in FW fishes or FW-adapted euryhaline species.

7.1.2.3 Growth Hormone – Insulin-Like Growth Factor I

Growth hormone (GH) has been demonstrated to play a major endocrine role in SW adaptation in many teleost species and therefore to have an osmoregulatory action antagonistic to that of PRL. Indeed, injection of GH in hypophysectomised tilapia, *Oreochromis mossambicus*, after transfer to SW, resulted in a decrease of the plasma osmolality through the stimulation of the gill Na^+, K^+-ATPase activity

(Borski et al. 1994; Sakamoto et al. 1997; Shepherd et al. 1997). In the same species, long-term treatment of GH produced an increase in chloride cells density in the opercular membrane (Flik et al. 1993). In the mummichog, *Fundulus heteroclitus*, injection of GH after transfer from BW to SW also increased the gill Na^+, K^+-ATPase activity (Mancera and McCormick 1998a). Similarly, GH could increase salinity tolerance in many salmonid species (Sakamoto et al. 1993). The injection of GH, in branchial chloride cells of the salmon, *Salmo salar*, resulted in an increase of α type and a decrease of β type cells (Prunet et al. 1994). In the rainbow trout, *Oncorhynchus mykiss*, and the Atlantic salmon, *Salmo salar*, GH administration increased chloride cell density in the gills (Pelis and McCormick 2001; Perry 1998). Thus, GH is strongly involved in the SW adaptation through its positive effect on the number and size of the chloride cells, the gill Na^+, K^+-ATPase activity and the ion transporters involved in the salt secretion (McCormick 2001).

Some of the osmoregulatory actions of GH are mediated by its interaction with another hormone, the insulin-like growth factor I (IGF-I). Several studies demonstrated the positive effect of exogenous IGF-I on salinity tolerance in teleosts (Madsen et al. 1995; Mancera and McCormick 1998b; McCormick 1996). SW transfer of rainbow trout resulted in a significant increase of the circulating levels of IGF-I (Shepherd et al. 2005). In the mummichog, *Fundulus heteroclitus*, GH treatment and IGF-I treatment both resulted in an increase in the gill Na^+, K^+-ATPase activity and the salinity tolerance. When the hormones were administered together, the effects observed were higher suggesting cooperation between GH and IGF-I in the SW adaptation (Mancera and McCormick 1998a). In the tilapia GH and IGF-I stimulated the number and size of the chloride cells in the gills (Xu et al. 1997). In *Salmo trutta* and *Salmo salar*, IGF-I has been shown to enhance Na^+, K^+-ATPase activity and GH the Na^+, K^+-$2Cl^-$ cotransporter expression in the branchial chloride cells (Pelis and McCormick 2001; Seidelin et al. 1999).

GH receptors have been mainly localized in the osmoregulatory organs including the gills (Kajimura et al. 2004; Lee et al. 2001; McCormick 2001; Nakao et al. 2004; Tse et al. 2003). The recent detection of GH in the same osmoregulatory organ suggests a possible paracrine/autocrine action of this hormone. In the same way, IGF-I and IGF-I receptors have both been found in the gill chloride cells of some salmonid species. The implication of GH and IGF-I in osmoregulation may involve both endocrine and paracrine/autocrine actions in teleost fishes (Sakamoto and McCormick 2006).

7.1.2.4 Cortisol

For a long time cortisol has been considered as the major SW endocrine factor in opposition to PRL, the major FW hormone. However, cortisol seems to be involved in both SW and FW adaptation, possibly through its interaction with PRL and GH and depending on environmental salinity (McCormick 2001; Mommsen et al. 1999; Wendelaar Bonga 1997).

In SW, cortisol acts in synergy with the GH/IGF-I axis to regulate ion excretion from the gills by increasing the gill Na^+, K^+-ATPase activity (Madsen 1990; McCormick 1996). This positive interaction between GH/IGF-I and cortisol axes on the adaptation to SW may also occur in part by the regulation of the cortisol receptor density in the gills (Evans 2002).

Opposite to its action in SW, cortisol has been shown to interact with PRL in low salinity environments mainly by acting on the ion uptake mechanisms in the gills. Both hormones together have been shown to exert a greater action on FW adaptation than either hormone alone (Eckert et al. 2001; Zhou et al. 2003).

Recently it has been suggested that GH and PRL could be involved in the control of cell turnover in the gills, possibly through their interaction with cortisol. Thus PRL/cortisol could act to increase, by differentiation or proliferation, the number of FW chloride cells, which are responsible for ion uptake, and GH/cortisol to increase the number of SW chloride cells involved in ion secretion (Sakamoto and McCormick 2006).

7.1.2.5 Natriuretic Peptides

Natriuretic peptides (NP) are multifunctional hormones which play important roles in the regulation of circulatory and fluid homeostasis in vertebrates (Loretz and Pollina 2000). Thus far, atrial, ventricular and C-type NP (ANP, VNP, CNP) have been identified in teleosts, mainly in salmonids and eels (Takei 2000). The major sites of production of NP in fishes are the heart and brain (Loretz and Pollina 2000).

ANP seems to be involved in the SW adaptation in fishes. In several teleost species, higher ANP levels were recorded after adaptation to elevated salinity and lower levels after adaptation to reduced salinity (Evans 1990; Smith et al. 1991; Westenfelder et al. 1988). Furthermore, ANP was shown to increase Na^+, K^+-ATPase activity in isolated gill cells from SW eels and no effect was observed in FW eels (Takei et al. 1990). In the SW acclimatized mummichog, *Fundulus heteroclitus*, ANP stimulated gill chloride secretion in isolated opercular membrane (Scheide and Zadunaisky 1988). In the SW-adapted flounder, *Platichthys flesus*, ANP was recorded to stimulate Na efflux (Arnold-Reed et al. 1991). Thus, the plasma ANP concentration in fishes is generally higher in SW than in FW. In the eel, however, ANP levels increase only transiently (Takei and Hirose 2002). For example, in the Japanese eel, *Anguilla japonica*, plasma ANP and VNP concentrations increase immediately after transfer from FW to SW and then gradually return to FW levels after long-term exposure to SW (Kaiya and Takei 1996).

ANP and VNP are circulating hormones secreted principally from the heart whereas CNP is a paracrine/autocrine factor mainly expressed in the brain and periphery (Inoue et al. 2005). However in FW eel, CNP must function as a circulating hormone as well as a paracrine factor (Takei and Hirose 2002). The expression of CNP mRNA, mainly in heart, and the expression of the CNP specific receptor in gills are enhanced in FW eels (Katafuchi et al. 1994; Takei and Hirose 2002).

Thus, ANP and CNP seem to have opposite functions in osmoregulation in the eel. CNP appears to be involved in FW adaptation by stimulating an uptake of Na^+ by the gills, whereas ANP would be involved in SW adaptation through stimulation of Na^+ excretion (Takei and Hirose 2002; Takei and Kaiya 1998).

The actions of NPs in osmoregulation may also be indirect by interactions with other hormones such as cortisol. In the flounder, *Platichthys flesus*, ANP elevated circulating cortisol levels in SW (Arnold-Reed et al. 1991). In 1994, Kloas et al. suggested that cortisol secretion could be under some combination of autocrine, paracrine and endocrine control by NP (Kloas et al. 1994). The exposure to high salinities in fishes would produce cortisol secretion and increased gill Na^+, K^+-ATPase activity, through a stimulation of ANP release (Loretz and Pollina 2000). The injection of ANP, CNP and VNP in eels demonstrated that only CNP was able to increase plasma cortisol concentration in FW, whereas only ANP had this effect on plasma cortisol concentration in SW (Li and Takei 2003). These results strengthen the possible role of ANP in SW adaptation and CNP in FW adaptation, by stimulating the release of cortisol. VNP and CNP were also shown to enhance a slow and lasting stimulation of GH release in tilapia (*Oreochromis mossambicus*), VNP being more potent than CNP, whereas ANP had no effect (Eckert et al. 2003).

7.1.2.6 Other Endocrine Hormones

Among the various other factors involved in the endocrine control of teleost osmoregulation we can mention some of the following hormones.

The neurohypophysial peptide arginine vasotocin (AVT), synthesized in the hypothalamus, seems to be involved in fish osmoregulation, particularly in salt and water regulation (Avella et al. 1999; Pierson et al. 1995; Warne and Balment 1995). This hormone could be involved in fishes in the initial response to both hyper and hypo-osmotic stress (Bond et al. 2002).

Studies concerning the involvement of thyroid hormones (TH) in osmoregulation in teleosts are contradictory. Although the direct role of T3 and T4 in osmoregulation remains unclear and conflicting, TH seem to be involved in SW adaptation through their interaction with the GH/IGF I and cortisol axes (Peter et al. 2000; Prunet et al. 1989; Mancera and McCormick 1999; McCormick 2001; Klaren et al. 2005).

Angiotensin II (ANG II) is a vasoactive hormone that affects renal function in both mammals and fishes (Brown et al. 1990; Yanagawa 1991; Marsigliante et al. 1996, 1997). This hormone seems to be involved in both SW and FW adaptation. In fact, although ANG II promotes SW adaptation in fishes it seems to be also indirectly involved in the FW adaptation by stimulating PRL release (Eckert et al. 2003; Leedom et al. 2003).

Urotensins (UI and UII), secreted in teleosts by the neurosecretory cells in the distal spinal cord, are neuropeptides involved in cardiovascular regulation. UI and UII seem to have both direct and indirect effects on fish osmoregulation namely by stimulating cortisol secretion (Arnold-Reed and Balment 1994; Kelsall and Balment 1998; Bond et al. 2002; Evans et al. 2005).

7.2 CGRP: A Possible New Hormone Involved in the Endocrine Control of Teleost Osmoregulation

In mammals the localization, synthesis and physiological roles of CT and CGRP have been widely described in the literature. Alternative splicing of the calcitonin gene (CALC I) generates two distinct peptides: calcitonin (CT), which is synthesized in the thyroid C cells and involved in the regulation of calcium metabolism, and calcitonin gene-related peptide (CGRP), which is a neuromediator and a potent vasodilatator synthesized in the peripheral and central nerves (Copp et al. 1962; Amara et al. 1982; Azria 1989; Van Rossum et al. 1997). In the non-mammalian phyla, sauropsid and anuran amphibians, both CT and CGRP have been identified. CT, synthesised in the ultimobranchial body (UBB) cells, seems to be involved in the regulation of calcium homeostasis and CGRP could play a role in the central and peripheral nervous system, as in mammals. In teleosts, the situation concerning these hormones remains to be defined in most groups. However CT could be involved in calcium regulation and CGRP in osmoregulation.

In this paper we have collected our data concerning the CT/CGRP family in the eel, *Anguilla anguilla* (Lafont et al. 2004, 2006a). This model was chosen for its biological and phylogenetic interests. In fact, in vertebrates most of the CT/CGRP family data concern the sarcopterygian group. The sarcopterygian sister group, the actinopterygian group, is mainly represented by teleosts. The presence and respective roles of CT and CGRP in teleosts remain to be defined in a lot of species. We chose to investigate their distribution and function in a representative species of a phylogenetically ancient group (Elopomorph) among teleosts species, the European eel, *Anguilla anguilla*. We also chose this species for its biological interest. The eel is an amphihaline fish, which will perform two oceanic migrations separated by a sedentary period in continental freshwater. These various changes in the external environment, from seawater to freshwater and from freshwater to seawater, set off a complex multi-hormonal system involved namely in osmoregulation, in which CT and CGRP could play a role.

Previous studies performed by our group on trout have demonstrated that CGRP interacts *in vitro* with fish branchial tissue and its action is mediated by a cyclic AMP dependent mechanism (Arlot-Bonnemains et al. 1991; Fouchereau-Peron et al. 1990; Lamharzi and Fouchereau-Peron 1994). Transfer of trout, *Oncorhynchus mykiss*, from FW to SW resulted in a significant increase in the circulating CGRP concentrations correlated with an increase in both CGRP binding and carbonic anhydrase activity in the gills (Lamharzi and Fouchereau-Peron 1996). The opposite transfer of rainbow trout, from SW to FW, also induced important modifications in the circulating levels of CGRP, in the interaction of CGRP with its specific gill receptors and in the carbonic anhydrase activity (Fouchereau-Peron 1998). Carbonic anhydrase is an enzyme involved in respiratory gas exchange, ion transport and acid base regulation processes, which is localised in the chloride cells of several teleost species (Conley and Mallatt 1988). Thus, CGRP, together with carbonic anhydrase, may be involved in the osmoregulation process in trout. In the flounder, *Paralichthys olivaceus*, a specific CGRP receptor mRNA has been detected in the gills whereas CGRP mRNA has not been detected in this organ (Suzuki et al. 2000). The same group demonstrated that the expression of this CGRP receptor mRNA in the gills decreased when flounders were transferred from SW to BW or FW (Suzuki et al. 2002). Therefore CGRP could represent one of the many endocrine factors involved in the gill osmoregulation.

Previous data obtained in salmonids and molluscs suggested that CGRP could represent the ancestral molecule of the CT/CGRP peptide family. In fact, both molecules could be detected in vertebrates, whereas only CGRP-like molecules could be characterized in invertebrates. In order to further understand the evolution of these two peptides and investigate the potential early role of CGRP in vertebrates, we initiated studies in the European eel, *Anguilla anguilla*, a representative species of a phylogenetically ancient group (Elopomorph) among teleosts.

Firstly, we investigated the characterization of both CT and CGRP related peptides in the eel. The presence of immunoreactive molecules was detected using radioimmunoassays specific for CT and CGRP. These molecules were further tested for their structural similarities with CT and CGRP, using their abilities to displace the binding of these peptides to their specific mammalian receptors. Finally, the molecules immunologically and biologically related to CGRP, detected in heart and spleen, were partially purified in order to determine their apparent molecular weight (Lafont et al. 2004).

In the second part of this chapter, we report on investigations into the possible role of the CGRP-like molecules in the eel by characterizing the presence and distribution of CGRP-like receptors. Screening of CGRP target organs was performed by studying specific binding of this peptide to various eel tissues. Binding properties of CGRP-like receptors were further characterized in the gills, one of the major organs involved in osmoregulation (Lafont et al. 2006a).

Finally, in the third part of this chapter, we describe studies on the possible role of calciotropic hormones (CT and CGRP) in osmoregulation. To that aim, we analyzed the effect of eel transfer from FW to SW on plasma osmolality, as well as on plasma CT and CGRP levels using radioimmunoassay. We also characterized the effect of this SW adaptation on the CGRP gill binding sites (Lafont et al. 2006a).

7.2.1 Characterization of CGRP and CT Related Peptides in the Eel

7.2.1.1 CGRP-Like Immunoreactivity in Various Tissues from Female and Male Silver Eels

In the first part of this work we obtained data concerning CGRP. These data represent the first reports on a study of this peptide in a representative species of a phylogenetically ancient group among teleosts, the eel. This peptide was partially characterized and its distribution was compared to that of CT, in order to provide new information concerning the potential role of these two peptides in this species.

We used specific radioimmunoassays to investigate the presence of molecules immunologically related to human CGRP (hCGRP) and to salmon CT (sCT). The specificity of each RIA was previously demonstrated by testing the cross reactivity between sCT and hCGRP, respectively, with various analogues of CT and CGRP (Lafont et al. 2004). With the specific CGRP RIA, we detected the presence of molecules immunologically related to this peptide in all the tissues tested (brain, gill, heart, intestine, kidney, liver, spleen and ultimobranchial body, ranging from 2 to 50 ng g^{-1} of organ) except for the stomach (Fig. 7.2). This ubiquitous distribution has also been described in the trout. In this species, CGRP-like molecules were found in brain, gill, heart, intestine and stomach ranging from 2 to 14 ng g^{-1} of organ (Fouchereau-Peron et al. 1990). In the eel, CGRP RIA was performed in both sexes and we observed significant differences between males and females in only two organs: heart and liver. In male, the CGRP-like concentration in these two organs was three times higher than in female.

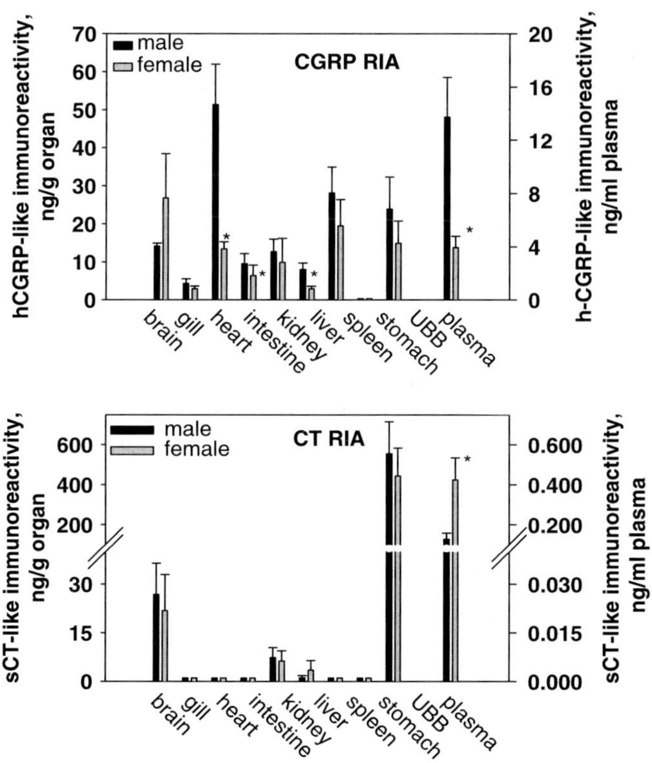

Fig. 7.2 Concentrations of CGRP- (A) and CT- (B) like immunoreactivity in tissues and plasma of male and female eels. These concentrations were determined by specific RIA. Each tissue was analysed at multiple dilutions. Results were expressed as ng immunoreactive peptide per g of organ or per ml of plasma. Each bar represents the mean ± SEM of five males and six female eels. $*p < 0.05$ when compared to male eels (Student's t-test)

7.2.1.2 CGRP-Biological Activity in Various Tissues from Female and Male Silver Eels

Molecules immunologically related to CGRP were then tested concerning their ability to interact with specific CGRP mammalian receptors. Thus, we tested whether they were able to displace the specific binding of labelled hCGRP in rat liver membranes. This specific radioreceptor assay (RRA) was used to determine if the molecules immunologically related to CGRP detected by RIA had also CGRP-like biological activities. The specificity of each RRA (CT and CGRP) was previously demonstrated by testing the ability of various analogues of CT and CGRP to displace the binding of the labelled hormone (hCGRP or sCT) from its specific binding sites (rat liver membranes or rat kidney membranes) (Lafont et al. 2004). Molecules biologically related to CGRP were detected in tissue extracts of brain, heart, kidney, liver, spleen and UBB (Fig. 7.3). The concentrations in these organs were measured at between 3 pg μg^{-1} of proteins in the kidney and 10 pg μg^{-1} of protein in the brain. In contrast, the molecules detected in gill and intestine extracts were not able to displace the binding of CGRP from its specific mammalian receptors. Therefore the immunoreactivity detected by RIA in these two tissues was not related to the presence of CGRP-like molecules. On one hand, this non-specific immunoreactivity can be explained by the fact that we used a heterologous system to detect the CGRP-like molecules (human hormone and antibody) in the various eel tissues. On the other hand, this non specific immunoreactivity can

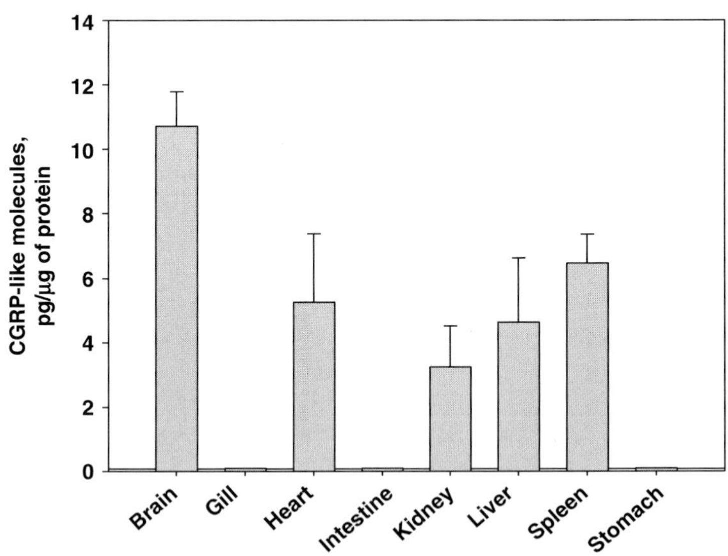

Fig. 7.3 Concentrations of biologically active CGRP molecules, in female silver eels, as determined by specific CGRP RRA, using specific mammalian receptor (rat liver membranes). Each tissue was analyzed at multiple dilutions. Results were expressed as pg CGRP-like per µg of protein. Each bar represents the mean ± SEM of four female eels

also be explained by the fact that in the RIA the antibody recognizes only an antigenic determinant of a short amino acid sequence. This is the reason why we performed both RIA and RRA in the various eel tissues, in order to determine more precisely the distribution of molecules related to the calciotropic hormones.

7.2.1.3 Partial Purification of the CGRP-Like Molecules Present in the Heart and Spleen Extracts

We then selected tissues containing the molecules immunologically and biologically related to CGRP to determine their apparent molecular weight by partial purification. Molecular sieving of heart and spleen extracts demonstrated that the CGRP-like molecules, bioactive in the specific RRA, corresponded to a molecular weight in the range of 3,300–3,950 Da, similar to that of hCGRP (3,750 Da). Eel spleen and heart extracts were partially purified by exclusion chromatography. Each fraction eluted from this column was tested by specific CGRP RIA and the profile of immunoreactivity is presented, for the spleen, in Fig. 7.4. Immunoreactivity was detected in most fractions eluting between 60 and 200 ml. These fractions were further tested by specific CGRP RRA. Only fractions eluting from 130 to 150 ml contained molecules able to displace the binding of labelled hCGRP from its specific mammalian

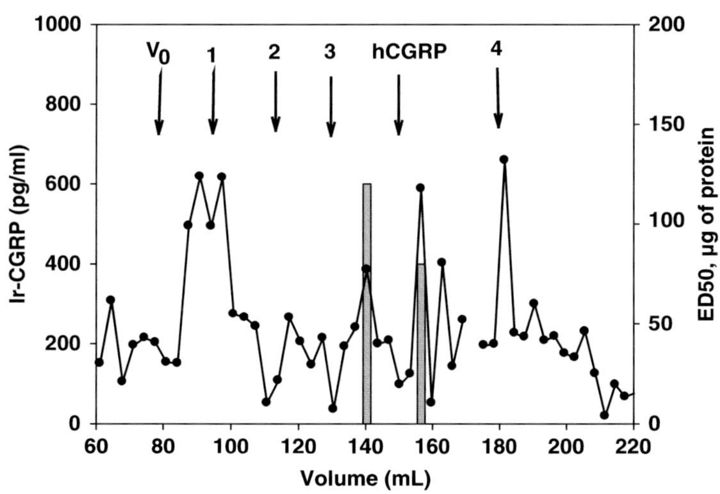

Fig. 7.4 Partial purification of an acid extract of spleen. Extract of eel spleen was subjected to gel exclusion chromatography. Eluted fractions were tested by specific CGRP RIA and RRA. Numbers indicate the elution position of the different markers: 1: bovine serum albumin, 2: lactalbumin C, 3: aprotinin, 4: bacitracin. The immunoreactive profile is indicated by closed circles. Each fraction was tested by specific hCGRP RRA. The grey bars symbolized the only two fractions able to interact in the hCGRP RRA and indicate the quantity of protein that induced a 50% displacement of the CGRP binding to its specific mammalian receptors (ED_{50})

receptors. Fifty-percent inhibition of the binding was obtained with protein levels between 80 and 120 µg. These molecules were immunologically and biologically related to CGRP and eluted for a volume similar to that of hCGRP (150 ml). No displacement of the CGRP binding to its specific mammalian receptors was observed with the other fractions up to 150 µg. Similar results were obtained with eel heart extracts (data not shown). The molecular weight of the heart and spleen fractions containing molecules immunologically and biologically related to CGRP was further characterized using SDS gel electrophoresis (data not shown). The apparent molecular weight determined for partially purified heart and spleen fractions was in the range of 3,300–3,950 Da, which approximates to that of hCGRP (3,750 Da). Therefore, the molecules detected in eel tissues were immunologically and biologically related to CGRP and with a close molecular weight.

The highest concentration of CGRP-like molecules was detected in the eel brain. The presence of CGRP in the central nervous system has been widely demonstrated in mammals, suggesting its involvement in various brain functions, namely in the sensorial, motor and integrative systems (Wimalawansa and El-Kholy 1993; Wimalawansa 1996; Van Rossum et al. 1997). It seems also that CGRP could play the same function in the central nervous system in sauropsids and amphibians (Kline et al. 1988, 1989; Lanuza et al. 2000; Martinez-Garcia et al. 2002). Therefore the role of CGRP in brain functions could represent a common feature among vertebrates.

We have also detected high concentrations of molecules immunologically and biologically related to CGRP in the eel heart. This situation is in agreement with the situation described in mammals. Immunocytochemistry has demonstrated the presence of CGRP related molecules in the nerve fibres of the cardiovascular system of the rat and the guinea pig (Mulderry et al. 1985; Gerstheimer and Metz 1986). This peptide was described as one of the most potent vasodilating substances inducing a decrease of vascular resistance and an increase in the rate and force of contraction of the heart (Girgis et al. 1985; Marshall et al. 1986). This vasodilatative effect in the arterial system described in mammals has also been demonstrated in amphibians (Kline et al. 1988).

The fact that we observed the presence of molecules related to CGRP in the UBB is not inconsistent with the situation described in mammals. In fact, a low concentration of CGRP is found in mammals in the thyroid C cells, around 5% of the CT concentration in this organ (Wimalawansa 1997). This CGRP detected in the thyroid is mainly hCGRP II, synthesized by a second form of the calcitonin gene (Höppener et al. 1987; Wimalawansa 1997). In some teleost species, such as salmon, several forms of CGRP have already been identified (Jansz et al. 1996). This could also be the situation in the eel, however this fact remains to be demonstrated in this species.

7.2.1.4 Comparison with CT Distribution

In order to compare the distribution of CGRP to that of CT in the eel, we also tested the same tissue extracts in the specific CT RIA. CT immunoreactivity was detected in brain, kidney, liver and UBB. No cross reactivity was observed in other tissues (gill, heart, intestine, stomach and spleen) (Fig. 7.2). The highest concentration was found in UBB in both male (554 ± 161 ng g^{-1} of organ) and female (443 ± 140 ng g^{-1} of organ) with no significant difference between sexes. In brain and kidney the CT-like immunoreactivities were measured at around 20 and 6 ng g^{-1} of organ, respectively, with no significant difference between male and female. In the liver, immunoreactivity could be measured only in females and was found to be around 3 ng g^{-1} of organ. As for CGRP, a specific CT RRA (rat kidney membranes) was used to discriminate biologically active CT. Among the four tissues that had shown CT-like immunoreactivity (brain, kidney, liver and UBB) only the ultimobranchial extract was able to displace the labelled sCT binding to its specific mammalian receptors (data not shown). The concentration of the molecules biologically related to CT accounted for 11 pg μg^{-1} of protein. Therefore, we could detect the presence of molecules immunologically and biologically related to CT only in UBB, which could represent in eel the only organ of production of biologically active CT. The high concentrations of CT in UBB are not surprising as this organ has been shown to be the site of CT synthesis in non-mammalian vertebrates, like the C cells of the thyroid in mammals (Copp et al. 1962, 1967; Pearse et al. 1967). The concentration detected in the eel (49 and 72 ng per gland of male and female, respectively) was lower than that previously found in the trout UBB where CT levels increased from 10 to 100 µg per gland at the time of spawning (Fouchereau-Peron et al. 1990). However, we must note that in the present experiment, both male and female silver eels were far from their complete maturity and that the CT level may increase in UBB during the reproduction period. Furthermore, these differences in the CT concentration of the UBB between eels and salmonids may be related to a species difference in the UBB physiology. In fact, no CGRP-like molecules had been detected in the UBB of trout (Fouchereau-Peron et al. 1990). Thus, the UBB could play an endocrine role, not only in CT production in the eel, but also in CGRP production and this function could represent a non-common feature among teleost species.

7.2.1.5 CT and CGRP Circulating Levels

The circulating levels of molecules related to CT and CGRP quantified by our group have shown significant differences not only between both peptides in the eel, but also in relation with the sex of the animal. In fact, in males we measured a hundred times more of molecules related to CGRP (13.7 ng ml^{-1}) than to CT (0.13 ng ml^{-1}), whereas in females there is only a ten times difference between both peptides (CGRP: 4.11 ng ml^{-1}, CT: 0.42 ng ml^{-1}). This variation observed between male and female can be related to the difference observed in some tissues, such as the heart. In this organ, the CGRP concentrations were five times higher in males than in females. This situation implies a release in the blood of a higher quantity of circulating peptides in males. Nevertheless, we observed in eels a higher concentration of CGRP-like molecules than of CT-like molecules in plasma, independently from the sex of the animal. This situation differs from that found in the trout where circulating CGRP levels of about 0.5 ng ml^{-1} are equivalent to those of CT: 0.6 ng ml^{-1}

(Lamharzi and Fouchereau-Peron 1996). In mammals, the situation is opposite to that observed in the eel with a ratio of circulating CGRP/CT of only 0.2 in humans, the circulating CGRP and CT (Roos et al. 1980; Trasforrini et al. 1991) levels were reported to be 7 and 35 pg ml^{-1}, respectively. This suggests an important endocrine role of CGRP in the eel, which would decline during vertebrate evolution.

7.2.2 Characterization of CGRP-Like Receptors in the Eel

7.2.2.1 Distribution of the Specific CGRP Binding Sites in the Eel

In order to determine the role of those molecules related to CGRP, we have in the second part of our study focused on the receptor characterization and on the possible involvement of these molecules in the complex system of teleost osmoregulation. We have identified the CGRP target organs by measuring the capacity of the peptide-specific binding to various eel tissues (brain, gill, heart, kidney, liver, spleen and stomach). The distribution of the ^{125}I-labeled hCGRP and the ^{125}I-labeled sCT specific binding in various eel tissues is presented in Fig. 7.5. The specific binding of sCT ranged from 0.14 ± 0.13 fmol mg^{-1} of protein in heart to 0.65 ± 0.5 fmol mg^{-1} of protein in gill. hCGRP binding could be measured in all the tissues except for the kidney. The highest values were observed in heart, brain and gill membranes

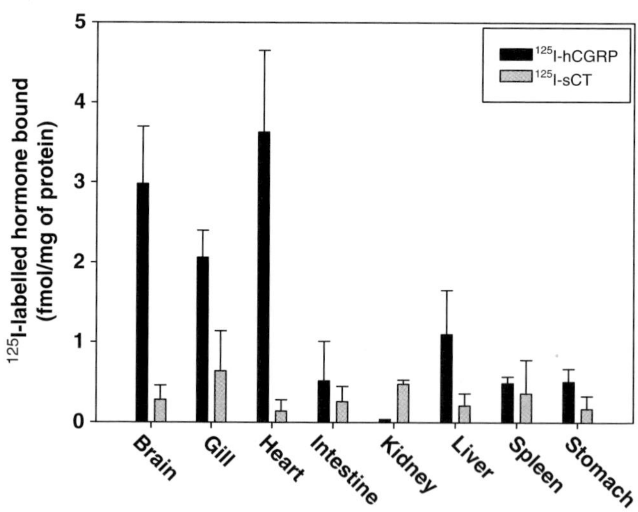

Fig. 7.5 Distribution of CT and CGRP receptors in different tissues of eels. ^{125}I-labelled hCGRP or ^{125}I-labelled sCT and 0.375 mg of membrane proteins per ml were incubated during 2 h at 11°C. Non-specific binding was measured with an excess of unlabelled hCGRP or sCT (2 µg ml^{-1}) and subtracted from the total binding to obtain the specific binding. Results were expressed as specific binding. Each point represents the mean ± SEM of four individual experiments performed in triplicate

with a specific binding of 3.6 ± 1.0, 3.0 ± 0.7 and 2.1 ± 0.3 fmol mg^{-1} of protein, respectively. The lowest hCGRP binding was observed in intestine, stomach and spleen membranes with a specific binding of about 0.5 fmol mg^{-1} of protein.

Maximum specific binding of CGRP was measured in brain, heart and gills. These three organs seem to represent important target organs in the eel. In mammals, CGRP plays a major role in the control of cerebral and cardiovascular functions, namely by its direct action on brain and hearts (Gennari and Fischer 1985; Goltzman and Mitchell 1985; Sigrist et al. 1986; Goto et al. 1992; Wimalawansa and El-Kholy 1993; Wimalawansa 1996; Van Rossum et al. 1997). This role in the central nervous system and heart has also been demonstrated in various non mammalian vertebrates and could represent a common feature among teleosts. In addition, high specific binding of CGRP was observed in gill. The characterization of the gill as a target organ for CGRP in the eel is in agreement with our previous data showing a high specific binding of CGRP in gill tissue of another teleost, the trout (Arlot-Bonnemains et al. 1991). In comparison with CGRP binding, specific CT binding was very low in these three tissues (heart, brain and gill).

7.2.2.2 Autocrine/Paracrine Function of CGRP

These results of CGRP binding in different eel tissues (Fig. 7.5) were compared to the concentrations of CGRP itself in these same tissues, presented in the first part (Fig. 7.3). The highest levels of biologically active CGRP were found in brain and heart, whereas none was detected in gill. These data are in agreement with the PCR study of the expression of CGRP and CGRP receptor in a recent teleost (Pleuronectiformes: flounder). These studies show an important expression of both the peptide and the receptor mRNA in the heart and brain, and only expression of the receptor mRNA in the gills (Suzuki et al. 2000, 2001).

In mammals, CGRP is a neuropeptide involved in several cerebral functions (Wimalawansa and El-Kholy 1993; Van Rossum et al. 1997). It plays a major role in pain perception probably through its participation to the nocioceptive transmission as neuromediator (Li et al. 2001; Yu et al. 2003; Winston et al. 2005). Its potent vasodilatative action in the brain circulation suggests that this peptide could also be strongly involved in migraine pathology (Brain and Grant 2004; Brain 2004). In sauropsids and amphibians, CGRP seems also to play an important role in the central and peripheral nervous system as a neuromediator or neuromodulator (Kuramoto and Fujita 1986; Kline et al. 1988, 1989; Lanuza et al. 2000; Martinez-Garcia et al. 2002). The presence of both CGRP and its specific receptors in the central nervous system of a phylogenetically ancient species among teleosts, the eel, suggests that this neuromediator or neuromodulator role could represent a conserved function among vertebrates. Furthermore, in the central nervous system of cephalopods (brain and optic lobes), molluscs presenting a real cephalisation, we have also detected the presence of both CGRP and its specific receptors (Lafont et al. 2006b). Similarly, in another cephalopod species, *Octopus vulgaris*, a previous study using immunohistochemical detection reported the presence of CGRP-like molecules in the centrifugal neurones of optic lobes (Suzuki and Yamamoto 2002). These data suggest that the neuromediator or neuromodulator role of CGRP, conserved among vertebrates, could represent an ancient role in metazoa, as it is already present in cephalopods.

We have also detected in the eel heart the presence of both specific CGRP binding sites and the peptide itself, suggesting an autocrine/paracrine role of CGRP in this organ. In mammals, CGRP is a potent vasodilatator (Brain et al. 1985; Girgis et al. 1985). It is widely distributed along the cardiovascular system in a sensory network innervating the arteries, veins and heart (Mulderry et al. 1985; McCulloch et al. 1986; Edvinsson et al. 1987). Several specific CGRP binding sites are also localised in the heart and blood vessels (Coupe et al. 1990; Knock et al. 1992). In amphibians this peptide could possess a vasodilatatory effect in the arterial system, similar to that in mammals (Kline et al. 1988). Thus the autocrine/paracrine role of CGRP could represent a common function in various species among vertebrates.

7.2.2.3 Characterization of the Specific CGRP Gill Binding Sites

The presence in gill of CGRP receptors but not of the peptide itself indicates that CGRP could exert an endocrine control on the gills, a major organ for osmoregulatory control in fishes. Therefore, we decided to focus on this organ to pursue the characterization of the CGRP binding sites. We demonstrated that specific binding of labelled hCGRP to eel gill membranes was time-and temperature-dependent (Fig. 7.6). At 8°C, the maximum binding was obtained after 90 min of incubation and decreased significantly after 120 min of incubation. At 11°C, the values of

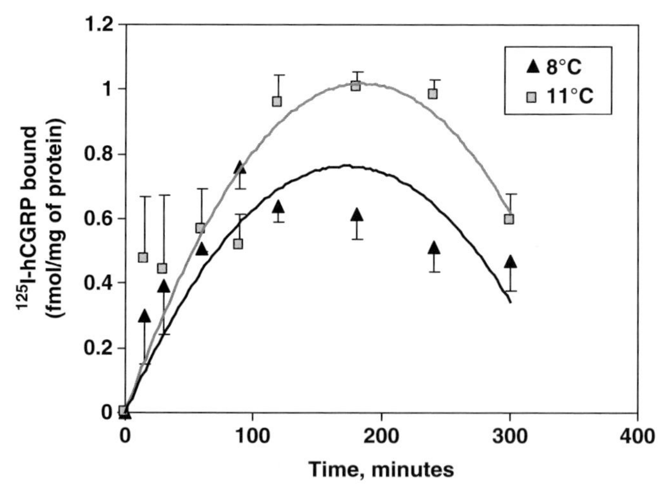

Fig. 7.6 Time course of specific ^{125}I-labelled hCGRP binding to eel gill membranes. 0.375 mg of membrane proteins per ml were incubated with 50 pM of ^{125}I-labelled hCGRP at 8°C and 11°C, for different times (0, 15, 30, 60, 90, 120, 180, 240 and 300 min), with or without 2 µg ml^{-1} unlabelled peptide, to determine the non-specific binding. Results were expressed as the quantity of specific binding in fmol per mg of protein. Each point represents the mean ± SEM of three individual experiments performed in triplicate

specific labelled hCGRP binding to gill membranes were higher, with a maximum observed after 200 min of incubation. Between 120 and 240 min of incubation the specific binding reached an apparent equilibrium. Different concentrations of gill membranes were tested at this temperature. The specific hCGRP binding was linear between 125 and 500 µg of protein per ml. The best specific binding was obtained for a concentration of 400 µg of gill membrane proteins per ml (data not shown). Thus, the parameters chosen for the following experiments were a 180 min incubation time at 11°C and a gill protein concentration of 400 µg per ml.

Increasing concentrations of unlabeled hCGRP (0.27–26.7 nM) displaced the binding of labelled hCGRP (0.05 nM) from gill membranes in a dose dependent fashion (data not shown). Fifty-percent of inhibition was obtained with 4.1 nM of unlabeled hCGRP. This interaction between CGRP and its specific gill binding sites was quantified with the method of Scatchard (Scatchard 1949) (data not shown). We obtained a linear relationship suggesting the presence of one single class of binding sites in the eel gill membranes. The affinity constant (Ka) was 1×10^8 M^{-1} and the binding capacity accounted for 343 fmol mg^{-1} of protein, i.e. a number of binding sites of 2.07×10^{11} per mg of protein. This affinity constant was of the same order of magnitude as that calculated in gill membranes from the trout (Fouchereau-Peron et al. 1994).

The specificity of the eel gill receptors was further characterized by analyzing the capacity of different CGRP (human I (hCGRP), human II, chicken, $CGRP_{8-37}$) and CT (human, chicken, salmon, eel) analogues to displace the binding of labelled hCGRP to the gill membranes. In each case, increasing peptide concentrations displaced the binding of labelled hCGRP to eel gill membranes in a dose-dependent fashion (data not shown). The quantity of each analogue needed to obtain a 50% inhibition of the binding was calculated and reported in Table 7.1. The CGRP analogues were able to displace the CGRP specific binding but with a higher affinity than CT analogues, indicating the presence of a CGRP-like receptor in the gill. Among the CGRP analogues, the best efficiency was exhibited by $CGRP_{8-37}$ suggesting that the CGRP receptors in eel gill would be related to the CGRP type 1 receptor described in rat brain (Quirion et al. 1992). The cross-reaction of CT analogues with eel gill CGRP receptors is in agreement with data in mammals which showed some cross-reactivity between CT and CGRP and their respective specific receptors, a phenomenon related to the strong structural similarity between the two peptides (Goltzman and Mitchell 1985; Henke et al. 1985). Among CT analogues,

Table 7.1 Comparative affinities of various CGRP and CT analogs for ^{125}I-labelled hCGRP binding to eel gill membranes

Peptide	ED_{50}	Peptide	ED_{50}
hCGRP I	4.1 nM	hCT	51.3 nM
hCGRP II	2.2 nM	chCT	36.8 nM
chCGRP	1.9 nM	sCT	29.2 nM
$CGRP_{8-37}$	4.2 nM	eelCT	12.0 nM

The values represent the concentrations of competitors required to displace 50% of specifically bound ^{125}I-labelled hCGRP (ED_{50}). Similar data were obtained with four other experiments.

the best cross-reactivity was exhibited by eel CT, suggesting a possible physiological interaction of both CT and CGRP with the same CGRP-like gill receptor.

7.2.3 *Possible Endocrine Role of CGRP in Osmoregulation: Seawater Transfer*

In order to investigate this possible role of CGRP in eel osmoregulation we performed a seawater transfer and analyzed the adaptive responses of the animals. Two transfer experiments from FW to SW were performed on two independent batches of eels. Tissues and blood sample were collected at various times after transfer (Exp1: 0, 2, 6, 14, 24, 38, 48, 96 and 504h; Exp2: 0, 2, 14, 38, and 504h). Radioimmunoassays of CGRP and CT circulating levels in FW control eels indicated much higher plasma levels of CGRP than CT, in agreement with our previous data. SW transfer resulted in a significant plasma hyperosmolality (FW: 286 ± 9mOsm kg^{-1}), from 2h into SW (337 ± 6mOsm kg^{-1}; $p<0.05$) to the end of the experiment (504h: 340 ± 2mOsm kg^{-1}; $p<0.05$) (Fig. 7.7). At the same time, the data showed a large increase in the circulating levels of CGRP, which was also significant from 2h after transfer (17.1 ± 1.0ng ml^{-1}; $p<0.05$) up to the end of the experiment (at 504h: 20.0 ± 6.5ng ml^{-1}; $p<0.05$) (Fig. 7.8). Plasma CGRP levels, observed after SW transfer, were up to eight times higher than in FW controls (FW: 3.4 ± 1.5ng ml^{-1}). In comparison, plasma CT levels showed a moderate increase, which became significant only after 24h in SW (1.5 ± 0.4ng ml^{-1}; $p<0.05$) and remained higher than in controls (0.5 ± 0.1ng.ml^{-1})

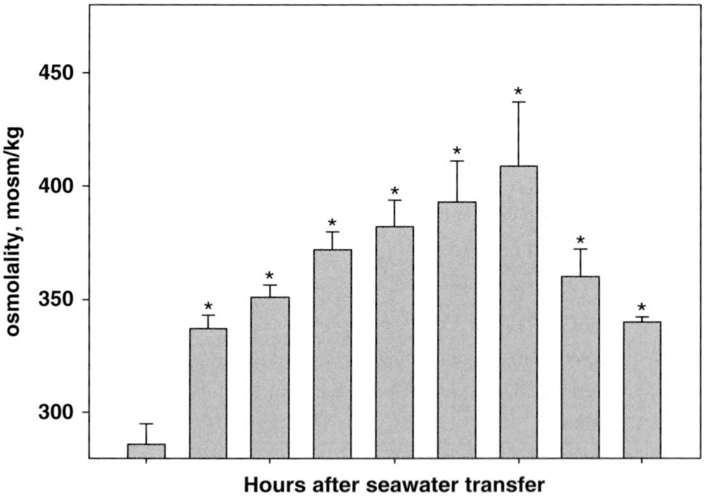

Fig. 7.7 Eel plasma osmolality after 0, 2, 6, 14, 24, 38, 48, 96 and 504h of SW transfer (experiment 2). Osmolality was measured in 10μl aliquots of plasma samples using a vapour pressure osmometer. Results were expressed as mOsm kg^{-1}. Each bar represents the mean ± SEM of eight eels for experiment 1 and seven eels for experiment 2. $^{*}p < 0.05$ when compared to the FW control value (0h in SW) (ANOVA)

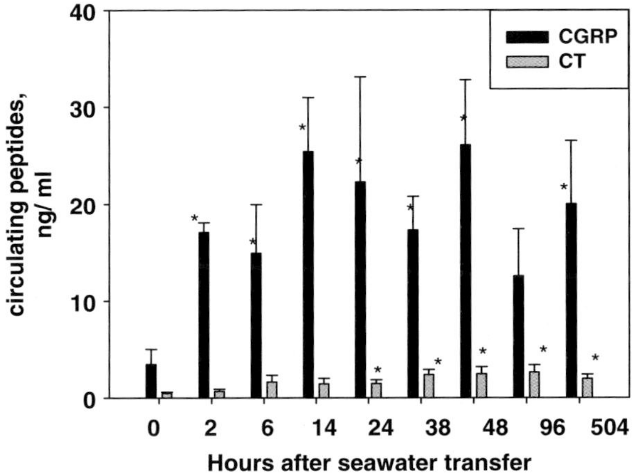

Fig. 7.8 Eel circulating concentrations of CT and CGRP after 0, 2, 6, 14, 24, 38, 48, 96 and 504 h of SW transfer (experiment 1). Plasma was assayed using RIAs specific for hCGRP and sCT. Results were expressed as ng immunoreactive peptide per ml of plasma. Each bar represents the mean ± SEM of eight eels for experiment 1 and seven eels for experiment 2. $*p < 0.05$ when compared to the FW control value (0 h in SW) (ANOVA)

until the end of the experiment (at 504 h: 2.0 ± 0.4 ng ml^{-1}; $p<0.05$). Accordingly, the circulating levels of CGRP were between 7 and 18 times higher than the circulating CT levels throughout the transfer experiment.

Concerning the UBB, the CT-like molecules measured during the transfer showed a significant difference from the control (336 ± 39 ng g^{-1} of UBB) at only one time, after 14 h in SW (44 ± 5 ng g^{-1} of UBB; $p < 0.05$) (Fig. 7.8).

We also measured the concentrations of molecules related to CGRP in the heart at various times following the SW transfer (Fig. 7.9). We observed a significant decrease after 14 h in SW (6.78 ± 0.14 ng g^{-1} of heart, $p < 0.05$), when compared to the controls (9.41 ± 0.63 ng g^{-1} of heart). After 48 h in SW (9.5 ± 0.19 ng g^{-1} of heart), and till the end of the experiment (504 h: 8.46 ± 0.51 ng g^{-1} of heart), the concentrations were no longer significantly different from the control values. From these results we can suggest a hypothesis. In fact, the increase in the plasma osmolality could set off the CGRP secretion by the heart, inducing the significant decrease of the CGRP concentration in this organ, observed after 14 h. This decrease would be followed by the synthesis of CGRP in the heart which would result in an increase in the concentration of this peptide concentration in this organ and a return to values not significantly different from the controls after 48 h. It would be interesting to confirm this hypothesis by a molecular approach, with the study of the CGRP mRNA expression in the heart during the SW transfer. In addition, we have observed that the variations measured in the heart were in the range of pg, whereas the variations observed in the plasma were in the range of ng. As we have demonstrated in the first part of the study, the molecules related to CGRP have been detected in several tissues in the eel. Therefore, the heart is probably not the only organ involved in the CGRP secretion.

Concerning the molecules related to CT, the variations in the circulating levels during the transfer were lower than that of CGRP and significant from 24 h in SW. In

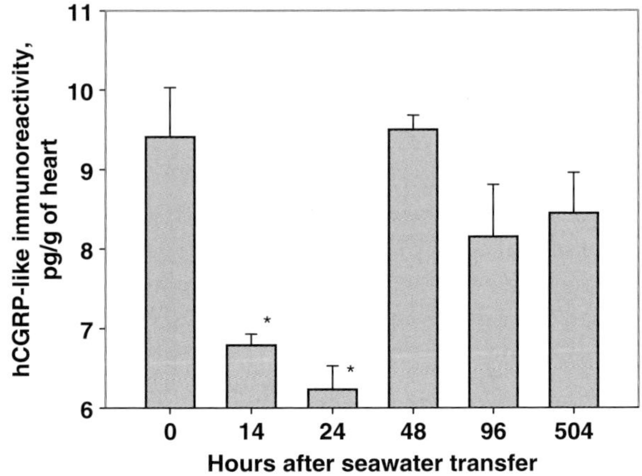

Fig. 7.9 Concentrations of CGRP-like molecules detected in the eel heart after 0, 2, 6, 14, 24, 38, 48, 96 and 504 h of SW transfer (experiment 1). Results are expressed as ng of immunoreactive peptide per g of heart. Each bar represents the mean ± SEM of eight eels for experiment 1 and seven eels for experiment 2. *$p < 0.05$ when compared to the FW control value (0 h in SW) (ANOVA)

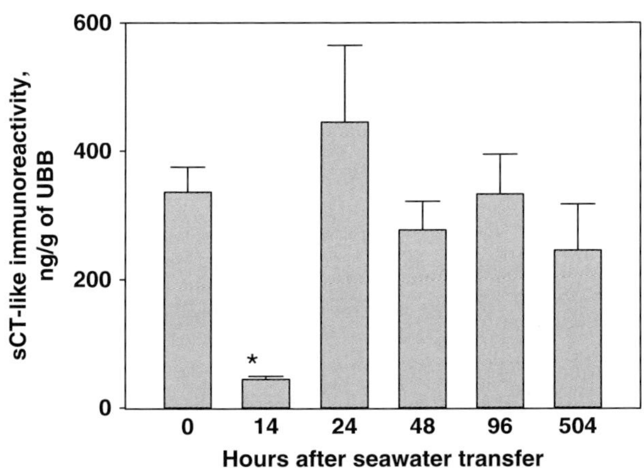

Fig. 7.10 Concentrations in CT-like molecules detected in the eel UBB after 0, 2, 6, 14, 24, 38, 48, 96 and 504 h of SW transfer (experiment 1). Results are expressed as ng of immunoreactive peptide per g of UBB. Each bar represents the mean ± SEM of eight eels for experiment 1 and seven eels for experiment 2. *$p < 0.05$ when compared to the FW control value (0 h in SW) (ANOVA)

the UBB, we observed a significant decrease in the CT-like concentrations after 14 h (44 ± 5 ng g^{-1} of UBB, $p < 0.05$) (Fig. 7.10). After 24 h and till the end of the experiment the concentrations were no longer significantly different from the controls (336 ± 39 ng g^{-1} of UBB). Therefore, we can suggest the same hypothesis as for CGRP, i.e. a secretion of CT from the UBB rapidly followed by new synthesis of this peptide in this organ. As in the case of CGRP, a molecular approach would be informative.

This finding together with the large variations of CGRP plasma levels under osmotic challenge, as shown in the present study, suggest an important endocrine role for this peptide in eels. In contrast, an inversion in the CGRP/CT ratio is found in mammals with higher circulating levels of CT than CGRP (CT levels five times higher than CGRP in human).

Finally, in order to see if there was any variation observed concerning the CGRP receptors during this SW adaptation, hCGRP binding to gill membranes was studied at different times after SW transfer (0, 2, 14, 38 and 504 h; Experiment 2), using the Scatchard method. In each case, a linear relationship was obtained (data not shown). No significant difference was found between the slopes indicating no significant variations in the affinity constant during the SW transfer (Table 7.2). In contrast, the data showed a twofold significant increase in the number of CGRP binding sites after 2 h in SW (Table 7.2). This increase was transient and, after 14 h, the number of sites returned to a level not significantly different from that of the FW controls. No further significant change was observed until the end of the experiment (504 h). This significant and transient increase, after 2 h of transfer, in the number of gill CGRP receptors with no change in their affinity constant may reflect an up regulation by CGRP of its specific receptors. However, it cannot be excluded that many other factors of osmoregulation could be involved in this process. Up regulation is a well known regulating system described for many peptide hormones: in particular, an up regulation of the receptors has also been described in the case of growth hormone, another peptide hormone involved in hyper-osmotic adaptation, as shown during the transfer of rainbow trout from FW to SW (Sakamoto and Hirano 1991). Furthermore, the transfer of flounders from SW to brackish water has shown a decrease in the CGRP receptor mRNA expression in the gills. This expression was even no more detectable in FW, suggesting also a possible role of CGRP in the flounder SW adaptation (Suzuki et al. 2002). Further studies should aim at identifying the physiological mechanism by which CGRP could regulate the osmotic stress. Studies could focus on the characterization of the gill cell type(s) in which the CGRP receptors are expressed and investigate the potential effects of CGRP on cell activity, such as Na, K-ATPase expression.

Table 7.2 Binding parameters of CGRP to eel gill membranes, after different times in SW (experiment 2)

Hours after seawater transfer	Maximal binding, fmol mg^{-1} of protein	Binding sites per mg of protein x 10^{10}	Affinity constant (Ka) $M^{-1} \times 10^8$
0	92.69 ± 5.36	5.58 ± 0.32	2.67 ± 0.28
2	167.05 ± 13.01[a]	10.06 ± 0.78[a]	2.00 ± 0.41
14	64.90 ± 19.47	3.91 ± 1.17	4.49 ± 1.02
38	70.52 ± 10.76	4.25 ± 0.65	2.80 ± 0.34
504	98.16 ± 16.51	5.91 ± 0.99	3.80 ± 0.47

Each value is the mean ± SEM of 5 individual experiments performed in triplicate.
[a]$p < 0.05$ when compared to the FW control value (0 h in SW) (ANOVA).

7.3 Conclusion

In conclusion, this study provides physiological evidence that the neuropeptide, CGRP, by its specific target on gill membranes and by the variations of its circulating levels in response to osmotic variation, may be one of the many regulators involved in osmoregulation. Osmoregulation represents a major challenge for aquatic animals implying the recruitment of many and various endocrine controls such as prolactin, growth hormone and cortisol. This study in the eel, a representative of an ancient group of teleosts, together with previous data from other teleost species, indicates that the involvement of CGRP in the control of the gill function may be an ancient and conserved regulatory mechanism among the wide group of teleosts, representative of the Actinopterygian lineage. Furthermore, we have demonstrated that in cephalopods, CGRP-like molecules and their specific receptors seem to be present in the gills (Lafont et al. 2006b). In the same way, in oyster, transfer from seawater to brackish water induced a similar decrease in gene expression of the gill CRLR receptor (Dubos et al. 2003). Other studies have indicated the absence of CGRP-like molecules in bivalve and gastropod mollusc gills, and in contrast the presence in this organ of specific CGRP binding sites (Fouchereau-Peron 1993, 1996; Duvail et al. 1997, 1999). Thus, the endocrine role of CGRP in the gill ionic regulation may constitute an ancestral function for this peptide activity among the aquatic animals, vertebrates and invertebrates. In addition, our data also demonstrate the presence of CGRP and its specific receptors in eel brain and heart, as shown in mammals, suggesting that the local roles of CGRP as an autocrine or paracrine neuromediator in the control of cardiovascular and neuronal functions may represent ancestral and conserved features among vertebrates.

Acknowledgements This work has been supported by the European contract EELREP N°Q5RS-2001-01836.

References

Amara SG, Jonas V, Rosenfeld MG, Ong ES, Evans RM (1982) Alternative RNA processing in calcitonin gene expression generates mRNAs encoding different polypeptide products. Nature 298: 240–244.

Amara SG, Arriza JL, Leff SE, Swanson LW, Evans RM, Rosenfend MG (1985) Expression in brain of a messenger RNA encoding a novel neuropeptide homologous to calcitonin gene related peptide. Science 229: 1094–1097.

Arlot-Bonnemains Y, Fouchereau-Peron M, Julienne A, Milhaud G, Moukhtar MS (1991) Binding sites of calcitonin gene related peptide (CGRP) to trout tissues. Neuropeptides 20(3): 181–186.

Arnold-Reed D, Balment J (1994) Peptide hormones influence in vitro interrenal secretion of cortisol in the trout, Oncorhynchus mykiss. Gen Comp Endocrinol 96: 85–91.

Arnold-Reed D, Hazon N, Balment J (1991) Biological actions of atrial natriuretic factor in flatfish. Fish Physiol Biochem 9: 271–277.

Auperin B, Rentier-Delrue F, Martial JA, Prunet P (1995) Regulation of gill prolactin receptors in tilapia (*Oreochromis niloticus*) after a change in salinity or hypophysectomy. J Endocrinol 145(2): 213–220.

Avella M, Part P, Ehrenfeld J (1999) Regulation of Cl^- secretion in seawater fish (*Dicentrarchus labrax*) gill respiratory cells in primary culture. J Physiol 516: 353–363.

Azria M (1989) The calcitonins: physiology and pharmacology. Karger, Basel.

Bond H, Winter MJ, Warne JM, McCrohan CR, Balment RJ (2002) Plasma concentrations of arginine vasotocin and urotensin II are reduced following transfer of the euryhaline flounder (*Platichthys flesus*) from seawater to fresh water. Gen Comp Endocrinol 125(1): 113–120.

Borski RJ, Yoshikawa JS, Madsen SS, Nishioka RS, Zabetian C, Bern HA, Grau EG (1994) Effects of environmental salinity on pituitary growth hormone content and cell activity in the euryhaline tilapia, *Oreochromis mossambicus*. Gen Comp Endocrinol 95(3): 483–494.

Brain SD (2004) Calcitonin gene-related peptide (CGRP) antagonists: blockers of neuronal transmission in migraine. Br J Pharmacol 142: 1053–1054.

Brain SD, Grant AD (2004) Vascular actions of calcitonin gene-related peptide and adrenomedullin. Physiol Rev 84: 903–934.

Brain SD, Williams TJ, Tippings JR, Morris HR, MacIntyre I (1985) Calcitonin gene related peptide is a potent vasodilator. Nature 313: 54–56.

Brown JA, Gray CJ, Taylor SM (1990) The renin-angiotensin system and glomerular function of teleost fish. Prog Clin Biol Res 342: 528–533.

Brown PS, Brown SC (1987) Osmoregulatory actions of prolactin and other adenohypophysial hormones. Vertebrate Endocrinology: Fundamentals and Biochemical Implications. Pang PKT and Schreibman MP (eds). Academic Press, San Diego, CA, Vol. 2, pp. 45–84.

Burden CE (1956) The failure of hypophysectomized *Fundulus heteroclitus* to survive in freshwater. Biol Bull 110: 8–28.

Chakrabarti P, Mukherjee D (1993) Studies on the hypocalcemic actions of salmon calcitonin and ultimobranchial gland extracts in the freshwater teleost *Cyprinus carpio*. Gen Comp Endocrinol 90: 267–273.

Chang IC, Lee TH, Wu HC, Hwang PP (2002) Effects of environmental Cl^- levels on Cl^- uptake and mitochondria-rich cell morphology in gills of the stenohaline goldfish, *Carassius auratus*. Zool Stud 41: 236–243.

Clark MS, Bendell L, Power DM, Warner S, Elgar G, Ingleton PM (2002) Calcitonin: characterisation and expression in a teleost fish, *Fugu rubripes*. J Mol Endocrinol 28: 111–123.

Conley DM, Mallatt J (1988) Histochemical localization of Na^+-K^+ATPase and carbonic anhydrase in gills of 17 fish species. Can J Zool 66: 2398–2405.

Copp DH, Cameron EC, Cheney BA, Davidson AG, Henze KG (1962) Evidence for calcitonin: a new hormone from the parathyroid that lowers blood calcium. Endocrinology 70: 638–649.

Copp DH, Cockcroft DW, Kueh Y (1967) Ultimobranchial origin of calcitonin, hypocalcemic effect of extracts of chicken gland. Can J Physiol Pharmacol 45: 1095–1099.

Cornish J, Callon KE, Bava U, Kamona SA, Cooper GJS, Reid IR (2001) Effects of calcitonin, amylin, and calcitonin gene-related peptide on osteoclast development. Bone 29(2): 162–168.

Coupe MO, Mak JC, Yacoub M, Oldershaw PJ, Barnes PJ (1990) Autoradiographic mapping of calcitonin gene-related peptide receptors in human and guinea pig hearts. Circulation 81: 741–747.

Dubos M-P, Badariotti F, Rodet F, Lelong C, Favrel P (2003) Molecular and physiological characterization of an invertebrate homologue of a calcitonin-related receptor. Biochem Biophys Res Commun 310: 972–978.

Duvail L, Lopez E, Fouchereau-Peron M (1997) Characterization of a calcitonin gene-related peptide-like molecule in the abalone, *Haliotis tuberculata*. Comp Biochem Physiol 116C: 155–159.

Duvail L, Lopez E, Fouchereau-Peron M (1999) Characterization of binding sites for calcitonin gene-related peptide in the abalone gill. Peptides 20: 361–366.

Eckert SM, Yada T, Shepherd BS, Stetson MH, Hirano T, Grau EG (2001) Hormonal control of osmoregulation in the channel catfish *Ictalurus punctatus*. Gen Comp Endocrinol 122(3): 270–286.

Eckert SM, Hirano T, Leedom TA, Takei Y, Gordon Grau E (2003) Effects of angiotensin II and natriuretic peptides of the eel on prolactin and growth hormone release in the tilapia, *Oreochromis mossambicus*. Gen Comp Endocrinol 130(3): 333–339.

Edvinsson L, Ekman R, Jansen I, McCulloch J, Uddman R (1987) Calcitonin gene-related peptide and cerebral blood vessels: distribution and vasomotor effects. J Cereb Blood Flow Metab 7: 720–728.

Evans DH (1990) An emerging role for a cardiac peptide hormone in fish osmoregulation. Annu Rev Physiol 52: 43–60.

Evans DH (2002) Cell signaling and ion transport across the fish gill epithelium. J Exp Zool 293(3): 336–347.

Evans DH, Piermarini PM, Choe KP (2005) The multifunctional fish gill: dominant site of gas exchange, osmoregulation, acid-base regulation, and excretion of nitrogenous waste. Physiol Rev 85(1): 97–177.

Flik G, Atsma W, Fenwick JC, Rentier-Delrue F, Smal J, Wendelaar Bonga SE (1993) Homologous recombinant growth hormone and calcium metabolism in the tilapia, *Oreochromis mossambicus*, adapted to fresh water. J Exp Biol 185: 107–119.

Fouchereau-Peron M (1993) Characterization of a molecule related to calcitonin gene related peptide (CGRP) in the scallop, *Pecten maximus*. Comp Biochem Physiol 105B: 707–711.

Fouchereau-Peron M (1996) Characterization of binding sites for calcitonin gene related peptide in the mollusc gill. Regul Pept 61: 57–61.

Fouchereau-Peron M (1998) Increased binding of calcitonin gene-related peptide to gill receptors in trout transferred from seawater to fresh water. J Mar Biotechnol 6: 95–98.

Fouchereau-Peron M, Arlot-Bonnemains Y, Milhaud G (1994) Mode of action of calcitonin gene related peptide in trout gill membranes: evidence for a GTP coupling process. Neuropeptides 26: 267–272.

Fouchereau-Peron M, Arlot-Bonnemains Y, Taboulet J, Milhaud G, Moukhtar MS (1990) Distribution of calcitonin gene-related peptide and calcitonin like immunoreactivity in trout. Regul Pept 27: 171–179.

Gennari C, Fischer JA (1985) Cardiovascular action of calcitonin gene related peptide in humans. Calcif Tissue Int 37: 581–584.

Gerstheimer FP, Metz J (1986) Distribution of calcitonin gene-related peptide-like immunoreactivity in the guinea pig heart. Anat Embryol (Berl) 175(2): 255–260.

Girgis SI, Stevenson JC, Lynch C, Srelf CH, Macdonald DWR, Bevis PJR, Wimalawansa SJ, Morris HR, MacIntyre I (1985) Calcitonin gene related peptide: potent vasodilator and major product of the calcitonin gene. Lancet 2(8445): 14–16.

Goltzman D, Mitchell J (1985) Interaction of calcitonin and calcitonin gene related peptide at receptor sites in target tissues. Science 227: 1343–1345.

Goto K, Miyauchi T, Homma S, Ohshima N (1992) Calcitonin gene-related peptide in the regulation of cardiac function. Calcitonin gene-related peptide. The first decade of a novel pleiotropic neuropeptide. Tache Y, Holzer P, Rosenfeld MG. Ann NY Acad Sci 657: 194–203.

Henderson IW, Hazon N, Hughes K (1985) Hormones, ionic regulation and kidney function in fishes. Symp Soc Exp Biol 39: 245–265.

Henke H, Tschopp FA, Fischer JA (1985) Distinct binding sites for calcitonin gene-related peptide and salmon calcitonin in rat central nervous system. Brain Res 360: 165–171.

Higashimoto Y, Nakao N, Ohkubo T, Tanaka M, Nakashima K (2001) Structure and tissue distribution of prolactin receptor mRNA in Japanese flounder (*Paralichtys olivaceus*): conserved and preferential expression in osmoregulatory organs. Gen Comp Endocrinol 123(2): 170–179.

Hirano T (1986) The spectrum of prolactin action in teleosts. Prog Clin Biol Res 205: 53–74.

Hirano T, Hasegawa S, Yamauchi H, Orimo H (1981) Further studies on the absence of hypocalcemic effects of eel calcitonin in the eel, *Anguilla japonica*. Gen Comp Endocrinol 43: 42–50.

Hirose S, Kaneko T, Naito N, Takei Y (2003) Molecular biology of major components of chloride cells. Comp Biochem Physiol B 136(4): 593–620.

Höppener JWM, Steenbergh PH, Slebos TRJC, Visser A, Lips CJM, Jansz HS, Bechet JM, Lenoir GM, Born W, Haller-Brem S, Petermann JB, Fisher JA (1987) Expression of the second calcitonin/calcitonin gene related peptide gene in ewing sarcoma cell lines. J Clin Endocrinol Metab 64: 809–817.

Inoue K, Sakamoto T, Yuge S, Iwatani H, Yamagami S, Tsutsumi M, Hori H, Cerra MC, Tota B, Suzuki N, Okamoto N, Takei Y (2005) Structural and functional evolution of three cardiac natriuretic peptides. Mol Biol Evol 22: 2428–2434.

Jansz H, Martial K, Zandberg J, Milhaud G, Benson AA, Julienne A, Moukhtar MS, Cressent M (1996) Identification of a new calcitonin gene in the salmon *Oncorhynchus gorbuscha*. Proc Natl Acad Sci 93(22): 12344–12348.

Jansz HS, Zandberg J (1992) Identification and partial characterization of the salmon calcitonin/CGRP gene by polymerase chain reaction. Calcitonin gene-related peptide. The first decade of a novel pleiotropic neuropeptide. Taché Y, Holtzer P, Rosenfeld MG. Ann NY Acad Sci 657: 63–69.

Kaiya H, Takei Y (1996) Changes in plasma atrial and ventricular natriuretic peptide concentrations after transfer of eels from freshwater to seawater or vice versa. Gen Comp Endocrinol 104(3): 337–345.

Kajimura S, Kawaguchi N, Kaneko T, Kawazoe I, Hirano T, Visitacion N, Grau EG, Aida K (2004) Identification of the growth hormone receptor in an advanced teleost, the tilapia (*Oreochromis mossambicus*) with special reference to its distinct expression pattern in the ovary. J Endocrinol 181(1): 65–76.

Katafuchi T, Takashima A, Kashiwagi M, Hagiwara H, Takei Y, Hirose S (1994) Cloning and expression of eel natriuretic-peptide receptor B and comparison with its mammalian counterparts. Eur J Biochem 222(3): 835–842.

Kelly SP, Chow IN, Woo NY (1999) Effects of prolactin and growth hormone on strategies of hypoosmotic adaptation in a marine teleost, *Sparus sarba*. Gen Comp Endocrinol 113(1): 9–22.

Kelsall CJ, Balment RJ (1998) Native urotensins influence cortisol secretion and plasma cortisol concentration in the euryhaline flounder, platichthys flesus. Gen Comp Endocrinol 112(2): 210–219.

Klaren PH, Guzman JM, Mancera JM, Geven EJ, Flik G (2005) The involvement of thyroid hormone metabolism in gilthead sea bream (*Sparus auratus*) osmoregulation. Ann NY Acad Sci 1040: 360–362.

Kline LW, Kaneko T, Chiu KW, Harvey S, Pang PKT (1988) Calcitonin gene-related peptide in the bullfrog *Rana catesbeiana* localization and vascular action. Gen Comp Endocrinol 72: 123–129.

Kline LW, Cooper CW, Harvey S, Pang PKT (1989) Calcitonin gene-related peptide in neural tissue: a phylogenetic study. Gen Comp Endocrinol 76: 408–414.

Kloas W, Reinecke M, Hanke W (1994) Role of the atrial natriuretic peptide for adrenal regulation in the teleost fish *Cyprinus carpio*. Am J Physiol 267: R1034–R1042.

Knock GA, Wharton J, Gaer JA, Yacoub MH, Taylor KM, Polak JM (1992) Regional distribution and regulation of (125I) calcitonin gene-related peptide binding sites in coronary arteries. Eur J Pharmacol 219: 415–425.

Kuramoto H, Fujita T (1986) An immunohistochemical study of calcitonin gene-related peptide (CGRP) containing nerve fibers in the sympathetic ganglia of bullfrogs. Biomed Res 7: 349–357.

Lafont AG, Dufour S, Fouchereau-Peron M (2004) Characterisation and distribution of calcitonin gene-related peptide in a primitive teleost, the eel, *Anguilla anguilla* and comparison with calcitonin. Regul Pept 117(2): 141–148.

Lafont AG, Fitzpatrick T, Cliff Rankin J, Dufour S, Fouchereau-Peron M (2006a) Possible role of calcitonin gene-related peptide in osmoregulation via the endocrine control of the gill in a teleost, the eel, *Anguilla anguilla*. Peptides 27(4): 812–819.

Lafont AG, Dufour S, Fouchereau-Peron M (2006b) Evidence for the presence of molecules related to the neuropeptide CGRP in two cephalopods, *Sepia officinalis* and *Nautilus macromphalus*: comparison with its target organ distribution. Neuroendocrinology 84(2): 138–150.

Laiz-Carrion R, Guerreiro PM, Fuentes J, Canario AV, Martin Del Rio MP, Mancera JM (2005) Branchial osmoregulatory response to salinity in the gilthead sea bream, *Sparus auratus*. J Exp Zool A 303(7): 563–576.

Lamharzi N, Fouchereau-Peron M (1994) Calcitonin gene related peptide stimulates carbonic anhydrase in trout gill membranes. Gen Comp Endocrinol 94: 166–170.

Lamharzi N, Fouchereau-Peron M (1996) Adaptation of rainbow trout to seawater: changes in calcitonin gene related peptide levels are associated with an increase in hormone receptor interaction in gill membranes. Gen Comp Endocrinol 102: 274–280.

Lanuza E, Davies DC, Landete JM, Novejarque A, Martinez Garcia F (2000) Distribution of CGRP-like immunoreactivity in the chick and quail brain. J Comp Neurol 421: 515–532.

Le Rouzic P, Sandra O, Grosclaude J, Rentier-Delrue F, Jolois O, Tujague M, Pakdel F, Sandowski Y, Cohen Y, Gertler A, Prunet P (2001) Evidence of rainbow trout prolactin interaction with its receptor through unstable homodimerisation. Mol Cell Endocrinol 172(1–2): 105–113.

Lee LT, Nong G, Chan YH, Tse DL, Cheng CH (2001) Molecular cloning of a teleost growth hormone receptor and its functional interaction with human growth hormone. Gene 270(1–2): 121–129.

Lee TH, Feng SH, Lin CH, Hwang YH, Huang CL, Hwang PP (2003) Ambient salinity modulates the expression of sodium pumps in branchial mitochondria-rich cells of Mozambique tilapia, *Oreochromis mossambicus*. Zool Sci 20(1): 29–36.

Lee Y, Takami K, Kawai Y, Girgis S, Hillyard CJ, MacIntyre I, Emson PC, Tohyama M (1985) Distribution of calcitonin gene-related peptide in the rat peripheral nervous system with reference to its coexistence with substance P. Neuroscience 15: 1227–1237.

Leedom TA, Hirano T, Grau EG (2003) Effect of blood withdrawal and angiotensin II on prolactin release in the tilapia, *Oreochromis mossambicus*. Comp Biochem Physiol A 135(1): 155–163.

Li N, Lundeberg T, Yu LC (2001) Involvement of CGRP and CGRP1 receptor in nociception in the nucleus accumbens of rats. Brain Res 901: 161–166.

Li YY, Takei Y (2003) Ambient salinity-dependent effects of homologous natriuretic peptides (ANP, VNP, and CNP) on plasma cortisol level in the eel. Gen Comp Endocrinol 130(3): 317–323.

Loretz CA, Pollina C (2000) Natriuretic peptides in fish physiology. Comp Biochem Physiol A 125(2): 169–187.

Madsen SS (1990) The role of cortisol and growth hormone in seawater adaptation and development of hypoosmoregulatory mechanisms in sea trout parr (*Salmo trutta trutta*). Gen Comp Endocrinol 79(1): 1–11.

Madsen SS, Bern HA (1992) Antagonism of prolactin and growth hormone: impact on seawater adaptation in two salmonids, *Salmo trutta* and *Oncorhynchus mykiss*. Zool Sci 9: 775–784.

Madsen SS, Jensen MK, Nohr J, Kristiansen K (1995) Expression of Na(+)-K(+)-ATPase in the brown trout, *Salmo trutta*: in vivo modulation by hormones and seawater. Am J Physiol 269(6 Pt 2): R1339–R1345.

Mancera JM, McCormick SD (1998a) Evidence for growth hormone/insulin-like growth factor I axis regulation of seawater acclimation in the euryhaline teleost *Fundulus heteroclitus*. Gen Comp Endocrinol 111(2): 103–112.

Mancera JM, McCormick SD (1998b) Osmoregulatory actions of the GH/IGF axis in non-salmonid teleosts. Comp Biochem Physiol B 121: 43–48.

Mancera JM, McCormick SD (1999) Influence of cortisol, growth hormone, insulin-like growth factor I and 3,3″,5-triiodo-l-thyronine on hypoosmoregulatory ability in the euryhaline teleost *Fundulus heteroclitus*. Fish Physiol Biochem 21(1): 25–33.

Mancera MJ, Laiz Carrion R, del Pilar Martin del Rio M (2002) Osmoregulatory action of PRL, GH, and cortisol in the gilthead seabream (*Sparus aurata* L.). Gen Comp Endocrinol 129(2): 95–103.

Manzon LA (2002) The role of prolactin in fish osmoregulation: a review. Gen Comp Endocrinol 125(2): 291–310.

Marshall I, Al-Kazwini S, Holman JJ, Craig RK (1986) Human and rat α CGRP but not calcitonin cause mesenteric dilatation in rats. Eur J Pharmacol 123: 217–222.

Marshall WS (2002) Na(+), Cl(−), Ca(2+) and Zn(2+) transport by fish gills: retrospective review and prospective synthesis. J Exp Zool 293(3): 264–283.

Marsigliante S, Muscella A, Vilella S, Nicolardi G, Ingrosso L, Ciardo V, Zonno V, Vinson GP, Ho MM, Storelli C (1996) A monoclonal antibody to mammalian angiotensin II AT1 receptor recognizes one of the angiotensin II receptor isoforms expressed by the eel (*Anguilla anguilla*). J Mol Endocrinol 16(1): 45–56.

Marsigliante S, Muscella A, Vinson GP, Storelli C (1997) Angiotensin II receptors in the gill of sea water- and freshwater-adapted eel. J Mol Endocrinol 18(1): 67–76.

Martinez Garcia F, Novejarque A, Landete JM, Moncho Bogani J, Lanuza E (2002) Distribution of calcitonin gene-related peptide-like immunoreactivity in the brain of the lizard *Podarcis hispanica*. J Comp Neurol 447(2): 99–113.

McCormick SD (1996) Effects of growth hormone and insulin-like growth factor I on salinity tolerance and gill Na^+, K^+-ATPase in Atlantic salmon (*Salmo salar*): interaction with cortisol. Gen Comp Endocrinol 101(1): 3–11.

McCormick SD (2001) Endocrine control of osmoregulation in teleost fish. Am Zool 41(4): 781–794.

McCulloch J, Uddman R, Kingam T, Edvinsson L (1986) Calcitonin gene-related peptide: functional role in cerebrovascular regulation. Proc Natl Acad Sci 83: 5731–5735.

Minvielle S, Cressent M, Delehaye MC, Segond N, Milhaud G, Julienne A, Moukhtar MS, Lasmoles F (1987) Sequence and expression of the chicken calcitonin gene. FEBS Lett 223: 63–69.

Mommsen TP, Vijayan MM, Moon TW (1999) Cortisol in teleosts: dynamics, mechanisms of action, and metabolic regulation. Rev Fish Biol Fish 9(3): 211–268.

Mukherjee D, Sen U, Bhattacharyya SP, Mukherjee D (2004) The effects of calcitonin on plasma calcium levels and bone metabolism in the fresh water teleost *Channa punctatus*. Comp Biochem Physiol 138A: 417–426.

Mulderry PK, Ghatei MA, Rodrigo J, Allen JM, Rosenfelg MG, Polak JM, Bloom SR (1985) Calcitonin gene related peptide in cardiovascular tissues of the rat. Neuroscience 14: 947–954.

Nakao N, Higashimoto Y, Ohkubo T, Yoshizato H, Nakai N, Nakashima K, Tanaka M (2004) Characterization of structure and expression of the growth hormone receptor gene of the Japanese flounder (*Paralichtys olivaceus*). J Endocrinol 182(1): 157–164.

Ogawa H, Takahashi T, Kuwayama T, Kawashima M (2003) Presence of calcitonin receptors in shell gland of the guinea fowl and changes in binding property during an oviposition cycle. Poultry Sci 82(8): 1302–1306.

Pearse AGE, Cavalheira AF (1967) Cytochemical evidence for an ultimobranchial origin of rodent thyroid C cells. Nature 214: 929–930.

Pelis RM, McCormick SD (2001) Effects of growth hormone and cortisol on Na(+)-K(+)-2Cl(−) cotransporter localization and abundance in the gills of Atlantic salmon. Gen Comp Endocrinol 124(2): 134–143.

Perry SF (1997) The chloride cell: structure and function in the gills of freshwater fishes. Annu Rev Physiol 59: 325–347.

Perry SF (1998) Relationships between branchial chloride cells and gas transfer in freshwater fish. Comp Biochem Physiol A 119(1): 9–16.

Peter MS, Lock RC, Wendelaar Bonga S (2000) Evidence for an osmoregulatory role of thyroid hormones in the freshwater mozambique tilapia *Oreochromis mossambicus*. Gen Comp Endocrinol 120: 157–167.

Pickford GE, Phillips JG (1959) Prolactin, a factor in promoting survival of hypophysectomized killifish in fresh water. Science 130(3373): 454–455.

Pickford GE, Griffith RW, Torretti J, Hendlez E, Epstein FH (1970) Branchial reduction and renal stimulation of (Na^+, K^+)-ATPase by prolactin in hypophysectomized killifish in fresh water. Nature 228(5269): 378–379.

Pierson PM, Guibbolini ME, Mayer-Gostan N, Lahlou B (1995) ELISA measurements of vasotocin and isotocin in plasma and pituitary of the rainbow trout: effect of salinity. Peptides 16(5): 859–865.

Pisam M, Rambourg A (1991) Mitochondria-rich cells in the gill epithelium of teleost fishes: an ultrastructural approach. Int Rev Cytol 130: 191–232.

Pisam M, Caroff A, Rambourg A (1987) Two types of chloride cells in the gill epithelium of a freshwater-adapted euryhaline fish: *Lebistes reticulatus*; their modifications during adaptation to saltwater. Am J Anat 179(1): 40–50.

Pisam M, Auperin B, Prunet P, Rentier-Delrue F, Martial J, Rambourg A (1993) Effects of prolactin on alpha and beta chloride cells in the gill epithelium of the saltwater adapted tilapia *Oreochromis niloticus*. Anat Rec 235(2): 275–284.

Prunet P, Boeuf G, Bolton JP, Young G (1989) Smoltification and seawater adaptation in Atlantic salmon (*Salmo Salar*): plasma prolactin, growth hormone and thyroid hormones. Gen Comp Endocrinol 74: 355–364.

Prunet P, Pisam M, Claireaux JP, Boeuf G, Rambourg A (1994) Effects of growth hormone on gill chloride cells in juvenile Atlantic salmon (*Salmo salar*). Am J Physiol 266: R850–R857.

Prunet P, Sandra O, Le Rouzic P, Marchand O, Laudet V (2000) Molecular characterization of the prolactin receptor in two fish species, tilapia *Oreochromis niloticus* and rainbow trout, *Oncorhynchus mykiss*: a comparative approach. Can J Physiol Pharmacol 78(12): 1086–1096.

Quirion R, Van Rossum D, Dumont Y, St-Pierre S, Fournier A (1992) Characterization of CGRP1 and CGRP2 receptor subtypes. Calcitonin gene-related peptide. The first decade of a novel pleiotropic neuropeptide. Tache Y, Holzer P, Rosenfeld MG. Ann NY Acad Sci 657: 88–105.

Raynaud A, Cohen R, Modigliani E (1994) Le peptide alternatif du gène de la calcitonine (CGRP). Presse Med 23(4): 171–175.

Roos BA, Lindall AW, Baylin SB, O'Neil J, Frelinger AL, Birnbaum RS, Lambert PW (1980) Plasma immunoreactive calcitonin in lung cancer. J Clin Endocrinol Metab 50: 659–696.

Sakamoto T, Hirano T (1991) Growth hormone receptors in the liver and osmoregulatory organs of rainbow trout: characterization and dynamics during adaptation to seawater. J Endocrinol 130(3): 425–433.

Sakamoto T, McCormick SD (2006) Prolactin and growth hormone in fish osmoregulation. Gen Comp Endocrinol 147(1): 24–30.

Sakamoto T, McCormick SD, Hirano T (1993) Osmoregulatory actions of growth hormone and its mode of action in salmonids: a review. Fish Physiol Biochem 11: 155–164.

Sakamoto T, Shepherd BS, Madsen SS, Nishioka RS, Siharath K, Richman NH, 3rd, Bern HA, Grau EG (1997) Osmoregulatory actions of growth hormone and prolactin in an advanced teleost. Gen Comp Endocrinol 106(1): 95–101.

Sandra O, Sohm F, de Luze A, Prunet P, Edery M, Kelly PA (1995) Expression cloning of a cDNA encoding a fish prolactin receptor. Proc Natl Acad Sci USA 92(13): 6037–6041.

Sandra O, Le Rouzic P, Cauty C, Edery M, Prunet P (2000) Expression of the prolactin receptor (tiPRL-R) gene in tilapia *Oreochromis niloticus*: tissue distribution and cellular localization in osmoregulatory organs. J Mol Endocrinol 24(2): 215–224.

Santos CR, Ingleton PM, Cavaco JE, Kelly PA, Edery M, Power DM (2001) Cloning, characterization, and tissue distribution of prolactin receptor in the sea bream (*Sparus aurata*). Gen Comp Endocrinol 121(1): 32–47.

Scatchard G (1949) The attraction of proteins for small molecules and ions. Ann NY Acad Sci 51: 660–672.

Scheide JI, Zadunaisky JA (1988) Effect of atriopeptin II on isolated opercular epithelium of *Fundulus heteroclitus*. Am J Physiol 254: R27–R32.

Seidelin M, Madsen SS, Byrialsen A, Kristiansen K (1999) Effects of insulin-like growth factor-I and cortisol on Na^+, K^+-ATPase expression in osmoregulatory tissues of brown trout (*Salmo trutta*). Gen Comp Endocrinol 113(3): 331–342.

Shepherd BS, Sakamoto T, Nishioka RS, Richman NH, 3rd, Mori I, Madsen SS, Chen TT, Hirano T, Bern HA, Grau EG (1997) Somatotropic actions of the homologous growth hormone and prolactins in the euryhaline teleost, the tilapia, *Oreochromis mossambicus*. Proc Natl Acad Sci USA 94(5): 2068–2072.

Shepherd BS, Drennon K, Johnson J, Nichols JW, Playle RC, Singer TD, Vijayan MM (2005) Salinity acclimation affects the somatotropic axis in rainbow trout. Am J Physiol 288(5): R1385–R1395.

Shikano T, Fujio Y (1998) Immunolocalization of Na^+, K^+-ATPase and morphological changes in two types of chloride cells in the gill epithelium during seawater and freshwater adaptation in a euryhaline teleost, *Poecilia reticulata*. J Exp Zool 281: 80–89.

Shiraishi K, Matsuda M, Mori T, Hirano T (1999) Changes in the expression of prolactin- and cortisol-receptor genes during early-life stages of euryhaline tilapia (*Oreochromis mossambicus*) in freshwater and seawater. Zool Sci 16: 139–146.

Sigrist S, Franco-Cereceda R, Muff R, Henke H, Lundberg JM, Fischer JA (1986) Specific receptor and cardiovascular effects of calcitonin-gene related peptide. Endocrinology 119: 381–390.

Skofitsch G, Jacobowitz DM (1985) Calcitonin gene-related peptide: detailed immunohistochemical distribution in the central nervous system. Peptides 6: 721–745.

Smith NF, Eddy FB, Struthers AD, Talbot C (1991) Renin, atrial natriuretic peptide and blood plasma ions in parr and smolts of Atlantic salmon, *Salmo salar* L. and rainbow trout *Oncorhynchus mykiss* (Walbaum) in freshwater and after short-term exposure to seawater. J Exp Biol 157: 63–74.

Suzuki H, Yamamoto T (2002) Centrifugal neurons of the octopus optic lobe cortex are immunopositive for calcitonin gene-related peptide. Neurosci Lett 324(1): 21–24.

Suzuki N, Eguchi C, Hirai T, Sasayama Y (1997) Nucleotide sequences of reptile calcitonins: their high homology to chicken calcitonin. Zool Sci 14: 833–836.

Suzuki N, Suzuki T, Kurokawa T (2000) Cloning of a calcitonin gene-related peptide receptor and a novel calcitonin receptor-like receptor from the gill of flounder, *Paralichthys olivaceus*. Gene 244(1–2): 81–88.

Suzuki N, Suzuki T, Kurokawa T (2001) Cloning of a calcitonin gene-related peptide from genomic DNA and its mRNA expression in flounder, *Paralichthys olivaceus*. Peptides 22(9): 1435–1438.

Suzuki N, Suzuki T, Kurokawa T (2002) Possible involvement of calcitonin gene-related peptide in seawater adaptation of flounder: expression analysis of its receptor mRNA in the gill. Fish Sci 68(2): 425–429.

Taché Y, Gunion M, Lauffenberger M, Goto Y (1984) Inhibition of gastric acid secretion by intracerebral injection of calcitonin gene-related peptide in rats. Life Sci 35: 871–878.

Takei Y (1993) Role of peptide hormones in fish osmoregulation. Fish Ecophysiology. Rankin JC and Jensen FB (eds). Chapman & Hall, London, pp. 136–160.

Takei Y (2000) Structural and functional evolution of the natriuretic peptide system in vertebrates. Int Rev Cytol 194: 1–66.

Takei Y, Hirose S (2002) The natriuretic peptide system in eels: a key endocrine system for euryhalinity? Am J Physiol 282: 940–951.

Takei Y, Kaiya H (1998) Antidiuretic effect of eel ANP infused at physiological doses in seawater-adapted eels, *Anguilla japonica*. Zool Sci 15: 399–404.

Takei Y, Takahashi A, Watanabe TX, Nakajima K, Sakakibara S, Takao T, Shimonishi Y (1990) Amino acid sequence and relative biological activity of a natriuretic peptide isolated from eel brain. Biochem Biophys Res Commun 170(2): 883–891.

Tippins JR (1986) CGRP: a novel neuropeptide from the calcitonin gene is the most potent vasodilator known. J Hypertens 4: S102–S105.

Trasforrini G, Margutti A, Portaluppi F, Menegatti M, Ambrosio MR, Bagni B, Pansini R, Uberti ECD (1991) Circadian profile of plasma calcitonin gene related peptide in healthy man. J Clin Endocrinol Metab 73: 945–951.

Tse DL, Tse MC, Chan CB, Deng L, Zhang WM, Lin HR, Cheng CH (2003) Seabream growth hormone receptor: molecular cloning and functional studies of the full-length cDNA, and tissue expression of two alternatively spliced forms. Biochim Biophys Acta 1625(1): 64–76.

Tse DLY, Chow BKC, Chan CB, Lee LTO, Cheng CHK (2000) Molecular cloning and expression studies of a prolactin receptor in goldfish (*Carassius auratus*). Life Sci 66: 593–605.

Utida S, Hirano T, Oide H, Ando M, Johnson DW, Bern HA (1972) Hormonal control of the intestine and urinary bladder in teleost osmoregulation. Gen Comp Endocrinol 3(Supplement 1): 317–327.

Van Rossum D, Hanisch UK, Quirion R (1997) Neuroanatomical localization, pharmacological, characterization and functions of CGRP related peptides and their receptors. Neurosci Biobehav Rev 21: 649–678.

Warne JM, Balment RJ (1995) Effect of acute manipulation of blood volume and osmolality on plasma [AVT] in seawater flounder. Am J Physiol 269: R1107–R1112.

Wendelaar Bonga SE (1997) The stress response in fish. Physiol Rev 77: 591–625.

Westenfelder C, Birch FM, Baranowski RL, Rosenfeld MJ, Shiozawa DK, Kablitz C (1988) Atrial natriuretic factor and salt adaptation in the teleost fish *Gila atraria*. Am J Physiol 255: F1281–F1286.

Wimalawansa SJ (1996) Calcitonin gene-related peptide and its receptors: molecular genetics, physiology, pathophysiology and therapeutic potentials. Endocrinol Rev 17: 533–584.

Wimalawansa SJ (1997) Amylin, calcitonin gene-related peptide, calcitonin, and adrenomedullin: a peptide superfamily. Crit Rev Neurobiol 11(2–3): 167–239.

Wimalawansa SJ, El-Kholy AA (1993) Comparative study of distribution and biochemical characterization of brain calcitonin gene-related peptide receptors in five different species. Neuroscience 54(2): 513–519.

Winston JH, He ZJ, Shenoy M, Xiao SY, Pasricha PJ (2005) Molecular and behavioral changes in nociception in a novel rat model of chronic pancreatitis for the study of pain. Pain 117: 214–222.

Xu B, Miao H, Zhang P, Li D (1997) Osmoregulatory actions of growth hormone in juvenile tilapia (*Oreochromis niloticus*). Fish Physiol Biochem 17: 295–301.

Yamauchi H, Orimo H, Yamauchi K, Takano K, Takahashi H (1978) Increased calcitonin levels during ovarian development in the eel, *Anguilla anguilla*. Gen Comp Endocrinol 36: 526–529.

Yanagawa N (1991) Angiotensin II and proximal tubule sodium transport. Ren Physiol Biochem 14(4–5): 208–215.

Yaoi Y, Suzuki M, Tomura H, Sasayama Y, Kikuyama S, Tanaka S (2003) Molecular cloning of otoconin-22 cDNA in the bullfrog endolymphatic sac: effect of calcitonin on otoconin-22 mRNA levels. Endocrinology 144: 3287–3296.

Yu LC, Weng XH, Wang JW, Lundeberg T (2003) Involvement of calcitonin gene-related peptide and its receptor in anti-nociception in the periaqueductal grey of rats. Neurosci Lett 349: 1–4.

Zhou B, Kelly SP, Ianowski JP, Wood CM (2003) Effects of cortisol and prolactin on Na^+ and Cl^- transport in cultured branchial epithelia from FW rainbow trout. Am J Physiol 285(6): R1305–R1316.

Chapter 8
Energy Requirements of European Eel for Trans Atlantic Spawning Migration

Guido van den Thillart, Arjan Palstra, and Vincent van Ginneken

8.1 Introduction

8.1.1 Migration

An important aspect of the reproduction of European silver eels is the huge distance they have to swim to reach their spawning grounds. After leaving the West European coast they still have to swim 5,000–6,000 km to the Sargasso Sea, the assumed spawning area. So, obviously long term swimming capacity is a major requirement for successful reproduction. Migrating eels don't feed; therefore they rely for their energy completely on fat stores (Tesch 2003), which can be as much as 30% of their body weight. Silver eels must swim across the Atlantic Ocean within 5–6 months, as this is the difference between the time they leave and the time the first larvae are observed in the Sargasso Sea. From the time needed to cross the ocean the minimal swimming speed of 0.4 m s^{-1} can be calculated. The long distance migration suggests two major questions: (1) Do they have enough energy reserves? (2) Are they built to swim long distances? To know whether they have enough energy left over for successful reproduction after arrival at the spawning site, it is important to know the energy consumption during long term swimming as well as the amount of the initial fat stores.

Long term swimming experiments were, to our knowledge, never carried out before with fishes. This requires the construction of special equipment, suitable to run continuously for at least several months; such as available at the Institute of Biology Leiden. Long term swimming may be a much heavier burden to animals than short term swimming, since under those conditions the experimental animals do not have the opportunity to recover. This may be a constant stress making them sensitive to otherwise harmless viral and bacterial infections. Thus far nothing was

---

G. van den Thillart, A. Palstra, and V. van Ginneken
Institute Biology Leiden, Leiden University, P.O. Box 9516, 2300 RA Leiden, The Netherlands

G. van den Thillart et al. (eds.), *Spawning Migration of the European Eel,*

known about the swimming and endurance capacity of eels. Swimming speeds, endurance capacity, and oxygen consumption rates have to be measured to answer the above questions. European eels migrate great distances to reach their spawning sites. As silver eels they leave the European west coast in the fall and are supposed to reach the Sargasso Sea after about 6 months in the spring (Tesch 2003). Although they leave for the spawning site to reproduce, they are still immature at that time. So, the gonads have to develop during or after their migration. Eels have much fat as energy stores, which are reserves for gonad development as well as for migration. For their long-distance migration to the Sargasso Sea the energy reserves may easily become critical particularly since the fat percentage varies largely (Svedäng and Wickström 1997a, b). An estimation of the energy required to cover 6,000-km was presented recently. Based on the oxygen consumption rates during a 10-day swim trial, the equivalent fat consumption extrapolated to 6,000-km was 120 g per kg or 40% of the initial fat reserve (Van Ginneken and Van den Thillart 2000). More extensive data were obtained from intermediate (1,000-km) to even long term (5,500-km) swim trials (Van den Thillart et al. 2004; Van Ginneken et al. 2005a), that showed the high endurance and low cost of swimming of the European eel.

Johannes Schmidt (1923) found the smallest eel larvae (leptocephali) of the European eel, *Anguilla anguilla* L., near the Sargasso Sea and the largest near the European coast. This is the only evidence to date that locates the spawning grounds in the Sargasso Sea (neither eggs nor mature adults have ever been found in this area). For Schmidt's theory to be supported, the following three conditions must be met: (a) Adult European eels must be able cover a distance of 6,000-km in a fasting state, implying that migrating eel must have sufficient energy reserves to cover this enormous distance (Tucker 1959); (b) Mature European eels and fertilized eggs must be found in the Sargasso Sea; and (c) Eel larvae must be shown to migrate towards the European coasts. On question (a) new information is available from our laboratory in Leiden, which will be elaborated in this chapter. As for question (c), the most recent observations on larval migration patterns were published by McCleave et al. (1987) indicating that the larvae are transported from the Sargasso Sea to the European coast. However, not resolved are questions concerning the large variation in age (Antunes and Tesch 1997) and the genetic make up (Wirth and Bernatchez 2001) of glass eels To test question (b), the group of Tesch (Post and Tesch 1982) tried, so far without success, to catch adult eels in the Sargasso. Until now there are only two reports of incidentally caught silver eels (*A. anguilla*) in the open Atlantic (Ernst 1977; Bast and Klinkhardt 1988).

Tucker (1959) raised severe doubts whether the European eel would be able to swim across the ocean and even suggested that all European eels are the offspring of the American eel. Tucker's "new solution to the Atlantic eel problem" provoked a long debate (D'Ancona and Tucker 1959; Deelder and Tucker 1960) and was finally rejected when a distinction could be made between the two Atlantic eel species based on allozymes (Williams and Koehn 1984), enzymes (Comparini and Rodino 1980), mitochondrial DNA (Avise et al. 1986, 1990; Tagliavini et al. 1995) and genomic DNA (Nieddu et al. 1998). However, it is still remarkable that the topic of energy requirements for long distance swimming of European eel was left open for so long, particularly since eel swimming was assumed by biomechanics to be very inefficient (Videler 1993). Remarkable indeed as eels swam across the Atlantic for millions of years since the opening of the Thetys Sea. Thus one would expect the opposite: an extreme strong selection pressure on swimming efficiency and capacity for long distance migration.

8.1.2 Environmental Factors

Parasites: Three former studies investigated the influence of the swim bladder parasite *A. crassus* on swimming of eels. Barni et al. (1985) found lower cruise speeds for infected eels. Sprengel and Lüchtenberg (1991) found a reduction of maximum swimming speed. Recently, Münderle et al. (2004) did not find any negative relation between swimming activity of eels and intensity of *A. crassus* infection, at least in the short-term. However, those studies were performed with elvers and not with silver eels. In fact for extrapolation to long distance migration to the Sargasso Sea only results on silver eels should be applied, as the other eel life stages are all sedentary. In addition oxygen consumption rates have not been measured in these studies, which makes it impossible to judge swimming efficiency and cost of transport of the infected eels. Our recent studies on the effect of *A. crassus* infection on swimming energetics in Leiden, revealed that infected eels and eels with damaged swim bladders (due to previous infections) swim at ~20% higher costs. Furthermore all infected eels in those studies were incapable of swimming longer that 1 month (Palstra et al. 2007a). So, it seems that *A. crassus* infection is a real threat for reproduction although the infection itself is not lethal (see Chapter 7).

Viruses: A new factor, that has not received much attention to date, is the worldwide occurrence of eel viruses (Van Ginneken et al. 2004, 2005c). Viruses are known to affect blood-forming tissues in fishes, and typically become virulent during stress (Wolf 1988). In salmon for example, Infectious Haematopoietic Necrosis Virus (IHNV) and Viral Hemorrhagic Septicemia Virus (VHSV), both rhabdoviruses, can affect haematopoietic tissues, leading to severe anaemia (Wolf 1988). For eels, long-term migration can certainly be considered as a major stressful event. Therefore, one may assume that an outbreak of a virus infection in infected individuals could take place during this journey (see Chapter 18).

PCBs: As eels have high fat levels (up to 30%), particularly when they undergo silvering, they accumulate lipophilic compounds from their environment. Eels have been even mentioned as suitable bio-indicators for the occurrence of PCBs (Van Leeuwen and Hermens 1995; Van Leeuwen et al. 2002). Of the different compounds the highest levels are always the PCBs (polychlor bi-phenyl), some of which are endocrine disruptors and some are interfering with gene-transcription factors (Pocar et al. 2003), and some are even direct toxic. The summated effect is in general expressed in TEQ (dioxin toxicity equivalent quantity), however it is rather unknown until what extent the effect holds for different species (species specific effects), and how they could interfere with the different processes of animals. Especially the effect on spawning migration could be complex as well as highly significant (see Chapter 17).

Trophic quality: It is well known that farmed female eels silver in 1–3 years, while in the wild it takes 8 to more than 20 years. This is likely related to growth rate, which in turn depends on temperature and food availability. There is not much information available on this item, but certainly we found in our studies younger female silver eels in high trophic areas like Lake Grevelingen and the Lek (lower branch of Rhine) than eels from the Loire. In the end of the growth phase, accumulation of high fat content is likely a trigger for silvering, however this question is thus far not resolved. We have only circumstantial evidence for it, like the higher fat content of silver eels vs. yellow eels.

8.2 Swimming Capacity

8.2.1 Swim Tunnel

The principle of the Blazka swim tunnel was explained in earlier publications (Blazka et al. 1960; Smith and Newcomb 1970; Van Dijk et al. 1993). The advantage of this swim tunnel is the compact design, which allowed us to construct 22 swimming tunnels in an air-conditioned room of 100 m^2. This type consists of two concentric tubes, with the inner circle and outer ring having the same surface area resulting in the same flow rate. The propeller pushes the water into the outer ring and further into a bundle of flow streamers to reduce the size of the vortices generating a semi-laminar flow.

Flow characteristics of swim tunnels of the Blazka-type were to our knowledge never described before. In a recent study we applied the very accurate Laser-Doppler system to demonstrate the homogeneity of the flow in the swim tunnels at three different cross sections (Van den Thillart et al. 2004). The actual flow in a 2-m long swimming tunnel was measured at three different cross-sections (11, 61, and 110 cm from the inflow) and at different distances from the wall (0.5, 1.0, 2.0, 4.0 and 9.5-cm). A linear relationship was observed between the number of revolutions per minute and the measured water velocity. The linearity existed up to 0.9 m s^{-1}, while the calibration allowed running the tunnels up to flows of 1.5 m s^{-1}.

Likely due to the semi-laminar flow type, the drop-off at the wall was very steep; at 2.0-cm the flow was about 70% and at 4.0-cm about 90% of the flow in the middle. So, fish with a width of >4.0-cm can not swim in the boundary layer. Furthermore eels need an even wider space because of the large amplitude of their tail beats. In addition, we observed that that the head of swimming eels remained between 5 and 10-cm from the wall (Fig. 8.1). Eels were rather easy to induce to swim, just placing them in streaming water is apparently sufficient to activate their swimming mode. The only crucial point is to avoid stress when forcing them to swim. Changing the flow rate up and down quietly during the first hour of starting up

Fig. 8.1 Swim tunnels for long distance migration studies. The set up consisted of 22 2-m long blazka-type swim tunnels. On the fore ground a tunnel with a 72-cm silver female eel swimming at 0.5 BL s^{-1}. The tunnels consist of two concentric perspex tubes of 2-m and two PVC endcaps. The propeller pushes water into the outer ring and 'sucks it' out from the inner tube. The cross section area of the inner tube and of the outer ring have the same surface area to obtain equal flow rates at both sides. The water is pushed through streamers with internal diameters of ~10-mm resulting in a semi laminar flow in the inner tube. Flow rates through different transects were found homogeneous as determined by Laser-Doppler tests

was almost always sufficient. Once they were swimming, eels could swim forever as long as the speed remained low (0.5–0.6 BL s^{-1}). Recently we published results from a 10 day swim trial with European Silver eel swimming at 0.5 BL s^{-1} in the same set up in sea water (Van Ginneken and Van den Thillart 2000).

To test the suitability of the set up for long distance migration, we have swum seven farmed silver eels for several months in fresh water (19°C) at 0.5 BL s^{-1}. Two eels stopped swimming during the first 2 months, the experiment was stopped after 3 months. Every day the oxygen consumption rate was measured during several hours by automatic control of the water inlet between 85% and 75% air saturation. Figure 8.2 shows the oxygen consumption data of the five eels swimming 95 days at 0.5 BL s^{-1}, corresponding to a distance of 2,850-km. These first long distance swimming results with European silver eels show their impressive endurance.

8.2.2 Swimming Fitness

The capacity of individual fish to swim can be tested in swim tunnels as described and used by many authors (e.g. Brett 1964) more or less in the same way as mammals can be tested on runways. The fish are usually pretrained to prevent interference by handling stress, followed by stepwise increase of swimming speed. At each speed the individual fish are usually left swimming for 1–2 h, as it takes about an

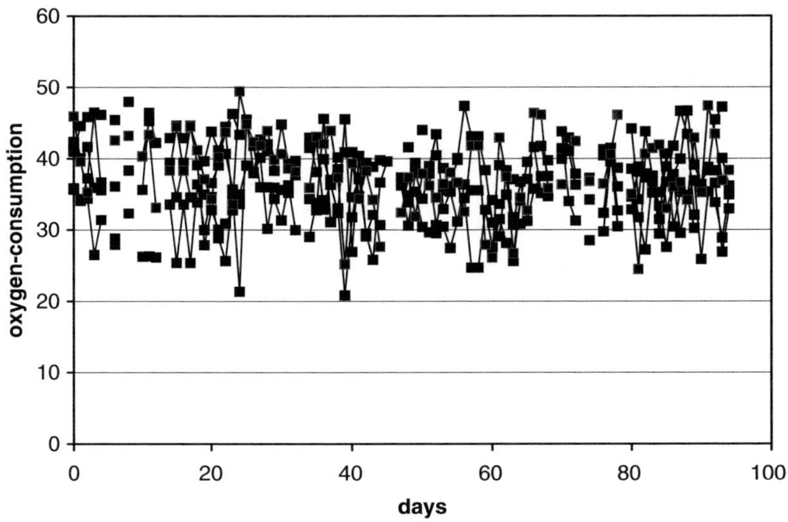

Fig. 8.2 The oxygen consumption (mg O_2.kg^{-1}.h^{-1}) rates of five adult silver eels. The eels (73 ± 21 g, 92 ± 64 cm) swam continuously at 0.5 BL s^{-1} (at 19°C) during a period of 95 days in the dark (red light invisible for eels). The oxygen consumption was measured every morning over 2 h. No changes were observed in swimming behavior (Van den Thillart et al. 2004. With permission from JFB)

hour to establish a new stable condition with respect to circulation and ventilation (Jones and Randall 1978). With each increase of speed the drag increases with the second power, and therefore the oxygen consumption increases also in the same way with speed. When the swimming speed approaches the maximal speed, anaerobic metabolism is activated, which ultimately results in high lactate levels and exhaustion of high-energy phosphates, leading to collapse. The collapse point is a reproducible parameter, as is the oxygen consumption rate at the collapse point. Likely the maximal oxygen consumption rate determines the maximal sustained swimming speed, which is the speed where the animal does not fatigue when swimming for a few hours. This is not necessarily the speed that the animal can sustain for much longer, as endurance is also determined by other factors.

Somewhere between resting and maximal oxygen consumption the optimal swimming speed can be found. This is the speed where the cost of transport is the lowest. The cost of transport (mg O_2 kg^{-1} km^{-1}) is by definition the ratio of oxygen consumption rate over swimming speed, which is high at low speed as well as at high speed. At low speed because the standard metabolic rate is a large part of energy consumption, which is not used for swimming, and at high speeds it is due to the drag that increases with second power of the speed. So, the lowest COT is found at an intermediate speed, which is than called the optimal swimming speed. Particularly for migrating animals the cost of transport is crucial as it determines the maximal distance that can be covered for the available energy. A low cost of transport is certainly a prerequisite for high endurance. Therefore maximal endurance should be expected at the optimal swimming speed. According to biomechanical criteria the best endurance is the maximal speed where the oxygen consumption rate remains stable and the mode of swimming does not change (Videler 1993). These two criteria are based on different assumptions, either lowest energy cost or stable swimming mode. It is likely though that both will come together at the same speed, as such would be selected out in the population over several generations. To determine maximal endurance conditions for eels, they have to be swum for extended periods at different swimming speeds. Those studies have been carried out recently in Leiden (Tables 8.1 and 8.2; Palstra et al. 2007c). Four different groups of eels were tested in a swimming fitness protocol over a range of 0.5 to 1.0 m s^{-1}, which corresponds roughly to 0.6 to 1.2 BL s^{-1}.

Swimming fitness can be described by a number of parameters such as maximal swimming speed, optimal swimming speed, and minimal cost of transport. These were recently determined for the European eel in a single day protocol, using the above described swim tunnel. In this study (Palstra et al. 2007c) eels were swum 2 h at each speed from 0.5 to 1.0 m s^{-1} in steps of 0.1 m s^{-1}. At each speed the oxygen consumption was measured continuously for 90 min. The maximal aerobic speed was interpolated according to the method of Brett (1964). A group of 40 farmed eels were tested twice (test 1 and test 2) with 2 h intervals, and in between at the same speeds with 12 h intervals (endurance). The endurance test lasted 6 days, so each

Table 8.1 Data of experimental female eels used for fitness tests. The tests included a stepwise increase in swimming speed starting at 0.5 BL s^{-1}, incrementing with 0.1 BL s^{-1} up to 1.0 m s^{1}. The eels swam for 2 h at each step until collapse or till last step at 1.0 m s^{-1}. Eels were placed in the swim tunnels ~24 h prior to swimming in the set up as described in Fig. 8.1. n = number experimental animals; y = yellow, s = silver; SW/FW: fresh or salt water (Data from Palstra et al. 2007a, b)

			Length (cm)		Weight (g)			
Origin	n	Status	Mean	SD	Mean	SD	Salinity	Temp (°C)
Farm	42	y	67.7	4.6	673	171	SW	18
Lake Balaton	81	y/s	66.4	6.3	464	146	FW	18
Lake Grevelingen	19	s	78.8	5.1	949	156	SW	18
Loire River	20	s	82.4	6.4	1,018	253	FW	18

Table 8.2 Fitness parameters of eels from different locations. Data and conditions of the eels are given in Table 8.1 (Data from Palstra et al. 2007a, b)

	Max speed		MO_2 (max)		MO_2 (rest)		Optimal speed		COT (min)	
	m s^{-1}		mg O_2 kg^{-1} h^{-1}		mg O_2 kg^{-1} h^{-1}		m s^{-1}		mg O_2 kg^{-1} km^{-1}	
Origin	Mean	SD	Mean	SD	Mean	SD	mean	SD	mean	SD
Farm	0.77	0.08	135	19	36	5	0.67	0.15	45	7
Lake Balaton	0.67	0.14	144	44	39	9	0.65	0.14	45	12
Lake Grevelingen	0.71	0.14	152	39	35	8	0.65	0.15	44	17
Loire River	0.66	0.18	141	35	43	13	0.61	0.05	44	11

MO_2 = oxygen consumption at rest or at maximal swimming speed; COT (min) = minimal cost of transport series took 8 days. Speed test 2 was a way to test whether the results of speed test 1 were repeatable. Results are summarized in Fig. 8.3 showing the relation between oxygen consumption and swimming speed for each of the three tests. The results show also that there were no statistical significant differences between the three tests; a 12 h run gave exactly the same result as the two 2 h speed tests. This implies that the 2 h speed test can be used for testing endurance as well. This corresponds with our observations; once the eels are swimming they don't change their mode of swimming nor their oxygen consumption at that speed. As these two remain the same then also the ratio remains the same, which is the cost of transport (mg O_2 h^{-1} kg^{-1}). When considering the COT at the different swimming speeds, we clearly see that the values remain almost the same, indicating that eels swim between 0.5 and 0.8 m s^{-1} at almost the same COT. As the results were the same for the 2 and 12 h speed tests of the farmed eels, three other eel groups were tested only in a 1 day speed test with 2 h intervals (Table 8.1).

Of all four groups the fitness parameters were calculated and listed in Table 8.2. The most remarkable result is that the COT (min) and the optimal swimming speed

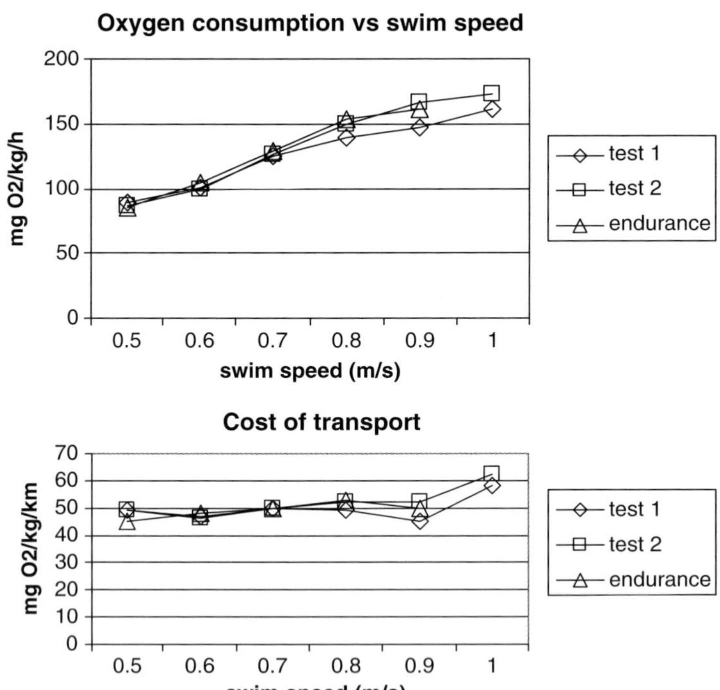

Fig. 8.3 Oxygen consumption (5 mg O_2 kg^{-1} h^{-1}) and cost of transport (mg O_2 kg^{-1} km^{-1}) of 40 farmed female eels (0.9 ± 0.1 kg) swimming at different swimming speeds. Eels were exposed to two swimming protocols: test 1 and test 2 with 2h intervals and the endurance test with 12 h intervals, each starting at 0.5 m s^{-1}. For the endurance test, eels were exposed each day to a different speed and rested overnight (Data modified after Palstra et al. 2007c)

values were almost the same for all four groups of female eels tested. The optimum swimming speeds, i.e. the speed where the oxygen consumption per km is the lowest, were found 0.61–0.67 m s^{-1}, which is ~60% higher than the generally assumed cruise speed of 0.4 m s^{-1}. This would imply that female eels may reach the Sargasso Sea within 3.5 months instead of 6 months.

8.2.3 Swim Bladder Parasites

Infection with the swim-bladder parasite, *Anguillicola crassus*, is suggested as one of the causes of the collapse of eel populations worldwide. This nematode has been introduced 20 to 30 years ago from Asia and in a short time parasitized various eel species in different geographical regions of the world. There are basically two kinds of adverse effects of *A. crassus* infection (Höglund et al. 1992): (1) energy drain due to sanguivorous activities of the parasites, and (2) mechanical damage of the swim-bladder wall. Concerning effect 1, Boon et al. (1989) found that the sanguivorous activities of the parasites decreased the number of circulating erythrocytes. So, highly infected eels will have a lower aerobic performance. Molnár et al. (1993) proved that in hypoxic water severely infected eels die first, while uninfected specimens endure the hypoxic condition for a long time. Concerning effect 2: The migratory activity of the larvae in the swim bladder wall and the direct invasion of the pre-adults and adults in blood vessels, result in extensive damage of the swim bladder wall (Molnár et al. 1993). Pathological changes include haemorrhages, formation of parasitic nodules, inflammatory cell proliferation, and hypertrophy of connective tissue, necrotic areas and oedema. These changes eventually cause substantial thickening of the swim bladder wall (Beregi et al. 1998) and shrinkage of the swim bladder. Thus cured swim bladders are mostly non-functional or even absent, resulting in negative buoyancy.

In a recent study (Palstra et al. 2007a) we attempted to estimate the effects of the *A. crassus* infection on swimming capacity and endurance. It was assumed that the parasite weight would impair the endurance mainly by energy drainage. On the other hand, damage of the swim-bladder function would impair buoyancy control. As the eels without buoyancy control will swim upwards, they must swim at a higher energy cost of transport. Most European eels are moderate to heavy infected by the swim bladder parasite.

We selected 80 eels, suffering from various degrees of infection, for a swimming fitness test. We found that infection and, even more, damage had serious effects on cruising ability and efficiency. Both infection and damage caused significantly higher O_2 consumption rates at all swimming speeds. Eels with damaged swim-bladders had a lower optimum swimming speed (U_{opt}). Infected eels have lower cruise speeds and higher cost of transport. Eels that were not infected, but with a damaged swim bladder from a previous infection, showed similar effects (Fig. 8.4). Almost half of those eels stopped swimming at low speeds <0.7 m s^{-1}. The effects thus seem to be associated with swim-bladder dysfunction and the resulting loss of neutral buoyancy.

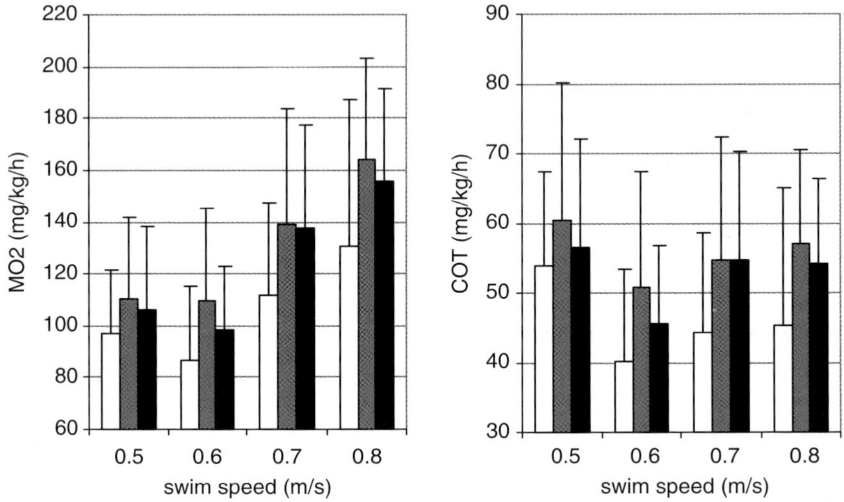

Fig. 8.4 The effect swim bladder parasite on swim performance of European eel. The oxygen consumption levels (MO_2) and cost of transport (COT) are presented for swim speeds 0.5, 0.6, 0.7, and 0.8 ($m\ s^{-1}$) of healthy eels (white bars), infected eels (grey bars) and eels with damaged swim bladders (black bars). MO_2 was higher (ANCOVA; $P < 0.01$) for infected and damaged eels at all swim speeds. The same applied for the COT which was higher for infected and damaged eels at all swim speeds. Hence, infection with swim bladder parasite impairs migration capacity (with permission; Palstra et al. 2007a)

Our results agree with Barni et al. (1985) who also found lower cruise speeds for infected eels. However, we did not find a significant reduction of the maximum swimming speed in contrast to Sprengel and Lüchtenberg (1991). Our results are also in contrast with Münderle et al. (2004), who did not find any negative effect of the infection on the swimming capacity of eels. However, their studies were performed with elvers and not with large female silver eels like in our study. Obviously only studies with silver eels should be used to extrapolate infection effects on spawning migration. We found that especially silver eels have significant higher infection levels, possibly related to a pre-silver state of increased food uptake. This leads to the conclusion that infected eels with damaged swim-bladders will likely fail to reach the spawning grounds.

8.3 Endurance Swimming

8.3.1 Simulated Migration

During long-distance migration, all animals are likely to maximize the distance covered per given fuel unit, thus they will try to migrate at the lowest cost of transport. The migration distance of the different eel species varies: the European eel (*A. anguilla*) 5,500-km

(Schmidt 1923), the American eel (*A. rostrata*) 4,000-km (Tucker 1959; McCleave et al. 1987); the Australian eel (*A. australis*) 5,000-km (Jellyman 1987) and Japanese eel (*A. japonica*) 4,000-km (Tsukamoto 1992). Thus all eel species migrate impressive distances, but European eels need to be the most efficient swimmers among eels.

The long-term swimming experiments with five eels of about 0.9-kg as discussed above (Fig. 8.2) indicate that eels can be stimulated to swim under laboratory conditions for a very long period without resting. Five out of seven eels were able to swim 3 months at 0.5 BL s^{-1}, covering a distance of 2,850-km (Van den Thillart et al. 2004). In the literature, limited data are available on swimming performance of eels or other anguilliform swimming teleost (Webb 1975; McCleave 1980). It is even suggested that the swimming movement of eel is less efficient than that of salmonids (Videler 1993; Bone et al. 1995). However, biomechanical efficiency of propulsion is different from overall swim efficiency. The latter is the total energy required by the animal to transport itself over a certain distance. This overall swim efficiency ultimately determines the amount of fuel stores i.e. grams of fat, that the eel needs to swim across the Atlantic Ocean.

Based on a 10 day swim trial with European silver eels we demonstrated previously that the energy cost of transport of those eels was extremely low: 0.137 cal g^{-1} km^{-1} (Van Ginneken and Van den Thillart 2000). This is 2.4–3.0 times lower than values reported in literature for other species (Schmidt-Nielsen 1972). In a more recent study we exposed female yellow eels of about 900-g to a 6 months swim trial at a mean swimming speed of 0.5 BL s^{-1} (Van Ginneken et al. 2005a). The eels swam in this experiment a complete simulated migration run of 5,500-km. Only a few eels fatigued during the journey and were unable to finish. The oxygen consumption rates were measured each day and shown in Fig. 8.5. The graphs demonstrate that the swimming eels have a twofold higher O_2 consumption than the resting eels, while over a period of almost 6 months the energy consumption remained almost constant. The slope shows an increase of about 10%, which is mainly due to weight loss. As the length of the animal did not change, the drag must have remained the same, so the energy per kg must increase correspondingly. Swimming eels had a higher weight loss than the resting eels (Table 8.3).

8.3.2 *Calorimetric Calculations*

From the oxygen consumption the energy consumption can be calculated based on fat combustion data given by Brafield and Llewellyn (1982). Thus we can estimate how much fat the eels would have used, assuming that fat is the only energy source, based on the oxicalorimetric data for fat combustion (Table 8.4). The values calculated for the two long term swim trials gave about the same oxygen consumption per km (28 vs. 31 mg kg^{-1} km^{-1}). This results in a cost of transport of about 0.4 kJ km^{-1} and 10 mg fat km^{-1}. As for a trans Atlantic journey of 6,000-km, this would then require about 60 g fat for an eel of 1-kg. The variance between individuals was 7% in experiment I and 14% in experiment II; meaning that the fat consumption ranged between 55 and 74 g fat kg^{-1} 6,000 km^{-1}. The calculated values for the cost

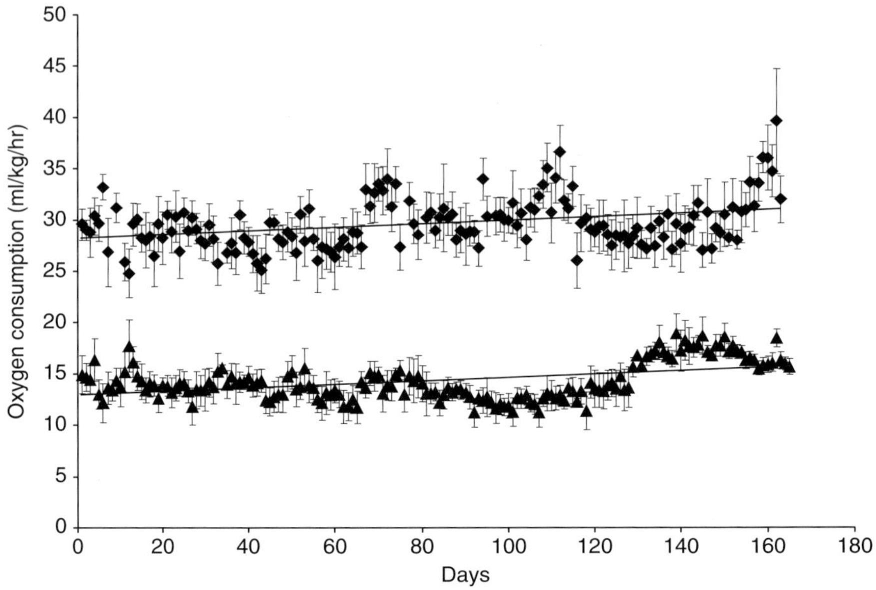

Fig. 8.5 Simulated migration of female eels (860 ± 82 g, 73.1 ± 3.8 cm). Oxygen consumption (ml O_2 kg^{-1} h^{-1}) is presented for a period of 6 months of female farmed eels swimming at 0.5 BL s^{-1} or resting. The swim group (n = 9) covered 5,533 ± 354 km in 173 days (at 18°C). The rest group (n = 6) remained at the same time under furthermore similar conditions. Regression lines: Rest-group: Y = 0.0326X + 25.294; Swim-group: Y = 0.0394X + 54.86. Diamonds: swimming; Circles: resting. (Van Ginneken et al. 2005a. With permission)

Table 8.3 Simulated migration of female eels. Energy consumption is presented of female yellow eels after a period of 6 months swimming at 0.5 BL s^{-1} or at rest. The swim group covered 5,533 ± 354 km at in 173 days (18°C). The rest group remained at the same time in tanks without swimming under furthermore similar conditions. Weight los, oxygen consumption, and energy consumption were significant difference (P < 0.01) between swim and rest groups (After Van Ginneken et al. 2005b)

	Swim group (n = 9)		Rest group (n = 6)	
Parameters	Mean	SD	Mean	SD
Length (cm)	74.7	3.4	70.6	3.6
Weight start (g)	915.0	58.0	795.0	72.0
Weight end (g)	734.0	45.0	691.0	71.0
Weight loss (g)	180.0	38.0	104.0	26.0
Oxygen consumption (mg kg^{-1} h^{-1})	30.0	4.2	14.3	1.8
Energy consumption (kJ kg^{-1} h^{-1})	2,317.0	221.0	1,123.0	108.0
Cost of transport (kJ kg^{-1} km^{-1})	0.42		–	

of transport (COT) correspond rather well for the two experiments. However, the COT values are some 4–5 fold lower than those obtained for salmonids (Schmidt-Nielsen 1972). If eels would swim at the same energy consumption rate as salmonids, they would need 300-g fat kg^{-1} for crossing the ocean instead of 60-g of fat per kg. Long term swim experiments with eels have been published for eels swimming

Table 8.4 Oxygen consumption of female European silver eels during simulated migration converted to fat use. Data were calculated based on the oxicaloric equivalent of fat (13.72 kJ gO_2^{-1} and 39.5 kJ g^{-1} fat; Brafield and Llewellyn 1982). Eels swam at 0.5 BL s^{-1}; length and weight (0.91 ± 0.06 kg, 0.73 m in exp I; 0.92 ± 0.06 kg, 74.7 ± 3.4 cm in exp II); data for experiment I are from Van den Thillart et al. (2004) and for experiment II from Van Ginneken et al. (2005b)

Experiment I		Three months		Distance 2,980 km	
Mean VO2	speed	COT	COT	Fat use	
mg O_2 kg^{-1} h^{-1}	km d^{-1}	mg O_2 km^{-1}	J km^{-1}	mg km^{-1}	g 6,000 km^{-1}
36.90	31.41	28.20	386.88	9.79	58.77

Experiment II		6 months		Distance 5,580 km	
Mean VO2	speed	COT	COT	Fat use	
mg O_2 kg^{-1} h^{-1}	km d^{-1}	mg O_2 km^{-1}	J km^{-1}	mg km^{-1}	g 6,000 km^{-1}
42.26	32.27	31.43	431.21	10.92	65.50

Table 8.5 Body composition of eels (% of dry weight) at the start, and after 6 months either at rest or swimming at 0.5 BL s^{-1} (After Van Ginneken et al. 2005b)

	Start (n = 15)		Swim (n = 9)		Rest (n = 15)	
	Mean	SD	Mean	SD	Mean	SD
Fat	67.9	1.9	68.1	2.5	68.1	2.2
Protein	28.2	1.8	28.3	2.2	28.0	2.2
Carbohydrate	0.9	0.4	0.6	0.5	0.9	0.5
Ash	3.0	0.6	3.0	0.6	3.0	0.6
Sum	100.0		100.0		100.0	
Dry matter %	49.6	2.4	50.3	2.9	50.7	2.2

at 0.5 BL s^{-1} for 3 months and for 5.5 months (Van den Thillart et al. 2004; Van Ginneken et al. 2005a).

During long term swimming the eels lost weight, which can be due to diminished energy stores, but also due to water loss. Therefore the total body composition was determined of the eels at the end of the experiment II. The body composition of the three eel groups in experiment II (control, resting, and swimming) remained the same in all three conditions, which is remarkable (Table 8.5). The weight fractions of water, fat, protein, carbohydrate, and ash remained constant; the fat and protein content of the dry weight were 68% and 28% respectively. This implies in the first place that the buoyancy of the animal did not change and that therefore no volume compensation is required by the swim bladder during swimming. Another implication is that the energy consumption calculated from oxygen consumption and weight loss cannot be based on fat alone, but needs to be compensated for protein oxidation as well. Thus weight loss for the resting and swimming eels can be converted to energy consumption.

On the other hand it is possible to calculate from the oxygen consumption data how much weight should have been lost. Therefore we need to recalculate the energy content of the dry weight according to the composition given in Table 8.5. The recalculated data are presented in Table 8.6. The energy content of the dry

Table 8.6 Calculation of the calorimetric value of eel whole body based on total body composition (Combustion energies of fat, protein, and carbohydrates are from Brafield and Llewellyn 1982)

		Combustion energy of eel			
		$kJ\ g^{-1}$	$kJ\ g^{-1}$	$kJ\ gO_2^{-1}$	$kJ\ gO_2^{-1}$
	% Dry weight	fuel	mix	fuel	mix
Fat	68.0	39.5	26.9	13.72	9.6
Protein	28.2	23.6	6.7	13.36	3.9
Carbohydrate	0.8	17.2	0.1	14.76	0.1
Ash	3.0	0.0	0.0	0.0	0.0
Sum	100.0		33.7		13.6

weight was found to be 33.7 kJ g^{-1}, which is only 15% lower than that of fat combustion (39.5 kJ g^{-1}), indicating that fat is by far the main energy source.

From the oxygen consumption per hour and the swimming speed (1.34 km h^{-1}), the cost of transport was found: 31.43 mg O_2 km^{-1} (Table 8.4). From this the total oxygen consumed over 6,000 km must have been 188.58 g O_2, which corresponds to 188.58 × 13.6 = 2,569.2 kJ. When dividing by 33.7 kJ g dry weight^{-1}, we find an estimate for the dry weight loss of 76.3 g, or 152.6 g wet weight. The observed wet weight loss over 5,533 km was 180.2 ± 38.2 g (Van Ginneken et al. 2005a). The calculated figure based on oxygen consumption is 84% of the observed value. Considering the errors occurring in the different measurements we can conclude that the respirometry data corroborate well with the calorimetric data.

Our new calculations based on 33.7 kJ g dry weight^{-1} and 13.6 kJ gO_2^{-1} (see Table 8.6) provide an estimation of the total energy consumption over 6,000-km of 2,569.2 kJ based on oxygen consumption and 3,294.5 kJ based on weight loss. The resultant COT for both methods is respectively 0.43 and 0.55 kJ km^{-1}. A break down of the fuel usage for a simulated migration over 6,000 km provides the following results: fat 52.1 g, protein 21.6 g, and carbohydrate 0.6 g. Thus the minimal fat requirement for migration is about 5.2% of total body weight. This value is lower than the first estimate of 6.5% (Table 8.4), which is due to the fact that also proteins are being used as fuel for energy generation. It is possible that the energy requirement gets even lower as migration possibly takes place at lower temperatures. When eels migrate at a depth of 500–1,000 m, the temperature is between 10°C and 4°C, which will certainly reduce the standard metabolic rate. On the other hand our calculations regard female silver eels of about 0.74 m and 0.9 kg. Values for male eels will be much higher as they are smaller and have to swim at higher speeds.

Recent data on salmon swimming confirmed earlier results from Schmidt-Nielsen (1972) that salmonids swim at much higher cost of transport. Lee et al. (2003) swam different salmon species at 0.5–2.2 BL s^{-1} (12–16°C), the salmon were about 64 cm (fork length) and 2.65 kg. At 0.5 BL s^{-1} the COT was 150–200 mg kg^{-1} km^{-1} falling to 100 mg kg^{-1} km^{-1} at 1.2 BL s^{-1}. These data are between four and seven times higher than those observed with female eels of even a smaller size, which clearly indicate again that eels are very efficient swimmers.

8.3.3 Hydrodynamic Performance

There are various levels of energy conversion in a swimming animal. The overall metabolic efficiency is made up of the efficiencies of various processes, e.g. propeller efficiency (how much momentum is gained by the animal and wasted in the wake) and muscle efficiency (how many ATP molecules are used for the muscle contraction). To a fish can alter its propeller efficiency by changing its structural design and its motion pattern. Both carangiform and anguilliform swimmers undulate their body, the former with a narrower amplitude envelope than the latter. How the shape of the body undulations affects locomotory efficiency has been estimated using analytical approximations. Lighthill's elongated body theory (EBT) concludes that efficient swimmers should undulate only the most posterior section of their body – in the ideal case only their trailing edge – to maximize propeller efficiency (Lighthill 1971, Tytell and Lauder 2004). Daniel's (1991) predictions differ in part: propeller efficiency of undulatory swimming decreases linearly as the rearward speed of the body wave increases relative to the swimming speed, and it is independent of the frequency and the amplitude of the body wave. Given that the swimming kinematics of trout and eel mainly differ in the amplitude envelope of their body wave, but have a similar range of body wave speeds (Videler 1993), it is unlikely that kinematical differences between trout and eel can explain the difference in their overall metabolic efficiency.

The combined effect of propeller shape and motion on performance can be studied by visualizing the flow generated by anguilliform and carangiform swimmers. The ratio of forward to total momentum of the entire wake provides the mean propeller efficiency over a complete tail beat. This approach, whether using experimental or computational flow fields, requires the quantification of the three-dimensional flow in the complete wake, which so far has not been done. The currently available two-dimensional slices through the wake suggest that eels generate considerable lateral momentum, which does not contribute to the forward motion and therefore reduces efficiency (Müller et al. 2001; Tytell and Lauder 2004). Tytell estimated a hydrodynamic efficiency of 0.5 to possibly up to 0.87 (Tytell and Lauder 2004). Equivalent estimates for carangiform fish are reported in the range from 0.74 to 0.97 (Drucker and Lauder 2001; Müller et al. 2001; Nauen and Lauder 2002). These values suggest that trout has a higher propeller efficiency than eel, which does not explain the higher overall metabolic efficiency of eels. Efficiency is also inversely related to thrust (Lighthill 1971; Daniel 1991). However, a 25% difference in swimming speed is insufficient to explain a fourfold difference in efficiency. So, the currently existing evidence on the hydrodynamics of undulatory swimming contradicts rather than explains the high swimming efficiency of eels.

8.3.4 Muscle Performance

The efficiency with which a muscle converts chemical energy into mechanical work is important in prolonged aerobic locomotion, such as migration. Cruising is characterized by cyclic contractions at a well-defined frequency. Swimming speed depends linearly on tail beat frequency, and tail beat frequency corresponds to contraction frequency. The mechanical efficiency of muscle contractions depends on contraction speed in a non-linear fashion. This relationship can be predicted from Hill's model of muscle contractions (McMahon 1984) and has also been documented in fish swimming muscles (Curtin and Woledge 1993). There is a narrow range of contraction frequencies over which efficiency remains high. At contraction frequencies above and below this range, efficiency drops off progressively (McMahon 1984; Curtin and Woledge 1993). McMahon (1984) calculations show that maximum efficiency occurs at a contraction speed at 13% of the maximum contraction speed of the muscle, which is at a slightly lower speed than the speed at maximum power. To swim at maximum muscle efficiency, the fish should maintain a tail beat frequency that allows the muscle to contract at this optimal speed. If we take the contraction frequency that maximizes power as a first approximation of the contraction speed that maximizes efficiency, we can compare eel aerobic swimming muscles to those of trout. Eel muscles deliver peak power at much lower contraction frequencies (0.5 to 0.8 Hz in silver eel; measured at 14°C; Ellerby et al. 2001) than the muscles of trout (2 to 3 Hz, measured at 11°C) The swimming speeds that correspond to these contraction frequencies are 0.5 BL s^{-1} for eel and 0.4 to 1.0 BL s^{-1} for trout (Webb 1971a, b). These values confirm that in our experiments both eel and trout were swimming close to their optimal swimming speed, and hence the much higher COT of trout is probably not due to the trout having been forced to swim under considerably suboptimal conditions for its swimming muscles.

At the low speeds used in this study, the eels will recruit only the posterior red muscle to swim continuously. As demonstrated in the work of Gillis (1998) muscle fiber type recruitment was clearly dependent upon swimming speed. A pattern of 'posterior-to-anterior' recruitment within a fibre type was observed as eels increased their swimming speed (Gillis 1998). For example, eels typically used mainly posteriorly located red muscle to power slow-speed swimming, but would then additionally recruit more anteriorly located red muscle to swim at the higher speeds (Gillis 1998). This unusual muscle activation pattern and kinematics may explain the low COT in eels compared with trout, in which most of the red muscle on each side of the body is stimulated during a tail-beat cycle – assuming that the European and American eels are similar in this regard. In contradiction to this theory/hypothesis of Gillis (1998) to explain the low swimming efficiency of eel by recruitment patterns of muscle, Wardle et al. (1995) have shown that the muscle activity pattern (% time active during one tail beat cycle) does not differ substantially between different undulatory swimmers. Wardle's values for eel agree with those mentioned by Gillis (1998). Compared with trout and other fish, recruitment in eel is certainly not less by a factor of 2 to 4. Hence it is not likely that more posterior muscle recruitment in eel can explain the many-fold difference in efficiency between eel and trout.

8.3.5 Metabolism

Overall metabolic efficiency is also influenced by the efficiency of the respiration and energy-conversion processes. The whole-organism locomotory performance is determined by its metabolic machinery, bringing us to the whole-body oxygen consumption (Routine Metabolic Rate) of the animal. In this study, we found a RMR of 29.55 ± 4.2 ml O_2 kg^{-1} h^{-1}, which corresponds to 42.21 ± 6.0 mg O_2 kg^{-1} h^{-1}. This value is similar to values reported in literature: 35 mg kg^{-1} h^{-1} (for the same-size animals at 18°C, Degani et al. 1989; McKenzie et al. 2000) for eel and also the routine metabolic rate (RMR) measurements of other fish species (Winberg 1956). Hence, we may conclude that, based on metabolic rate comparisons with other fish species, the mitochondrial capacity remains the same. However, in the wild, eel do not migrate at the surface but in the deep sea: a migrating eel has been photographed at the Bahamas at a depth of 2,000 m (Robins et al. 1979). There, they experience considerably greater pressures that might further increase metabolic efficiency at the mitochondrial level by increasing the efficiency of their oxidative phosphorylation (Theron et al. 2000). In a laboratory study, exposing eels for 21 days at a hydrostatic pressure of 10.1 MPa, Theron et al. (2000) demonstrated that the ADP/O ratios, calculated from mitochondrial respiration measurements, were significantly increased. Eels actually performing the migration will not only experience higher pressures, but also lower temperatures, which will also affect their efficiency. Furthermore, eels might adapt their migratory route to take advantage of favourable sea currents, which would further reduce the energy requirements. However, with the migratory routes unknown, nothing can be said about the possible energy savings from pressure, temperature and sea-current effects.

8.4 Conclusions

Our respiratory measurements as well as the carcass analyses suggest that eels have a much higher swimming efficiency than trout. In the eel, the COT values obtained from oxygen consumption data and carcass analyses are 0.42 and 0.62 kJ kg^{-1} km^{-1}, respectively, whereas trout has a much higher COT value of around 2.7 kJ kg^{-1} km^{-1}. The COT in trout matches the value measured by Webb (1971b), and is similar to other salmonids (Brett 1973) and many adult fish species (Videler 1993). This means that eel swim four to six times more efficiently than other fish species, even across swimming styles. European eel is able to swim 5,500-km, a distance corresponding to their supposed spawning area in the Sargasso Sea at a remarkably low energy costs. The amount of energy for swimming 5,500-km corresponds to

60-g fat. Remarkably, as eels consume proteins and fat in the same ratio as present as in their body the energy losses have no effect on buoyancy. So, we can conclude that healthy well fed eels are able to reach the Sargasso leaving enough reserves for reproduction.

Swimming fitness and the ability to migrate are negatively influenced by infections with *A. crassus*. High infection levels result in a damaged swim bladder; both the parasite weight as well as the damaged swim bladder interferes with swimming performance. Both swim-bladders with high parasite weight causes as well as damaged swim bladders have reduced volume causing negative buoyancy. Thus in both cases the chance to reach the spawning site is very low.

Speed tests with four groups of eels showed that the swimming performance between the groups were rather similar. There were no differences between 2 and 12 h speed tests, suggesting that eels swim at the same low efficiency from the start. COT values changed little with speed; the lowest COT over 0.5–1.0 m s^{-1} was found at 0.65 m s^{-1}, a swimming speed 60% higher than the minimal cruising speed. At that speed (0.8 BL s^{-1}) female silver eels can reach the spawning site within 3.5 months, leaving ample time for final maturation and finding mates.

Acknowledgements Funded by grants from EU (EELREP #Q5RS-2001-01836), and STW – Dutch technology foundation (LBI66.4199).

References

Antunes C, Tesch FW (1997) Eel larvae (*Anguilla anguilla* L.) caught by RV "Heincke" at the European continent slope in autumn 1991. Ecol Fresh Fish 6, 50–52.

Avise JC, Helfman GS, Saunders NC, Hales LS (1986) Mitochondrial DNA differentiation in North Atlantic eels: Population genetics consequences of an unusual life history pattern. Proc Natl Acad Sci USA 83, 4350–4354.

Avise JC, Nelson WS, Arnold J, Koehn RK, Williams GC, Thorsteinsson V (1990) The evolutionary genetic status of Icelandic eels. Evolution 44, 1254–1262.

Barni S, Bernocchi G, Gerzeli G (1985) Morphohistochemical changes during the life cycle of the European eel. Tissue Cell 17: 97–109.

Bast HD, Klinkhardt MB (1988) Fang eines Silberaales (*Anguilla anguilla* (L. 1758)) im Iberischen Becken (Nordostatlantik) (Teleostei: *Anguillidae*). Zool Anz 221, 386–398.

Beregi A, Molnár K, Békési L, Székely CS (1998) Radiodiagnostic method for studying swimbladder inflammation caused by *Anguillicola crassus* (Nematoda: Dracunculoidea). Dis Aquat Organ 34, 155–160.

Blazka P, Volf M, Ceplea M (1960) A new type of respirometer for determination of the metabolism of fish in an active state. Physiol Bohemoslov 9, 553–560.

Bone Q, Marshall NB, Blaxter JH (1995) Biology of Fishes. 2nd ed, Chapman & Hall, London. ISBN 0-7514-022. The eels *Anguilla* and *Histiobranchus*, photographed on the floor of the deep Atlantic in the Bahamas. Bull Mar Sci 29, 401–405.

Boon JH, Lokin CJA, Ceusters R, Ollevier F (1989) Some properties of the blood of European eel (*Anguilla anguilla*) and the possible relationship with *Anguillicola crassus* infestations. Aquaculture 76, 203–208.

Brafield AE, Llewellyn MJ (1982) Animal Energetics. Blackie, Glasgow.

Brett JR (1964) The respiratory metabolism and swimming performance of young sokeye salmon. J Fish Res Board Can 21, 1183–1226.

Brett JR (1973). Energy expenditure of sockeye salmon, *Oncorhynchus nerka*, during sustained performance. *J Fish Res Bd Can* 30, 1799–1809.

Comparini A, Rodino E (1980) Electrophoretic evidence for two species of *Anguilla* leptocephali in the Sargasso Sea. Nature 287, 435–437.

Curtin NA, Woledge R (1993) Efficiency of energy conversion during sinusoidal movement of red muscle fibres from the dogfish *Scyliorhinus canicula*. J Exp Biol 185, 195–206.

D'Ancona U, Tucker DW (1959) Old and new solutions to the eel problem. Nature 183, 1405–1406.

Daniel TL (1991) Efficiency in aquatic locomotion: Limitations from single cells to animals. In: Blake RW (ed) Efficiency and Economy in Animal Physiology. Cambridge University Press, Cambridge, pp. 83–95.

Deelder CL, Tucker DW (1960) The Atlantic eel problem. Nature 185, 589–592.

Degani G, Gallagher ML, Metzler A (1989) The influence of body size and temperature on oxygen consumption of the European eel, *Anguilla anguilla*. J Fish Biol 34, 19–24.

Drucker EG, Lauder GV (2001) Locomotor function of the dorsal fin in teleost fishes: Experimental analysis of wake forces in sunfish. J Exp Biol 204, 2943–2958.

Ellerby DJ, Spierts ILY, Altringham JD (2001) Slow muscle power output of yellow-and silver-phase European eels (Anguilla anguilla L.): Changes in muscle performance prior to migration. J Exp Biol 204, 1369–1379.

Ernst P (1977) Catch of an eel (*Anguilla anguilla*) northeast of the Faroe Islands. Ann Biol 32, 175.

Gillis GB (1998) Neuromuscular control of anguilliform locomotion: Patterns of red and white muscle activity during swimming in the American eel *Anguilla rostrata*. J Exp Biol 201, 3245–3256.

Höglund J, Andersson J, Härdig J (1992) Haematological responses in the European eel, *Anguilla anguilla* L., to sublethal infestation by *Anguillicola crassus* in a thermal effluent of the Swedish Baltic. J Fish Dis 15, 507–514.

Jellyman D (1987) Review of the marine life history of Austral-Asian temperate species of Anguilla. Am Fish Soc Symp 1, 276–285.

Jones DR, Randall DJ (1978) The respiratory and circulatory systems during exercise. In: Hoar WS, Randall DJ (eds) Fish Physiology VII: Locomotion. Academic Press, London, pp. 425–502.

Lee CG, Farrell AP, Lotto A, MacNutt MJ, Hinch SG, Healey MC (2003a) The effect of temperature on swimming performance and oxygen consumption in adult sockeye (Oncorhynchus nerka) and coho (O. kisutch) salmon stocks. J Exp Biol 206, 3239–3251.

Lighthill MJ (1971) Large-amplitude elongated-body theory of fish locomotion. Proc R Soc Lond A 179, 125–138.

McCleave JD (1980) Swimming performance of European eel (Anguilla anguilla L.) elvers. J Fish Biol 16, 445–452.

McCleave JD, Kleckner RC, Castonguay M (1987) Reproductive sympatry of American en European eels and implications for migration and taxonomy. Am Fish Soc Symp 1, 286–297.

McKenzie DJ, Piraccini G, Piccolella M, Steffensen JF, Bolis CL, Taylor EW (2000) Effects of dietary fatty acid composition on metabolic rate and responses to hypoxia in the European eel (Anguilla anguilla). Fish Physiol Biochem 22, 281–296.

McMahon TA (1984) Muscles, Reflexes and Locomotion. Princeton University Press, Princeton, NJ.

Molnár K, Baska F, Csaba GY, Glávits R, Székely CS (1993) Pathological and histopathological studies of the swimbladder of eels Anguilla anguilla infected by Anguillicola crassus (Nematoda: Dracunculoidea). Dis Aquat Organ 15, 41–50.

Müller UK, Smit J, Stamhuis EJ, Videler JJ (2001) How the body contributes to the wake in undulatory fish swimming: Flow fields of a swimming eel (Anguilla anguilla). J Exp Biol 204, 2751–2762.

Münderle M, Sures B, Taraschewski H (2004) Influence of Anguillicola crassus (Nematoda) and Ichthyophthirius multifiliis (Ciliophora) on swimming activity of European eel Anguilla anguilla. DAO 60, 133–139.

Nauen JC, Lauder GV (2002) Hydrodynamics of caudal fin locomotion by chub mackerel, Scomber japonicus (Scombridae). J Exp Biol 205, 1709–1724.

Nieddu M, Pichiri G, Coni P, Salvadori S, Deiana AM, Mezzanotte R (1998) A comparative analysis of European and American eel (Anguilla anguilla and Anguilla rostrata) genomic DNA: 5S rDNA polymorphism permits the distinction between the two populations. Genome 41, 728–732.

Palstra AP, Heppener DFM, Van Ginneken VJT, Székely C, Van den Thillart GEEJM (2007a) Swimming performance of silver eels is severely impaired by the swim-bladder parasite Anguillicola crassus. J Exp Mar Biol Ecol, 352, 244–256.

Palstra A, Curiel D, Fekkes M, De Bakker M, Székely C, Van Ginneken V, Van den Thillart G (2007b) Swimming stimulates oocyte development in European eel (Anguilla anguilla L.). Aquaculture 270, 321–332.

Pocar P, Brevini TA, Fischer B, Gandolfi F (2003) The impact of endocrine disruptors on oocyte competetence. Reproduction 125, 313–325.

Post A, Tesch FW (1982) Midwater trawl catches of adolescent and adult anguilliform fishes during the Sargasso Sea eel expedition 1979. Helgoländer Meeresun 35, 341–356.

Robins CR, Cohen DM, Robins CH (1979) The eels Anguilla and Histiobranchus, photographed on the floor of the deep Atlantic in the Bahamas. Bull Mar Sci 29, 401–405.

Schmidt J (1923) Breeding places and migration of the eel. Nature 111: 51–54.

Schmidt-Nielsen K (1972) Locomotion: Energy cost of swimming, flying and running. Science 177, 222–228.

Smith LS, Newcomb TW (1970) A modified version of the Blazka respirometer and exercise chamber for large fish. J Fish Res Board Can 27, 1321–1324.

Sprengel G, Lüchtenberg H (1991) Infection by endoparasites reduces maximum swimming speed of European smelt Osmerus eperlanus and European eel Anguilla anguila. Dis Aquat Organ 11, 31–35.

Svedang H, Wickström H (1997a) Maturation patters in female European eel: Age and size at the silver eel stage. J Fish Biol 48, 342–351.

Svedang H, Wickstrm H (1997b) Low fat contents in female silver eels: Indications of insufficient energetic stores for migration and gonadal development. J Fish Biol 50, 475–486.

Tagliavini J, Harrison IJ, Gandolfi G (1995) Discrimination between Anguilla anguilla and Anguilla rostrata by polymerase chain reaction-restriction fragment length polymorphism analysis. J Fish Biol 47, 741–743.

Tesch WW (2003) "The eel", 5-th edition. JE Thorpe (ed), Blackwell, Oxford, pp 408.

Theron M, Guerrero F, Sébert P (2000) Improvement in the efficiency of oxidative phosphorylation in the freshwater eel acclimated to 10.1 MPa hydrostatic pressure. J Exp Biol 203, 3019–3023.

Tsukamoto K (1992) Discovery of the spawning area for Japanese eel. Nature 356, 789–791.

Tucker DW (1959) A new solution to the Atlantic eel problem. Nature 183, 495–501.

Tytell ED, Lauder GV (2004) The hydrodynamics of eel swimming. I. Wake structure. J Exp Biol 207, 1825–1841.

Van den Thillart G, van Ginneken V, Körner F, Heijmans R, van der Linden R, Gluvers A (2004) Endurance swimming of European eel. J Fish Biol 65, 1–7.

Van Dijk PL, Van den Thillart G, Balm P, Wendelaar Bonga S (1993) The influence of gradual water acidification on the acid/base status and plasma hormone levels in the carp. J Fish Biol 42, 661–671.

Van Ginneken V, Van den Thillart G (2000) Eel fat stores are enough to reach the Sargasso. Nature 403, 156–157.

Van Ginneken V, Haenen O, Coldenhoff K, Willemze R, Antonissen E, Van Tulden P, Dijkstra S, Wagenaar F, Van den Thillart G (2004) Presence of eel viruses in eel species from various geographic regions. Bull Eur Ass Fish Pathol 24(5), 268–271.

Van Ginneken V, Vianen G, Muusze B, Palstra AP, Verschoor L, Lugten O, Onderwater M, Van Schie S, Niemansteverdriet P, Van Heeswijk R, Eding E, Van den Thillart G (2005a) Gonadal development and spawning behavior of artificially-matured European eel (Anguilla anguilla L.). Anim Biol 55, 203–218.

Van Ginneken V, Antonissen E, Muller UK, Booms R, Eding E, Verreth J, Van den Thillart G (2005b) Eel migration to the sargasso: Remarkably high swimming efficiency and low energy costs. J Exp Biol 208, 1329–1335.

Van Ginneken V, Ballieux B, Willemze R, Coldenhoff K, Lentjes E, Antonissen E, Haenen O, Van den Thillart G (2005c) Hematology patterns of migrating European eels and the role of EVEX virus. Comp Biochem Physiol C 140, 97–102.

Van Leeuwen C, Hermens J (1995) Risk Assessment of Chemicals: An introduction. Kluwer, Dordrecht, The Netherlands.

Van Leeuwen SPJ, Traag WA, Hoogenboom LAP, De Boer J (2002) Dioxins, furans and dioxin-like PCBs in wild, farmed, imported and smoked eel from the Netherlands. RIVO Rapport C034/02.

Videler JJ (1993) Fish Swimming. Chapman & Hall, London, Fish and Fisheries Series 10, 260 pp.

Wardle CW, Videler JJ, Altringham JD (1995) Tuning in to fish swimming waves: Body form, swimming mode and muscle function. J Exp Biol 198, 1629–1636.

Webb PW (1971a) The swimming energetics of trout. 1. Thrust and power at cruising speeds. J Exp Biol 55, 489–520.

Webb PW (1971b) The swimming energetics of trout. II. Oxygen consumption and swimming efficiency. J Exp Biol 55, 521–540.

Webb PW (1975) Hydrodynamics and energetics of fish propulsion. J Fish Res Board Can 190, 77–105.

Williams GC, Koehn K (1984) Population genetics of North Atlantic catadromous eels (Anguilla). In: Turner BJ (ed) Evolutionary Genetics of Fishes. Plenum Press, New York, pp. 529–560.

Winberg GG (1956) Rate of metabolism and food requirements of fishes. Trans Fish Res Board Can 194, 202 pp.

Wirth T, Bernatchez L (2001) Genetic evidence against panmixia in the European eel. Nature 409, 1037–1040.

Wolf K (1988) Fish Viruses and Fish Viral Diseases. Cornell University Press/Comstock Publishing Associates, Ithaca, NY/New York, 476 pp.

Chapter 9
Impact of the Swim-Bladder Parasite on the Health and Performance of European Eels

Csaba Székely, Arjan Palstra, Kálmán Molnár, and Guido van den Thillart

9.1 Parasitic Infections in Eels

A growing number of diseases including infections by parasites are thought to play an important role in the drastic reduction of eel in European rivers and lakes. Until now, the occurrence of about forty parasite species has been reported for the European eel. Bykhovskaya-Pavlovskaya et al. (1962) listed 30 species in this fish. Since then, several new species were recorded and among them highly pathogenic ones. Several of these species arrived by Pacific eels (*Anguilla japonica, A. australis*) that were introduced into Europe for experimental purposes. Parasitic infections leading to severe symptoms and eventually death of the fish are caused primarily by the highly pathogenic species. The common occurrence of some helminths like *Proteocephalus macrocephalus, Bothriocephalus claviceps, Acanthocephalus anguillae, Paraquimperia tenerrima* has been known for a long time (Murai 1971; Moravec 1994). There were also data on the pathogenic effect of some well-known myxosporeans as *Myxidium giardi, Hoferellus gilsoni* (Copland 1983; Lom et al. 1986), but these parasites rarely caused any diseases. The first real pathogenic disease of parasitic origin in Europe was reported by Molnár (1983) who discovered *Pseudodactylogyrus anguillae* and *P. bini* in cultured eels. These monogeneans caused heavy losses in intensively cultured eel stocks (Buchmann et al. 1987; Buchmann 1993) but no losses were recorded in natural waters. From time to time reports appeared about severe infections caused by unknown or less known parasites affecting eels in natural water, but none of these parasites caused measurable losses in eel populations. Besides monogeneans, Moravec and Køie (1987) described *Daniconema anguillae*, a skrjabillanid nematode from the abdominal cavity of the European eel. Molnár and Moravec (1994) found a heavy infection by its larval stages in the fins of eels of Lake Balaton. *Myxobolus portucalensis* as

C. Székely and K. Molnár
Veterinary Medical Research Institute, Hungarian Academy of Sciences, H-1581 Budapest, Hungary

A. Palstra and G. van den Thillart
Integrative Zoology, Institute of Biology Leiden, van der Klaauw Laboratories, P.O. Box 9516, Kaiserstraat 63, 2300 RA Leiden, The Netherlands

G. van den Thillart et al. (eds.), *Spawning Migration of the European Eel*,

described by Saraiva and Molnár (1990) seems also a common and frequent parasite but due to its specific location on the fins, losses of the hosts were not recorded. On the other hand an introduced nematode species, *Anguillicola crassus* proved to be highly pathogenic for eels and its role in the permanent decrease of eel populations in Europe should be seriously considered. A general review on the pathogenic helminth parasites of eels was presented by Kennedy (2007).

9.2 Anguillicola Infection in Eels

Five species of the genus *Anguillicola* are currently known. Four of these five species; *Anguillicola globiceps* Yamaguti 1935, *A. crassus* Kuwahara, Niimi et Itagaki 1974, *A. australiensis* Johnston et Mawston 1940 and *A. novaezelandiae* Moravec et Taraschewski 1988, are native and specific to Pacific eels, while the fifth one, *A. papernai* Moravec and Taraschewski 1988 is a parasite of the African species *Anguilla mossambica* (Taraschewski et al. 2005). In their original biotope, all of these species seem to be well adapted to hosts and they cause no apparent damage on them. Adults of all *Anguillicola* spp. live in the lumen of the swim-bladder and feed on blood (Fig. 9.1). As far as is known, none of the Pacific species can employ paratenic hosts in their life cycle (Puquin and Yuru 1980; Wang and Zhao 1980; Nagasawa et al. 1994). Introduced specimens of *A. crassus* in Europe however develop in several fishes as in paratenic hosts (Thomas and Ollevier 1992; Székely 1994; Moravec and Škoriková 1998), and cause heavy infections in the latter. This means that in the life cycle of the European eels, a more frequent indirect infection route through paratenic hosts and a less frequent direct route similar to the one of Far-East eels can be distinguished.

9.3 Life Cycle of *Anguillicola crassus*

The basic life cycle of *A. crassus* was studied by several researchers in the Far East and in Europe (Nagasawa et al. 1994; Kennedy 1994; Moravec et al. 1994a). In the Pacific Region adults (Fig. 9.2) are found in the swim-bladder of the Japanese eel where they feed on blood. Eggs are released into the swim-bladder and contain the second larval stage encased in the sheath of the first stage larva (Fig. 9.3). The larvae may hatch in the swim-bladder, or they may stay inside the eggs while they are passed out through the pneumatic duct of the eel and get into water via the intestine. The free-living larvae float for a while in the water and sink to the bottom, where they undulate in response to tactile or other similar stimuli. Larvae are captured and consumed by copepods either in water or on the bottom of the water basin. They are ingested by different species of cyclopoid copepods and pass into the hemocoel, where they molt to the third larval stage (Fig. 9.4).

Eels usually become infected in a direct way by eating infected copepods. Infected copepods can also enter the intestine of the eel through the gut content of small prey fishes. Larvae in the intestine of eel rapidly leave the copepods and migrate through

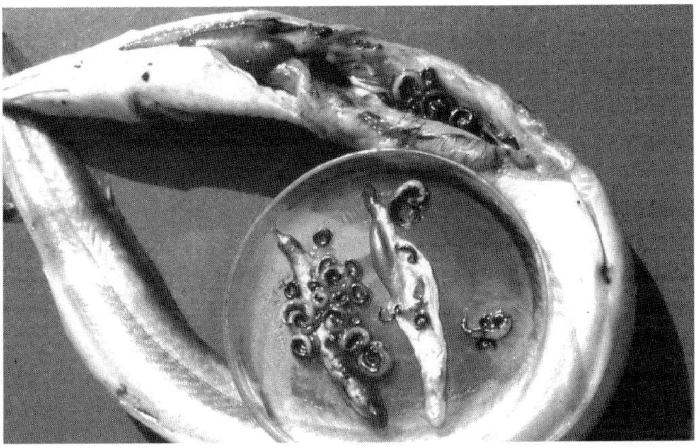

Fig. 9.1 Anguillicola-infected swim-bladders of Lake Balaton eels in 1991

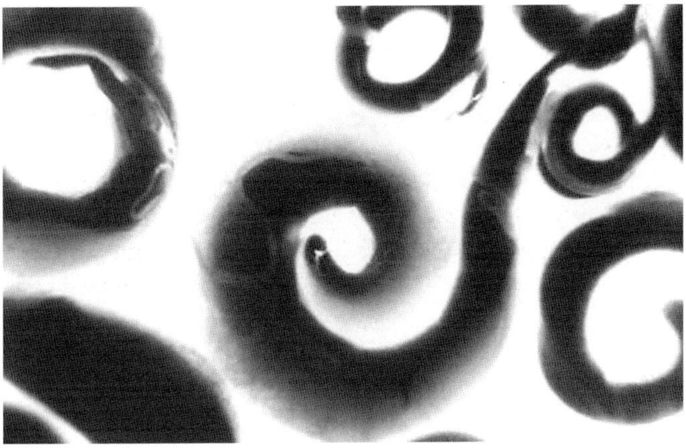

Fig. 9.2 Adult *Anguillicola* females containing L2 larvae in their uterus

the intestinal wall towards the swim-bladder (Wang and Zhao 1980). In the wall of the swim-bladder they molt to the fourth larval stage and then develop into adults. In the Pacific region *A. crassus* usually infects its host in low numbers and is not pathogenic to the natural host (Hine and Boustead 1974; Wang and Zhao 1980; Kennedy 1994).

The life cycle of *A. crassus* in Europe was studied by several authors (Haenen et al. 1989, 1994 1996; De Charleroy et al. 1990; Kennedy and Fitch 1990; Thomas and Ollevier 1992, 1993; Höglund and Thomas 1992; Moravec et al. 1993, 1994b). The above authors proved by their field and experimental studies that *A. crassus* in Europe develops using, in most cases, small prey fishes as paratenic hosts. At least 20 species of freshwater fishes as well as a snails, amphibians and insects have been identified as paratenic hosts for the third larval stage (Thomas and Ollevier 1992; Moravec 1996; Moravec and Konecny 1994; Székely 1994, 1995; Moravec and Škoriková 1998)

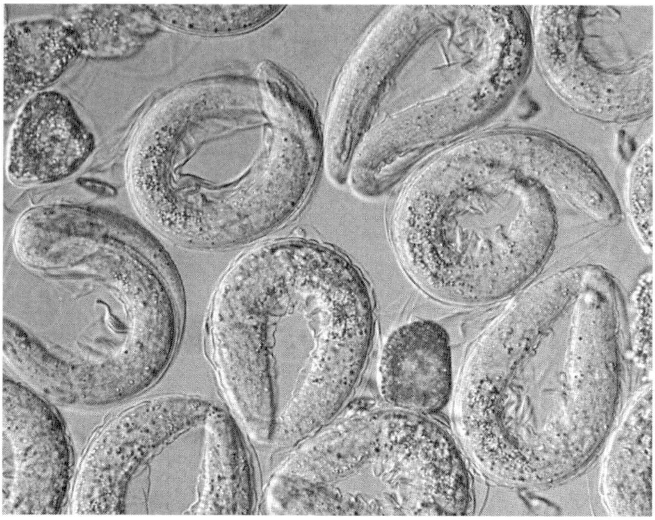

Fig. 9.3 Eggs of *Anguillicola crassus* containing second stage larvae

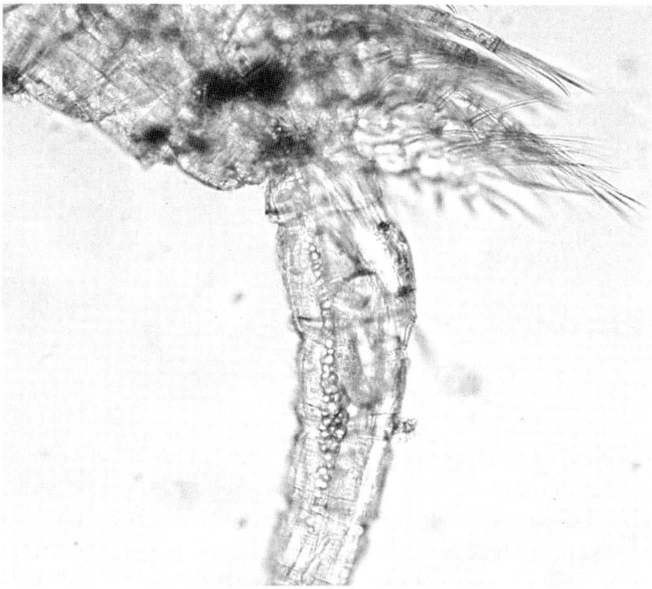

Fig. 9.4 Third stage *Anguillicola*-larva in a Cyclops intermediate host

(Fig. 9.5). Not all of them are of similar suitability: in some species only a few larvae survive as the host reacts against them, whereas in others, especially in species of Perciformes, larvae preserve their vitality for several months (Pazooki and Székely 1994; Székely et al. 1996). These authors concluded, that young eels may become

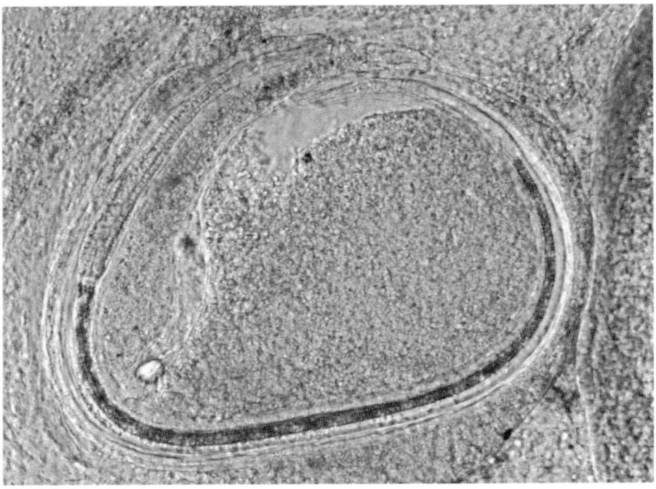

Fig. 9.5 Encapsulated third stage *Anguillicola*-larva from a paratenic fish host

infected by ingesting infected copepods but older and larger eels by ingesting infected fish. These studies also revealed that larvae in Pacific eels move rapidly through the intestinal and swim-bladder walls, but in the European eel they do not, and in case of host reaction they stay in the intestinal and swim-bladder wall before moving into the lumen of the swim-bladder and molting to the adult stage.

9.4 Transcontinental and Intracontinental Spread of *Anguillicola* Spp.

Contrary to the Pacific eels, which are infected with at least four *Anguillicola* spp., before intercontinental eel transmissions, no *Anguillicola* infection was reported in the European eel *A. anguilla* and the American eel *A. rostrata*. Little is known about the occurrence of parasites in the less studied Pacific eels (*A. celebesiensis* or *A. marmorata*). *Anguillicola globiceps* appears to occur only in wild *A. japonica* and may cause some thickening of the swim-bladder wall. It is more tolerant to cold temperatures than *A. crassus* (Nagasawa et al. 1994). In Japan, *A. crassus* is found in both wild and farmed specimens of *A. japonica*, but it does not appear to cause any damage to this host. On the other hand, Egusa (1979) and Egusa and Hirose (1983) reported that when *A. anguilla* was introduced into Japanese aquaculture ponds to increase stock production, it was found that both the prevalence and intensity of infection were higher in this introduced species than in *A. japonica*. Furthermore, *A. crassus* was pathogenic to, and caused mortality in, *A. anguilla*. Their observation was confirmed by experiments by Knopf and Manhke (2004) who described that recoveries of *A. crassus* are higher in *A. anguilla* (33.2%) than in *A. japonica* (13.8%). In the latter, development of the parasite is slower and a greater proportion of larvae is encapsulated in the swim-bladder wall and dies. Based on Japanese experiences, Egusa (1979) warned fish culturists that transmission of the Japanese eel to other continents might be a hazard. Despite Japanese experiences and Egusa's warning, two species of the Pacific eels were transported to Europe. *A. crassus* first appeared in Europe around 1980 (Peters and Hartmann 1986, Hartmann 1987; Taraschewski et al. 1987; Moravec and Taraschewski 1988; Kennedy and Fitch 1990) and has spread through the eel trade throughout the continent and into North Africa (El-Hilali et al. 1996), and subsequently to North America (Barse and Secor 1999; Moser et al. 2001). The source of this *Anguillicola* infection was possibly Japanese eels brought to Europe for experimental purposes. Before this introduction to Western Europe, another shipment was transported for a comparative experiment to the Soviet Union and held in heated water. This shipment, however, transmitted only a monogenean (*Pseudodactylogyrus*) infection to the European eel (Golovin 1977), but no *Anguillicola crassus* infection was reported.

Other species have also been introduced into Europe when a stock of Australian eel (*A. australis*) was imported to Italy in 1975 and placed into Lake Bracciano. By these fishes, *Anguillicola novaezelandiae* (originally identified as *A. australiensis* by Paggi et al. (1982)) was introduced. It survived there for several years, reaching prevalence levels of 80% and intensities of up to 27 individuals per eel, but it caused no damage to the native *A. anguilla* and never spread from this isolated habitat to other lakes. Following the introduction of *A. crassus* into the lake in 1993, its numbers have declined (Moravec et al. 1994a). Though the chance for introduction of *A. globiceps* by Pacific eels is reasonable, there are no reports of its establishment. *A. crassus* spread very fast in Europe towards the East and North. It was found in Hungary in 1990 (Székely et al. 1991), in the Czech Republic (Baruš 1995), in Denmark (Køie 1988; Boetius 1990) and in Sweden (Höglund and Thomas 1992). The parasitosis spread to Africa; El-Hilali et al. (1996) and Maamouri et al. (1999) found it in Morocco and Tunisia, respectively. Fries et al. (1996) and Barse and Secor (1999) discovered this parasite in the American eel (*Anguilla rostrata*). The spread of the parasite in each case could be attributed to human movements of infected eel stocks around the European continent and across the seas as a consequence of the unconsidered eel trade (Kennedy and Fitch 1990). Therefore, *A. crassus* appears to be an important pathogen for both species of Atlantic eels.

9.5 Differences of *Anguillicola* Infection Between Pacific Eels and Atlantic Eels

Anguillicola spp seem to be specific parasites for the genus *Anguilla* and they seem to easily infect other eel species than the original hosts. When European eels were first transmitted for experimental purpose to Japan (Egusa 1979; Egusa and Hirose 1983), this stock became quickly and heavily infected. Moreover, its infection rates surpassed the rate of their Japanese counterparts cultured in the same way and density. At a similar way, infected Japanese and Australian eel stocks transported to Europe readily transmitted their *Anguillicola* infection to the European eel and an *Anguillicola novaezelandiae* (identified as *A. australiensis*) infection appeared in Italy (Paggi et al. 1982) and *A. crassus* infection in Western Europe (Hartmann 1987). This latter species became infective also for the American eel (*A. rostrata*) (Barse and Secor 1999). Of the two species, *A. novaezelandiae* seem to be less adaptable to the new biotope and host, and showed no ability for intensive spread. *A. crassus*, however, proved to be an aggressively disseminating species and rapidly colonized Europe, some African countries and Northern America (Peters and Hartmann 1986; El-Hilali et al. 1996; Barse et al. 2001; Kirk 2003). Besides colonization in the European eel, *A. crassus* demonstrated new qualities unknown in the original biotope, the Pacific region. Its much wider specificity for its first intermediate host, as it is able to infect several species of freshwater cyclopoid copepods as well as estuarine copepods, seems to be one of the most significant changes in its life cycle. This enables the parasite to complete its life cycle in freshwater and estuarine biotopes. The other major difference between the life cycle of *A. crassus* in Pacific eels and the Atlantic eels is the role of paratenic hosts. No paratenic hosts were found in the Pacific region and a direct way of infection route is supposed (Puquin and Yuru 1980; Wang and Zhao 1980; Nagasawa et al. 1994). On the other hand the heavy infection of the European eel with *A. crassus* is promoted by the large number of paratenic host, like prey fishes and their intensive infection with third stage *A. crassus* larvae (Thomas and Ollevier 1992; Moravec and Konecny 1994; Székely 1994, 1995). Differences in immune response to infection of the eels of the two regions may be a major factor to regulate the infection rate and balance between parasite and host. Nielsen (1999) proved that *A. japonica* mounts more effective immune responses to *A. crassus* than does *A. anguilla*.

9.6 Biology and Ecology of *Anguillicola crassus*

The life cycle of *A. crassus* is temperature related. It grows faster at higher temperatures and is retarded at low temperatures. Studies show that in most cases eggs containing second stage larvae and less frequently hatched larvae, leave the fish via the ductus pneumaticus of the swim-bladder and the gut. Eggs of *A. crassus* released from the worms do not hatch below 10°C and the rate of hatching increases with temperature up to 25–30°C. Hatching rate is also related to salinity, but the percentage of eggs hatched, and survival and infectivity of the second larval stage declines as salinity increases (Kirk et al. 2000a). A part of the eggs and larvae might be captured and eaten by copepods. The majority, however, sinks to the bottom. Second stage larvae attach to the substrate within 2 or 3 days. Kennedy and Fitch (1990) reported L_2 survival of 160 days at 10°C, up to 8 months at 7°C and 5 months at 24°C. In contrast, Thomas and Ollevier (1993) found that they survive only up to 45 days at 23°C. Survival and infectivity decrease exponentially with time (Kennedy and Fitch 1990) and with increasing salinity even though the parasite can survive in some saline lagoons (Di Cave et al. 2001). Third stage larvae in some paratenic host fish species might survive for more than a year (Székely 1996). Third stage larvae infecting the swim-bladder of eels may survive in the wall of this organ for 4 months at 4°C. Over time they become unable to enter the lumen. Adult mortality increases over 4 months at the same temperature (Knopf et al. 1998). *A. crassus* prefers warmer temperatures and its life cycle is hindered at lower temperatures. It may explain why *A. crassus* is uncommon in, or absent from, the more northern boreal regions (Höglund and Thomas 1992; Thomas and Ollevier 1992, 1993; Knopf et al. 1998) where it may be restricted to thermal effluents (Höglund et al. 1992). The whole cycle can be completed in 90 days at suitable temperatures, but will normally take longer; at least 4 months (Haenen et al. 1989). Eels become vulnerable to infection as soon as they commence feeding in rivers or estuaries. Nimeth et al. (2004) found that even glass eels and elvers are susceptible to infection. Infection can continue throughout life, and in general infection levels tend to be higher in older and larger eels. Eels can recover from infection, but they are not immune to re-infection. Molnár et al. (1994) proved that in Lake Balaton eels there is a permanent reinfection, with recovery and intensification periods. There is no relationship between primary and secondary infections (Haenen et al. 1996) and higher doses of infection will normally produce more severe symptoms. It was initially thought that there was no antibody response to the parasites, but later studies by Sures and Knopf (2004) using ELISA have shown that the body wall of the parasite is a good antigen and significant levels of antibodies can be detected in the blood after 61 days. The infection rate might be affected by the antibody response. The response is suppressed by the initial rise in cortisol levels in all eels due to handling stress, which assists parasite establishment. Knopf et al. (2000a, b) has also reported a humoral response. Nielsen (1999) has shown that the antibody response of *A. japonica* to *A. crassus* is higher than that of *A. anguilla*, which suggests that an immune response may be involved in specificity and control of numbers.

Ashworth et al. (1996) and Ashworth and Kennedy (1999) have identified three density-dependent regulatory processes that may be responsible for this situation. Parasite induced copepod mortality is the first factor. Uninfected copepods might survive for 30 days post infection (pi), whereas the equivalent survival time for infected copepods is only 12 days pi. Mortality of infected copepods is also density-dependent, and heavily infected copepods may die before larvae reach the infective third stage. The second factor is the intensity of gravid females per infrapopulation, which remains relatively constant over time and is independent of the overall infrapopulation density. The third factor is the relation of gravid females to third and fourth stage larvae in the swim-bladder wall. The authors suggested that the presence of adult males and females in the swim-bladder could inhibit the movement of fourth stage larvae from the swim-bladder wall into the lumen and larvae were arrested in development in a density dependent manner. Density-dependent regulatory processes within the definitive host may be affected by environmental factors such as endocrine disruptors. Fazio et al. (2008a) showed that the steroid hormone 11-ketotestosterone induced a significant male-biased ratio in the nematode infrapopulations.

Kennedy and Fitch (1990) determined that adult parasites could survive in *A. anguilla* for up to 4 weeks when the eel was kept in 100% seawater. Survival declined in coastal lagoons of increasing salinity (Di Cave et al. 2001). Kirk et al. (2000a, b) showed that some adults could survive and continue to produce eggs in eels in 50% and 100% seawater for up to 6 months, but around 10% of the adult parasites were damaged following exposure to high salinity. Kirk et al. (2002) showed that the parasites are osmoconformers, achieving this by feeding on eel blood. Still, around 20% of the parasites could not tolerate the osmotic stress of living in eels in 100% seawater but died and disintegrated. It is thus possible that parasites of freshwater origin can survive in eels in coastal lagoons and estuaries as well as during the eel's migration to the Sargasso Sea. The life cycle could also be completed in waters of enhanced (up to 50%) salinity by using estuarine copepods such as *Eurytemora affinis* as an intermediate host, but it was considered unlikely that the parasites could transmit in sea water, as most marine copepods were of the wrong size to serve as intermediate hosts. The ability of the parasite to survive in eels in the Baltic Sea (Höglund and Thomas 1992; Reimer et al. 1994) could thus be due to transmission there or to the survival of freshwater infections.

9.7 Histopathological Changes Caused by *A. crassus* Infection

It is generally known for parasitic infections that during a low infection there is a kind of balance between parasites and hosts. At severe infection, however, parasitic diseases may have an important impact on the health of the hosts. As for *Anguillicola* infection, in the Pacific region this balance works well and heavy infections develop neither in Japanese eels (Egusa and Hirose 1983; Knopf and Mahnke 2004) nor in the Australian species (Kennedy 1994). On the other hand in the European eel, the newly colonized *A. crassus* caused heavy infections from the first years following its first detection (Hartmann 1987; Haenen et al. 1994). At a similar way in Hungary, heavy infection was recorded already at the time when the parasite was first found (Székely et al. 1991).

Severe pathological symptoms can develop in all sizes of eels. The effect of the parasite on its host will relate to the number of parasites present and the size of the eel. Changes are caused by matured worms and migrating larvae. The most evident visual effects can be observed on the swim-bladder but disease symptoms may develop in the intestinal wall as well. The major symptoms of infection include dilated blood vessels and congestion of blood vessels, haemorrhages, inflammation, thickening and fibrosis of the swim-bladder wall (Van Banning and Haenen 1990; Haenen et al. 1989, 1994; Molnár 1994; Molnár et al. 1991, 1993, 1994). There may be an increase in the spleen mass (Lefebvre et al. 2004), and finally the swim-bladder adheres totally to surrounding organs, such as kidney and intestine (van Banning and Haenen 1990).

The number of worms found in the eel is a vital, but not the only factor affecting the health of the fish. Van Banning and Haenen (1990), Haenen et al. (1996) and Csaba et al. (1993) proved that changes in the swim-bladder wall have higher impact on the health of the host than the number of worms. Molnár et al. (1993) pointed out that migrating *Anguillicola* larvae showed up higher pathogenic impact on the changes in swim-bladder wall than blood sucking imago stages in the lumen. Van Banning and Haenen (1990), Csaba et al. (1993) and Molnár et al. (1993) supposed that pathological alterations in the swim-bladder and a general decrease in the host's resistance promote development of bacterial, fungal and other parasitic infections. Therefore heavy infection with *A. crassus* appears in most cases as a disease complex. Liewes and Schaminee-Main (1987) and Kamstra (1990) described seven stages of infection by the alterations in the swim-bladder:

Stage 1 Swim-bladder is normal and without nematodes.
Stage 2 Swim-bladder is normal but contains a few nematodes.
Stage 3 Swim-bladder enlarged and partly filled with red-brown fluid. Swim-bladder wall can be inflamed.
Stage 4 Swim-bladder much enlarged and filled with red-brown fluid. In this stage actively moving nematode larvae (L2) can be noticed.
Stage 5 Rupture of the swim-bladder wall or the ductus pneumaticus is highly irritated. Secondary infections of surrounding tissues are externally visible as a swollen and inflamed abdomen.
Stage 6 In this stage the swim-bladder wall (possible after rupture) is replaced by a thick layer of connective tissue. In the swim-bladder remainders of the nematodes can be found.
Stage 7 The swim-bladder is replaced by a hard brown-black mass in which remainders of the nematodes can be found.

Csaba et al. (1993) examined eels during the massive eel mortality in Lake Balaton in 2001 and observed three major changes in the swim-bladder: (1) At the first stage of infection developing and mature worms were found in the lumen of the swim-bladder with a transparent wall, (2) In a progressed stage of infection, decayed, fragmented worms and a red-brown fluid were found in the lumen of the swim-bladder with an inflammated wall, and (3) With the most serious cases, the serosa and subserosa of the swim-bladder became as thick as 5 mm due to fibrous changes, but no worms inhabited the lumen. These authors found also a secondary bacterial infection at the final stage of anguillicolosis, but they recorded no increase of pesticides or heavy metals in the flesh of the eels. Molnár (1994) stated that changes in the swim-bladder wall were caused by the third and fourth stage larvae migrating in this organ. Larvae were detectable in the oedematous connective tissue of the subserosa and in the gas glands. Tissue proliferation consisting of epithelioid cells started to develop around migrating larvae blocking their route towards the lumen and finally most of the larvae became surrounded by a fibrous capsule. A similar nodule formation around invading larvae was observed in the wall of the intestine in which several hundreds of encapsulated or decayed larvae were found. Würtz and Taraschewski (2000) suggested that the histotrophic larvae did not create a severe cellular reaction. The leucocytes gathering around larvae seemed to be attracted rather to cellular debris resulting from the parasites' movements. Molnár (1994) however, observed that a granulation tissue consisting of epitheloid cells and

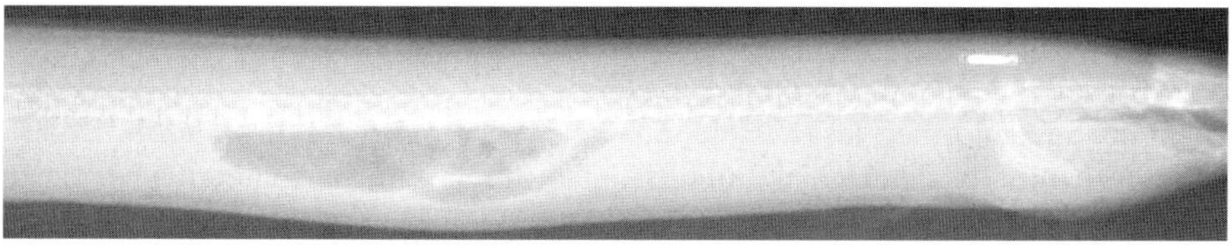

Fig. 9.6 X-ray of an *Anguillicola*-infected eel. The contours of the large *Anguillicola*-nematodes in the swim-bladder are well visible

macrophages was formed around larvae which restricted their movement of the larvae. Changes found in the epithelium are better attributed to the effect of blood sucking imago stages. Würtz and Taraschewski (2000) described that the epithelium of the heavily infected eel is characterized by large folds and cauliflower-like proliferation. Molnár (1994) found that during severe infection, serum filled cysts appeared in the tunica propria of the swim-bladder wall. These cysts were lined with epithelium and the serum filling their cavity contained pycnotic, rounded cells.

Beregi et al. (1998) and Székely et al. (2004) examined the anguillicolosis in the swim-bladder parallel with x-ray and CT-methods and with dissections (Fig. 9.6). They differentiated five major stages of changes in the pathological process:

At grade 1 the swim-bladder gave a homogeneous radiographic shadow. When opening the lumen, no worms were found and the wall thickness did not exceeded 0.3 mm.

At grade 2 radiographic shadow was not homogeneous, the contours of small worms were discernible. When opening the lumen of the mildly thickened wall, small worms were found in it.

At grade 3 radiographic shadow of the bladder was deformed and contours of large worms were observed. By dissection, large worms and some fluid were found in the lumen of the dilated swim-bladder, which had a thickened portion at both ends.

At grade 4 the swim-bladder gave a narrowed radiographic shadow. By opening the bladder the lumen was narrow, almost completely devoid of air but contained some exudates and dead worms. The wall of the swim-bladder in this stage of changes might reach 2 to 3 mm.

At grade 5 no radiographic shadow of the bladder was found. By dissecting the abdominal cavity a degenerated small swim-bladder with thickened wall and atelectised lumen was found.

The long term change in the pathological status of a given *Anguillicola* infection was studied also by a radiodiagnostic method (Székely et al. 2005). Naturally infected eels caught in Lake Balaton were x-rayed at 0, 4, 8 and 12 weeks and changes in radiographic shadows of the different grades of infected swim-bladders were followed in 78 eels. Results obtained from individually tagged eels dissected after 3 months observations showed that the pathological status of the swim-bladder had deteriorated in 55% and remained the same in 37% of the cases. Tendency of improvement (one eel) and variable findings (7%) were recorded in a low percentage of cases only.

9.8 Effects of *A. crassus* Infection on the Physiology of Eel

There are several other, less visible, effects of *A. crassus* on *A. anguilla*. Boon et al. (1989) initially detected no significant difference in haematocrit and plasma proteins in infected eels but later (Boon et al. 1990a, b), using more sophisticated controls, these authors found a significant difference. Barus et al. (1998, 1999a) demonstrated lower level of methionine and aspartic acid in the muscles of infected eels, and significantly lower levels of muscle Ca, P, Fe and Mn. Scholz and Zerbst-Boroffka (1994) have determined that *A. crassus* is iso-osmotic with the eel body fluids, but that there are ionic differences in eel chloride levels in sea water composition. These ionic and osmotic changes impose stress on the parasites which are ionic and osmotic conformers.

Kelly et al. (2000) could not find significant differences in stress hormones, metabolic hormones or osmoregulatory status of infected eels and concluded that eels were able to adapt to chronic infection levels. However, Gollock et al. (2004) found that infected eels were more stressed under aquaculture conditions, when netted and exposed to air. They found that the cortisol response did not differ between infected and uninfected eels but that plasma glucose levels were higher in infected eels, and that glucose metabolism and utilization was increased i.e. stress of infection elevated glucose turnover. In a later study Gollock et al. (2005a) showed that acute temperature alone had little effect as an eel stressor. Under such conditions there was a lag in glucose metabolism in infected eels. There was no significant increase in hemoglobin levels when compared to the responses of uninfected eels as both groups showed a significant increase in hemoglobin. Gollock et al. (2005b) went on to demonstrate that infected eels exhibited a more pronounced stress response to hypoxia than uninfected individuals. Finally, Fazio et al. (2008b) found that the expression level of deep-sea rod opsin (DSO) gene in the eyes was significantly greater in infected wild eels. The authors suggested that the parasite may have an effect on the eel's silvering process.

9.9 Effect of *A. crassus* on the Survival of Eel

Despite intensive pathological changes and morphological degenerations in the swimbladder, no outer symptoms are observable on infected eels. Køie (1991) could not find evidence of lack of appetite in infected European eels or difference in condition factor, but confirmed a greater mortality of infected eels during storage or transport due to stress and possibly to secondary bacterial infections. Other workers have found that the length/weight relationship does not differ between infected and uninfected European eels and that any difference in weight between infected and uninfected eels is not significant even if, paradoxically, the eels are suffering mortality due to the parasite (Barus and Prokeš 1996). Koops and Hartmann (1989) also found no difference in condition factor between infected and uninfected eels or a relationship between condition factor and parasite intensity. Möller et al. (1991) reported a higher condition factor in infected eels but no change in the hepatosomatic index. The differences in estimation of the condition factor by the above authors might relate to the fact that in infected eels the enlarged swim-bladder filled by worms or fluid, as well as general fibrosis and serous infiltration of the abdominal serosa and inner organs, seemingly increase the body weight.

Contrary to the lack of outer symptoms, there are a series of observations on the effect of *A. crassus* infection on the physiological performance and survival of the eel. Thomas and Ollevier (1992) found that heavily infected eels were more easily sucked into power station intakes. Molnár and Székely (1995 unpublished) experienced that the escape reaction of eels harvested by commercial fishermen differs significantly. They observed that less active eel specimens, first taken off by a landing-net from a densely filled fishing smack, were more infected than actively hiding specimens harvested from the smack at the end. Gollock et al. (2004) who examined the effect of netting and aerial exposure to the plasma glucose of eels infected with *A. crassus* concluded that these stressors potentially could result in decreased growth.

Molnár (1993) demonstrated that when uninfected and infected eels were deprived of oxygen, the severely infected eels died first. The impact of the oxygen shortage was temperature-dependent and the effect on individual eels related more closely to the thickening of the swim-bladder wall rather than to the number of parasites present. Infected eels had an increased demand for oxygen, but the presence of *A. crassus* impaired the functioning of the swim-bladder and this in turn could result in eel death.

The swim-bladder is an important but not a vital organ for the fish, at least in freshwater where they habitually live on (or in) the bottom. The fish can compensate most of its dysfunction relatively well. Nevertheless, Würtz et al. (1996) were able to demonstrate the importance of this organ. They proved, that there was in fact a significant correlation between the oxygen concentration in the swim-bladder and the level of *A. crassus* infection. The contribution of oxygen to the swim-bladder gas was reduced between 36% and 60% in naturally infected eels, and this related to the changes in the swim-bladder wall. Overall, the presence of parasites impeded the function of the swim-bladder as a buoyancy and hydrostatic organ by impairing the functioning of the gas gland.

9.10 Mortalities of Eel Caused by *A. crassus*

Since Egusa (1979) first observed the death of introduced European eels due to *Anguillicola* infection in Japan, data obtained in Europe proved that mortalities among intensively infected eel stocks can also occur. Eel mortalities in lakes in Central Europe, where eels are stocked to form the basis of fisheries, also suggest that *A. crassus* plays a role in these mortalities. The best documented of these mortalities occurred in Lake Balaton in Hungary (Molnár et al. 1991). The mass mortality of eels occurred during the summer of 1991, with an estimated loss of

250 t of eels in the western basin. In 1992 losses were lower, when 40 t were lost in the central basin as conditions improved in the western basin of the lake and the infection spread to the eastern basin. The last eel mortality was recorded in 2005 when the eastern part of the lake was affected but the losses in this region were more severe than in the other two. No other fish species was involved. Initially it was suggested that *A. crassus* alone might have been the cause. Subsequently, following a detailed examination of ichthyological and physico-chemical conditions in the lake, the opinion on the exclusive role of the parasite lessened. Effects of population density, the temperature and oxygen content in the water, co-infections with other pathogenic agents and a supposed intoxication with pesticides were considered. Infection levels were very high in the lake at the time of fish mortality and virtually all eels were infected, with 30–50 adults per eel and up to 200 larvae. Eel population densities were also very high. The effects of the parasites on eel swim-bladders were typical, with eels showing haemorrhagic swim-bladder walls that were thickened up to ten times in comparison with the normal condition, and with the swim-bladder filled with fluid, eggs and decaying and live adults. This, together with the known ability of the parasite to cause mortalities in eel farms, suggested the major role of anguillicolosis in fish mortality. It was clear that water temperature levels in the lake were unusually high during those summers while oxygen levels were correspondingly very low. Later on, it turned out that the lake was heavily overpopulated due to the repeated eel introductions and the extremely good survival of the glass eels. At the time when about 300 t of eel died, the mass of the eel population was calculated to be 1,000 to 1,200 t (Tátrai et al. 2002). Contradictory to the calculations, in the last 16 years (1991–2006) 2,584 t of eel was harvested and the yearly catch is still reasonable, although since 1991 no new introduction of eels happened. These conditions caused stress to the eels, and it now seems more likely that the combined effects of this stress together with heavy infection with adult and larval *A. crassus* invasion were the causes of the mortality (Molnár et al. 1991, 1994). On the other hand, the role of pesticides forwarded by Bálint et al. (1997) can be excluded as no other fish species was affected. In laboratory experiments the eel proved to be more resistant to piretroid-pesticides than cyprinid fishes (own experiences and Csaba G: personal communication 1995).

Similar mass mortalities have been reported from other water bodies which have been stocked with eels, for example in the Vranov Reservoir in the Czech Republic (Barus et al. 1999). Here there was a loss of some 3–5 t in 1994, and the mortalities occurred also in summer when water temperatures were high, water oxygen levels low and eel densities were high. No other fish species were involved in losses. The conditions in Vranov Reservoir were similar to those of Lake Balaton. Barus and Prokeš (1996) stated that the conditions in both lakes were ideal for such epizootics as in these closed, shallow and overpopulated lakes the density of copepods and paratenic hosts facilitated the rapid increase of *Anguillicola* population levels. It is likely, therefore, that it was a combined effect of environmental and parasite induced stresses that caused the mortalities.

Conditions in the shallow, productive lakes of central Europe in which eels are stocked for commercial fisheries can clearly result in eel mortalities from time to time, but these are not regular or inevitable occurrences. Schabuss et al. (2005) have noted a similarity in physical and chemical conditions between Neusiedler See in Austria and Lake Balaton. In both lakes a stocked eel fishery is going on. Infection levels with *A. crassus* were also high in Neusiedler See, but there was no occurrence of mass mortalities there over a period of many years. Despite similarities, however, there were great differences between the two lakes. Before there was a regular fisheries activity in Neusiedler See, in Lake Balaton eels were exclusively harvested by a trap at the outlet of the lake until the first eel mortality occurred and thus the density of the eel population was only moderately decreased. Ashworth and Kennedy (1999) suggested that parasite-induced mortalities might occur in all natural waters but that the eel density in these systems is never as high as in stocked lakes and that density dependent regulatory processes work against massive eel mortality in the former.

A. crassus can cause host mortalities in eel farms as well. Egusa and Hirose (1983) and Nagasawa et al. (1994) reported that *A. crassus* in *A. anguilla* in Japanese eel farms caused severe mortality of infected eels at their first experimental colonization to Japan and these were largely responsible for mortalities. Eventually, the mortalities in *A. anguilla* in Japanese eel farms due to the parasite were largely responsible for the abandonment of cultivation of *A. anguilla* in Japan (Taraschewski et al. 1987). In Europe, Liewes and Schaminee-Main (1987), Kamstra (1990), van Banning and Haenen (1990) and Mellergaard (1988) reported increased mortality of infected eels in Dutch and Danish eel farms. Contrary, however, to Japanese farms, eels in Europe were already infected in natural waters but disease symptoms developed later on during intensive culture.

9.11 Population Effects and Control

The experimental evidence that the functioning of the swim-bladder is impaired in infected eels and that they are more susceptible to stress and to human activities such as netting (Gollock et al. 2004) and transport (Køie 1991) suggests that infected eels in natural populations may respond differently to conditions than uninfected eels. It would seem very likely, for example, that infected eels would be more susceptible to natural avian predators. There is no direct evidence that this is the case, and such evidence would be very difficult to obtain, but even the reduction in swimming speed reported by Sprengel and Lüchtenberg (1991) must surely affect their escape response to predators.

Control measures are clearly impracticable in natural water bodies, and indeed may be unnecessary if *A. crassus* populations are regulated in a density-dependent manner at levels below those which cause mortality. If stocking densities can be reduced to a level at which the fishery remains commercial but the eel population does not rise to a density at which the eels are stressed, then mortalities are unlikely to occur. In the Lake Balaton where eel fisheries is depend on stocking with glass eels, the cessation of new stockings and intensive fisheries activity stopped the chance of a massive eel kill and reduced the intension of infection, although a relatively high level infection is still continues.

In eel ponds, control of the parasite levels may be difficult. A reduction of copepods' densities as the intermediate host can help. Egusa and Hirose (1983) suggested an intensive flow through the pond. Using chemicals to eliminate copepods might be effective but is not environmentally acceptable, as the effluent of the ponds would contaminate natural water bodies. In Europe, usually there are no problems in fish farms as the eel is cultured in intensive recirculation systems where no intermediate hosts exist. Eel farms are usually free from the parasite, or they harbor a low level infection due to the higher water salinity (Køie 1991). Taraschewski (2006) thinks that for farmed European eels, *A. crassus* is no more an economic threat, as reported from Japan in the 1970s (Egusa 1979) and Europe in the 1980s (Liewes and Schaminee-Main 1987; Van Banning and Haenen 1990; Kamstra 1990), due to progress in chemotherapy (Taraschewski et al. 1988; Hartmann 1987; Geets et al. 1992). The above authors reported about a number of nematocidal drugs that proved to be effective against *A. crassus*. Of the latter the most effective are levamisole and metrifonate in freshwater baths (Taraschewski et al. 1988). Levamisole appears to be the better of the two drugs as the cumulative lethal dose ratio is more favorable than with metrifonate (Taraschewski et al. 1988). Although these drugs have an excellent effectiveness against adult parasites in eels, the larvae in the swim-bladder wall are not affected. Ashworth and Blanc (1997) suggested applying levamisole to increase the success of treatment.

9.12 The Effects of Swim-Bladder Parasite Infection on Swimming Performance of Silver Eels

The reviewed effects of *A. crassus* infection on physiology and survival of the European eel mainly concerned the continental yellow eels stage. However, since the first appearance of *A. crassus* in Europe, concern has been expressed about possible effects of the parasite upon the migration of adult eels back to the Sargasso Sea to spawn. Yellow eels experience stress in their movements from freshwater to the sea and in their transition to silver eels. The silver eels themselves depend on their swim-bladder as a hydrostatic organ in the course of their migration to the Sargasso. Knowledge of the effects of *A. crassus* on the gas gland and oxygen concentration in the swim-bladder suggests very strongly that its ability to function as a hydrostatic organ will be impaired in infected eels and this must surely affect their vertical movements on migration as well as their swimming performance (see also Chapter 5).

Swim-bladder parasites drain energy due to their sanguivorous feeding and they cause mechanical damage on the swim-bladder wall. These two effects are hypothesized to impair the spawning migration of the European eel. Two earlier studies investigated the influence of *A. crassus* on the swimming of eel

(Sprengel and Lüchtenberg 1991; Münderle et al. 2004). However, those studies were performed with small yellow eels (≤45 cm). Obviously large silver eels had to be tested over long distances and periods, as not only swimming speed but particularly a low cost of transport and a high endurance are crucial for long distance migration. Therefore, in a recent study (Palstra et al. 2007a), we have investigated both effects on swimming performance. We hypothesized that parasitic sanguivorous activities – related to parasite weight – reduce swimming endurance, while mechanical damage of the swim-bladder impairs buoyancy control. Recently, we have developed an experimental test to quantify swimming performance (Palstra et al. 2006) using 22 swim-tunnels suitable for exercising many large female silver eels at the same time. This test and set-up was used to investigate the relation between swimming endurance and the adverse effects of *A. crassus* infection. The relation between *A. crassus* infection and swimming efficiency was measured for 80 large female silver eels suffering various degrees of infection and swimming at various swimming speeds. It was found that:

- Oxygen consumption in rest and at critical aerobic swimming speeds were unaffected by infection (Fig. 9.7a, b).
- Critical aerobic swimming speed values tended to be negatively correlated with infection and damage levels (Fig. 9.7c).
- Eels with damaged swim-bladders dropped out early in comparison with the healthy eels; most of them could not swim faster than 0.7 m s^{-1} (Fig. 9.7d).
- Infected and damaged eels had at all speeds higher O_2 consumption levels and corresponding cost of transport (COT) levels.
- The optimum swimming speed of infected eels was significantly lower by 18% and that of eels with damaged swim-bladders by 21% in comparison with healthy eels (Fig. 9.7e).
- These eels also showed a higher (non significant) COT at their optimum swimming speeds by respectively 21% and 18% (Fig. 9.7f).

Hence, it could be concluded that both infected eels as well as eels without any parasites but with a damaged swim-bladder showed a considerable loss of swimming endurance. Simulated migration trials confirmed that eels with a high parasite level or with damaged swim-bladders show early migration failure (< 1,000-km; Palstra et al. 2007b). Since swimming performance in both groups of eels was reduced, effects thus seem to be associated with swim-bladder dysfunction.

9.13 Implications for Eel Migration and Reproductive Success

The swim-bladder parasite has high impact during migration since (a) silver eels have much higher infection levels than yellow eels (Palstra et al. 2007a), (b) the infection level does not diminish during longer periods of up to 6 months (Székely et al. 2005; Palstra et al. 2007a), and (c) infection continues under salt water conditions (Kennedy and Fitch 1990; Kirk et al. 2000a, b, 2002; Palstra et al. 2007a).

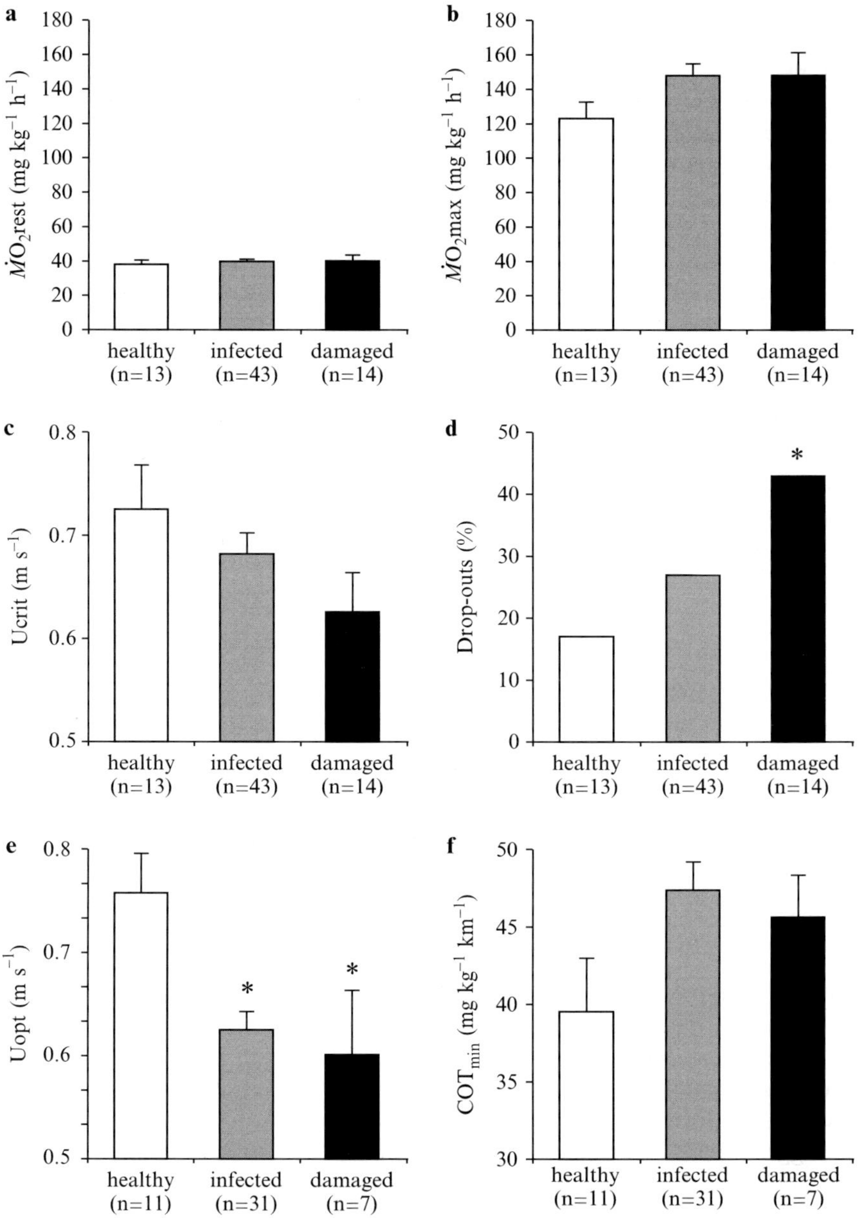

Fig. 9.7 Swimming parameters of healthy eels (white bars), infected eels (grey bars) and damaged eels (black bars). Healthy eels had large swim-bladders (SBI ≥ 10) without parasites. Infected eels had all-sized swim-bladders with parasites. Damaged eels had small swim-bladders (SBI < median 10.3%) without parasites. Significant differences ($P < 0.05$) are indicated by asterisk. No significant differences were found for (a) O_2 consumption in rest ($\dot{M}O_{2\,rest}$) (b) and maximal O_2 consumption ($\dot{M}O_{2\,max}$), (c) critical swimming speeds (U_{crit}) tended to decrease with increasing damage and (d) 43% of these eels dropped out before reaching U_{opt} (Mann-Whitney; $P = 0.03$), (e) eels with small swim-bladders had lower optimum swimming speeds U_{opt} (ANCOVA; $P = 0.01$) and (f) cost of transport (COT_{min}) tended to increase with increasing damage (Reproduced from Palstra et al. 2007a. With permission from Elsevier)

When we assume that eels continuously cruise at optimum swimming speeds, the trip would take only 3.5 months (Palstra et al. 2006). The 20% lower optimum swimming speed of infected and damaged eels may cause them to cruise slower, which extents the swimming period to about 4.2 months. They may arrive too late at the spawning grounds for final maturation and the spawning event itself.

The cost of transport at optimum swimming speeds was about 20% higher in heavily infected and damaged eels than in healthy eels, making it likely that they will spend at least 20% more of their energy reserves on migration. This leaves less fat for egg production. In a study on the spawning characteristics of downstream migrating silver eels from the River Rhine (A. Palstra et al. 2005, unpublished results), a negative correlation was found between the number of parasites (up to n = 46 parasites per swim-bladder) and the relative gonad mass suggesting a direct effect of infection on maturation. Müller et al. (2001) concluded that *Anguillicola crassus* infection is not a barrier factor for the artificial induction of maturation. Egg quality may however be lower in these eels.

From the study of Palstra et al. (2007a), it can be concluded that fewer eels will be able to reach the spawning grounds. These eels may arrive too late and egg quality of these eels may be lower. With these effects on spawning migration and gamete quality, the swim-bladder parasite is a serious threat for the overall reproductive success of the European eel.

9.14 *Anguillicola crassus* Infection and Eel Decline

It was thought at one time that the decline in population levels of *A. anguilla* throughout Europe during the 1980s might be directly related to the spread and increase of *A. crassus* over the same period. A decline in eel populations and elver runs has been well documented throughout the continent (Moriarty and Dekker 1997), and a number of factors including overfishing of elvers and adults and global warming have been considered to be wholly or partially responsible. The correlation between the increase in *A. crassus* infection levels and decrease in host population levels might suggest a causal relationship, but doubt was thrown upon this suggestion when it was realized that a similar decline in magnitude of recruitment (98%) was taking place simultaneously in the *A. rostrata* population in North America at a time before *A. crassus* had spread to that continent. This decline was also blamed on overfishing and pollution. However, it has been suggested that the co-incidence in timing of the declines on both sides of the Atlantic implies an Atlantic-wide cause, e.g. changes in climate or Gulf Stream (Castonguay et al. 1994). Nevertheless, given the effects on basic physiology and survival during the yellow eel phase, and the impaired swimming performance of silver eels, it is very hard to believe that *A. crassus* is not at least partially responsible for, or does not contribute to the decline in eel populations and many workers believe that this is in fact the case (Køie 1991; Sures and Knopf 2004).

9.15 Conclusion

Anguillicola crassus is a pathogenic nematode of *Anguilla anguilla*. This parasite infecting originally pacific eels without obvious pathogenic symptoms was introduced into the Atlantic region in the years of 1980s, where it caused heavy infections and even massive fish dies in the European eel. Here the parasite has altered its life cycle by infecting a wide range of intermediate hosts, employing paratenic hosts and surviving as larvae for months in the swim-bladder wall. Pathogenic symptoms are partially caused by adult worms living in the lumen of the swim-bladder and feeding on blood, but major degeneration changes are caused by migrating larvae boring through the intestinal wall and causing proliferative, oedematous, fibrous and degenerative changes in the swim-bladder wall. Major pathogenic effects on eels result from haemorrhaging in, and thickening of the swim-bladder wall. At heavy infections no worms and air are formed in the swim-bladder thickened ten times compared to the original thickness. The process of development and changes in the swim-bladder wall were followed by x-ray and CT methods. The loss of oxygen concentration in the swim-bladder seems to reduce function as a hydrostatic organ, and increases the stress response of eels. In shallow lakes at warm temperatures this can result in mass mortalities. Experiments performed by the authors proved that swim-bladder infection had an adverse effect on swimming performance. It is, therefore feared that the parasite negatively affects the ability of eels to migrate to the Sargasso Sea and so contributes to the decline in eel populations.

Acknowledgements This work has been supported by the European contract EELREP N°Q5RS-2001-01836 and partially by the Hungarian Scientific Research Fund (OTKA project No. T 45891 and K71837).

References

Ashworth ST, Blanc G (1997) *Anguillicola crassus*, un colonisateur agressif récemment introduit dans les stocks européeans d'anguilles. Bulletin Francais de Pêche et Pisciculture 1997: 342–345.

Ashworth ST, Kennedy CR (1999) Density-dependent effects on *Anguillicola* crassus (Nematoda) within its European eel definitive host. Parasitology 118: 289–296.

Ashworth ST, Kennedy CR, Blanc G (1996) Density-dependent effects of *Anguillicola crassus* (Nematoda) within and on its copepod intermediate hosts. Parasitology 113: 303–309.

Balint T, Ferency J, Kåtåi F, Kiss I, Kråczer L, Kufesák O, Láng G, Polyhos C. Szabó I, Szegleres T, Nemcsók J (1997) Similarities and differences between the massive eel (*Anguilla anguilla* L.) devastations that occurred in Lake Balaton in 1991 and 1995. Ecotoxicology and Environmental safety 37: 17–23

Barse AM, Secor DH (1999) An exotic nematode parasite of the American eel. Non-native Species Management 24: 6–10.

Barse AM, McGuire SA, Vinores MA, Eierman LE, Weeder JA (2001) The swimbladder nematode *Anguillicola crassus* in American eels (*Anguilla rostrata*) from middle and upper regions of Chesapeake Bay. Journal of Parasitology 87: 1366–1370.

Barus V (1995) First record of *Anguillicola crassus* (Nematoda) in the Morava River drainage basin. Helminthologia (Bratislava) 32: 89.

Baruš V, Prokeš M (1996) Length-weight relations of uninfected and infected eels (*Anguilla anguilla*) by *Anguillicola crassus* (Nematoda). Folia Zoologica 45: 183–189.

Baruš V, Kráčmar S, Tenora F, Prokeš M (1998) Amino acid spectrum of uninfected and infected eels (*Anguilla anguilla*) by *Anguillicola crassus* (Nematoda). Folia Zoologica 47: 231–234.

Baruš V, Moravec F, Prokeš M (1999) Anguillicolosis of the European eel (*Anguilla anguilla*) in the Czech Republic. Czechoslovak Journal of Animal Science 44: 423–431.

Baruš V, Tenora F, Kráčmar S, Dvořáček J (1999a) Contents of several inorganic substances in European eel infected and uninfected by *Anguillicola crassus* (Nematoda). Diseases of Aquatic Organisms 37: 135–137.

Beregi A, Molnár K, Békési L, Székely CS (1998) Radiodiagnostic method for studying swim-bladder inflammation caused by *Anguillicola crassus* (Nematoda: Dracunculiodea). Diseases of Aquatic Organisms 34: 155–160.

Boetius I (1990) Preliminary report on the occurrence of anguillicola in some Danish fresh-water and seawater areas. Internationale Revue der Gesamten Hydrobiologie 75: 889–889.

Boon JH, Lokin CJA, Ceusters R, Ollevier F (1989) Some properties of the blood of European eel (*Anguilla anguilla*) and the possible relationship with *Anguillicola crassus* infestations. Aquaculture 76: 203–208.

Boon JH, Augustijn H, Cannaerts VMH, Lokin CJA, Machiels MAM, Ollevier F (1990a) The suitability of experimental inoculations with infective larvae of *Anguillicola crassus* and their effects on the growth and mortality of the European eel (*Anguilla anguilla*). Aquaculture 87: 111–120.

Boon JH, Cannaerts VMH, Augustijn H, Machiels MAM, De Charleroy D, Ollevier F (1990b) The effects of different infection levels with infective larvae of *Anguillicola crassus* on haematological parameters of European eel (*Anguilla anguilla*). Aquaculture 87: 243–253.

Buchmann K (1993) Epidémiologie et contrôle des infections à *Pseudodactylogyrus* dans les elevages intensifs d'anguilles: Données actuelles. Bulletin Francais Peche Pisciculture 328: 66–73.

Buchmann K, Mellergaard S, Køie M (1987) *Pseudodactylogyrus* infections in eel: A review. Diseases of Aquatic Organisms 3: 51–57.

Bykhovskaya-Pavlovskaya IE, Gussev AV, Dubinina MN, Izyumova NA, Smirnova TS, Sokolovskaya IL, Sthein GA, Shulman SS, Epsthein VM (1962) Key to the parasites of freshwater fish of the USSR. Moscov-Leningrad, Izdatelstvo Akademii Nauk SSSR. 776 pp. (in Russian).

Castonguay M, Hodson PV, Moriarty C, Drinkwater KF, Jessop BM (1994) Is there a role of ocean environment in American and European eel decline? Fisheries Oceanography 3: 197–203.

Copland JW (1983) The pathology of *Myxidium giardi* Cépéde, 1906 infections in wild and cultured eels *Anguilla anguilla* L. Journal of Fish Diseases 6: 209–217.

Csaba G, Láng M, Sályi G, Ramotsa J, Glávits R, Rátz F (1993) The nematode *Anguillicola crassus* (Nematoda, Anguillicolidae), and its role in the death of eels in the Lake Balaton during 1991. Magyar Állatorvosok Lapja 48: 11–21.

De Charleroy D, Grisez I, Thomas K, Belpaire C, Olivier F (1990) The life cycle of *Anguillicola crassus*. Diseases of Aquatic Organisms 8: 77–84.

Di Cave D, Berrilli F, De Liberato C, Orecchia P, Kennedy CR (2001) Helminth communities in eels *Anguilla anguilla* from adriatic coastal lagoons in Italy. Journal of Helminthology 75: 7–13.

Egusa S (1979) Notes on the culture of the European eel (*Anguilla anguilla* L.) in Japanese eel farming ponds. Rapports et Proces-Verbaux des Réunions. Conseil International pour l'Éxploration de la Mer 174: 51–58.

Egusa S, Hirose H (1983) Anguillicolosis of eels. In: Egusa S (ed) Fish Pathology (Infectious Diseases and Parasitic Diseases). Koseisha Koseikaku, Tokyo, pp. 308–310.

El-Hilali M, Yahyaoui A, Sadak A, Macchi M, Taghy Z (1996) Data on anguillicolosis observed for the first time in Morocco. Bulletin Francais de la Peche et de la Pisciculture 340: 57–60.

Fazio G, Moné H, Mouahid G, Sasal P (2008a) Biased sex ratio in the European eel (*Anguilla anguilla*) swimbladder parasite *Anguillicola crassus*, experimentally induced by 11-ketotestosterone. The Journal of Parasitology 94: 956–958.

Fazio G, Moné H, Lecomte-Finiger R, Sasal P (2008b) Differential gene expression analysis in European eels (*Anguilla anguilla*, l. 1758) naturally infected by macroparasites. The Journal of Parasitology 94: 571–577.

Fries LT, Williams DJ, Johnson SK (1996) Occurrence of *Anguillicola crassus*, an exotic parasitic swimbladder nematode of eels, in south-western United States. Transactions of the American Fisheries Society 125: 794–797.

Geets A, Liewes EW, Ollevier F (1992) Efficacy of some anthelmintics against the swimbladder nematode *Anguillicola crassus* of eel *Anguilla anguilla* under saltwater conditions. Diseases of Aquatic Organisms 13: 123–128.

Gollock MJ, Kennedy CR, Quabius ES, Brown JA (2004) The effect of parasitism of European eels with the nematode, *Anguillicola crassus* on the impact of netting and aerial exposure. Aquaculture 233: 45–54.

Gollock MJ, Kennedy CR, Brown JA (2005a) Physiological responses to acute temperature increase in European eels *Anguilla anguilla* infected with *Anguillicola crassus*. Diseases of Aquatic Organisms 64: 223–228.

Gollock MJ, Kennedy CR, Brown JA (2005b) European eels, *Anguilla anguilla* (L.), infected with *Anguillicola crassus* exhibit a more pronounced stress response to severe hypoxia than uninfected eels. Journal of Fish Diseases 28: 429–436.

Golovin PP (1977) Monogeneans in eel during its culture using heated water. In: Skarlato OA (ed) Investigation of Monogenea in USSR. Academy of Sciences, Leningrad, pp. 144–150 (in Russian).

Haenen OLM, Grisez L, De Charleroy D, Belpaire C, Ollevier F (1989) Experimentally induced infections of the European eel *Anguilla anguilla* with *Anguillicola crassus* (Nematoda, Dracunculoidea) and subsequent migration of larvae. Diseases of Aquatic Organisms 7: 97–101.

Haenen OLM, van Banning P, Dekker W (1994) Infection of eel *Anguilla anguilla* (L.) and smelt *Osmerus eperlanus* (L.) with *Anguillicola crassus* (Nematoda, Dracunculoidea) in the Netherlands from 1986 to 1992. Aquaculture 126: 219–229.

Haenen OLM, van Wijngaarden TAM, van der Heijdedn MHT, Höglund J, Cornelissen JBJW, van Leengoed LAMG, Borgsteede FHM, van Muiswinkel WB (1996) Effects of experimental infections with different doses of *Anguillicola crassus* (Nematoda, Dracunculoidea) on European eel (*Anguilla anguilla*). Aquaculture 141: 41–57.

Hartmann S (1987) Schwimmblasenwürmer beim aal. Fischer und Teichwirt 1: 2–3.

Hine PM, Boustead NC (1974) A guide to disease in eel farms. Fisheries Research Division Occasional Publication 6: 1–26.

Höglund J, Thomas K (1992) The black goby *Gobius niger* as a potential paratenic host for the parasitic nematode *Anguillicola crassus* in a thermal effluent of the Baltic. Diseases of Aquatic organisms 13: 175–180.

Höglund J, Anderrson J, Wickström H, Reizenstein M (1992) The distribution of *Anguillicola* in Sweden and its association with thermal discharge areas. Irish Fisheries Investigations series A (Freshwater) 36: 143–150.

Kamstra A (1990 vagy 1991) *Anguillicola* in Dutch eelfarms; Current state. International Revue der gesamten Hydrobiologie 75: 867–874.

Kelly CE, Kennedy CR, Brown JA (2000) Physiological status of wild European eels (*Anguilla anguilla*) infected with the parasitic nematode, *Anguillicola crassus*. Parasitology 120: 195–202.

Kennedy CR (1994) The distribution and abundance of the nematode *Anguillicola australiensis* in eels *Anguilla reinhardtii* in Queensland, Australia. Folia Parasitologica 41: 279–285.

Kennedy CR (2007) The pathogenic helminth parasites of eels. Journal of Fish Diseases 30: 319–334.

Kennedy CR, Fitch DJ (1990) Colonization, larval survival and epidemiology of the nematode *Anguillicola crassus*, parasitic in the eel, *Anguilla anguilla*, in Britain. Journal of Fish Biology 36: 117–131.

Kirk RS (2003) Impact of *Anguillicola crassus* on European eels. Fisheries Management and Ecology 10: 385–394.

Kirk RS, Lewis JW, Kennedy CR (2000a) Survival and transmission of *Anguillicola crassus* Kuwahara, Niimi and Itagaki, 1974 (Nematoda) in seawater eels. Parasitology 120: 289–295.

Kirk RS, Kennedy CR, Lewis JW (2000b) Effect of salinity on hatching, survival and infectivity of *Anguillicola crassus* (Nematoda: Dracunculoidea) larvae. Diseases of Aquatic Organisms 40: 211–218.

Kirk RS, Morritt D, Lewis JW, Kennedy CR (2002) The osmotic relationship of the swim-bladder nematode *Anguillicola crassus* with seawater eels. Parasitology 124: 339–347.

Knopf K, Mahnke M (2004) Differences in susceptibility of the European eel (*Anguilla anguilla*) and the Japanese eel (*Anguilla japonica*) to the swimbladder nematode *Anguillicola crassus*. Parasitology 129: 491–496.

Knopf K, Würtz J, Sures B, Taraschewski H (1998) Impact of low water temperature on the development of *Anguillicola crassus* in the final host *Anguilla anguilla*. Diseases of Aquatic Organisms 33: 142–149.

Knopf K, Naser K, Van der Heijden MHT, Taraschewski H (2000a) Humoral immune response of European eel *Anguilla anguilla* experimentally infected with *Anguillicola crassus*. Diseases of Aquatic Organisms 42: 61–69.

Knopf K, Naser K, Van der Heijden MHT, Taraschewski H (2000b) Evaluation of an ELISA and immunoblotting for studying the humoral immune response in *Anguillicola crassus* infected European eel (*Anguilla anguilla*). Diseases of Aquatic Organisms 43: 39–48.

Koops H, Hartmann F (1989) Infection of eels from different regions with *Anguillicola*. Journal of Applied Ichthyology 1: 41–45.

Køie M (1988) Parasites in eels, *Anguilla anguilla* (L.) from eutrophic Lake Esrum (Denmark). Acta Parasitologica Polonica 33: 89–100.

Køie M (1991) Swimbladder nematodes (*Anguillicola* spp.) and gill monogeneans (*Pseudodctylogyrus* spp.) parasitic on the European eel (*Anguilla anguilla*). Journal de la Conseil International pour Exploration de la Mer 47: 391–398.

Lefebvre F, Mounaix B, Poizat G, Crivelli AJ (2004) Impacts of the swimbladder nematode *Anguillicola crassus* on *Anguilla anguilla*: Variations in liver and spleen masses. Journal of Fish Biology 64: 435–447.

Liewes EW, Schaminee-Main S (1987) Onderzoek aalparasiet vordert. Aquacultuur 2: 5–17.

Lom J, Molnár K, Dyková I (1986) *Hoferellus gilsoni* (Debaisieux,1925) comb. n. (Myxozoa, Myxosporea): Redescription and mode of attachment to the epithelium of the urinary bladder of its host, the European eel. Protistologica 22: 405–413.

Maamouri F, Gargouri L, Daddah MO, Bouix G (1999) Occurrence of *Anguillicola crassus* (Nematode, Anguillicolidae) in th Ichkeul Lake (Northern Tunisia). Bulletin of the European Association of Fish Pathologists 19: 17–19.

Mellergaard S (1988) The swimbladder nematode of eel – a new parasite in the European eel stock Norges. Aquaculture 4: 50–54.

Möller, H, Holst S, Lüchtenberg H, Petersen F (1991) Infection of eel *Anguilla anguilla* from the River Elbe estuary with two nematodes, *Anguillicola crassus* and *Pseudoterranova decipiens*. Diseases of Aquatic Organisms 11: 193–199.

Molnár K (1983) Das Vorkommen von parasitären Hakensaugwurmern bei der Aalaufzucht in Ungarn. Zeitschrift für Binnenfischerei der DDR 30: 341–345.

Molnár K (1993) Effects of decreased oxygen content on eels (*Anguilla anguilla*) infected by *Anguillicola crassus* (Nematoda: Dracunculoidea). Acta Veterinaria Hungarica 41: 349–360.

Molnár K (1994) Formation of parasitic nodules in the swimbladder and intestinal walls of the eel *Anguilla anguilla* due to infections with larval stages of *Anguillicola crassus*. Diseases of Aquatic Organisms 20: 163–170.

Molnár K, Moravec F (1994) Third-stage larvae of *Daniconema anguillae* (Nematoda: Dracunculoidea) in the subcutaneous tissue of eel *Anguilla anguilla*. Folia Parasitologica 41: 215–219.

Molnár K, Székely C, Baska F (1991) Mass mortality of eel in Lake Balaton due to *Anguillicola crassus* infection. Bulletin of the European Association of Fish Pathologists 11: 211–212.

Molnár K, Baska F, Csaba G, Glávits R, Székely C (1993) Pathological and histopathological studies of the swimbladder of eels *Anguilla anguilla* infected by *Anguillicola crassus* (Nematoda: Dracunculoidea). Diseases of Aquatic Organisms 15: 41–50.

Molnár K, Székely C, Perényi M (1994) Dynamics of *Anguillicola crassus* (Nematoda: Dracunculoidea) infection in eels of Lake Balaton, Hungary. Folia Parasitologica 41: 193–202.

Moravec F (1994) Parasitic Nematodes of Freshwater Fishes of Europe. Academia Press, Praha, 473 pp.

Moravec F (1996) Aquatic invertebrates (snails) as new paratenic hosts of *Anguillicola crassus* (Nematoda: Dracunculoidea) and the role of paratenic hosts in the life cycle of this parasite. Diseases of Aquatic Organisms 27: 237–239.

Moravec F, Køie M (1987) *Daniconema anguillae* gen. et sp. n., a new nematode of a new family Daniconematidae fam. n. parasitic in European eels. Folia Parasitologica 34: 335–340.

Moravec F, Konecny R (1994) Some new data on the intermediate and paratenic hosts of the nematode *Anguillicola crassus* Kuwahara, Niimi et Itagaki, 1974 (Dracunculoidea), a swimbladder parasite of eels. Folia Parasitologica 41: 65–70.

Moravec F, Škoriková B (1998) Amphibians and larvae of aquatic insects as new paratenic hosts of *Anguillicola crassus* (Nematoda: Dracunculoidea), a swimbladder parasite of eels. Diseases of Aquatic Organisms 34: 217–222.

Moravec F, Taraschewski H (1988) Revision of the genus *Anguillicola* Yamaguti, 1935 (Nematoda: Anguillicolidae) of the swimbladder of eels, including descriptions of two new species, *A. novaezelandiae* sp.n. and *A. papernai* sp.n. Folia Parasitologica 35: 125–146.

Moravec F, Di Cave D, Orecchia P, Paggi L (1993) Studies on the development of *Anguillicola crassus* Kuwahara, Niimi et Itagaki, 1974 (Nematoda: Dracunculoidea) in the intermediate host. Folia Parasitologica 40: 39–48.

Moravec F, Di Cave D, Orecchia P, Paggi L (1994a) Present occurrence of *Anguillicola novaezelandiae* (Nematoda: Drancunculoidea) in Europe and its development in the intermediate host. Folia Parasitologica 41: 203–208.

Moravec F, Di Cave D, Orecchia P, Paggi L (1994b) Experimental observations on the development of *Anguillicola crassus* (Nematoda: Dracunculoidea) in its definitive host, *Anguilla anguilla*, Pisces. Folia Parasitologica 41: 138–148.

Moriarty C, Dekker W (1997) Management of the European eel. Fisheries Bulletin (Dublin) 15: 1–110.

Moser ML, Patrick WS, Crutchfield JU (2001) Infection of American eels, *Anguilla rostrata*, by the introduced nematode parasite, *Anguillicola crassus*, in North Carolina. Copeia 3: 848–853.

Müller T, Binder T, Váradi B, Bercsényi M (2001) Induced maturation and successful egg release in the European eel (*Anguilla anguilla* L.). Halászat 94: 115–118 (in Hungarian with English summary).

Münderle M, Sures B, Taraschewski H (2004) Influence of *Anguillicola crassus* (Nematoda) and *Ichthyophthirius multifiliis* (Ciliophora) on swimming activity of European eel *Anguilla anguilla*. Diseases of Aquatic Organisms 60: 133–139.

Murai É (1971) Tapeworms (Cestodes) parasiting eels introduced into Lake Balaton. Parasitologica Hungarica 4: 145–156.

Nagasawa K, Kim Y-G, Hirose H (1994) *Anguillicola crassus* and *A. globiceps* (Nematoda: Dracunculoidea) parasitic in the swimbladder of eels (*Anguillicola japonica* and *A. anguilla*) in East Asia: A review. Folia Parasitologica 41: 127–137.

Nielsen ME (1999) An enhanced humoral immune response against the swimbladder nematode, *Anguillicola crassus*, in the Japanese eel, *Anguilla japonica*, compared with the European eel, *A. anguilla*. Journal of Helminthology 73: 227–232.

Nimeth K, Zwerger P, Würtz J, Salvenmoser W, Pelster B (2004) Infection of the glass-eel swimbladder with the nematode *Anguillicola crassus*. Parasitology 121: 75–83.

Paggi L, Orecchia P, Minervini R, Mattiucci S (1982) Sulla comparsa de *Anguillicola australiensis* Johnston and Mawson, 1940 (Dracunculoidea: Anguillicolidae) in *Anguilla anguilla* del Lago di Bracciano. Parassitologia 24: 139–144.

Palstra A, Van Ginneken V, Van den Thillart G (2006) Swim performance of European silver eels (*Anguilla anguilla*). In: Palstra AP (ed) Energetic Requirements and Environmental Constraints of Reproductive Migration and Maturation of European Silver Eel (*Anguilla anguilla* L.). Ph.D. thesis, University of Leiden, Leiden, The Netherlands.

Palstra AP, Heppener DFM, Van Ginneken VJT, Székely C, Van den Thillart GEEJM (2007a) Swimming performance of silver eels is severely impaired by the swim-bladder parasite *Anguillicola crassus*. Journal of Experimental Marine Biology and Ecology 352(1): 244–256.

Palstra AP, Curiel D, Fekkes M, De Bakker M, Székely C, Van Ginneken VJT, Van den Thillart GEEJM (2007b) Swimming stimulates oocyte development in European eel. Aquaculture 270: 321–332.

Pazooki J, Székely C (1994) Survey of the paratenic hosts of *Anguillicola crassus* in Lake Velence, Hungary. Acta Veterinaria Hungarica 42: 87–97.

Peters G, Hartmann F (1986) *Anguillicola*, a parasitic nematode of the swimbladder spreading among eel populations in Europe. Diseases of Aquatic Organisms 1: 229–230.

Puquin W, Yuru Z (1980) Observations of the life history of *Anguillicola globiceps* (Nematoda: Anguillicolidae). Acta Zoologica Sinica 26: 243–249.

Reimer LW, Hildebrand A, Scharberth D, Walter U (1994) *Anguillicola crassus* in the Baltic Sea: Field data supporting transmission in brackish waters. Diseases of Aquatic Organisms 18: 77–79.

Saraiva A, Molnár K (1990) *Myxobolus portucalensis* n. sp. in the fins of European eel *Anguilla anguilla* (L.) in Portugal. Revista Iberica Parasitologica 50: 31–35.

Schabuss M, Kennedy CR, Konecny R, Grillitsch B, Reckendorfer W, Schiemer F, Herzig A (2005) Dynamics and predicted decline of *Anguillicola crassus* infection in European eels, *Anguilla andhilla*, in Neusiedler See, Austria. Journal of Helminthology 79: 159–167.

Scholz F, Zerbst-Boroffka I (1994) Osmo-und ionenegulation des aalparasiten *Anguillicola crassus* (Nematoda) und seines wirtes (*Anguilla anguilla*). Zoologische Beiträgen 35: 103–117.

Sprengel G, Lüchtenberg H (1991) Infection by endoparasites reduces maximum swimming speed of European smelt *Osmerus eperlanus* and European eel *Anguilla anguilla*. Diseases of Aquatic Organisms 11: 31–35.

Sures B, Knopf K (2004) Individual and combined effects of cadmium and 3, 3′, 4, 4′, 5-pentachlorobiphenyl (PCB 126) on the humoral immune response in European eel (*Anguilla anguilla*) experimentally infected with larvae of *Anguillicola crassus* (Nematoda). Parasitology 128: 1–10.

Székely C (1994) Paratenic hosts for the parasitic nematode *Anguillicola crassus* in Lake Balaton, Hungary. Diseases of Aquatic Organisms 18: 11–20.

Székely C (1995) Dynamics of *Anguillicola crassus* (Nematoda: Dracunculoidea) larval infection in paratenic host fishes of Lake Balaton, Hungary. Acta Veterinaria Hungarica 43: 401–422.

Székely C (1996) Experimental studies on the infectivity of *Anguillicola crassus* third-stage larvae (Nematoda) from paratenic hosts. Folia Parasitologica 43: 305–311.

Székely C, Láng M, Csaba G (1991) First occurrence of *Anguillicola crassus* in Hungary. Bulletin of the European Association of Fish Pathologists 11: 162–163.

Székely C, Pazooki J, Molnár K (1996) Host reaction in paratenic fish hosts against 3rd stage larvae of *Anguillicola crassus*. Diseases of Aquatic Organisms 26: 173–180.

Székely C, Molnár K, Müller T, Szabó A, Romvári R, Hancz C, Bercsényi M (2004) Comparative study of x-ray computed tomography and conventional x-ray methods in the diagnosis of swimbladder infection of eel caused by *Anguillicola crassus*. Diseases of Aquatic Organisms 58: 157–164.

Székely C, Molnár K, Rácz O (2005) Radiodiagnostic method for studying the dynamics of *Anguillicola crassus* (Nematoda: Dracunculoidea) infection and pathological status of the swim-bladder in Lake Balaton eels. Diseases of Aquatic Organisms 64: 53–61.

Taraschewski H (2006) Hosts and parasites as aliens. Journal of Helminthology 80: 99–128.

Taraschewski H, Moravec F, Lamah T, Anders K (1987) Distribution and morphology of two helminths recently introduced into European eel populations: *Anguillicola crassus* (Nematoda,

Dracunculoidae) and *Paratenuis ambiguus* (Acanthocephala: Tenuisentidae). Diseases of Aquatic Organisms 3: 167–176.

Taraschewski Hk, Renner C, Mehlhorn H (1988) Treatment of fish parasites 3. Effects of levamisole HCl, metrifonate, fenbenzadole, mebenzadole, and invermectin on *Anguillicola crassus* (nematodes) pathogenic in the air bladder of eels. Parasitology Research 74: 281–289.

Taraschewski H, Boomker J, Knopf H, Moravec F (2005) *Anguillicola papernai* (Nematoda: Anguillicolidae) and other helminths parasitizing the African longfin eel *Anguilla mossambica*. Diseases of Aquatic Organisms 63: 185–195.

Tátrai I, Paulovits G, Szabó I (2002) The stock of introduced fish species in Lake Balaton. In: Mahunka S, Banczerowski J (eds) Results of Balaton Research in 2001. Magyar Tudományos Akadémia, Budapest, pp. 149–159 (in Hungarian).

Thomas K, Ollevier F (1992) Paratenic hosts of the swimbladder nematode *Anguillicola crassus*. Diseases of Aquatic Organisms 13: 165–174.

Thomas K, Ollevier F (1993) Hatching, survival, activity and penetration efficiency of second-stage larvae of *Anguillicola crassus* (Nematoda). Parasitology 107: 211–217.

Van Banning P, Haenen O (1990) Effect of the swimbladder nematode *Anguillicola crassus* in wild and farmed eel, *Anguilla anguilla*. In: Perkins FO, and Cheng TC (eds) Pathology in Marine Science. Academic Press, New York, pp. 317–330.

Wang P, Zhao Y (1980) Observations on the life history of *Anguillicola globiceps* (Nematoda: Anguiillicolaidae). Acta Zoologica Sinica 26: 243–249.

Würtz J, Taraschewski H (2000) Histopathological changes in the swimbladder wall of the European eel *Anguilla anguilla* due to infection with *Anguillicola crassus*. Diseases of Aquatic Organisms 39: 121–134.

Würtz J, Taraschewski H, Pelster B (1996) Changes in the gas composition of the swimbladder of the European eel (*Anguilla anguilla*) infected with *Anguillicola crassus* (Nematoda). Parasitology 112: 233–238.

Part IV
Maturation and Reproduction

Chapter 10
Effects of Swimming on Silvering and Maturation of the European Eel, *Anguilla anguilla* L.

Arjan Palstra, Vincent van Ginneken, and Guido van den Thillart

Abbreviations 11-KT: 11-ketotestosteron; ACTH: adrenocorticotropic hormone; Ca: calcium; DHP: dihydroxy progesterone (17, 20β-dihydroxy-4-pregnen-3-one); E2: 17β-estradiol; EI: eye index; ERα: estradiol receptor α; FSH: follicle stimulating hormone; FSH-β: FSH specific subunit β; FW: fresh water; GnRH: gonadotropin releasing hormone; GPα: gonadotropin common α subunit; GSI: gonadosomatic index (relative gonad mass); LH: luteinising hormone; LH-β: LH specific subunit β; NEFA: non-esterified fatty acids; q-rtPCR: quantitative real time polymerase chain reaction; StAR: steroidogenic acute regulatory protein; SW: salt water; T: testosteron; VTG: vitellogenin; αMSH: melanophore-stimulating hormone α

10.1 Introduction

Every year at the end of the growth season in autumn, the majority of the large European eels cease feeding, become restless and start to mature. Only those with sufficient lipid stores (Larsson et al. 1990; Svedäng and Wickström 1997) will start their reproductive migration to the spawning grounds in the Sargasso. They leave in a prepubertal condition with less than 2% relative gonad mass (GSI). The smaller males leave by August (Usui 1991). The large and fatty females leave between October and December to arrive an estimated 3.5 months later (Palstra et al. 2008a) at the spawning grounds. There, in early spring, males reach gonad masses of 10% and females of 40% to 60% of the body mass (Palstra et al. 2005).

A. Palstra, V. van Ginneker, and G. van den Thillart
Integrative Zoology, Institute of Biology Leiden, Van der Klaauw Laboratories, P.O. Box 9516, Kaiserstraat 63, 2300 RA Leiden, The Netherlands

G. van den Thillart et al. (eds.), *Spawning Migration of the European Eel*,

They are sexually mature and ready to spawn more than one million eggs (Van Ginneken et al. 2005).

As European eels have to swim about 5,500-km to reach their spawning site, swimming is supposed to play a crucial role in natural triggering and stimulation of maturation. Fine-tuning between migration and maturation of fishes is a non-elucidated research topic that deserves much more attention. It is peculiar that the influence of swim exercise on maturation has never been thoroughly investigated, especially since fatty migrant fishes like tuna and eel are of major commercial interest but very difficult or even impossible to reproduce in captivity.

10.2 Environmental Triggers of Silvering and Maturation

10.2.1 The Continental Phase: Depressed Lipid Mobilization and Pre-pubertal Blockage of Maturation

The continental phase in the life cycle of the European eel can be considered as a feeding phase to attain the lipid reserves to fuel migration and to provide the future offspring with sufficient energy. In order to store the required amount of lipid before maturing, this phase is characterised by a severely depressed lipid mobilization (EELREP 2005) and blockage of maturation. Pre-pubertal blockage of maturation is due to a deficient gonadotropin releasing hormone (GnRH) stimulation and a simultaneous inhibition of the pituitary gonadotropes FSH and LH by dopamine (Chapter 12). These gonadotropes control gonad development directly, or indirectly by acting on steroid metabolism in the production of estrogens and androgens in both male and female eels. This dual neuroendocrine control is extreme, but not specific for eels and occurs in various adult teleosts (Vidal et al. 2004). However, dopamine only counteracts regulation of the last steps of gametogenesis in these species while in eel, dopamine seems to play a role in earlier stages (Vidal et al. 2004). The extreme blockage of maturation is the main reason why eel still cannot be bred naturally in captivity.

10.2.2 Possible Environmental Triggers of Maturation

Temperature and light do not show a clear effect on the hypothalamo-pituitary-gonad axis in silver eels (Boëtius and Boëtius 1967; Nilsson et al. 1981). Salinity seems to play a role as trigger since GSI and oocyte diameters increased after rearing Japanese farmed eels for 3 months in seawater up to values that are similar for wild silver females (Kagawa et al. 1998; Kagawa 2003). Also water pressure has been demonstrated to have a positive effect on maturation (Dufour and Fontaine 1985), also suggested by the observation of a migrating silver eel with swollen belly at the Bahamas at 2,000-m depth (Robins et al. 1979). After 3 weeks exposure of eels to high pressure in hyperbaric chambers, Sébert et al. (2005) found a decrease in thyroxin hydroxylase mRNA levels; the rate limiting enzyme in the biosynthesis of dopamine, suggesting that high pressure is involved in lifting the dopaminergic inhibition and triggering maturation. Recently, Sébert et al. (2007) indeed found that females submitted to high pressure of 101 ATA showed a significant increase in oocyte diameter and plasma levels of E2 and VTG, while both sexes showed increased plasma levels of 11-KT.

10.2.3 Is Swimming the Trigger for Lipid Mobilization and Maturation?

Silvering marks the start of lipid mobilization and maturation. It appears that silvering is not a true metamorphosis e.g. a marked and abrupt developmental change in the form or structure of an animal, but a mere initiation of maturation; the start of puberty (Chapter 3). Silvering is more flexible than generally presumed (Svedäng and Wickström 1997) and may well be arrested at various stages as occurs for Atlantic salmon *Salmo salar* (Mills 1989). Durif et al. (2005) identified such intermediate phases and found that they were correlated to migration. The most advanced stages of maturation we know are from migrating silver eels *Anguilla* spp. caught closest to the spawning grounds. Moreover, a negative correlation seems to exist between migration distance to the spawning grounds and GSI at the start of oceanic migration of the various *Anguilla* species (Todd 1981).

From this we could assume that lipid mobilization and early maturation are linked to migration and that swimming itself may be the natural trigger for these processes. Indeed, no change in lipid mobilization was found between yellow and silver eels from the same location without having swum (EELREP 2005). We therefore hypothesized that lipolysis becomes activated during and due to sustained swimming. Furthermore, we hypothesized that swimming triggers silvering, the start of maturation, but that blockage of more advanced stages; vitellogenesis, is likely required in order to allow the long spawning migration.

10.3 Evidence from Experimental Swim Trials: Swimming Induces Silvering and Maturation

10.3.1 Experimental Swim Trials

At Leiden University (The Netherlands), numerous swim trials have been performed in 22 Blazka-type calibrated swim-flumes described in detail by Van den Thillart et al. (2004), elucidating aspects of swim performance of eels (Chapters 8, 9 and 16) and swimming induced silvering and maturation (this chapter). In 2006, a new stream-gutter (Fig. 10.1) was built to allow group-wise swimming of males and females, expected to lead to lower stress levels that would have negative effects on maturation.

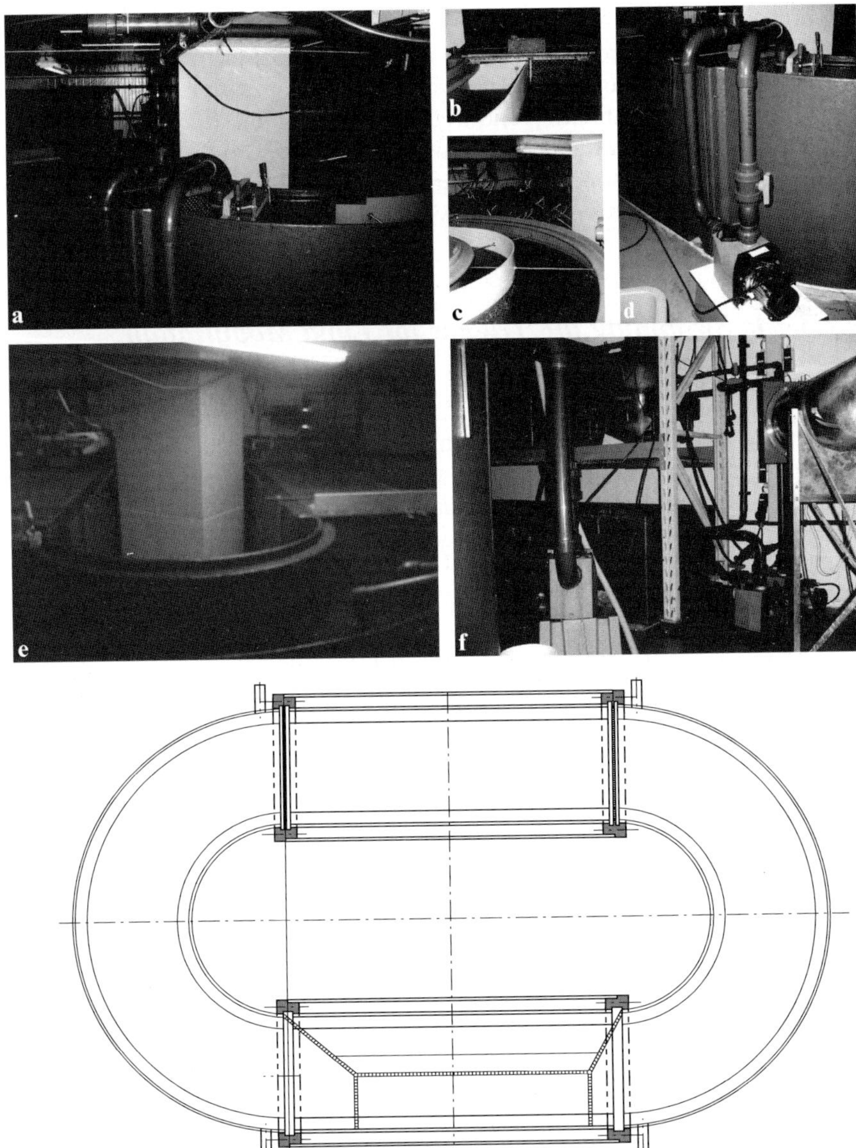

Fig. 10.1 (**a**) The new stream gutter at Leiden University to allow group-wise swimming with (**b**) the front fence of the female compartment, (**c**) the curve before the female compartment with a constructed curved fence to create a faster and linear water velocity profile, (**d**) one of the two SPECK pumps creating a stream, (**e**) the stream gutter running during the experiment under conditions of far-red 670-nm light (bandwidth 20-nm), (**f**) the recirculating filter system with precipitation filter, sump, protein skimmer and biological filter and (**g**) experimental design of the round-shaped stream gutter. To allow a large amount of eels, the straight ends were extended creating the shape of a racetrack. Two compartments allow swimming of males (top) and females (bottom). The female compartment is only half the width of the male compartment to create a two-fold harder stream

10.3.2 Swimming Induced Changes in External Appearance

Swimming induces an increase of eye diameter (Table 10.1). This has been observed repeatedly in different swim trials. Continuous swimming at 0.5 body-lengths per second of eels from the landlocked Lake Balaton resulted in an increase of the eye index (EI) that was already apparent after 2 weeks and occurred in all eels exposed to the swim trial (Palstra et al. 2007a). The observed changes appeared

Table 10.1 Experimental swim-trials performed at Leiden University in (a) fresh water (FW) and (b) salt water (SW)

	Origin	n	Sex (m/f)	Age (year)	Migr. period (weeks)	Migr. distance (km)	EI	GS	Reference
(a)	FW trials Lake Balaton (Hu)	40	f	16*	1	<300	–	Stage 3	Palstra et al. 2007a
	River Loire (Fr)	20	f	16 ± 4	1	<300	+		Palstra et al. 2008a
	Lake Balaton (Hu)	6	f	16 ± 1	2	350	+	Stage 3	Palstra et al. 2007a
	Lake Balaton (Hu)	9	f	16 ± 3	6	1,100	+	Stage 3	Palstra et al. 2007a
	Farm	6	f	5	6	2,200	+	Stage 3	Palstra et al. 2006a
	Farm	15	f	3	25	5,500	–	Stage 3	Van Ginneken et al. 2007a
(b)	SW trials								
	Lake Grevelingen (NL)	20	f	11 ± 4	0.6	<300	–		Palstra et al. 2008a
	Lake Grevelingen (NL)	6	f	8 ± 2	4	1,200	–	Stage 3	Palstra et al. 2006a
	Lake Grevelingen (NL)	11	f		4	1,000	–	Stage 3	A. Palstra et al., 2000, unpublished data
	Greece	6	f		6	900	–	Stage 3	Palstra et al. 2007c
	Greece	6	f		12	1,400	–	Stage 3	Palstra et al. 2007c
	Greece	6	m		6	500	–	Stage 2	A. Palstra et al., 2007, unpublished data
	Greece	6	m		12	900	–	Stage 2	A. Palstra et al., 2007, unpublished data

Columns represent the origin of experimental eels, the number (n), their sex male (m) or female (f), their age in years, the experimental simulated migration period in weeks, the migrational distance in km, occurrence of changes in eye index EI (+) or not (–), the dominant gonad stage after swimming and the literature reference. The asterisk marks an estimated age corresponding with findings for other Lake Balaton eels.

even stronger after 6 weeks of swimming. Significant increases were also apparent in swim-trials with migrating eels from the River Loire and with older farmed eels (Table 10.1). Since younger farmed eels did not show swimming – induced increase of the eyes, age – dependent maturation sensitivity is suggested (see later in this chapter). It also appears that this phenomenon solely occurs in fresh-water since no changes were detected in salt-water trials (Table 10.1), in either males or females.

Since the enlargement of the eyes is used for discriminating between the yellow and silver phase (Pankhurst 1982), it can thus be stated that swimming induces silvering. Changes in the length (Durif et al. 2005) and shape (Tesch 2003) of the pectoral fins are also considered as indicative of the degree of silvering. However, such changes were not detected in any of the swim-trials (Palstra et al. 2007a).

10.3.3 Swimming Induced Changes in Oocyte Histology

The ovaries of European silver eels show oocytes after transformation of the oogonia in the first developmental stages (stage 1–2; Adachi et al. 2003). Further progression requires incorporation of lipids (stage 3) and vitellogenin (stage 4). Already after 1 week of swimming of Lake Balaton eels (Palstra et al. 2007a), the GSI increased and oocytes became larger with large numbers of lipid droplets (Figs. 10.2 and 10.3). After 6 weeks of swimming, changes were much more pronounced than after 2 weeks of swimming, both GSI and oocyte diameter were significantly higher. In contrast to resting eels the swimming eels had oocytes in the lipid droplet stage 3. These results indicate that a high level of lipid mobilization induced by swimming is required not only for fuel but also for a natural incorporation of lipid droplets in the oocytes. Oocytes of eels that had swum contained more than 100 large droplets. Most developed oocytes had lipid droplets that covered >50% of the cytoplasm and formed a complete ring around the circumference of the developing oocyte (Couillard et al. 1997), which is typical for previttelogenic oocytes (Colombo et al. 1984).

However, we did not observe any yolk globuli in the oocytes. Also the oocytes did not reach sizes that are characteristic for vitellogenesis (stage 4). Adachi et al. (2003) showed for *A. japonica* that vitellogenesis begins when oocytes are about 250 μm in diameter. In our studies we found maximum oocyte diameters of 236 μm, which are quite close to the onset of vitellogenesis.

10.3.4 Swimming – Induced Changes in Testis Histology

Male European silver eels have a GSI <0.2%. Swimming male eels had a higher GSI after 1.5 months swimming and even higher after 3 months swimming (Fig. 10.4; Palstra et al., 2008b). These changes in GSI reflected histological changes in testis development and the onset of spermatogenesis. The testes of European silver eels show thick layers of connective tissue. Spermatogonial stem cells are present with clear large homogenous nuclei containing one or two nucleoli (Fig. 10.4). Spermatogonia occur independently (type A) or in cysts of two or four germ cells (type B; Miura et al. 2003). Swimming male eels had higher percentages of stage 2 and late

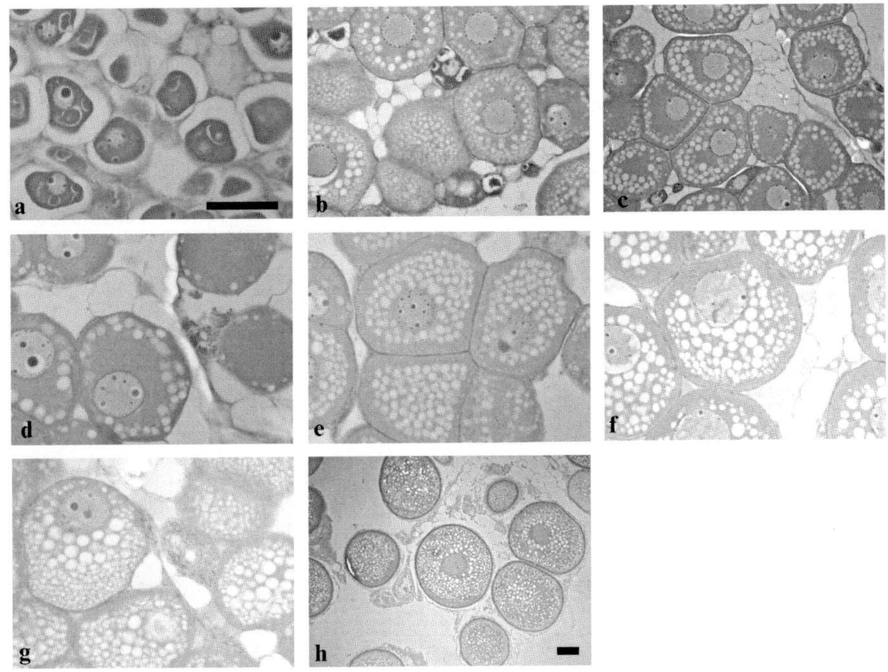

Fig. 10.2 Oocyte development in swimming eels, representing subsequent stages in (a) a Lake Balaton eel before swimming with stage 1 and 2 oocytes, (b) a Lake Balaton eel after 2 weeks swimming (350-km) with many stage 3 oocytes but also still less developed oocytes, (c) a Lake Balaton eel after 6 weeks swimming (1,100-km) with mainly stage 3 oocytes with many lipid droplets and decreasing lipid stores, (d) a 5 year old farmed eel with mainly early stage 3 oocytes, (e) a 5 year old farmed eel after 2,200-km of swimming with solely stage 3 oocytes, (e) a migrating silver eel from Lake Grevelingen with stage 3 oocytes and fat reserves, (f) a migrating silver eel from lake grevelingen wtih stage 300 cytes and fat reserves, (g) a silver eel from Lake Grevelingen after swimming 1,200-km with stage 3 oocytes fully covered with large lipid droplets, and (h) an artificially matured Lake Grevelingen eel with oocytes containing yolk globuli during final maturation. Swimming stimulates incorporation of large amounts of lipid and synchronizes development up to late stage 3 oocytes. The scale bar represent 100 μm; the scale bar in a accounts for a to g, h is on a smaller scale

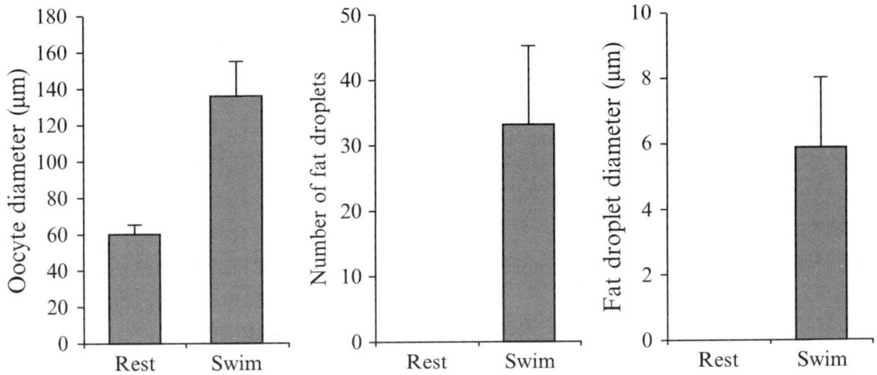

Fig. 10.3 Histological differences in oocyte development between Lake Balaton eels that rested or swam for 2 weeks in (a) oocyte diameter, (b) the number of lipid droplets in the oocytes and (c) the diameter of lipid droplets (Based on data from Palstra et al. 2007a)

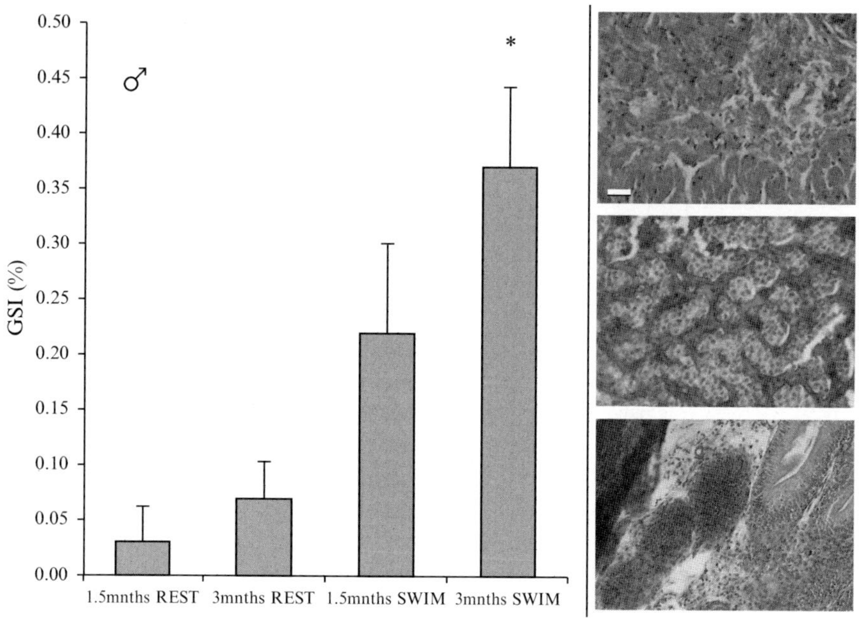

Fig. 10.4 Relative gonad mass (mean GSI + standard error) and testis histology in male eels. Experimental groups were sampled after swimming or resting after 1.5 and 3 months. The asterisk indicates a significant difference ($P < 0.05$) in GSI after 3 months swimming. Pictures from top to bottom show testis histology of a resting eel with stage 1 spermatogonia (top), a swimming eel for 1.5 months with a GSI = 0.32 with clear clusters of stage 2 spermatogonia (middle) and a swimming eel for 3 months with a GSI = 0.59 with stage 3 spermatocytes (bottom). The scale bar represent 100 μm

type b spermatogonia (Fig. 10.4). An increase of organization in clusters occurred with spermatogonia grouped in spermatic tubules, and a strong reduction of connective tissue and lipid tissue. The testis of one male eel that swam for 3 months even contained spermatocytes. Our latest results show that male eels that swam for 3 months in salt water and that were subsequently treated with human chorionic gonadotropin started their spermiation earlier, and they produced more sperm of higher density.

10.3.5 Swimming – Induced Changes in Pituitary and Blood Plasma Maturation Parameters

After swimming 5,500-km in fresh-water, young farmed eels showed increased LH levels in the pituitary (Fig. 10.5; Van Ginneken et al. 2007a). The same occurred in wild migratory eels from Lake Grevelingen after swimming 1,000-km in salt water (A. Palstra et al., 2000, unpublished data). Swimming thus stimulates gonadotrope production of the pituitary but it is still unclear whether secretion is also stimulated.

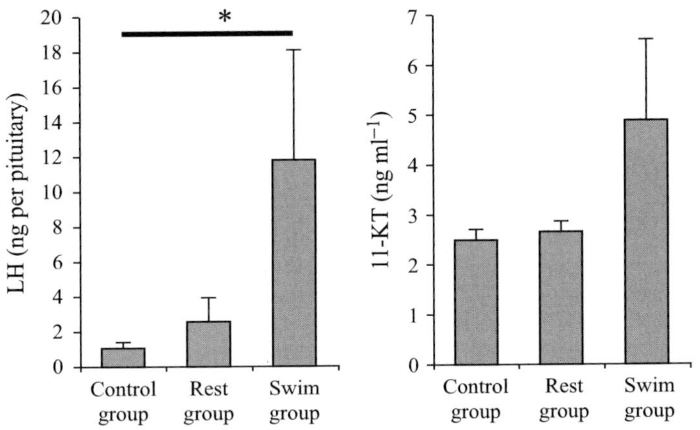

Fig. 10.5 Differences (means + standard error) in pituitary LH and 11-KT in young farmed eels that were sampled at the start of the experiment (control), that rested during the experimental period (rest) or swam 5,500-km (swim). Pituitary LH is increased in swimmers and 11-KT tends to increase but individual variation is high. Based on data from Van Ginneken et al. 2007a

After swimming 5,500-km in fresh-water, the young farmed eels also showed a tendency to an increased level of plasma 11-KT (Fig. 10.5), while plasma-VTG, pituitary-adrenocorticotropic hormone (ACTH), plasma-ACTH, pituitary- melanophore-stimulating hormone (MSH) α and plasma-αMSH were unaffected (Van Ginneken et al. 2007a). 11-KT is considered as the major hormonal mediator of silvering in female eels (Lokman et al. 2003), as evidenced by the numerous silvering effects, like enlargement of the eyes, that Rohr et al. (2001) observed after implanting 11-KT in female *A. australis*. Just recently, Endo et al. (2007) showed that 11-KT has an important role in lipid transfer and deposition in the oocytes which agrees with the histological observations. The young farmed eels in Van Ginneken's study however did not show increase in eye size but other swimming eels did after swimming in fresh-water (Table 10.1).

VTG levels in swimming Lake Balaton eels were below detection limits (A. Palstra et al. 2005, unpublished data). Wild migratory silver eels even showed decreased levels of plasma VTG after swimming 1,000-km in salt water while testosterone (T) and E2 were unaffected (A. Palstra et al. 2005, unpublished data). Recently, we measured blood plasma E2 and VTG indirectly through plasma calcium (Ca) (Palstra et al. 2007c). A significant positive correlation and similar sensitivity to VTG has been demonstrated for rainbow trout, *Oncorhynchus mykiss*, by Verslycke et al. (2002) and used on eels by Versonnen et al. (2004). Swimming, but also resting, increased E2 levels but only in first instance. Ca levels were found lower in swimming eels. Results thus show that swimming does not stimulate vitellogenesis which corresponds with histological findings; the absence of yolk globuli in the oocytes of swimmers.

10.3.6 Swimming – Induced Changes in Expression Profiles of Maturation Parameters

Recently, we have cloned four different genes from eel tissue extracts (Palstra et al. 2007c): E2 receptor α (ERα), VTG 1, VTG 2 and β-actin. We have applied the developed molecular probes for housekeeping gene β-actin and targeted genes for ERα-, VTG 1- and VTG 2-expression on liver samples of female silver eels that swam for 1.5 or 3 months in salt water. In swimming eels, the expression of ERα was slightly lower and in resting eels higher than in the control group (Fig. 10.6a). This reduction in expression probably caused the reduced expression of VTG 1 and VTG 2. From this we can conclude that hepatic vitellogenesis is indeed reduced in swimming silver eels in salt water.

FSH and LH play separate roles (Suetake et al. 2003) during maturation of teleost fishes. In general, FSH is related to early maturation (vitellogenesis/spermatogenesis) and LH related with late maturation processes (maturation/spermiation). During artificially induced maturation, an immediate FSH-β decrease and a LH-β over-expression occur in female Japanese (Nagae et al. 1993; Saito et al. 2003; Suetake et al. 2002) and European eels (Schmitz et al. 2005). Their relatives, naturally maturing New Zealand longfinned eels *Anguilla dieffenbachii* (Suetake et al. 2002) and common Japanese congers (Suetake et al. 2003), however show high FSHβ expression levels suggesting that this decrease is abnormal. The question is now whether a natural stimulator of maturation, like swimming, causes an increase in FSH expression. At this moment we are investigating the expression of the common α-subunit (GPα) and the specific β-subunits (FSH-β and LH-β) in the pituitary and the expression of their receptors in the gonads of swimmers. The latest results (Palstra et al. 2008b) have shown that swimming did not have a significant effect on FSHB expression. Swimming did cause a two- to three-fold higher LHB expression but only in males and not in females.

10.4 Hypothesis for a Mechanism

10.4.1 Swimming Activates Lipid Metabolism

Swimming eels were found to have large oocytes in the lipid droplet stage containing large amounts of lipid droplets (Palstra et al. 2007a). These results indicate that a high level of lipid mobilization induced by swimming is required not only to fuel migration but also for a natural incorporation of lipid droplets in the oocytes. This is regulated by swimming increased 11-KT levels and represents a crucial step in oocyte maturation since the amount of lipid droplets influences the following developmental events before and after fertilization (Palstra et al. 2005), and provides the necessary reserves for the offspring. For comparison, in artificially matured eels, 57 ± 22 g lipid is incorporated into the oocytes corresponding to 28% of the lipid reserves of the average silver eel (Palstra et al. 2006a).

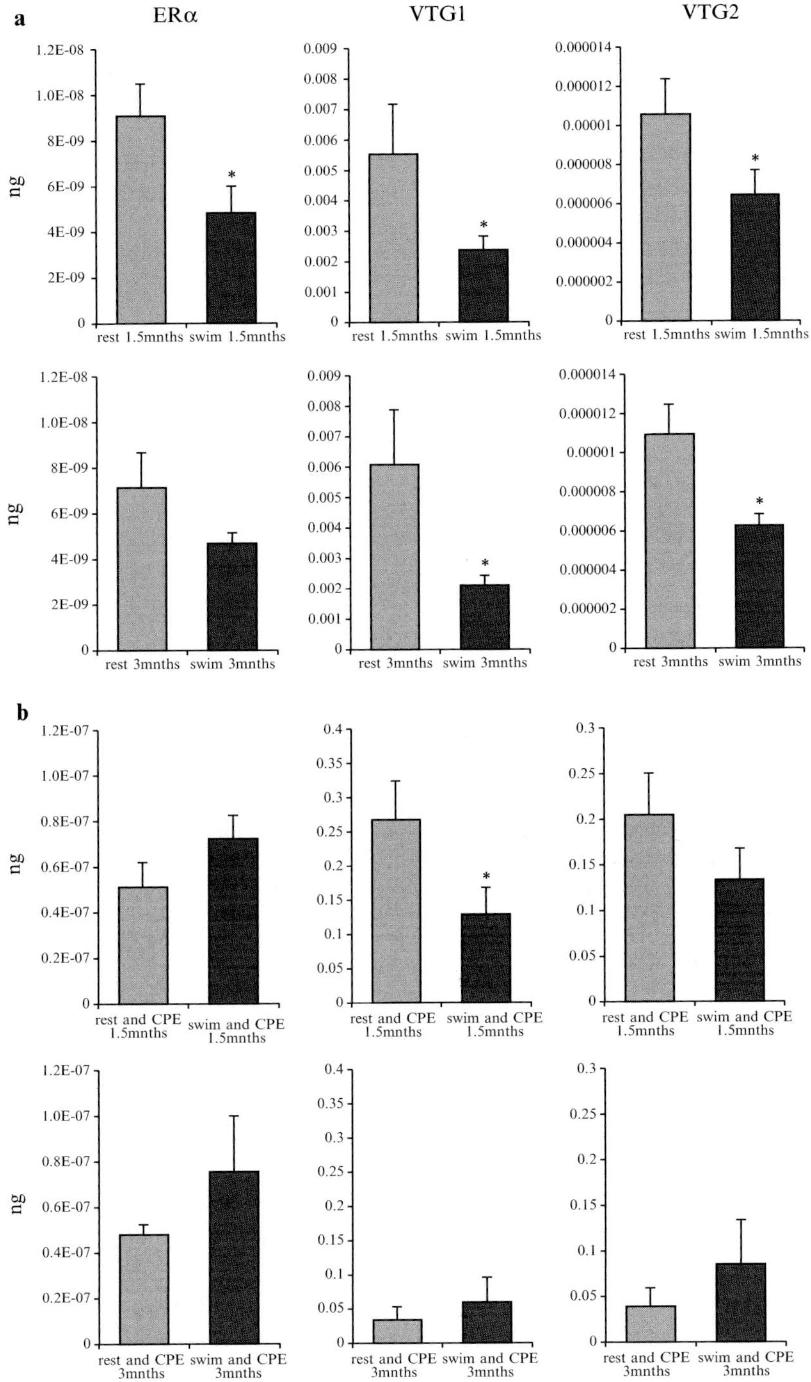

Fig. 10.6 Erα-, VTG 1- and VTG 2-expression (mean number of copies of mRNA per ngram of total RNA with standard error) in livers after (a) resting or swimming for 1.5 and 3 months, and (b) of 1.5 and 3 months rest and swim groups that were subsequently stimulated with three weekly CPE injections. Asterisks indicate a significant difference ($P < 0.05$) of a rest group vs. the control (Based on data from Palstra et al. 2007c)

10.4.2 Swimming Inhibits Vitellogenesis

On the basis of evidence from different angles, it can be concluded that swimming inhibits the whole process of vitellogenesis, at least in first instance. Firstly, ERα-, VTG 1- and VTG 2-expression were reduced in the livers of swimming females (Palstra et al. 2007c). Secondly, plasma VTG and Ca were repeatedly determined as not detectable and not elevated in swimming females (Van Ginneken et al. 2007a, Palstra et al. 2007c). Thirdly, oocytes of swimming females from Lake Balaton (Palstra et al. 2007a), Lake Grevelingen (A. Palstra et al., 2002, unpublished) and River Loire (A. Palstra et al., 2004, unpublished) did not contain any yolk globuli and were all smaller than 250 μm, the border for switching to stage 4 vitellogenic oocytes.

10.4.3 Linking Metabolism with Maturation: A Central Role for Cortisol

We hypothesize that cortisol plays a major role in the endocrinological connection between metabolism and maturation. Silver eels have higher cortisol levels (Van Ginneken et al. 2007b) and higher cortisol levels have been measured in swimming eels of Lake Grevelingen (A. Palstra et al., 2000, unpublished data) and Lake Balaton although individual variance is very high (Fig. 10.7). Cortisol is known for mobilization of lipids. Cortisol peaks lead to lysis of muscle and hepatic lipids (Freeman and Idler 1973; Davis et al. 1985; Barton et al. 1987; Mommsen et al. 1999) releasing fatty acids into the blood.

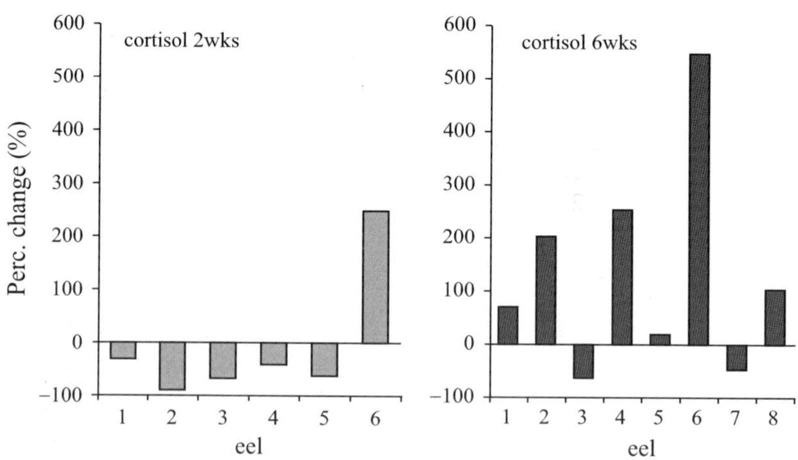

Fig. 10.7 Individual percentual change (%) in cortisol levels before and after swimming for (a) 2 weeks and (b) 6 weeks of Lake Balaton eels. After swimming 2 weeks, only one eel showed increase of cortisol level (>200%). After swimming 6 weeks, six out of eight eels showed increase of cortisol level with three eels showing an increase >200%

Cortisol has a less defined role in maturation of fishes. DiBattista et al. (2005) showed that cortisol treatment in rainbow trout significantly decreased dopaminergic activity in the telencephalon. Epstein et al. (1971) and Fontaine (1994) showed that successive high concentrations of plasma cortisol, lasting for at least 7 days, triggers silvering in the eel. Cortisol is also known as stimulator of LH synthesis in vitro and in vivo (Huang et al. 1999; Dufour et al. 2003) and as inhibitor of vitellogenin synthesis (Sbaihi et al. 2001). Cortisol was shown to inhibit E2-induced vitellogenin synthesis in the rainbow trout, an effect mediated by a decrease in ERα mRNA levels (Lethimonier-Desdoits et al. 2000). These observations perfectly match our observations on swimming eels.

Swimming increases cortisol levels chronically and with this, key steps in steroid metabolism may be stimulated. By activating lipid metabolism, cortisol may increase cholesterol transport in plasma by lipoproteins. In the cell, cholesterol is transported from the outer to the inner mitochondrial membrane by steroidogenic acute regulatory protein StAR (Li et al. 2003) and then transformed into pregnenolone by cytochrome P-450 cholesterol sidechain cleavage enzyme; the rate-limiting step (Stocco 2001) before transformations into for instance 11-ketotestosteron, estradiol and dihydroxy progesterone (DHP). Recently, Kazeto et al. (2006) found that transcript levels of this enzyme increased in the ovary of Japanese eels during artificially induced ovarian development, suggesting that expression is induced by gonadotrope stimuli. Precursor 17α hydroxyprogesteron can be transformed back into cortisol creating a possible regulatory mechanism.

10.5 The Necessity of Swimming for Reproduction

10.5.1 Age Dependent Sensitivity for Maturation

Otolith analysis has revealed the old age of Lake Balaton eels (13–21 years) and migratory Loire eels (10–28 years) at the time of experimenting (Palstra et al. 2008a, 2006a, 2007a). These eels show extensive changes after just 1 week swimming while young, 3 year old, farmed eels only showed minor changes under identical conditions (Table 10.1). These results suggest that older eels are more sensitive to stimulation of maturation and that maturation is an age – dependent process (Palstra et al. 2007a). Indeed, older farmed eels did show increase of the eyes within 1 week swimming (Table 10.1). Arguments for age – dependent maturation also come from other observations: (1) older eels showed increased capacity to incorporate more lipid from the muscle into the oocytes (Palstra et al. 2006b), and (2) older eels were more sensitive to hormonal stimulation (Palstra et al. 2006a; also Durif et al. 2006). Repeated yearly silvering and subsequent regression (Durif et al. 2005) might unlock the strong inhibition of sexual maturation in eels. Age might thus be a key factor for successful maturation.

10.5.2 Does Swimming Increase the Maturation Sensitivity?

In a recent study (Palstra et al. 2007c), we have tried to find time points during swimming where maturation sensitivity increases. Groups of females that had either swum or rested for 1.5 months and for 3 months were subjected to a maturation sensitivity test (EELREP 2005; Durif et al. 2006). In this test, eels were injected weekly IP with 20-mg carp pituitary extract for a period of 3 weeks since at this point eels show elevated expression of VTG 1 and VTG 2, elevated plasma calcium levels and yolk globuli appearing in the oocytes (A. Palstra et al., 2005, unpublished data). In this study, blood plasma E2 and Ca levels increased when swimming and resting eels were subsequently stimulated by three weekly CPE injections. Expression of ERα-, VTG 1- and VTG 2-strongly increased after hormonal stimulation. Female eels that has swim for three months did however not show a higher sensitivity (Fig. 10.6b). Recently (A. Palstra et al., 2007, unpublished results) we even found that swimmers evaluated on average 2-3 weeks later than the resters. In contrast, male swimmers were more sensitive and spermiated earlier in response to hormonal treatment than resters.

10.5.3 Does Swimming Increase Gamete Quality?

Although current protocols for artificial reproduction of eels by hormonal injections are successful to a certain extent (Kagawa et al. 2005; Chapter 15), the induced process of maturation can be considered abnormal in many aspects. Abnormality of maturation is evidenced by limited reproductive success (Pedersen 2003, 2004; Palstra et al. 2005 and references therein), and observed phenomena like variations in yolk accumulation, egg membrane formation, differences in the process of oocyte maturation and plasma hormone levels (Adachi et al. 2003; Kagawa et al. 2005). Oocytes of non-exercised silver eels are probably still too premature for hormonal stimulation by pituitary extract. Two probable causes of abnormality may be prevented by swimming: (1) incomplete lipid incorporation, and (2) incomplete vitellogenesis.

As for point 1, during artificial induction by hormonal injections, hepatic vitellogenesis is immediately induced (A. Palstra et al., 2005, unpublished data) Lipid and VTG incorporation occur simultaneously in artificially matured Japanese eel (Adachi et al. 2003) and European eel (A. Palstra et al., 2005, unpublished data), which suggests an unnatural situation. By stimulating incorporation of lipids in the oocytes and inhibiting vitellogenesis, swimming may optimize the natural sequence of these processes.

As for point 2, the effects of FSH in European eel are still largely unclear. Plasma FSH levels are higher in migrating silver eels (Dufour et al. 2003) indicating a relation with swimming. Kamei et al. (2005, 2006) recently showed for Japanese eel that FSH stimulated in vitro: (1) T and 11-KT secretions in a dose-dependent manner in immature testis, and (2) T and E2 secretion in a dose-dependent manner from mid-vitellogenic oocytes. Early oocyte stages lacked fully developed theca and granulosa cells and respective secretion of T and E2. The action of both T and E2 may be required for hepatic vitellogenin synthesis (Kwon et al. 2005). Vittelogenesis through administration of solely estrogens to silver eels has never succeeded (Olivereau and Olivereau 1979; Petersen and Korsgaard 1989; Peyon et al. 1993 as reviewed by Lokman et al. 2003). As stated before, artificially induced maturation may lack FSH effects and the observed phenomena of abnormal maturation may well be consequences of a lack of (swimming increased) FSH. Sato et al. (2003) reported that long term treatment of FSH (21 to 23 weekly injections) followed by weekly LH injection lead to rapid maturation of oocytes that ovulated as eggs of the highest quality with highest fertilization rates. Long term swimming may induce such FSH effects and may, if followed by LH treatment, reflect the natural variant of these results.

Recently, Patterson et al. (2004) recognized this issue also for salmon and stated that '*currently, exercise associated with migration is presented as a potential obstacle to successful reproduction. There has been no attempt, however, to reverse this paradigm and examine exercise as an integral part of normal reproductive development for long distance migrators*'. Indeed they found that non-exercised females had delayed maturity, lower egg deposition rates, and were more likely to die prior to egg ovulation than exercised females and natal spawners. Pre-treatment by swimming in current protocols for artificial reproduction will likely result in higher gamete quality and general reproductive success. These are topics of our current investigations.

10.6 Extrapolation to the Field

10.6.1 Fresh Water Migration: Silvering and Lipid Mobilization

When we ignore effects of other possible triggers and results from the laboratory are extrapolated to the field situation, it can be assumed that migratory eels do not necessarily silver before, but especially during, their fresh-water migration. Old swimming silver eels showed increases of eye size but it appears that this occurs only in fresh-water trials (Table 10.1). As far as we know, it is unknown if eels from salt or brackish water generally have larger eyes than eels from fresh water so that only the last mentioned show this swimming-induced change in eye size.

During fresh water migration, lipid mobilization occurs. Extensive lipid incorporation in the oocytes was apparent during fresh water swim-trials. In the field, Cottril et al. (2001) found that the concentration of total plasma non-esterified fatty acids (NEFA) appeared to follow the trend of E2 and GSI increasing with sexual maturity in migrating *A. rostrata*.

Vitellogenesis seems to be inhibited during fresh water migration. VTG and Ca levels are low in wild silver eels. Versonnen et al. (2004) measured Ca as indicator for VTG levels at 20 locations and found very low levels. In a recent study (A. Palstra et al., 2005, unpublished data), we measured E2 and Ca in blood plasma in migrating silver eels in the River Rhine (Germany) in August, September and

October 2005 and investigated the gonads histologically. E2 levels were higher in October but Ca levels stayed low over the months and oocytes were still smaller than 250 μm without yolk globuli. A. Palstra et al., 2000, unpublished data measured VTG in 104 large female silver eels from the brackish Lake Grevelingen (The Netherlands). Of these eels, 96% showed low VTG levels <0.5 μg ml^{-1}.

10.6.2 Oceanic Migration: Inhibiting Vitellogenesis

Salt water swim trials revealed no changes in eye diameter (Table 10.1). The question remains whether silver eels migrating in the ocean continue the increase in eye size. Probably, eyes will increase during progressive maturation since the eye diameter of silver eels from brackish waters increases further when stimulated by hormonal injections (Palstra 2006).

Also during oceanic migration, or at least during the first part, vitellogenesis seems to be reduced in migrating silver eels since eels did not show an increase in GSI and still showed reduced ERα-, VTG 1- and VTG 2-expression after swimming 3 months in salt water. This may be because (1) lifting the severely depressed lipid mobilization and extensive lipid incorporation requires long-term swim exercise, and (2) preventing undesired effects of vitellogenesis during swimming. VTG synthesis is associated with mobilization of phospho-calcium reserves coming from skeletal vertebrae resorption (Sbaihi 2002; Sbaihi et al. 2007) which is naturally undesired during swimming. Growth of oocytes is most pronounced during vitellogenesis and subsequent maturation. The oocyte diameter will increase two-fold up to 400 μm during vitellogenesis and again two-fold up to 800 μm due to hydration during final maturation (Palstra et al. 2005). These increases result in similar increases of gonad mass and with that of the body diameter. This then increases drag during swimming and with that increases the cost of transport (reviewed by Van Ginneken and Maes 2005). Also this situation is considered as undesirable during migration.

10.6.3 Vitellogenesis and Maturation at the Spawning Grounds?

When we extrapolate lab results to the field, it can be hypothesized that vitellogenesis and maturation occur near or at the spawning grounds. Besides other triggers that were discussed before, spawning ground specific triggers may be involved during vitellogenesis and final maturation. Spawning ground specific triggering of the final stages of maturation has also been considered for other homing fishes (Palstra et al. 2004). These triggers may involve area-specific odour, the intersex pheromonal communication (Liley and Stacey 1983; Lam 1983; Van Ginneken et al. 2005; Huertas et al. 2006) or triggering by an increase in water temperature by rising in the water column. A rise in water temperature is known to increase the responsiveness of the liver to estrogen in the production of vitellogenins (Yaran et al. 1980).

10.7 Conclusions

During their continental phase, yellow eels have depressed lipid mobilization and pre-pubertal blockage of maturation (Fig. 10.8). Since the start of spawning migration marks the onset of lipid mobilization and maturation, swimming may be a crucial trigger of these processes. Swim trials of older eels in fresh water show that eyes were enlarged in all swimmers and that swimming thus induced silvering. Swimming stimulates early oocyte development and deposition of lipids in the oocytes of female eels (Fig. 10.8), most probably regulated by increased 11-ketotestosteron (11-KT) levels. Effects appear stronger in males where swimming induces spermatogenesis. Increased LH levels in the pituitary show that swimming stimulates its gonadotrope production. Swimming also increases plasma 17β-estradiol (E2) but hepatic vitellogenesis is however not initiated: plasma vitellogenin (VTG) levels remain low and yolk globuli do not appear in the oocytes (Fig. 10.8). Swim trials of silver eels in salt water show that swimming inhibits vitellogenesis; expression of the E2 receptor α (ERα), VTG 1 and VTG 2 were lower in the liver of swimmers and plasma VTG levels were reduced. Lipid mobilization by swimming for fuel and deposition in the oocytes occurs most probably through the action of cortisol (Fig. 10.8) that is well-known as activator of lipid mobilization and has numerous positive effects on maturation parameters. Both the induction of lipid deposition in the oocytes and the obstruction of vitellogenesis by swimming may allow a natural sequence of events leading to a higher gamete quality in contrast to stimulation with pituitary extract injections. Such a natural sequence may reflect long term swimming – induced FSH effects followed by short term LH effects to finish maturation. The effect of long term swimming on gamete quality is subject of our current research. When these results are extrapolated to the field and linked to observations in the field, it appears that fresh water migration triggers silvering and lipid mobilization. Vitellogenesis is inhibited by swimming and probably occurs only near or at the spawning grounds. Our latest results show that male swimmers were more sensitive and spermiated earlier in response to hormonal treatment than resters.

Acknowledgements The authors would like to thank Prof. H. Spaink, Drs. D. Schnabel, C. Székely, C. Durif, F. Daverat, J. Klein-Breteler, G. Van der Laak, Ing. M. Nieveen, Ing. M. de Bakker, P. Niemantsverdriet, M. Fekkes, W. Spoor, E. Antonissen, M. Casteijn, R. Van der Linden, R. Heijmans, E. de Kuyper, J. Bij, M. Brittijn, S. Van Schie, D. Curiel, E. Clavero, J. Van Rijssel and M.A. Guerrero. The LU research was funded by the Dutch Technology Foundation (STW-project #LBI66.4199), the European Union (EELREP #Q5RS-2001–01836) and the Dutch Ministry of Agriculture, Nature and Food Quality (project #320187).

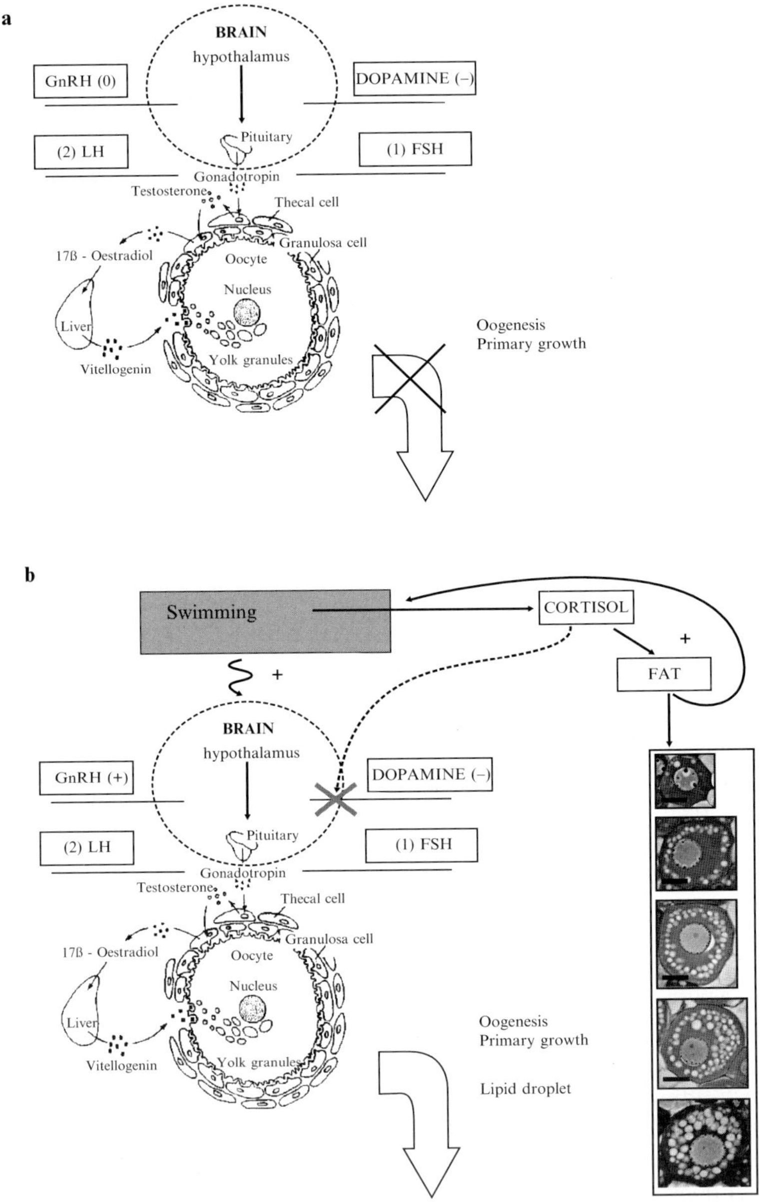

Fig. 10.8 Model for swimming as trigger and stimulator of maturation and lipid incorporation with (**a**) During their continental feeding stage, maturation of eels is blocked at a pre-pubertal stage due to a deficient gnrh stimulation and a simultaneous inhibition of the pituitary gonadotropes FSH and LH by dopamine (Dufour et al. 2003), and (**b**) During their fresh-water and oceanic reproductive migration, swimming lifts the dopaminergic inhibition probably through the action of cortisol. Cortisol mobilises lipid stores for fuel and for incorporation in the oocytes. Swimming induces oocyte development and lipid deposition during an extensive lipid droplet stage. Vitellogenesis is inhibited by swimming in first instance. Includes illustration from Palstra et al. 2007a (Reproduced from Palstra et al. 2007a. With permission of Elsevier)

References

Adachi S, Ijiri S, Kazeto Y, Yamauchi K (2003) Oogenesis in the Japanese Eel, *Anguilla japonica*. In: Aida K, Tsukamoto K, Yamauchi K (eds) Eel Biology. Springer, Tokyo, pp 502–518

Barton BA, Schreck CB, Barton LD (1987) Effects of chronic cortisol administration and daily acute stress on growth, physiological conditions, and stress responses in juvenile rainbow trout. Dis Aquat Organ 2: 173–185

Boëtius I, Boëtius J (1967) Studies in the European Eel, *Anguilla anguilla* (L.). Experimental induction of the male sexual cycle, its relation to temperature and other factors. Meddelelser fra Danmarks Fiskeri- og Havundersogelser 4: 339–405

Colombo G, Grandi G, Rossi R (1984) Gonad differentiation and body growth in *Anguilla anguilla* L. J Fish Biol 24: 215–228

Cottril RA, McKinley RS, Van der Kraak G, Dutil J-D, Reid KB, McGrath KJ (2001) Plasma non-esterified fatty acid profiles and 17β-oestradiol levels of juvenile immature and maturing adult American eels in the St Lawrence River. J Fish Biol 50: 364–379

Couillard CM, Hodson PV, Castonquay M (1997) Correlation between pathological changes and chemical contamination in American eels, *Anguilla rostrata*, from the St Lawrence River. Can J Fish Aquat Sci 54: 1916–1927

Davis K., Torrance P, Parker NC, Shuttle MA (1985) Growth, body composition and hepatic tyrosine aminotransferase activity in cortisol-fed channel catfish, *Ictalurus punctatus*. J Fish Biol 29: 177–184

DiBattista JD, Anisman H, Whitehead M, Gilmour KM (2005) The effects of cortisol administration on social status and brain monoaminergic activity in rainbow trout *Oncorhynchus mykiss*. J Exp Biol 208: 2707–2718

Dufour S, Fontaine YA (1985) La migration de reproduction de l'anguille européenne (*Anguilla anguilla* L.): un rôle probable de la pression hydrostatique dans la stimulation de fonction gonadotrope. B Soc Zool Fr 110: 291–299

Dufour S, Burzawa-Gerard E, Le Belle N, Sbaihi M, Vidal B (2003) Reproductive endocrinology of the European eel, *Anguilla anguilla*. In: Aida K, Tsukamoto K, Yamauchi K (eds) Eel Biology. Springer, Tokyo, pp 373–386

Durif C, Dufour S, Elie P (2005) The silvering process of the eel: a new classification from the yellow resident stage to the silver migrating stage. J Fish Biol 66: 1–19

Durif CMF, Dufour S, Elie P (2006) Impact of silvering stage, age, body size and condition on the reproductive potential of the European eel. Mar Ecol-Prog Ser, 327:171–181

EELREP (2005) Estimation of the reproduction capacity of European eel. Final report. http://www.fishbiology.net/eelrepsum.html

Endo T, Takashi T, Lokman MP, Ijiri S, Adachi S, Kohei Y (2007) *In vitro* induction of oil droplet accumulation into previtellogenic oocytes of Japanese eel, *Anguilla japonica*. Abstract book 8th International Symposium on Reproductive Physiology of Fish, June 3–8, Saint Malo, France

Epstein FH, Cynamon M, McKay W (1971) Endocrine control of Na-K-ATPase in seawater adaptation in *Anguilla rostrata*. Gen Comp Endocrinol 16: 323–328

Fontaine YA (1994) L'argenture de l'anguille: métamorphose, anticipation, adaptation. B Fr Peche Piscic 335: 171–186

Freeman HC, Idler DR (1973) Effects of corticosteroids on liver transaminases in two salmonids, the rainbow trout (*Salmo gairdneri*) and the brook trout (*Salvelinus fontinalis*). Gen Comp Endocrinol 20: 69–76

Huang Y, Rousseau K, Sbaihi M, Le Belle N, Schmitz M, Dufour S (1999) Cortisol selectively stimulates pituitary gonadotropin β-subunit in a primitive teleost, *Anguilla anguilla*. Endocrinology 130: 1228–1235

Huertas M, Scott AP, Hubbard PC, Canario AV, Cerda J (2006) Sexually mature European eels (*Anguilla anguilla* L.) stimulate gonadal development of neighbouring males: possible involvement of chemical communication. Gen Comp Endocrinol 147 (3): 304–313

Sbaihi M, Fouchereau-Perron M, Meunier F, Elie P, Mayer I, Burzawa-Gerard E, Vidal B, Dufour S (2001) Reproductive biology of the conger eel from the south coast of Brittany, France and comparison with the European eel. J Fish Biol 59: 302–318

Sbaihi M (2002) Interaction des steroïdes sexuels et du cortisol dans le contrôle de la fonction de reproduction et du métabolisme calcique chez un Téléostéen migrateur, l'anguille (*Anguilla anguilla* L.). Cybium 2002: 26 (1): 64

Sbaihi M, Kacem A, Aroua S, Baloche S, Rousseau K, Lopez E, Meunier F, Dufour S (2007) Thyroid hormone-induced demineralisation of the vertebral skeleton of the eel, *Anguilla anguilla*. Gen Comp Endocrinol 151: 98–107

Schmitz M, Aroua S, Vidal B, Le Belle N, Elie P, Dufour S (2005) Differential Regulation of Luteinizing Hormone and Follicle-Stimulating Hormone Expression during Ovarian Development and under Sexual Steroid Feedback in the European Eel. Neuroendocrinology 81: 107–119

Sébert M-E, Vettier A, Dufour S, Amerand A, Baloche S, Weltzien FA, Sébert P (2005) Mimicking environmental conditions of European eel oceanic migration: effects of exposure to high hydrostatic pressure on the brain-pituitary-gonad axis. Fish and Diadromy in Europe. Ecology, Management, Conservation. Bordeaux 29 March–1 April 2005

Sébert M-E, Amerand A, Vettier A, Weltzien FA, Pasqualini C, Sébert P, Dufour S (2007) Effects of high hydrostatic pressure on the pituitary-gonad axis in the European eel, *Anguilla anguilla* (L.). Gen Comp Endocrinol. 153: 289–298

Stocco DM (2001) StAR protein and the regulation of steroid hormone biosynthesis. Ann Rev Physiol 63: 193–213

Suetake H, Okubo K, Sato N, Yoshiura Y, Suzuki Y, Aida K (2002) Differential expression of two gonadotropin (GTH) β subunit genes during ovarian maturation induced by repeated injection of salmon GTH in the Japanese eel *Anguilla japonica*. Fish Sci 68: 290–298

Suetake H, Okubo K, Yoshiura Y, Aida K (2003) GTH and GnRH molecules and their expression in the Japanese eel. In: Aida K, Tsukamoto K, Yamauchi K (eds) Eel Biology. Springer, Tokyo, pp. 351–372

Svedäng H, Wickström H (1997) Low fat contents in female silver eels: indications of insufficient energetic stores for migration and gonadal development. J Fish Biol 50: 475–486

Tesch FW (2003) The Eel. Blackwell Science, Oxford

Todd PR (1981) Morphometric changes, gonad histology, and fecundity estimates in migrating New Zealand fresh-water eels (*Anguilla* spp.). N Z J Mar Freshw Res 15: 155–170

Usui A (1991) Eel Culture. Blackwell Scientific Publications Ltd, London

Van den Thillart G, Van Ginneken V, Körner F, Heijmans R, Van der Linden R, Gluvers A (2004) Endurance swimming of European eel. J Fish Biol 65: 1–7

Van Ginneken VJT, Maes GE (2005) The European eel (*Anguilla anguilla* Linnaeus), its lifecycle, evolution and reproduction: a literature review. Rev Fish Biol Fisher 15: 367–398

Van Ginneken V, Vianen G, Muusze B, Palstra A, Verschoor L, Lugten O, Onderwater M, Van Schie S, Niemantsverdriet P, Van Heeswijk R, Eding E, Van den Thillart G (2005) Gonad development and spawning behaviour of artificially-matured European eel (*Anguilla anguilla* L.). Anim Biol 55 (3): 203–218

Van Ginneken V, Durif C, Dufour S, Sbaihi M, Boot R, Noorlander K, Doornbos J, Palstra A, Goos H, Antonissen E, Mayer I, Van den Thillart G (2007a) A 5,500-km swim trial stimulates gonad maturation in the European eel (*Anguilla anguilla* L.). Comp Biochem Physiol. Part A 147: 1095–1103

Van Ginneken V, Durif C, Balm SP, Boot R, Verstegen MWA, Antonissen E, Van den Thillart G (2007b) Silvering of European eel (*Anguilla anguilla* L.): Seasonal changes of morphological and metabolic parameters. Anim Biol 57 (1): 63–77

Verslycke T, Vandenbergh GF, Versonnen B, Arijs K, Janssen CR (2002). Induction of vitellogenesis in 17α-ethinylestradiol-exposed rainbow trout (*Oncorhynchus mykiss*): a method comparison. Comp Biochem Physiol. Part C 132: 483–492

Versonnen BJ, Goemans G., Belpaire C, Janssen CR (2004). Vitellogenin content in European eel (*Anguilla anguilla*) in Flanders, Belgium. Environ Pollut 128 (3): 363–371

Vidal B, Pasqualini C, Le Belle N, Claire M, Holland H, Sbaihi M, Vernier P, Zohar Y, Dufour S (2004) Dopamine inhibits luteinizing hormone synthesis and release in the juvenile European eel: A neuroendocrine lock for the onset of puberty. Biol Reprod 71: 1491–1500

Weltzien F-A, Pasqualini C, Sébert M-E, Vidal B, Le Belle N, Kah O, Vernier P, Dufour S (2006) Androgen-Dependent Stimulation of Brain Dopaminergic Systems in the Female European Eel *(Anguilla anguilla*). Endocrinology 147 (6): 2964–2973

Yaran Z, Cocos M, Salzer H (1980) Effects of temperature and photoperiod on ovarian recrudescence in the cyprinid fish *Mirogrex terrae-sanctae*. J Fish Biol 16: 371–382

Chapter 11
The Gonadoliberin(S)-Gonadotropin(S) Axis in the Eel: Expression and Regulation Under Induced Maturation and Sex Steroid Feedbacks

Salima Aroua, Karine Rousseau, Monika Schmitz, Ching-Fong Chang, and Sylvie Dufour

Abbreviations 11-KT: 11-ketotestosterone; AR: androgen receptor; CPE: carp pituitary extract; DHT: dihydrotestosterone; E2: estradiol; ER: estrogen receptor; ERE: estrogen responsive element; FSH: follicle-stimulating hormone; FSHβ: β subunit of FSH; FSH-R: FSH receptor; GAP: GnRH-Associated Peptide; GnRH: gonadotropin-releasing hormone; cGnRH-II: chicken-II form of GnRH; cfGnRH: catfish form of GnRH; dfGnRH: dogfish form of GnRH; hrGnRH: herring form of GnRH; mGnRH: mammalian form of GnRH; mdGnRH: medaka form of GnRH; pjGnRH: pejerrey form of GnRH; sGnRH: salmon form of GnRH; sbGnRH: sea bream form of GnRH; wfGnRH: whitefish form of GnRH; GnRH-R: GnRH receptor; GPα: common subunit of gonadotropins; GTH: gonadotropins; hCG: human chorionic gonadotropin; HPLC: high performance liquid chromatography; LH: luteinizing hormone; LHβ: β subunit of LH; LH-R: LH receptor; RIA: radioimmunoassay; RT-PCR: reverse-transcription polymerase chain reaction; T: testosterone

11.1 Introduction

At the start of the reproductive migration towards the Sargasso Sea, silver eels are still immature and remain blocked at this prepubertal stage as long as migration is prevented. To date, adult mature eels have never been caught and only experimental treatments of silver eels with gonadotropic hormones have led to the observation of sexually mature animals (Fontaine 1936; Fontaine et al. 1964). The lack of sexual maturation at the silver stage is due to a deficient production of pituitary gonadotropins (GtHs) (Dufour et al. 1983a, b). Stimulation of synthesis and release of pituitary

---

S. Aroua, K. Rousseau, and S. Dufour
Museum National d'Histoire Naturelle, DMPA, UMR CNRS 5178 «Biology of Marine Organisms and Ecosystems», 7 rue Cuvier, CP 32, 75231 Paris cedex 05, France

M. Schmitz
Department of Biology, Karlstad University, S-651 88 Karlstad, Sweden

C.-F. Chang
Department of Aquaculture, National Taiwan Ocean University, Keelung 202, Taiwan, ROC

G. van den Thillart et al. (eds.), *Spawning Migration of the European Eel*,

gonadotropin in the silver eel can be induced by combined treatments with a GnRH-agonist and a dopamine-antagonist, indicating that a double neuroendocrine block was responsible for the deficient pituitary gonadotropic function: a lack of endogenous stimulation by GnRH due to a deficient production of GnRH and a strong dopaminergic inhibition of GnRH action (Dufour et al. 1988, 1991; Vidal et al. 2004).

In vertebrates, it is well known that the gonadotropins (luteinizing hormone, LH and follicle stimulating hormone, FSH) are secreted by the gonadotrophs in the anterior pituitary under the control of the gonadoliberin (GnRH) produced by the brain. The gonadotropins act on the ovaries and testes to promote gametogenesis and reproductive function, and to stimulate the production of sex steroids. In turn, gonadotropin secretion and subunit gene expression are regulated by sex steroids acting either directly on the gonadotrophs or indirectly by alterations of GnRH from the hypothalamus (Gharib et al. 1990). This review will mainly focus on the differential feedbacks exerted by sex steroids on the gonadoliberins/gonadotropins system in teleost fishes (especially the European eel).

11.2 GnRH System

To date, a total of 24 forms of GnRH have been isolated, 14 in vertebrates (12 in gnathostomes and 2 in lampreys) and 10 in invertebrates (9 in tunicates and 1 in molluscs) (for review: Gorbman and Sower 2003; Tsai 2006). Although GnRH peptides have not been isolated and sequenced from invertebrate neural structures, their presence has been suggested in cnidarians (for review: Rastogi et al. 2002; Gorbman and Sower 2003; Tsai 2006; Twan et al. 2006). In common with other neuropeptides GnRH peptides are first synthesized as a large precursor (pre-proGnRH), which includes a signal peptide (around 20–25 residues), the biologically active decapeptide GnRH, a cleavage tripeptide (Glys-Lys-Arg) and the GnRH-Associated Peptide (GAP; around 40–50 residues) (for review: Somoza et al. 2002; Lethimonier et al. 2004).

The majority of gnathostomes possess two forms of GnRH: the form in the preoptic-hypothalamic system is species-specific and highly variable, while the

Table 11.1 Amino-acid sequence of the nine GnRH variants identified in fish. C-II = chicken-II form; s = salmon form; m = mammalian form; hr = herring form; cf = catfish form; sb = sea bream form; pj = pejerrey form; wf = whitefish form

GnRH form	1	2	3	4	5	6	7	8	9	10
Mammalian	p-Glu	His	Trp	Ser	Tyr	Gly	Leu	Arg	Pro	Gly-NH2
Chicken II	p-Glu	His	Trp	Ser	His	Gly	Trp	Tyr	Pro	Gly-NH2
Salmon	p-Glu	His	Trp	Ser	Tyr	Gly	Trp	Leu	Pro	Gly-NH2
Catfish	p-Glu	His	Trp	Ser	His	Gly	Leu	Asn	Pro	Gly-NH2
Seabream	p-Glu	His	Trp	Ser	Tyr	Gly	Leu	Ser	Pro	Gly-NH2
Herring	p-Glu	His	Trp	Ser	His	Gly	Leu	Ser	Pro	Gly-NH2
Pejerrey	p-Glu	His	Trp	Ser	Phe	Gly	Leu	Ser	Pro	Gly-NH2
Whitefish	p-Glu	His	Trp	Ser	Tyr	Gly	Met	Asn	Pro	Gly-NH2

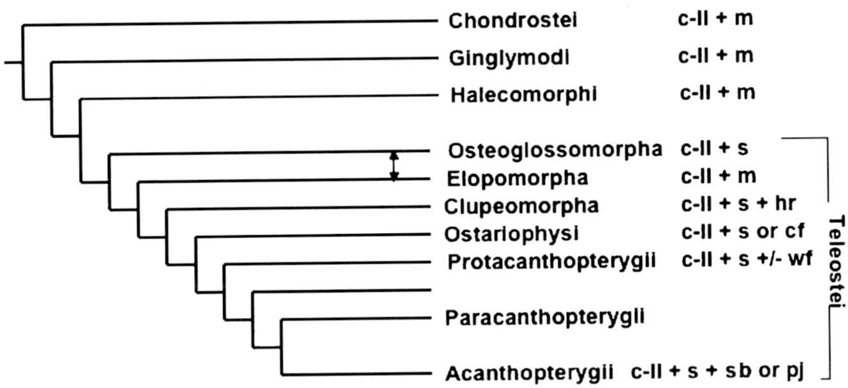

Fig. 11.1 Phylogeny of the different forms of GnRH present in teleosts. The double-arrow represents the hypothesis from some authors (O'Neill et al. 1998) that Elopomorpha are more ancient than Osteoglossomorpha. C-II = chicken-II form; s = salmon form; m = mammalian form; hr = herring form; cf = catfish form; sb = sea bream form; pj = pejerrey form; wf = whitefish form

form in the hindbrain is consistently the chicken II form (cGnRH-II), considered as an ancestral peptide (Muske 1993). The highest diversity has been observed in teleosts (eight GnRHs isolated) (Table 11.1; Fig. 11.1). In early teleosts such as the eel (Elopomorphs), chicken GnRH-II coexists with mammalian GnRH (mGnRH) (Japanese eel *Anguilla japonica*: Nozaki et al. 1985; European eel *Anguilla anguilla*: King et al. 1990), a situation similar to that found in other primitive Actinopterygii (for instance in Chondrostei such as sturgeon). In most other teleosts, cGnRH-II coexists with salmon GnRH (sGnRH) (for instance: Salmoniforms, Cypriniforms and Osteoglossiforms), with the exception of Siluriforms where catfish GnRH (cfGnRH) is present. In recent teleosts, a third GnRH form (perciforms and rockfish: sbGnRH; medaka: mdGnRH = pejerrey: pjGnRH) coexists with cGnRH-II and sGnRH. Interestingly, in more primitive species, the herring (clupeiform) and the whitefish (salmoniform), a third form was also identified (herring: hrGnRH; whitefish: wfGnRH), which also coexists with cGnRH-II and sGnRH.

In the European eel, a first immunocytochemical study of the brain distribution of GnRH neurons has been performed using antibodies recognizing all forms of GnRH (Kah et al. 1989). Two GnRH molecular forms, similar to mammalian GnRH (mGnRH) and to chicken GnRH-II (cGnRH-II), were then demonstrated by High Performance Liquid Chromatography (HPLC) and specific radioimmunoassay (RIA) (King et al. 1990). A differential distribution of these two forms was shown in the brain and the pituitary of silver eels (RIA using specific antisera: Dufour et al. 1993; immunocytochemistry: Montero et al. 1994). By RIA, we first showed that mGnRH levels were higher than cGnRH-II levels in the pituitary, olfactory lobes and tel-, di- and mes-encephalon, while the opposite was found in the posterior part of the brain (met- and myel-encephalon) (Dufour et al. 1993). These distributions were confirmed by our immunocytochemistry study in which mGnRH and cGnRH-II appeared to be produced by distinct neurons: mGnRH by neurons from olfactory bulbs, the nucleus olfactoretinalis, the ventral telencephalon, the preoptic

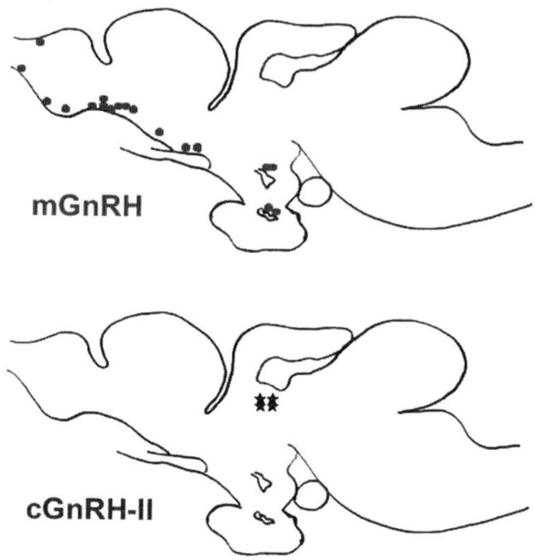

Fig. 11.2 Distribution of the two different forms of GnRH present in the European eel (after immunocytochemical studies from Montero et al. 1994). mGnRH = mammalian form of GnRH; cGnRH-II = chicken II form of GnRH

area and the hypothalamus, while cGnRH-II was produced in a few neurons in the midbrain tegmentum (Montero et al. 1994) (Fig. 11.2). All these data suggest differential physiological roles for the two GnRH forms in the eel. The occurrence of two GnRH forms in the eel brain was confirmed by the isolation of their cDNAs and genes in the Japanese eel (Okubo et al. 1999a, b). In addition, this study demonstrated the occurrence of three splicing variants of the messenger RNA coding for mGnRH, revealing further diversity of mGnRH potential roles and regulation (Okubo et al. 1999a, b). We are currently investigating the distribution and regulation of mGnRH and cGnRH-II precursors in the European eel (Weltzien, Dufour and coworkers, in the frame of the Norway-France international cooperation).

The mGnRH molecular form, as found in the eel, is thought to have first appeared in evolution in early-emerged bony fish (Osteichthyes) based on evidence from two species of Chondrostei (reedfish and sturgeon), one species of Finglymodi (gar) and one species of Halecomorphi (*Amia*) (for review: O'Neill et al. 1998), as well as two species of primitive Teleostei (eels and butterfly fish). The question of the existence of an undiscovered third GnRH form in the eel, in addition to mGnRH and cGnRH-II, has been addressed using phylogenetic approaches. Some authors (Okubo and Aida 2001) postulate that if only two GnRH forms occur in primitive teleosts such as eel and arowana (osteoglossiform), then the eel prepro-mGnRH and the arowana prepro-sGnRH should share high homology because they would be orthologues. In contrast, as they share low homology meaning they would rather be paralogues, the authors suggest that a gene duplication giving rise to mGnRH and sGnRH occurred before the emergence of teleosts and that a third form of GnRH may still be found in the eel (Okubo and Aida 2001). Other authors (O'Neill et al. 1998) postulate that there was a substitution/replacement in the mGnRH gene after the eels evolved resulting in the sGnRH gene, based on the presence of sGnRH in four members of osteoglossiforms (*Osteoglossum bicirrosum, Xenomystus nigri, Gnathonemus petersii, Chitala chitala*).

GnRH peptides bind to protein G-coupled receptors (GnRH-R) composed of an extracellular N-terminal region (30–40 amino acids), a large seven amino acid transmembrane domain (280–290 amino acids), and a short cytoplasmic C-terminal tail (30–50 amino acids). Within the teleost lineage, two main types of GnRH-R (termed type I and type II) could exist, each of which may include two or three subtypes (Lethimonier et al. 2004). While the two types of GnRH-R in mammals have distinct selectivity, all the teleost types of GnRH-R have a higher affinity for cGnRH-II, followed by sGnRH and a third endogenous GnRH form (Lethimonier et al. 2004). The two types of GnRH receptors have been identified in the Japanese eel (Okubo et al. 2000), and also in goldfish (Illing et al. 1999; Peter et al. 2003), African catfish (Tensen et al. 1997; Bogerd et al. 2002) and pufferfish (Yumoto et al. 2001). The situation may be still more complex, as recently, five different subtypes of GnRH-R were detected in masu salmon (Jodo et al. 2003), in the spotted green pufferfish *Tetraodon nigroviridis* (Ikemoto and Park 2005) and in European sea bass (Moncaut et al. 2005), while three forms have been reported in medaka (Okubo et al. 2001, 2003) and tilapia (Soga et al. 2005). In other species, up to now, only one type of GnRH-R was reported (rainbow trout: type II: Madigou et al. 2000; striped bass: type I: Alok et al. 2000). Further studies aiming at cloning and characterizing GnRH-R types and eventual subtypes in the European eel are clearly needed.

11.3 Gonadotropins

As in other vertebrates, the teleost pituitary secretes two gonadotropins (GTHs), follicle-stimulating hormone (FSH, formerly designated as GTH I in fish) and luteinizing hormone (LH, formerly GTH II in fish). GTHs are heterodimeric glycoproteins composed of a common α subunit and a hormone-specific β subunit. Our recent study in the eel using *in situ* hybridization demonstrated that LH and FSH were expressed by separate cells in the proximal pars distalis of the pituitary (Schmitz et al. 2005). In other teleost species, they are also produced in different cells in the pituitary (salmonids: Nozaki et al. 1990a; Naito et al. 1993; tuna: Kagawa et al. 1998; tilapia: Melamed et al. 1998; gilthead sea bream: Garcia Ayala et al. 2003; halibut: Weltzien et al. 2004; zebrafish: So et al. 2005). This distinct cellular source for the two GTHs in teleosts may facilitate the distinct regulation of gonadotropin synthesis and secretion. In contrast, in mammals, it was demonstrated that FSH and LH were expressed by the same gonadotropic cells (Childs et al., 1986; Liu et al., 1988). LH and FSH β subunits as well as the common glycoprotein α subunit (Gpα) have been cloned in the European eel (Quérat et al. 1990a, b; Schmitz et al. 2005) and in the Japanese eel (Nagae et al. 1996a). Measurement of their mRNA indicated that FSHβ increased at the early steps of the silvering process, while LHβ increased strongly later in the silvering process (European eel: Aroua et al. 2005; Rousseau et al. this book).

Fish gonadotropins act via binding to two gonadal gonadotropin receptors, homologous to mammalian LH receptor (LH-R) and FSH receptor (FSH-R). In contrast to the situation in mammals, the interactions are not highly specific (for review: Bogerd et al. 2005). Indeed, the catfish FSH-R is highly responsive to both catfish LH and FSH (Bogerd et al. 2001; Vischer et al. 2003), while the LH-R is rather specific to LH (Vischer and Bogerd 2003). The same situation is observed in coho salmon (Miwa et al. 1994; Yan et al. 1992). However, studies in zebrafish (Kwok et al. 2005), amago salmon (Oba et al. 1999a, b) and sea bass (Rocha et al. 2007) showed that FSH-R was specific to FSH, while LH-R was activated by both LH and FSH. Two types of gonadotropin receptors, respectively homologous to other teleost LH-R and FSH-R have been recently cloned in the Japanese eel (Jeng et al. 2007). Measurement of their ovarian mRNA levels by absolute quantitative real time RT-PCR indicated that FSH-R expression was much higher (50-fold) than that of LH-R in the previtellogenic eel (Jeng et al. 2007). In immature male Atlantic salmon, FSH-R transcripts are also more abundant (8-fold) than LH-R ones (Maugars and Schmitz 2007). Maturation experiments in the eel indicated that human chorionic gonadotropin, a LH-like hormone, was unable to induce ovarian development, which can be triggered by fish pituitary extract. This suggests that eel FSH-R may have a strong specificity and does not recognize mammalian gonadotropin. Our future investigations will aim at cloning European eel gonadotropin receptors and further characterizing gonadotropin receptor selectivity in the eel (Dufour, Chang and coworkers in the frame of the Taiwan-France international cooperation).

11.4 Sex Steroids

As in other vertebrates, the androgen, testosterone (T) and the estrogen, estradiol (E2) are present in teleost fishes. In addition, 11-oxygenated androgens, especially 11-ketotestosterone (11-KT) are also detected (Borg 1994). In mature male salmon parr and mature anadromous males, 11-ketotestosterone was even found to be the predominant androgen in the plasma (Mayer et al. 1990). Compared with adult mammals, the brain of most teleost fishes is characterized by an extremely high capacity to aromatize androgens into estrogens, because of exceptionally high levels of aromatase and of high aromatase activity (Callard et al. 1978; Pasmanik and Callard 1985; for review: Pellegrini et al. 2005). Unique among vertebrates, teleost fishes possess in fact three estrogen receptor (ER) subtypes (ERalpha, ER beta1, ER beta2) (Atlantic croaker, *Micropogonias undulatus*: Hawkins et al. 2000; zebrafish: Bardet et al. 2002; rainbow trout: Menuet et al. 2002; goldfish: Choi and Habibi 2003; European sea bass: Halm et al. 2004; fathead minnow, *Pimephales promelas*: Filby and Tyler 2005; sea bream: Pinto et al. 2006). Two androgen receptors (AR alpha and AR beta) have been identified in the eel (Ikeuchi et al. 1999) as in many teleosts (for review: Douard et al. 2004). Different studies showed the interdependence between androgen or estrogen receptors and aromatase in the brain of teleosts (for review: Pellegrini et al. 2005), with similar temporal patterns.

Low levels of androgens (mainly testosterone and 11-KT) are detected in the plasma of male silver eels (European eel: Khan et al. 1987; Japanese eel: Miura et al. 1991). Androgen production by eel testis is greatly stimulated during experimental maturation induced by hCG (European eel: Khan et al. 1987; Japanese eel: Ohta and Tanaka 1997; Japanese eel: Miura et al. 1991). In the female eel, plasma levels of T, 11-KT and E2 significantly increase between the pre-vitellogenic (yellow) and early vitellogenic (silver) stages as shown in *Anguilla anguilla*, (Sbaihi et al. 2001; Aroua et al. 2005) as well as in other eel species (*A. australis* and *dieffenbachi*: Lokman et al. 1998; *A. rostrata*: Cottrill et al. 2001; *A. japonica*: Han et al. 2003). Further increase in androgens and in E2 levels are observed during experimental maturation induced by gonadotropic (fish pituitary extract) treatments (Leloup-Hatey et al. 1988; Peyon et al. 1997). The similarity in plasma levels of androgens and estradiol is a remarkable feature in the female eel, likely related to androgen-specific regulations, as discussed in Chapter 12. Our recent study in Japanese eel showed that eel brain aromatase has a low activity compared to enzymatic activity in other teleosts (Jeng et al. 2005). This allows, in the eel, androgen-specific actions to be exerted, not only by non-aromatizable androgens such as 11-KT but also by aromatizable androgens, such as testosterone. Accordingly, testosterone-specific and estradiol-specific actions were found in the eel (see below). For the moment, no data are available concerning which androgen and/or estrogen receptor(s) is involved in the sex steroid feedbacks observed in the eel, on brain GnRHs and pituitary gonadotropins. However, concerning androgen receptors, in the Japanese eel only AR alpha is expressed in the hypothalamus (brain) (Ikeuchi et al. 1999).

11.5 Effects of Sexual Maturation on Eel Endogenous Brain-Pituitary Gonadotropic Axis

11.5.1 Effects on Gonadoliberins

As we mentioned before, silver eels remain blocked at a prepubertal stage as long as migration towards the Sargasso Sea is prevented, and to date, only experimental treatments of silver eels have led to the observation of sexually mature animals (male European eel: Fontaine 1936; female European eel: Fontaine et al. 1964). Based on these pioneer experiments in the European eel, similar treatments (hCG in males and fish pituitary extract in females) have been since currently employed to induce sexual maturation (gametogenesis and steroidogenesis) in various eel species (*Anguilla japonica*: Yamauchi et al. 1976; Ohta et al. 1997; *Anguilla rostrata*: Edel 1975; Sorensen and Winn 1984; *Anguilla dieffenbachii*: Todd 1979; Lokman and Young 2000). In the female, long-term treatment with carp pituitary extract stimulated ovarian vitellogenesis, leading to a gradual increase in gonado-somatic index, which reached up to 30–40% after several months, an index much higher than in control eels (1.5–2%) (Fontaine et al. 1964; Dufour et al. 1989, 1993; Schmitz et al. 2005; Durif et al. 2005).

Our first studies, employing antibodies recognizing all forms of GnRH, indicated a positive effect of experimental sexual maturation on total GnRH level in the brain of female or male silver eels (males treated with hCG and females with estradiol: Dufour et al. 1985; females treated with pituitary extract: Dufour et al. 1989). This effect was even more marked in the pituitary, reflecting the accumulation of GnRH in the axonal endings, which are directly innervating the adenohypophysis in the eel as in other teleosts. These data were confirmed by immunocytochemical observation (Kah et al. 1989), which indicated a strong accumulation of GnRH peptide in the pituitary and, in particular, in the axonal endings of the hypophysiotropic neurons. Moreover, castration was able to abolish the increase in brain and pituitary GnRH content, which indicates that gonadal hormones are responsible for this positive effect (Dufour et al. 1989). Later on, using specific RIAs for each native form of GnRH in the eel, we could perform more specific analyses of the effect of experimental maturation on mGnRH and cGnRH-II. We were able to demonstrate an opposite regulation of the two forms with an increase in mGnRH levels in the brain and pituitary, whereas a decrease in cGnRH-II levels in the brain was found, cGnRH-II levels being not detectable by RIA in the pituitary (Dufour et al. 1993). This opposite regulation suggests that mGnRH and cGnRH-II play drastically different roles during eel sexual maturation, and that mGnRH would play a major role in the neuroendocrine control of pituitary gonadotropins.

A differential regulation of the two GnRH forms was also observed in the goldfish and the salmon, with an increase in sGnRH but not cGnRH-II in the anterior brain and in the pituitary during natural sexual maturation (Amano et al. 1992; Rosenblum et al. 1994). In (masu) salmon brain, sGnRH genes are activated long before sexual maturation (Ando et al. 2001). In the striped bass, the levels of the two most abundant forms in the pituitary, sbGnRH and cGnRH-II, increased during the autumn and peaked prior to (for cGnRH-II) and during (for sbGnRH) the natural breeding season in March to May (Holland et al. 2001).

In sea bass and striped bass, pituitary GnRH-R gene expression increases according to maturation (Alok et al. 2000). Similarly, in masu salmon, the different GnRH-R genes were shown to vary with the season and after a GnRH analog treatment (Jodo et al. 2005). Even though we may also expect an increase in pituitary GnRH-R during eel induced maturation, direct data on the regulation of eel GnRH-R are missing.

11.5.2 Effects on Gonadotropins

In the European eel, our early studies, using heterologous radioimmunoassay for carp LH β subunit, showed a large increase in pituitary LH content in artificially matured eels, namely in females treated with carp pituitary extract or in males treated with human chorionic gonadotropin (Dufour 1985). The effect of carp pituitary extract on pituitary LH content was prevented by ovariectomy (Dufour et al. 1989), and the production of sex steroids was stimulated after GtH treatment in male (Khan et al. 1987) and experimentally matured female eels (Leloup-Hatey et al. 1988). These data suggested the involvement of gonadal hormones in the stimulation of endogenous pituitary LH during experimental maturation. In contrast, no change in plasma LH level was found in experimentally matured male eels, in spite of greatly elevated pituitary LH content (Dufour 1985). This indicates that endogenous LH synthesis but not release is stimulated during experimental maturation. The situation is likely the same in experimentally matured female eels, but the recognition of exogenous carp LH by RIA prevented the determination of endogenous LH plasma levels. Recently, the cloning of eel FSHβ and LHβ subunits allowed us to demonstrate that during experimental maturation induced by carp pituitary extract in females, LH and FSH undergo an opposite regulation with a large increase in LHβ mRNA levels but a decrease in FSHβ mRNA levels (Schmitz et al. 2005). Similarly, in males matured using hCG injections, LHβ mRNA levels increased while FSHβ mRNA levels decreased (Aroua S et al. 2007, unpublished).

In the Japanese eel, repeated treatment with salmon gonadotropin (Sato et al. 1996) is required for the artificial induction of ovarian maturation, and stimulation of LH synthesis is only observed after treatment with salmon pituitary homogenate (Nagae et al. 1996a, b, 1997), salmon GtH (Yoshiura et al. 1999) or sex steroids (Lin et al. 1998). An increase in both LHβ and GPα mRNA levels was observed during the induction of ovarian development (Nagae et al. 1996a, b, 1997; Yoshiura et al. 1999; Suetake et al. 2002; Saito et al. 2003a). A dramatic decrease in FSHβ mRNA levels was reported in experimentally matured male and female Japanese eels, and FSHβ mRNA levels were undetectable after 14 weeks of gonadotropic treatment (Yoshiura et al. 1999). Saito and colleagues found profound differences in FSHβ and LHβ mRNA profiles between artificially maturing Japanese eels and naturally maturing New Zealand longfinned eels *Anguilla dieffenbachii* that they correlated to differences in steroid hormone profiles (Saito et al. 2003a). Indeed, FSHβ mRNA level was high at the previtellogenic stage in Japanese eels, but low in New Zealand longfinned eels and then, at the mid-vitellogenic stage, increased in New Zealand longfinned eels but decreased in Japanese eels; LHβ mRNA level increased considerably at the mid-vitellogenic stage in Japanese eels, but only slightly in New Zealand longfinned eels. This reveals that the opposite variations in FSHβ and LHβ mRNA pituitary levels observed in artificially maturing Japanese as well as European eels, may differ considerably from their natural profiles.

Different profiles of gonadotropin expression have been observed during gametogenesis among teleost species. In salmonids, pituitary contents (rainbow trout: Suzuki et al. 1988; Nozaki et al. 1990a, b; Naito et al. 1991; amago and chum salmon: Suzuki et al. 1988) and plasma levels (rainbow trout: Suzuki et al., 1988; Prat et al., 1996; amago and chum salmon: Suzuki et al. 1988; coho salmon: Swanson et al. 1989; Atlantic salmon: Oppen-Berntsen et al. 1994; chinook salmon: Slater et al. 1994) of FSH and LH vary differently during gametogenesis, with FSH being elevated during vitellogenesis and spermatogenesis and LH increasing during final maturation. Similar results were found for FSH and LH mRNA levels (rainbow trout: Weil et al. 1995; Gomez et al. 1999; Atlantic salmon: Maugars and Schmitz 2007). These results suggest that, in salmonids, LH and FSH play separate roles during gonadal development with FSH controlling the first stages of gametogenesis and LH acting during the final steps (Swanson 1991). Salmonids are annual spawners and possess an ovary that shows synchronous oocyte development. In other teleost species such as the goldfish *Carassius auratus* (Yoshiura et al. 1997; Sohn et al. 1998), the blue gourami *Trichogaster trichopterus* (Jackson et al. 1999), the Japanese flounder *Paralichthys olivaceus* (Kajimura et al. 2001) and red seabream *Pagrus major* (Gen et al. 2000), both LHβ and FSHβ mRNA levels increase in parallel during gonadal maturation and spawning. These species are all multiple spawners, showing asynchronous development of oocytes throughout their spawning season. However, in another repeat spawner, the three-spined stickleback (*Gasterosteus aculeatus*), FSHβ expression peaks earlier than LHβ expression (Hellqvist et al. 2006), indicating that the patterns in expression of GtHs vary considerably, even between species that have similar patterns in ovarian development. Among perciforms, various profiles have been observed. In striped bass, *Morone saxatilis*, FSHβ mRNA levels increase quickly at the onset of gonadal development and then decline to basal levels, while LHβ levels remain elevated at the final stages of vitellogenic growth (Hassin et al. 1999). In contrast, in the sea bass, *Dicentrarchus labrax*, mRNA levels for FSH and LH increase simultaneously throughout sexual maturation and decline sharply at post-spermiation (Mateos et al. 2003).

Concerning gonadotropin receptors, in catfish ovaries, FSH-R mRNA peaks after spawning, while LH-R shows the highest expression at ovulation, suggesting that FSH-R is associated with ovarian recrudescence while LH-R plays a role in final oocyte maturation and ovulation (Kumar et al. 2001a, b). In zebrafish, while FSH-R expression increases during vitellogenesis of the first cohort of developing follicles, LH-R level becomes detectable at the beginning of vitellogenesis and steadily increases afterward with the peak level reached at the full-grown stage (Kwok et al. 2005). In male fish, FSH-R mRNA levels show an increase during early spermatogenesis, while at spermiation, the transcript levels either decrease or fluctuate (yellowtail: Rahman et al. 2003; rainbow trout: Kusakabe et al. 2006; Atlantic salmon: Maugars and Schmitz 2007). For LH-R mRNA, a steady increase is observed as testicular maturation advances (yellowtail: Rahman et al. 2003; rainbow trout: Kusakabe et al. 2006; Atlantic salmon: Maugars and Schmitz 2007). Our recent data on LH-R and FSH-R expression in the Japanese eel, using quantitative real time RT-PCR indicate an increase in the ovarian expression of both receptors during induced maturation (Jeng et al. 2007).

11.6 Effects of Exogenous Sex Steroid Treatment on Endogenous Brain-Pituitary Gonadotropic Axis

Castration experiments in the European eel demonstrated that the increase in endogenous mGnRH and LH levels, as observed during induced maturation, resulted from a positive feedback by gonadal hormones (Dufour et al. 1991).

Similarly, the involvement of gonadal hormones in the stimulation of pituitary gonadotropin during sexual maturation was demonstrated by castration experiments and gonadotropic treatments in the juvenile male rainbow trout (Crim et al. 1982; Gielen et al. 1982). Steroid measurement in various eel species have shown that the production of sex steroids (estradiol in females and androgens in both sexes) is significantly increased during eel artificial maturation (European eel: Khan et al. 1987; Japanese eel: Ijiri et al. 1995; Ohta and Tanaka 1997; New Zealand longfinned eel: Lokman et al. 2001). Their potential effect on the brain pituitary gonadotropic axis could be assessed by *in vivo* treatments with various sex steroids, as reported below.

11.6.1 Effects on Gonadoliberins

The first demonstration of the estradiol stimulatory effect on GnRH in the European female silver eel was performed by Dufour et al. (1985), using a RIA for mammalian GnRH. Later, using specific RIA for each native GnRH, we could reveal a differential regulation of mGnRH and cGnRH-II by sex steroids. We observed a stimulatory effect of estradiol on mGnRH brain and pituitary levels, while an inhibitory effect of testosterone on brain cGnRH-II level was found (Montero et al. 1995). The combined treatment with estradiol and testosterone enhanced both regulations. This indicated that steroids exert differential feedbacks on the two GnRH forms, with an estrogen-dependent control of mGnRH, an androgen-dependent control of cGnRH-II, and the possibility of a potentialization of the two types of steroids in the combined treatment (Montero et al. 1995). The increase in mGnRH levels and the decrease in cGnRH-II levels found after exogenous sex steroid treatments were in good agreement with the differential regulation of mGnRH and cGnRH-II observed in experimentally matured female European eel (Dufour et al. 1993). This suggests a prominent role of endogenous sex steroids feedbacks in the opposite regulation of mGnRH and cGnRH-II during experimental maturation.

In the Japanese eel, contradictory results using yellow (juvenile) eels have been obtained. Indeed, while no increase in pituitary mGnRH content was observed after injection of E2 or T in female or male eels (Jeng et al. 2002), Okubo et al. (2002) showed an up-regulation of mGnRH mRNAs by testosterone in males. These variations may depend upon age and stage of the eels, as well as mode and duration of sex steroid treatments.

Data from other teleosts revealed various effects and specificities of sexual steroids on GnRHs according to species and possibly physiological stage and mode of investigation, as illustrated below. However, as for mGnRH in the eel, altogether these studies indicate a stimulatory role of sex steroids on the GnRH form of the anterior brain, likely to be involved in the regulation of pituitary gonadotropins in the other teleosts (sGnRH, cfGnRH, sbGnRH).

In juvenile salmonids, T stimulated sGnRH content (*Oncorhynchus masou*: Amano et al. 1994; *Salmo trutta*: Breton et al. 1986; *Oncorhynchus mykiss*: Goos et al. 1986;

Breton and Sambroni 1996). In yearling masu salmon, oral 17α-methyltestosterone was able to activate sGnRH neurons (increase of the number of cells expressing sGnRH mRNA in the preoptic area) in future precocious males, but not in immature females (Amano et al. 1994). T and 11-KT induced an increase of GnRH levels in the brain of platyfish (*Xiphophorus maculatus*: Schreibman et al. 1986). In juvenile African catfish (*Clarias gariepinus*), steroid effects were rather selective, in that only T, but not E2 or 11-oxygenated androgens, induced a precocious maturation of the cfGnRH-producing neurones in the ventral forebrain, while the mesencephalic cGnRH-II-producing neurones did not appear to be affected by these treatments (Dubois et al. 1998). In contrast, T and 11-KT were reported to induce an increase in GnRH immunoreactive cells in African catfish (Goos 1987). Dubois et al. (2001) demonstrated that the onset of puberty in the male African catfish coincided with the completion of the steroid-dependent structural maturation of the cfGnRH system in the brain, but that T and E2 were also able to exert a positive influence on the amounts of cfGnRH during the later stages of pubertal development. In goldfish (*Carassius auratus*), sex steroids did not have any significant effects on the GnRH neuron activity in brain (Parhar et al. 2001). Nevertheless, other authors found that the ratio between salmon GnRH and cGnRH-II changed with sexual maturation (Rosenblum et al. 1994). In sexually immature male tilapia (*Oreochromis niloticus*), estrogen significantly increased preoptic seabream GnRH neuronal numbers, while KT failed to do so (Parhar et al. 2000). In juvenile tilapia, estrogen had no effects on mRNA levels of salmon- and chicken II-GnRH (Parhar et al. 1996). In black porgy, sex steroids (E2, T and 11-KT) significantly stimulated pituitary sbGnRH levels *in vivo*. Furthermore, the authors were able to establish primary culture of brain neurons and to demonstrate that sex steroids also stimulated the content and release of sbGnRH *in vitro* (Lee et al. 2004).

11.6.2 Effects on Gonadotropins

Early histological studies by Olivereau and co-workers suggested that the administration of estradiol in the eel stimulated gonadotropin synthesis but not release, leading to a large accumulation of secretory granules in the pituitary gonadotropic cells (Olivereau and Chambolle 1978; Olivereau and Olivereau 1979a, b). This assumption was confirmed by RIA studies, showing a large increase in pituitary LH content in E2-treated eels, while plasma LH levels remained as low as in control fish (Dufour et al. 1983a, b). An immunocytochemical study of the cultured pituitary cells of the eel indicated an increase in the gonadotropic cell size and LH content after an *in vivo* estradiol pre-treatment of the animals (Montero et al. 1996). In male eels, both estradiol and testosterone were effective in stimulation of pituitary LH content *in vivo*, while in the female testosterone had no or only moderate effect as compared to estradiol (Dufour et al. 1983a, b; Montero et al. 1995; Vidal et al. 2004). The effect of estrogen on pituitary LH in female European eel was shown to be mediated by an increase in mRNA coding GP α (Counis et al. 1987) and LH β subunits (Quérat et al. 1991). A synergistic effect of estradiol and testosterone on the α subunit mRNA and on the β subunit mRNA of LH (Quérat et al. 1991), as well as on pituitary radioimmunoassayable LH content (Montero et al. 1995) has been reported. The synergistic effect of testosterone on estradiol stimulation of pituitary LH expression can be compared to the similar synergy exerted on mGnRH (as discussed in Section 11.5.1). Recent cloning of European eel FSHβ cDNA (Schmitz et al. 2005) let us decipher the differential regulations of LH and FSH by sex steroids *in vivo*. Chronic *in vivo* treatment of female eels with E2 induced an increase in LHβ and GPα mRNA levels but not in FSHβ, while T induced a decrease in FSHβ mRNA levels (Schmitz et al. 2005). We have not yet investigated the possible synergistic effects of sex steroids on FSHβ. The opposite regulation of LH and FSH expression by sexual steroid treatments is in agreement with the opposite profiles of LH and FSH expression during induced maturation. This suggests that feedbacks by endogenous steroids may be responsible for the opposite changes in LH and FSH during experimental maturation.

Development of primary culture of eel pituitary cells (Montero et al. 1996) allowed us to investigate the direct effects of sexual steroids on pituitary gonadotropins. Surprisingly, these studies revealed a stimulatory effect of androgens (T and non-aromatizable androgens) but not E2 on the synthesis of LH, mediated by a specific increase in LHβ but not GPα mRNA levels (Huang et al. 1997). The discrepancy between the E2 large positive effect on LH expression *in vivo* but not *in vitro*, suggests that E2 action *in vivo* is indirect, possibly via its stimulatory effect on mGnRH. In contrast, the direct stimulatory effect of androgens on LHβ, as shown *in vitro*, may account for their moderate positive effect *in vivo*. Besides this direct effect, we recently demonstrated that androgens enhance *in vivo* the activity of dopaminergic neurons involved in the inhibition of eel pituitary LH (Weltzien et al. 2006). Altogether, this inhibition may counteract the direct stimulatory action of androgens on LH expression and explain their lack of, or moderate, effect *in vivo* (Aroua et al. 2007). Investigation of the regulation of FSHβ expression by pituitary cells *in vitro* indicated a moderate stimulation by E2 and no significant effect of androgens (Aroua et al. 2007). These effects cannot account for the inhibition of FSH expression observed *in vivo* in androgen-treated eels. A challenging hypothesis may come from the comparison between the effects of steroids on gonadotropins and gonadoliberins *in vivo*. While an estradiol-dependent positive regulation is exerted on LH and mGnRH fitting with the hypothesis of an mGnRH-LH axis, an androgen-dependent inhibition is observed on FSH and cGnRH-II, suggesting the possibility of a cGnRH-II-FSH axis. Further investigations are clearly needed to test this hypothesis.

In the Japanese eel, recent *in vitro* studies demonstrated that sex steroid treatment (T, E2 and 11-KT) suppressed FSHβ subunit mRNA levels (cultured eel pituitary: Mitsuhashi et al. 2000), while E2 was able to induce LH synthesis (organ-cultured pituitary glands: Saito et al. 2003b). For LH, this discrepancy in the data could be explained by the differences in the methods used (organ cultures still possess neuronal endings) or in the species studied.

In other teleosts, there is evidence for both positive and negative effects of sex steroids on gonadotropin production, depending on the reproductive stages of the fish. For instance, in goldfish, a stimulatory effect of T on LH mRNA levels has been shown (Huggard et al. 1996; Sohn et al. 1998). This stimulatory effect was less effective in fish in which gonadal maturation had been initiated in comparison to those in immature fish (Kobayashi and Stacey 1990; Huggard et al. 1996; Sohn et al. 1998), which demonstrates a difference in sensitivity of the pituitary gland to sex steroids depending on the maturity of the fish. *In vitro*, the levels of LHβ mRNA were stimulated by T (both sexually immature and mature fish: Huggard et al. 1996; immature fish: Sohn et al. 2001). Concerning FSH, *in vivo* investigations showed that sex steroids exerted a strong inhibitory effect on FSHβ mRNA levels in sexually immature and maturing goldfish (Sohn et al. 1998; Kobayashi et al. 2000), while no effect was observed *in vitro* on cells of immature, recrudescent, mature and regressed fish (Sohn et al. 2001), which indicates a possible indirect action of T on FSH. In juvenile salmonids, a number of studies have demonstrated that E2 increased pituitary levels of LH and LHβ mRNA (Crim et al. 1981; coho salmon: Dickey and Swanson 1998; rainbow trout: Trinh et al. 1986; Xiong et al. 1994). In contrast, negative or no effects of sex steroids on LH have been reported in mature fish (Billard et al. 1977; Larsen and Swanson 1997; Xiong et al. 1994). Xiong et al. (1994) studied the effect of sex steroids on LHβ gene expression in cultured pituitary cells of chinook salmon: several estrogen-responsive elements (EREs) were characterized and a proximal ERE was suggested to be involved in the positive feedback of steroids in juvenile fish, but not in mature ones. In contrast, *in vitro* treatment of pituitary cells from immature rainbow trout with E2 had no effect on the amounts of FSHβ mRNA (Xiong et al. 1994). Concerning the regulation of FSH *in vivo*, steroids had mostly negative but also some positive effects, which could be observed depending on the reproductive stage of fish. T and E2 reduced FSH plasma levels in immature coho salmon (Larsen and Swanson 1997; Dickey and Swanson 1998) and Atlantic salmon (Borg et al. 1998). In goldfish, a stimulatory effect of T on LH mRNA levels has been shown (Huggard et al. 1996; Sohn et al. 1998). This stimulatory effect was less effective in fish in which gonadal maturation has been initiated in comparison to those in immature fish (Kobayashi and Stacey 1990; Huggard et al. 1996; Sohn et al. 1998), which demonstrates a difference in sensitivity of the pituitary gland to sex steroids depending on the maturity of the fish. *In vitro*, the levels of LHβ mRNA were stimulated by T (both sexually immature and mature fish: Huggard et al. 1996; immature fish: Sohn et al. 2001). Concerning FSH, *in vivo* investigations showed that sex steroids exert a strong inhibitory effect on FSHβ mRNA levels in sexually immature and maturing goldfish (Sohn et al. 1998; Kobayashi et al. 2000), while no effect was observed *in vitro* on cells of immature, recrudescent, mature and regressed fish (Sohn et al. 2001), which indicates a possible indirect action of T on FSH. In the sea bass, like in goldfish, implantation with E2, T or DHT suppressed basal FSHβ expression, while slightly increased LHβ expression (Mateos et al. 2002).

Altogether, these data indicate that sex steroids have in general a stimulatory effect on LH and negative one on FSH in immature fishes from various teleost species, including the eel, even though the steroid specificity and mechanism of these actions may differ according to the species.

11.7 Conclusions and Applied Perspectives

In the eel as in other teleosts, but differently from mammals, LH and FSH are expressed by different cells in the proximal pars distalis of the pituitary. Their differential regulation by gonadal steroids, as shown during experimental maturation or after sex steroid treatment, with a large increase in LH and a decrease in FSH, suggests that LH and FSH play different roles in eel reproduction. We hypothesize (see Chapter 3) that FSH may be responsible for the initiation of lipidic vitellogenesis ("endogenous vitellogenesis"), LH and FSH may participate in the induction of vitellogenin production and "exogenous vitellogenesis", and LH may control the final steps (oocyte maturation and ovulation). Similarly, LH could be mostly involved in the final stages of spermatogenesis in male.

The differential distribution and regulation of eel GnRH native forms (mGnRH and cGnRH-II) in the brain also leads to the conclusion of different physiological roles. The large increase in pituitary mGnRH levels in experimentally matured eels or steroid-treated eels suggest the prime implication of mGnRH in the neurohormonal control of LH surge.

However, one should keep in mind that these opposite regulations (mGnRH-LH/cGnRH-II-FSH) (Fig. 11.3) are observed in experimental conditions that bypass the natural brain-pituitary control. They may therefore differ from the natural pat-

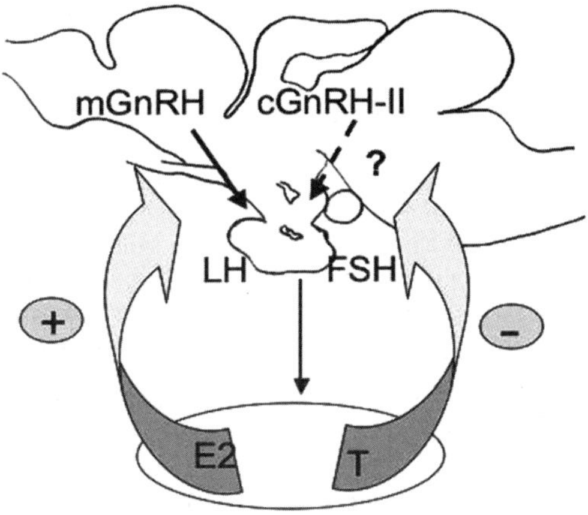

Fig. 11.3 Schematic representation of sex steroid feedbacks on the GnRH-gonadotropin axis in the European eel, deduced from *in vivo* experiments. mGnRH = mammalian form of GnRH; cGnRH-II = chicken II form of GnRH; FSH = follicle-stimulating hormone; LH = luteinizing hormone; E2 = estradiol; T = testosterone

tern of GnRH peptides and pituitary gonadotropins during spontaneous eel sexual maturation. In particular, the drop of FSH may be an artefactual consequence of sex steroid feedbacks in the absence of other natural triggering factors. Whatever the situation is, we could already take advantage of the large increase in LH and mGnRH induced by the sex steroid feedback by suggesting innovative protocols for inducing the last steps of gametogenesis in the eel. For instance, in experimentally-matured eels, short term treatments with GnRH agonists and dopamine antagonists could be applied to induce the release of pituitary LH and mimic the endogenous LH ovulatory peak. Similarly, factors able to trigger GnRH release, such as pheromones, could be applied to induce mGnRH-LH ovulatory peak. These alternative methods could lead to more physiological levels of hormones inducing final oocyte maturation and ovulation, favouring a better quality of gametes and larvae. Concerning FSH, future studies should aim at investigating which factors could maintain its expression in spite of the negative steroid feedback. Our ongoing studies indicate a strong stimulatory role of activin on eel FSHβ expression. One other major issue would be to study the localization and expression of GnRH-R and gonadotropin receptors during experimental maturation as well as their regulation by sex steroids. All this knowledge would give new perspectives for reproductive control and long-term aquaculture of eels.

Acknowledgements We thank S. Baloche, N. Le Belle, B. Vidal and Drs. O. Kah, Y-S Huang, M. Montero, Y. Zohar, M. Sbaihi and F-A. Weltzien for their contributions to the studies presented in this chapter. This work has been supported by the European contract EELREP N°Q5RS-2001-01836.

References

Alok D, Hassin S, Sampath KR, Trant JM, Yu K, Zohar Y (2000) Characterization of a pituitary GnRH-receptor from a perciform fish, *Morone saxatilis*: functional expression in a fish cell line. Mol Cell Endocrinol 168:65–75.

Amano M, Aida K, Okumoto N, Hasegawa Y (1992) Changes in salmon GnRH and chicken GnRH-II contents in the brain and pituitary, and GTH contents in the pituitary in the female Masu salmon, *Oncorhynchus masou*, from hatching through ovulation. Zool Sci 9:375–386.

Amano M, Hyodo S, Urano A, Okumoto N, Kitamura S, Ikuta K, Suzuki Y, Aida K (1994) Activation of salmon gonadotropin-releasing hormone synthesis by 17α-methyltestosterone administration in female masu salmon. Gen Comp Endocrinol 95:374–380.

Ando H, Sasaki Y, Okada H, Urano A (2001) Prepubertal increases in the levels of two salmon gonadotropin-releasing hormone mRNAs in the ventral telencephalon and preoptic area of masu salmon. Neurosci Lett 307:93–96.

Aroua S, Schmitz M, Baloche S, Vidal B, Rousseau K, Dufour S (2005) Endocrine evidence that silvering, a secondary metamorphosis in the eel, is a pubertal rather than a metamorphic event. Neuroendocrinology 82:221–232.

Aroua S, Weltzien FA, Le Belle N, Dufour S (2007) Development of real-time rt-pcr assays for eel gonadotropins and their application to the comparison of *in vivo* and *in vitro* effects of sex steroids. Gen Comp Endocrinol 153:333–343.

Bardet PL, Horard B, Robinson-Rechavi M, Laudet V, Vanacker JM (2002) Characterization of estrogen receptors in zebrafish (*Danio rerio*). J Mol Endocrinol 28:153–163.

Billard R, Richard M, Breton B (1977) Stimulation of gonadotropin secretion after castration in rainbow trout. Gen Comp Endocrinol 33:163–165.

Bogerd J, Li KW, Janssen-Dommerholt C, Goos H (1992) Two gonadotropin-releasing hormones fom African catfish (*Clarias gariepinus*). Biochem Biophys Res Commun 187:127–134.

Bogerd J, Blomenröhr M, Andersson E, ven derPutten HHAGM, Tensen CP, Vischer HF, Grannenman JCM, Janssen-Dommerholt C, Goos HJTh, Schulz RW (2001) Discrepancy between molecular structure and ligand selectivity of a testicular follicle-stimulating hormorne receptor of the African catfish (*Clarias gariepinus*). Biol Reprod 64:1633–1643.

Bogerd J, Diepenbroek WB, Hund E, van Oosterhout F, Teves AC, Leurs R, Blomenrohr M (2002) Two gonadotropin-releasing hormone receptors in the African catfish: no differences in ligand selectivity, but differences in tissue distribution. Endocrinology 143:4673–4682.

Bogerd J, Granneman JCM, Schulz RW, Vischer HF (2005) Fish FSH receptors bind LH: how to make the human FSH receptor to be more fishy? Gen Comp Endocrinol 142:34–43.

Borg B (1994) Androgens in teleost fishes. Comp Biochem Physiol 109C:219–245.

Borg B, Antonopoulou E, Mayer I, Andersson E, Berglund I, Swanson P (1998) Effects of gonadectomy and androgens treatments on pituitary and plasma levels of gonadotropins in mature male Atlantic salmon, *Salmo salar*, parr: positive feedback control of both gonadotropins. Biol Reprod 58:814–820.

Breton B, Sambroni E (1996) Steroid activation of the brain-pituitary complex gonadotropic function in the triploid trout *Oncorhynchus mykiss*. Gen Comp Endocrinol 101:155–164.

Breton B, Motin A, Billard R, Kah O, Geoffre S, Percigoux G (1986) Immunoreactive gonadotropin releasing hormone like material in the brain and pituitary gland during the periovulatory period in the brown trout (*Salmo trutta* L.): relationships with the plasma and pituitary gonadotropin. Gen Comp Endocrinol 61:109–119.

Callard GV, Petro Z, Ryan KJ (1978) Phylogenic distribution of aromatase and other androgen-converting enzymes in the central nervous system. Endocrinology 103:2283–2290.

Childs GV, Hazum E, Amsterdam A, Limor R, Naor Z (1986) Cytochemical evidence for different routes of gonadotropin-releasing hormone processing by large gonadotropes and granulosa cells. Endocrinology 119:1329–338.

Choi CY, Habibi HR (2003) Molecular cloning of estrogen receptor alpha and expression pattern of estrogen receptor subtypes in male and female goldfish. Mol Cell Endocrinol 204:169–177.

Cottrill RA, McKinley RS, van der Kraak G, Dutil J-D, Reid KB, McGrath KJ (2001) Plasma non-esterified fatty acid profiles and 17β-oestradiol levels of juvenile immature and maturing adult American eels in the St Lawrence River. J Fish Biol 59:364–379.

Counis R, Dufour S, Ribot G, Quérat B, Fontaine YA, Jutisz M (1987) Estradiol has inverse effects on pituitary glycoprotein hormone α-subunit messenger ribonucleic acid in the immature European eel and the gonadectomized rat. Endocrinology 121:1178–1184.

Crim LW, Peter RE, Billard R (1981) Onset of gonadotropic hormone accumulation in the immature trout pituitary gland in response to estrogen or aromatizable androgen steroid hormones. Gen Comp Endocrinol 44:374–381.

Crim LW, Billard R, Genge PD, Idler DR (1982) The influence of immature gonads on onset of gonadotropic hormone accumulation in the juvenile rainbow trout pituitary gland. Gen Comp Endocrinol 48:161–166.

Dickey JT, Swanson P (1998) Effects of sex steroids on gonadotropin (FSH and LH) regulation in coho salmon (*Oncorhynchus kisutch*). J Mol Endocrinol 21:291–306.

Douard V, Robinson-Rechavi M, Laudet V, Guiguen Y (2004) Molecular evolution of androgen receptors in fish. Fish Physiol Biochem 28:207–208.

Dubois EA, Florijn MA, Zandbergen MA, Peute J, Goos HJTh (1998) Testosterone accelerates the development of the catfish GnRH system in the brain of immature African catfish (*Clarias gariepinus*). Gen Comp Endocrinol 112:383–393.

Dubois EA, Slob S, Zandbergen MA, Peute J, Goos HJ (2001) Gonadal steroids and the maturation of the species-specific gonadotropin-releasing hormone system in brain and pituitary of the male African catfish (*Clarias gariepinus*). Comp Biochem Physiol 129B:381–387.

Dufour S (1985) La fonction gondotrope de l'anguille européenne, *Anguilla anguilla* L., au stade argenté (au moment du depart pour la migration de reproduction): les mécanismes de son blocage et sa stimulation expérimentale. Ph.D. thesis University Paris VI.

Dufour S, Delerue-Le Belle N, Fontaine YA (1983a) Development of a heterologous radioimmunoassay for eel (*Anguilla anguilla*) gonadotropin. Gen Comp Endocrinol 49:404–413.
Dufour S, Delerue-Le Belle N, Fontaine YA (1983b) Effects of steroid hormones on pituitary immunoreactive gonadotropin in European freshwater eel, *Anguilla anguilla* L. Gen Comp Endocrinol 52:190–197.
Dufour S, Fontaine YA, Kerdelhué B (1985) Increase in brain and pituitary radioimmunoassayable gonadotropin releasing hormone (GnRH) in the European silver eel treated with sexual steroid or human chorionic gonadotropin. Neuropeptides 6:495–502.
Dufour S, Lopez E, Le Menn F, Le Belle N, Baloche S, Fontaine YA (1988) Stimulation of gonadotropin release and ovarian development, by the administration of a gonadoliberin agonist and of dopamine antagonists, in female silver eel pretreated with estradiol. Gen Comp Endocrinol 70:20–30.
Dufour S, Le Belle N, Baloche S, Fontaine YA (1989) Postive feedback control by the gonads on gonadotropin (GTH) and gonadoliberin (GnRH) levels in experimentally matured female silver eels, *Anguilla anguilla*. Fish Physiol Biochem 7:157–162.
Dufour S, Bassompierre M, Montero M, Le Belle N, Baloche S, Fontaine YA (1991) Stimulation of pituitary gonadotropic function in female silver eels treated by a gonadoliberin agonist and dopamine antagonist. In: Scott AP, Sumpter JP, Kime DE, Rolfe MS (Eds), Proceedings of the Fourth International Symposium on the Reproductive Physiology of Fish, FishSymp 91, Sheffield, pp 54–56.
Dufour S, Montero M, Le Belle N, Bassompierre M, King JA, Millar RP, Peter RE, Fontaine YA (1993) Differential distribution and response to experimental sexual maturation of two forms of brain gonadotropin-releasing hormone (GnRH) in the European eel, *Anguilla anguilla*. Fish Physiol Biochem 11:99–106.
Durif C, Dufour S, Elie P (2005) The silvering process of the European eel: a new classification from the yellow resident stage to the silver migrating stage. J Fish Biol 66:1–19.
Edel RK (1975) The induction of maturation of female American eels through hormone injections. Helgoland Mar Res 27:131–138.
Filby AL, Tyler CR (2005) Molecular characterization of estrogen receptors 1, 2a and 2b and their tissue and ontogenic expression profiles in fathead minnow (*Pimephales promelas*). Biol Reprod 73:648–662.
Fontaine M (1936) Sur la maturation complète des organes génitaux de l'anguille mâle et l'émission spontanée de ses produits sexuels. CR Acad Sci Paris 202:1312–1315.
Fontaine M, Bertrand E, Lopez E, Callamand O (1964) Sur la stimulation des organes génitaux de l'anguille femelle (*Anguilla anguilla* L.) et l'émission spontanée des œufs en aquarium. CR Acad Sci, Paris, 259:2907–2910.
Garcia Ayala A, Villaplana M, Garcia Hernandez MP, Chaves Pozo E, Aqulleiro B (2003) FSH-, and LH-, and TSH-expressing cells during development of *Sparus aurata* L (Teleostei). An immunocytochemical study. Gen Comp Endocrinol 134:72–79.
Gen K, Okuzawa K, Senthilkumaran B, Tanaka H, Moriyama S, Kagawa H (2000) Unique expression of gonadotropin-I and –II subunit genes in male and female red seabream (*Pagrus major*) during sexual maturation. Biol Reprod 63:308–319.
Gharib SD, Wierman ME, Shupnik Man Chin WW (1990) Molecular biology of the pituitary gonadotropins. Endocrine Rev 11:177–199.
Gielen JT, Goos HJ, Peute J, Van den Bosch RA, Van Oordt PG (1982) The brain-pituitary-gonadal axis in the rainbow trout, *Salmo gairdneri*: gonadal hormones and the maturation of gonadotropic cells. Cell Tissue Res 225:45–56.
Gomez JM, Weil C, Ollitrault M, Le Bail PY, Breton B, Le Gac FL (1999) Growth hormone (GH) and gonadotropin subunit gene expression and pituitary and plasma changes during spermatogenesis and oogenesis in rainbow trout (*Oncorhynchus mykiss*). Gen Comp Endocrinol 113:413–428.
Goos HJTh (1987) Steroid feedback on pituitary gonadotropin secretion. In : Idler DR, Crim LW, Walsh JM (Eds), Proceeding of Third International Symposium on the Reproductive Physiology of Fish. St. John's, Newfoundland, Canada, pp 16–20.

Goos HJTh, De Leeuw R, Cook H, Van Oordt PGWJ (1986) Gonadotropic hormone-releasing hormone (GnRH) bioactivity in the brain of immature rainbow trout, *Salmo gairdneri*: the effect of testosterone. Gen Comp Endocrinol 64:80–84.

Gorbman A, Sower SA (2003) Evolution of the role of GnRH in animal (Metazoan) biology. Gen Comp Endocrinol 134:207–213.

Halm S, Martinez-Rodriguez G, Rodriguez L, Prat F, Mylonas CC, Carrillo M, Zanuy S (2004) Cloning, characterisation, and expression of three oestrogen receptors (ER alpha, ER beta 1 and ER beta 2) in the European sea bass, *Dicentrarchus labrax*. Mol Cell Endocrinol 223:63–75.

Han Y-S, Liao I-C, Tzeng W-N, Huang Y-S, Yu J Y-L (2003) Serum estradiol-17β and testosterone levels during silvering in wild Japanese eel *Anguilla japonica*. Comp Biochem Physiol 136B: 913–920.

Hassin S, Holland MCH, Zohar Y (1999) Ontogeny of follicle-stimulating hormone and luteinizing hormone gene expression during pubertal development in the female striped bass, *Morone saxatilis* (Teleostei). Biol Reprod 61:1608–1615.

Hawkins MB, Thornton JW, Crews D, Skipper JK, Dotte A, Thomas P (2000) Identification of a third distinct estrogen receptor and reclassification of estrogen receptors in teleosts. Proc Natl Acad Sci USA 97:10751–10756.

Hellqvist A, Schmitz M, Mayer I, Borg B (2006) Seasonal changes in expression of LH-β and FSH-β in male and female three-spined stickleback, *Gasterosteus aculeatus*. Gen Comp Endocrinol 145:263–269.

Holland MC, Hassin S, Zohar Y (2001) Seasonal fluctuations in pituitary levels of the three forms of gonadotropin-releasing hormone in striped bass, *Morone saxatilis* (Teleostei), during juvenile and pubertal development. J Endocrinol 169:527–538.

Huang YS, Schmitz M, Le Belle N, Chang CF, Quérat B, Dufour S (1997) Androgens stimulate gonadotropin-II β-subunit in eel pituitary cells in vitro. Mol Cell Endocrinol 131:157–166.

Huggard D, Khakoo Z, Kassam G, Mahmoud SS, Habibi HR (1996) Effect of testosterone on maturational gonadotropin subunit messenger ribonucleic acid levels in the goldfish pituitary. Biol Reprod 54:1184–1191.

Ijiri S, Kazeto Y, Takeda N, Chiba H, Adachi S, Yamauchi K (1995) Changes in serum steroid hormones and steroidogenic ability of ovarian follicles during artificial maturation of cultivated Japanese eel, *Anguilla japonica*. Aquaculture 135:3–16.

Ikemoto T, Park MK (2005) Identification and molecular characterization of three GnRH ligands and five GnRH receptors in the spotted green pufferfish. Mol Cell Endocrinol 242:67–79.

Ikeuchi T, Todo T, Kobayashi T, Nagahama Y (1999) cDNA cloning of a novel androgen receptor subtype. J Biol Chem 274:25205–25209.

Illing N, Troskie BE, Nahorniak CS, Hapgood JP, Peter RE, Millar RP (1999) Two gonadotropin-releasing hormone receptor subtypes with distinct ligand selectivity and differential distribution in brain and pituitary in the goldfish (*Carassius auratus*). Proc Natl Acad Sci USA 96:2526–2531.

Jackson K, Goldberg D, Ofir M, Abraham M, Degani G (1999) Blue gourami (*Trichogaster trichopterus*) gonadotropic β subunits (I and II) cDNA sequences and expression during oogenesis. J Mol Endocrinol 23:177–187.

Jeng SR, Chen GR, Lai JY, Huang YS, Dufour S, Chang CF (2002) Regulation of pituitary gonadotropin II and growth hormone content by sex steroids and pituitary extract in the aquacultured Japanese eel, *Anguilla japonica*. Aquaculture 209:319–332.

Jeng SR, Dufour S, Chang CF (2005) Differential expression of neural and gonadal aromatase enzymatic activities in relation to gonadal development in Japanese eel, *Anguilla japonica*. J Exp Zoolog 303A:802–812.

Jeng SR, Yueh WS, Chen GR, Lee YH, Dufour S, Chang CF (2007) Differential expression and regulation of gonadotropins and their receptors in the Japanese eel, *Anguilla japonica*. Gen Comp Endocrinol 154:161–173.

Jodo A, Ando H, Urano A (2003) Five different types of putative GnRH receptor gene are expressed in the brain of masu salmon (*Oncorhynchus masou*). Zool Sci 20:1117–1125.

Jodo A, Kitahashi T, Taniyama S, Ueda H, Urano A, Ando H (2005) Seasonal changes in expression of genes encoding five types of gonadotropin-releasing hormone receptors and responses to GnRH analog in the pituitary of masu salmon. Gen Comp Endocrinol 144:1–9.

Kagawa H, Kawazoe I, Tanaka H, Okuzawa K (1998) Immunocytochemical identification of two distinct gonadotropic cells (GtH I and GtH II) in the pituitary of bluefin tuna, *Thunnus thunnus*. Gen Comp Endocrinol 110:11–18.

Kah O, Dufour S, Baloche S, Breton B (1989) The GnRH systems in the brain and pituiatry of normal and hCG-treated European silver eels. Fish Physiol Biochem 6:279–284.

Kajimura S, Yoshiura Y, Suzuki M, Aida K (2001) cDNA cloning of two gonadotropin β subunits (GTH-Iβ and -IIβ) and their expression profiles during gametogenesis in the Japanese flounder (*Paralichthys olivaceus*). Gen Comp Endocrinol 122:117–129.

Khan IA, Lopez E, Leloup-Hatey J (1987) Induction of spermatogenesis and spermiation by a single injection of human chorionic gonadotropin in intact and hypophysectomized immature European eel (*Anguilla anguilla* L.). Gen Comp Endocrinol 68:91–103.

King JA, Dufour S, Fontaine YA, Millar RP (1990) Chromatographic and immunological evidence for mammalian GnRH and chicken II GnRH in eel (*Anguilla anguilla*) brain and pituitary. Peptides 11:507–514.

Kobayashi M, Stacey NE (1990) Effects of ovariectomy and steroid homone implantation on serum gonadotropin levels in female goldfish. Zool Sci 7:715–721.

Kobayashi M, Sohn YC, Yoshiura Y, Aida K (2000) Effects of sex steroids on the mRNA levels of gonadotropin subunits in juvenile and ovariectomized goldfish, *Carassius auratus*. Fish Sci 66:223–231.

Kumar RS, Ijiri S, Trant JM (2001a) Molecular biology of the channel catfish gonadotropin receptors: 1. Cloning of a functional luteinizing hormone receptor and preovulatory induction of gene expression. Biol Reprod 64:1010–1018.

Kumar RS, Ijiri S, Trant JM (2001b) Molecular biology of the channel catfish gonadotropin receptors: 2. Complementary DNA cloning, functional expression, and seasonal gene expression of the follicle-stimulating hormone receptor. Biol Reprod 65:710–717.

Kusakabe M, Nakamura I, Evans J, Swanson P, Young G (2006) Changes in mRNAs encoding steroidogenic regulatory protein, steroidogenic enzymes and receptors for gonadotropins during spermatogenesis in rainbow trout testes. J Endocrinol 189:541–554.

Kwok HF, So WK, Wang Y, Ge W (2005) Zebrafish gonadotropins and their receptors : I. Cloning and characterization of zebrafish follicle-stimulating hormone and luteinizing hormone receptors-evidence for their distinct functions in follicle development. Biol Reprod 72:1370–1381.

Larsen DA, Swanson P (1997) Effects of gonadectomy on plasma gonadotropins I and II in coho salmon, *Oncorhynchus kisutch*. Gen Comp Endocrinol 108:152–160.

Lee YH, Du JL, Shih YS, Jeng SR, Sun LT, Chang CF (2004) In vivo and in vitro sex steroids stimulate seabream gonadotropin-releasing hormone content and release in the protandrous black porgy, *Acanthopagrus schlegeli*. Gen Comp Endocrinol 139:12–19.

Leloup-Hatey J, Hardy A, Nahoul K, Quérat B, Zohar Y (1988) Influence of gonadotrophic treatment upon the ovarian steroidogenesis in European silver eel (*Anguilla anguilla* L.). INRA Paris, Les colloques de l'INRA, N°44, Reproduction chez les poissons, Bases fondamentales et appliquées en endocrinologie et génétique. Tel-Aviv, November 10–12, pp 127–130.

Lescheid DW, Powell JFF, Fischer WH, Park M, Craig A, Bokovskaya O, Barannikova IA, Sherwood NM (1995) Mammalian gonadotropin-releasing hormone (GnRH) identified by primary structure in Russian sturgeon, *Acipenser gueldenstaedti*. Regul Peptides 55:299–309.

Lethimonier C, Madigou T, Munoz-Cueto JA, Lareyre JJ, Kah O (2004) Evolutionary aspects of GnRHs, GnRH neuronal systems and GnRH receptors in teleost fish. Gen Comp Endocrinol 135:1–16.

Lin HR, Xie G, Zhang LH, Wang XD, Chen LX (1998) Artificial induction of gonadal maturation and ovulation in the Japanese eel (*Anguilla japonica* T & S). Bull Fr Peche Piscic 349:163–176.

Liu YC, Kato Y, Inoue K, Tanaka S, Kurosumi K (1988) Co-localization of LH β and FSH β mRNAs in the porcine anterior pituitary by *in situ* hybridization with biotinylated probes. Biochem Biophys Res Commun 154:80–84.

Lokman PM, Young G (2000) Induced spawning and early ontogeny of New Zealand freshwater eels (*Anguilla dieffenbachii* and *A. australis*). New Zeal J Mar Fresh Res 34:135–145.
Lokman PM, Vermeulen GI, Lambert JGD, Young G (1998) Gonad histology and plasma steroid profiles in wild New Zealand freshwater eels (*Anguilla dieffenbachii* and *A. australis*) before and at the onset of the natural spawning migration. I. Females. Fish Physiol Biochem 19:325–338.
Lokman PM, Wass RT, Suter HC, Scott SG, Judge KF, Young G (2001) Changes in steroid hormone profiles and ovarian histology during salmon pituitary-induced vitellogenesis and ovulation in female New Zealand longfinned eels, *Anguilla dieffenbachii*. J Exp Zool 289:119–129.
Madigou T, Mananos-Sanchez E, Hulshof S, Anglade I, Zanuy S, Kah O (2000) Cloning, tissue distribution, and central expression of the gonadotropin-releasing hormone receptor in the rainbow trout (*Oncorhynchus mykiss*). Biol Reprod 63:1857–1866.
Mateos J, Mananos E, Carrillo M, Zanuy S (2002) Regulation of follicle-stimulating hormone (FSH) and luteinizing hormone (LH) gene expression by gonadotropin-releasing hormone (GnRH) and sexual steroids in the Mediterranean Sea bass. Comp Biochem Physiol B 132:75–86.
Mateos J, Mananos E, Martinez-Rodriguez G, Carrillo M, Quérat B, Zanuy S (2003) Molecular characterization of sea bass gonadotropin subunits (alpha, FSHbeta and LHbeta) and their expression during the reproductive cycle. Gen Comp Endocrinol 133:216–232.
Maugars G, Schmitz M (2006) Molecular cloning and characterization of FSH and LH receptors in Atlantic salmon (*Salmo salar* L.). Gen Comp Endocrinol 149:108–117.
Maugars G, Schmitz M (2007) Expression of gonadotropins and gonadotropin receptor genes during early sexual maturation in male Atlantic salmon parr. Mol Reprod Dev. Epub ahead of print.
Mayer I, Berglund I, Rydevik M, Borg B, Schulz R (1990) Plasma levels of five androgens and 17α-hydroxy-20β-dihydroprogesterone in immature and mature male Baltic salmon (*Salmo salar*) parr, and the effects of castration and androgen replacement in mature parr. Can J Zool 68:263–267.
Melamed P, Rosenfeld H, Elizur A, Yaron Z (1998) Endocrine regulation of gonadotropin and growth hormone gene transcription in fish. Comp Biochem Physiol 119:325–338.
Menuet A, Pellegrini E, Anglade I, Blaise O, Laudet V, Kah O, Pakdel F (2002) Molecular characterization of three estrogen receptor forms in zebrafish: binding characteristics, transactivation properties, and tissue distributions. Biol Reprod 66:1881–1892.
Mitsuhashi E, Saito K, Matsubara H, Hirai T, Adachi S, Yamauchi K (2000) Effects of sex steroids on FSHβ and LHβ mRNA levels in cultured pituitary of Japanese eel. Proc Jpn Soc Comp Endocrinol 15:14.
Miura T, Yamauchi K, Nagahama Y, Takahashi H (1991) Induction of spermatogenesis in male Japanese eel, *Anguilla japonica*, by a single injection of human chorionic gonadotropin. Zool Sci 8:63–73.
Miwa S, Yan L, Swanson P (1994) Localization of two gonadotropin receptors in the salmon gonad by *in vitro* autoradiography. Biol Reprod 50:629–642.
Moncaut N, Somoza G, Power DM, Canario AV (2005) Five gonadotropin-releasing hormone receptors in a teleost fish: isolation, tissue distribution and phylogenetic relationships. J Mol Endocrinol 34:767–779.
Montero M, Vidal B, King JA, Tramu G, Vandesande F, Dufour S, Kah O (1994) Immunocytochemical localization of mammalian GnRH (gonadotropin-releasing hormone) and chicken GnRH-II in the brain of the European silver eel (*Anguilla anguilla* L.). J Chem Neuroanat 7:227–241.
Montero M, Le Belle N, King JA, Millar RP, Dufour S (1995) Differential regulation of the two forms of gonadotropin-releasing hormone (mGnRH and cGnRH-II) by sex steroids in the European female silver eel (*Anguilla anguilla*). Neuroendocrinology 61:525–535.
Montero M, Le Belle N, Vidal B, Dufour S (1996) Primary cultures of dispersed pituitary cells from estradiol-pretreated female silver eels (*Anguilla anguilla* L.): immunocytochemical characterization of gonadotropic cells and stimulation of gonadotropin release. Gen Comp Endocrinol 104:103–115.
Muske LE (1993) Evolution of gonadotropin-releasing hormone (GnRH) neuronal systems. Brain Behav Evol 42:215–230.

Nagae M, Todo T, gen K, Kato Y, Young G, Adachi S, Yamauchi K (1996a) Molecular cloning of the cDNAs encoding pituitary glycoprotein hormone α- and gonadotropin IIβ-subunits of the Japanese eel, *Anguilla japonica*, and increase in their mRNAs during ovarian development induced by injection of chum salmon pituitary homogenate. J Mol Endocrinol 16:171–181.
Nagae M, Adachi S, Yamauchi K (1996b) Changes in transcription of pituitary glycoprotein hormone α and gonadotropin IIβ subunits during ovarian development induced by repeated injections of salmon pituitary homogenate in the Japanese eel, *Anguilla japonica*. Fish Physiol Biochem 17:179–186.
Nagae M, Adachi S, Yamauchi K (1997) Changes in transcription of pituitary glycoprotein hormones α and gonadotropin IIβ subunits during ovarian development induced by repeated injections of salmon pituitary homogenate in the Japanese eel, *Anguilla japonica*. Fish Physiol Biochem 17:179–186.
Naito N, Hyodo S, Okumoto N, Urano A, Nakai Y (1991) Differential production and regulation of gonadotropins (GTH I and GTH II) in the pituitary gland of the rainbow trout, *Oncorhynchus mykiss*, during ovarian development. Cell Tissue Res 266:457–467.
Naito N, De Jesus EG, Nakai Y, Hirano T (1993) Ontogeny of pituitary cell types and the hypothalamo-hypophysial relationship during early development of chum salmon, *Oncorhynchus keta*. Cell Tissue Res 272:429–437.
Ngamvongchon S, Lovejoy DA, Fischer WH, Craig AG, Nahorniak CS, Peter RE, Rivier JE, Sherwood NM (1992) Primary structures of two forms of gonadotropin-releasing hormone, one distinct and one conserved, from catfish brain. Mol Cell Neurosci 3:17–22.
Nozaki M, Fujita I, Saito N, Tsukahara T, Kobayashi H, Ueda K, Oshima K (1985) Distribution of LHRH-like immunoreactivity in the brain of the Japanese eel (*Anguilla japonica*) with special reference to the nervous terminals. Zool Sci 2:537–548.
Nozaki M, Naito N, Swanson P, Miyata K, Nakai Y, Oota Y, Suzuki K, Kawauchi H (1990a) Salmonid pituitary gonadotropes. I. Distinct cellular distributions of two gonadotropins GTH I and GTH II. Gen Comp Endocrinol 77:348–357.
Nozaki M, Naito N, Swanson P, Dickhoff WW, Nakai Y, Suzuki K, Kawauchi H (1990b) Salmonid pituitary gonadotrophs. II. Ontogeny of GTH I and GTH II cells in the rainbow trout (*Salmo gairdneri irideus*). Gen Comp Endocrinol 77:358–367.
Oba Y, Hirai T, Yoshiura Y, Yoshikuni M, Kawauchi H, Nagahama Y (1999a) Cloning, functional characterization, and expression of a gonadotropin receptor cDNA in the ovary and testis of amago salmon (*Oncorhynchus rhodurus*). Biochem Biophys Res Commun 263:584–590.
Oba Y, Hirai T, Yoshiura Y, Yoshikuni M, Kawauchi H, Nagahama Y (1999b) The duality of fish gonadotropin receptors: cloning and functional characterization of a second gonadotropin receptor cDNA expressed in the ovary and testis of amago salmon (*Oncorhynchus rhodurus*). Biochem Biophys Res Commun 265:366–371.
Ohta H, Tanaka H (1997) Relationship between serum levels of human chorionic gonadotropin (hCG) and 11-ketotestosterone after a single injection of hCG and induced maturity in the male Japanese eel, *Anguilla japonica*. Aquaculture 153:123–134.
Ohta H, Kagawa H, Tanaka H, Okuzawa K, Hirose K (1997) Artificial induction of maturation and fertilization in the Japanese eel, *Anguilla japonica*. Fish Physiol Biochem 17:163–169.
Okubo K, Aida K (2001) Gonadotropin-releasing hormones (GnRHs) in a primitive teleost, the arowana: phylogenetic evidence that three paralogous lineages of GnRH occurred prior to the emergence of teleosts. Gen Comp Endocrinol 124:125–133.
Okubo K, Suetake H, Aida K (1999a) Expression of two gonadotropin-releasing hormone (GnRH) precursor genes in various tissues of the Japanese eel and evolution of GnRH. Zool Sci 16:471–478.
Okubo K, Suetake H, Aida K (1999b) A splicing variant for the prepro-mammalian gonadotropin-releasing hormone (prepro-mGnRH) mRNA is present in the brain and various peripheral tissues of the Japanese eel. Zool Sci 16:645–651.
Okubo K, Suetake H, Usami T, Aida K (2000) Molecular cloning and tissue-specific expression of a gonadotropin-releasing hormone receptor in the Japanese eel. Gen Comp Endocrinol 119:181–192.

Okubo K, Nagata S, Ko R, Kataoka H, Yoshiura Y, Mitani H, Kondo M, Naruse K, Shima A, Aida K (2001) Identification and characterization of two distinct gonadotropin-releasing hormone receptor subtypes in a teleost, the medaka *Oryzias latipes*. Endocrinology 142:4729–4739.

Okubo K, Suetake H, Aida K (2002) Three mRNA species for mammalian-type gonadotropin-releasing hormone in the brain of the eel *Anguilla japonica*. Mol Cell Endocrinol 192:17–25.

Okubo K, Ishii S, Ishida J, Mitani H, Naruse K, Kondo M, Shima A, Tanaka M, Asakawa S, Shimizu N, Aida K (2003) A novel third gonadotropin-releasing hormone receptor in the medaka *Oryzias latipes*: evolutionary and functional implications. Gene 314:121–131.

Olivereau M, Chambolle P (1978) Ultrastructure of gonadotropic cells of the normal eel and after estradiol injection. CR Acad Sci 287:1409–1412.

Olivereau M, Olivereau J (1979a) Effect of estradiol-17 beta on the cytology of the liver, gonads and pituitary, and on plasma electrolytes in the female freshwater eel. Cell Tissue Res 199:431–454.

Olivereau M, Olivereau J (1979b) Estradiol-positive feedback on gonadotropic (GTH) cells in freshwater male silver eels. Gen Comp Endocrinol 39:247–261.

O'Neill DF, Powell JF, Standen EM, Youson JH, Warby CM, Sherwood NM (1998) Gonadotropin-releasing hormone (GnRH) in ancient teleosts, the bonytongue fishes: putative origin of salmon GnRH. Gen Comp Endocrinol 112:415–425.

Oppen-Berntsen DO, Olsen SO, Rong CJ, Taranger GL, Swanson P, Walther BT (1994) Plasma levels of eggshell Zr-proteins, estradiol-17β, and gonadotropins during an annual reproductive cycle of Atlantic salmon (*Salmo salar*). J Exp Zool 268:59–70.

Parhar IS, Soga T, Sakuma Y (1996) In situ hybridization for two differentially expressed GnRH genes following estrogen and triidothyronine treatment in the brains of juvenile tilapia (cichlid). Neurosci Lett 218:135–138.

Parhar IS, Soga T, Sakuma Y (2000) Thyroid hormone and estrogen regulate brain region-specific messenger ribonucleic acids encoding three gonadotropin-releasing hormone genes in sexually immature male fish, *Oreochromis niloticus*. Endocrinology 141:1618–1626.

Parhar IS, Tosaki H, Sakuma Y, Kobayashi M (2001) Sex differences in the brain of goldfish: gonadotropin-releasing hormone and vasotocinergic neurons. Neuroscience 104:1099–1110.

Pasmanik M, Callard GV (1985) Aromatase and 5 alpha-reductase in the teleost brain, spinal cord, and pituitary gland. Gen Comp Endocrinol 60:244–251.

Pellegrini E, Menuet A, Lethimonier C, Adrio F, Gueguen MM, Tascon C, Anglade I, Pakdel F, Kah O (2005) Relationships between aromatase and estrogen receptors in the brain of teleost fish. Gen Comp Endocrinol 142:60–66.

Peter RE, Prasada Rao PD, Baby SM, Illing N, Millar RP (2003) Differential brain distribution of gonadotropin-releasing hormone receptors in the goldfish. Gen Comp Endocrinol 132:399–408.

Peyon P, Baloche S, Burzawa-Gérard E (1997) Investigation into the possible role of androgens in the induction of hepatic vitellogenesis in the European eel: *in vivo* and *in vitro* studies. Fish Physiol Biochem 16:107–118.

Pinto PIS, Passos AL, Martins RS, Power DM, Canario AVM (2006) Characterization of estrogen receptor βb in sea bream (*Sparus auratus*): phylogeny, ligand-binding and comparative analysis of expression. Gen Comp Endocrinol 145:197–207.

Prat F, Sumpter JP, Tyler CR (1996) Validation of radioimmunoassays for two salmon gonadotropins (GTH I and GTH II) and their plasma concentrations throughout the reproductive cycle in male and female rainbow trout (*Oncorhynchus mykiss*). Biol Reprod 54:1375–1382.

Quérat B, Jutisz M, Fontaine YA, Counis R (1990a) Cloning and sequence analysis of the cDNA for the pituitary glycoprotein hormone alpha-subunit of the European eel. Mol Cell Endocrinol 71:253–259.

Quérat B, Moumni M, Jutisz M, Fontaine YA, Counis R (1990b) Molecular cloning and sequence analysis of the cDNA for the putative beta subunit of the type-II gonadotropin from the European eel. J Mol Endocrinol 4:257–264.

Quérat B, Hardy A, Fontaine YA (1991) Regulation of the type-II gonadotropin α and β subunit mRNAs by oestradiol and testosterone in the European eel. J Mol Endocrinol 7:81–86.

Yoshiura Y, Suetake H, Aida K (1999) Duality of gonadotropin in a primitive teleost, Japanese eel (*Anguilla japonica*). Gen Comp Endocrinol 114:121–131.

Yumoto S, Suetake H, Suzuki Y, Nakamura O, Watanabe T (2001) Expression of GnRH and GnRH receptor genes in fugu *Takifugu rubripes*. Proc Jpn Soc Comp Endocrinol 16:20.

Chapter 12
Dopamine Inhibition of Eel Reproduction

Finn-Arne Weltzien, Marie-Emilie Sébert, Bernadette Vidal, Catherine Pasqualini, and Sylvie Dufour

12.1 Introduction

The silvering process includes various physiological and morphological changes that prepare the future genitors (silver eels) for the oceanic reproductive migration (e.g. Lokman et al. 1998). Silvering also marks the beginning of puberty (Aroua et al. 2005, and Chapter 11). However, silver eels are still sexually immature when they leave the continental habitats, and they remain blocked at this prepubertal stage if prevented from their reproductive migration. Because no maturing or spawning eels have ever been observed in the wild, the silver prepubertal stage is the last known stage of the eel biological cycle (for review, see Dufour et al. 2003). Accordingly, the regulatory mechanisms of puberty, sexual development, migration, and finally spawning are still not well understood.

Puberty can be defined as the transformation from a sexually immature juvenile into a mature adult by providing the brain-pituitary-gonad (BPG) axis with its full hormonal and gametogenetic capacity (Norris 1997; Schulz et al. 2000). Puberty is marked by the onset of gametogenesis, and the age of puberty and sexual maturation is determined by genetic factors as well as controlled by the nutritional status and/or body growth rate. Both in teleosts and mammals it seems that activation of the brain neuroendocrine system(s) is the key event to initiate puberty. What leads to this activation is, however, not understood.

The BPG axis consists of three physiologically connected constituents: brain, pituitary and gonads (Fig. 12.1). Stimulatory and inhibitory inputs merge in the

F.-A. Weltzien, M.-E. Sébert, B. Vidal, and S. Dufour (✉)
Museum National d'Histoire Naturelle, DMPA, UMR CNRS 5178 BOME, CP 32, 75231 Paris cedex 05, France

F.-A. Weltzien and C. Pasqualini
CNRS, UPR 2197, DEPSN, INAF, 91198 Gif-sur-Yvette Cedex, France

F.-A. Weltzien
University of Oslo, Department of Molecular Biosciences, PB 1041 Blindern, 0316 Oslo, Norway

G. van den Thillart et al. (eds.), *Spawning Migration of the European Eel*,

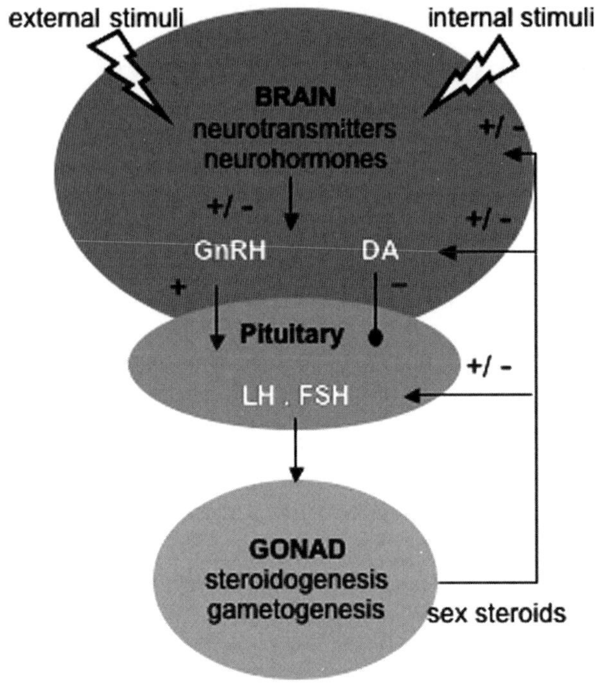

Fig. 12.1 The brain-pituitary-gonad (BPG) axis. Gonadotropin-releasing hormone (GnRH) originating mainly in the ventral forebrain stimulates the synthesis and release of pituitary gonadotropins (LH, luteinizing hormone and FSH, follicle-stimulating hormone), which in turn activate gonadal gametogenesis and steroidogenesis. An additional inhibitory brain control by dopamine (DA) is important in some teleosts, inhibiting the synthesis and release of gonadotropins. Sex steroids and growth factors produced in the gonads are important regulators of the BPG axis through various feedback mechanisms. A multitude of environmental and endogenous (e.g. sex steroids) factors are integrated by the central nervous system and modulates the activity of GuRH and DA neurons

forebrain on neuroendocrine neurons that produce an integrated output in the form of decapeptides known as gonadotropin-releasing hormones (GnRH). These phylogenetically ancient neuropeptides bind to GnRH receptors (GnRH-R) expressed by gonadotropin-producing cells (gonadotropes) in the pituitary, thereby stimulating production and release of two gonadotropins, follicle-stimulating hormone (FSH) and luteinizing hormone (LH). In some species, notably in several teleosts, the gonadotropes are subjected to a dual neuroendocrine control: stimulatory by GnRH and inhibitory by doparmine (DA). In teleosts as in other vertebrates, FSH and LH are released into the general circulation and bind to gonadal receptors where they stimulate gametogenesis and steroidogenesis. The products of steroidogenesis are important for the regulation of the BPG axis, either directly in the gonads or through feedback mechanisms on the brain and pituitary (see Weltzien et al. 2004).

Its striking life cycle with a long delay before sexual maturation makes the European eel a relevant model to study the regulatory controls of puberty (Huang et al. 1998,

1999; Dufour et al. 2003; Vidal et al. 2004): Contrary to most other models where pubertal development will proceed unless artificially inhibited, the eel will remain at a pre-pubertal stage unless stimulatory factors are given. Furthermore, its phylogenetical position, as a member of the group of Elopomorphs, an early branching teleost order (Lauder and Liem 1983), may provide information on ancestral regulations of puberty in vertebrates.

In this chapter, we will focus on the brain neurotransmitter DA, and how this is implicated in eel reproduction. We start by a general introduction to the role of DA in teleosts, before we discuss the special case of the eel. See also recent reviews by Dufour et al. (2003, 2005) and Sébert et al. (2008).

12.2 Dopaminergic Inhibition of Gonadotropes – Comparative View

DA is a widespread modulatory neurotransmitter in the central and peripheral nervous system of vertebrates. The physiological roles of DA range from regulation of pituitary function, sensorimotor control, thermoregulation, and modulation of appetite, to regulation of reproductive and maternal behaviour. DA is enzymatically synthesized from the amino acid tyrosine (Fig. 12.2). The rate-limiting enzyme in this process is tyrosine hydroxylase (TH), and the expression of this enzyme is often used as measure of dopaminergic activity.

In the teleost brain, dopaminergic cell bodies are distributed in various regions, most notably in the olfactory bulbs, in several telencephalic nuclei, in hypothalamic and preoptic regions, and in association with several cranial nerve nuclei of the brain stem. In contrast, the corpus cerebellum and optic tectum are devoid of DA cell bodies. Based on immunohistological data from various sources (Lefranc et al. 1969, 1970; L'Hermite and Lefranc 1972; Fremberg et al. 1977; Roberts et al. 1989; Kapsimali et al. 2000; Weltzien et al. 2006; S., Dufour et al. 2006 unpublished data), we have reconstructed the brain distribution of dopaminergic cell bodies in the European eel (Fig. 12.3; adapted from Sébert et al. 2008). This shows a similar distribution to other teleost species (e.g. Kaslin and Panula 2001). Quantification of TH mRNA in the eel brain gives a similar distribution pattern (Fig. 12.4; Weltzien et al. 2005b). In teleosts, various reports have shown that DA is involved in locomotion (Mok and Munro 1998), reproduction (Dufour et al. 2003, 2005), and aggressive and dominant behavior (Winberg and Nilsson 1993).

In several pituitary cell types, control of hormone synthesis and release is regulated through a duality of hypothalamic factors (a main positive and a main negative). However, in most textbooks you will find that gonadotropes are mainly controlled by a single neuroendocrine factor, the stimulatory GnRH. This seems to be the case for the rat and several other mammalian species, but should not be generalized for vertebrates as a group. Anatomical and physiological investigations performed mainly in teleosts, have shown that gonadotropes may also be submitted

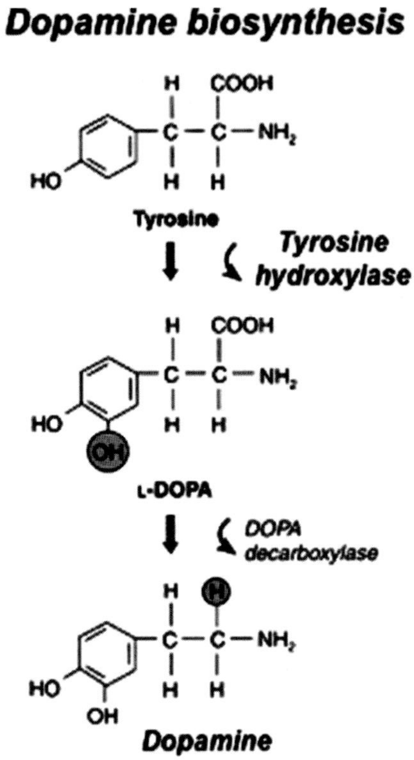

Fig. 12.2 Enzymatic biosynthesis of dopamine (DA) from the amino acid tyrosine. Quantification of mRNA expression of tyrosine hydroxylase, the rate-limiting enzyme, is often used as a measure of dopaminergic activity

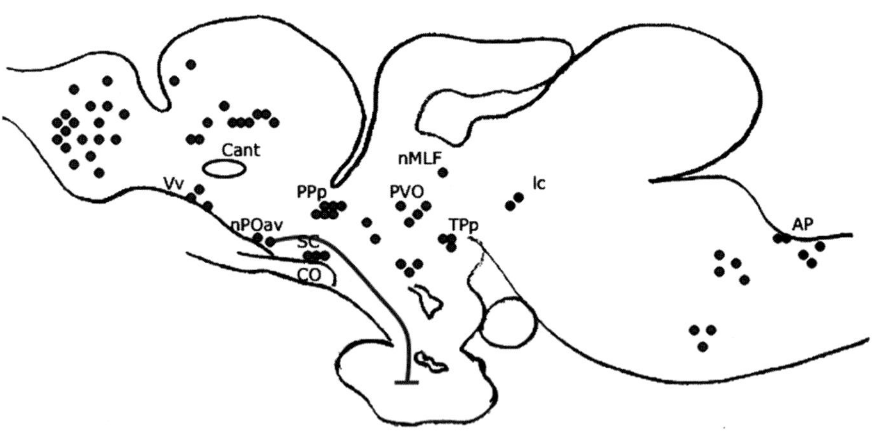

Fig. 12.3 Schematic organization of dopaminergic brain systems in the European eel. Abbreviations: AP, area postrema; Cant, anterior commissure; CO, optic chiasm; lc, locus coeruleus; nmlf, nucleus of medial longitudinal fascicle; Ob, olfactory bulbs; ppp, posterior parvocellular preoptic nucleus; PVO, paraventricular organ; Vv, ventral nucleus of the ventral telencephalic area; SC, suprachiasmatic nucleus; tpp: periventricular posterior tuberculum. Grey points: dopaminergic neurons (DA-ir/TH-ir); dark points: hypothetic noradrenergic neurons (Modified from M.E., Sébert et al. 2005 unpublished results)

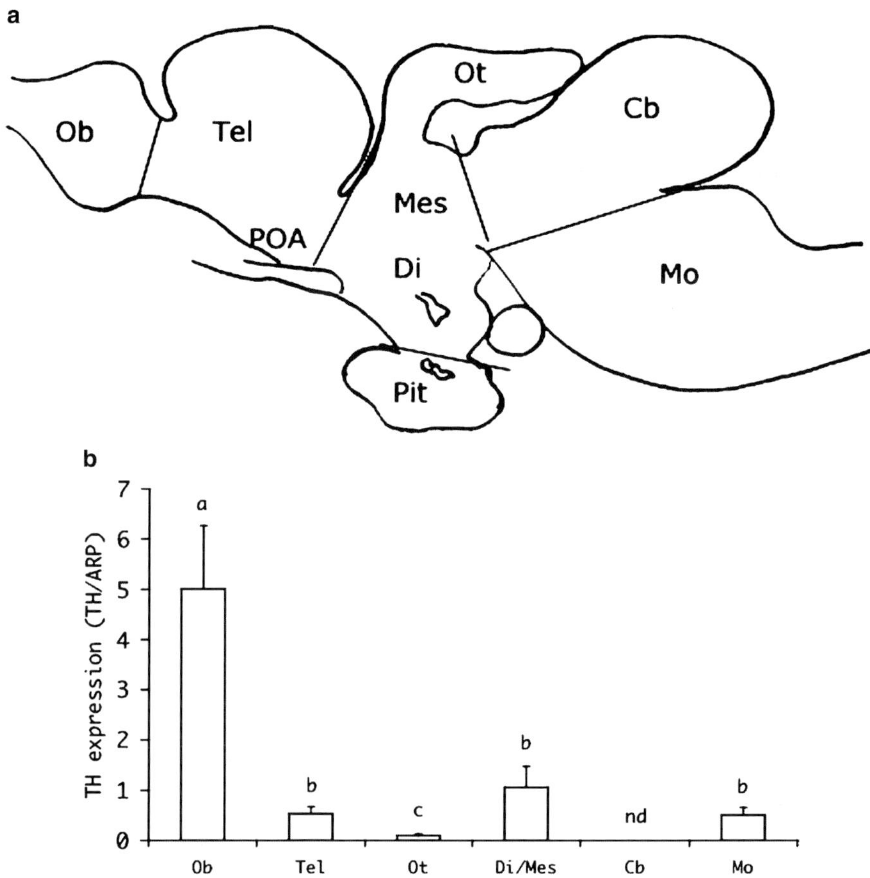

Fig. 12.4 (**a**) Side-view of a European eel brain with cut sites for qrt-PCR analyses indicated. Ob, olfactory bulbs; Tel, telencephalon (including preoptic area, POA); Ot, optic tectum (including pretectum); Di/Mes, di- and mesencephalic areas; Cb, corpus cerebellum; Mo, medulla oblongata; Pit, pituitary. (**b**) Relative expression of tyrosine hydroxylase (TH) in different brain regions of female silver eels (mean ± S.E.M.; n = 5–7). Data are normalized to eel acidic ribosomal phosphoprotein P0 (ARP) and expressed as the fold difference from a total brain cDNA calibrator control. Different letters indicate significant difference ($p < 0.05$). nd, not detectable

to a dual neuroendocrine control, stimulatory by GnRH and inhibitory by DA, challenging the dogma of a single control of gonadotropes by GnRH.

12.2.1 Initial Evidence of Inhibition of Final Maturation in Goldfish

Pioneer studies by Richard Peter and colleagues revealed the presence of a gonadotropin-release inhibiting factor in female goldfish (*Carassius auratus*) (Peter et al. 1978). Such a factor was identified, apparently by serendipity, during investigations on the regulatory control of gonadotropin release by a gonadotropin-releasing factor (Peter and Crim 1978), which was later identified as GnRH. Initial studies showed that gonadal recrudescence was prevented and gonadal regression induced following experimental lesions of the hypothalamus, thus revealing the neuronal pathways of GnRH in teleosts. However, during the same series of experiments, when performing larger hypothalamic lesions, the opposite result was sometimes observed; an elevated gonadotropin secretion, and subsequent ovulation in mature females. Similar evidence for a gonadotropin-release inhibiting factor was later shown also in male goldfish. More subtle lesions specifically in the preoptic area, hypothalamus, or pituitary stalk demonstrated that the neurons responsible for the observed inhibition of LH secretion and subsequent ovulation or spermiation originated in the preoptic area with axonal projections to the pituitary (Peter and Paulencu 1980).

Following these anatomical demonstrations, Peter's group performed in vivo and in vitro experiments to reveal the biochemical identity of the gonadotropin-release inhibiting factor. After studying the effects of various known pharmacological agonists and antagonists on LH secretion and ovulation in vivo in mature female goldfish, they suggested that DA could be the gonadotropin-release inhibiting factor (Chang and Peter 1983). This was supported by immunohistochemical and neuroanatomical studies identifying DA cell bodies in a specific region of the preoptic area termed the nucleus preopticus anteroventralis (NPOav). These dopaminergic neurons directly innervate the pituitary proximal pars distalis (PPD), the region where the gonadotropes are located (Kah et al. 1984). Final evidence was provided by Kah et al. (1987), showing that lesions specifically in the NPOav lead to a surge in LH release and subsequent ovulation in mature female goldfish.

In vitro experiments using agonists and antagonists specific for mammalian DA receptors showed that DA inhibits both basal and GnRH-stimulated LH release by acting directly on gonadotropes through binding of receptors that are pharmacologically related to mammalian D2-like receptors (Chang et al. 1990). In addition to these direct effects on pituitary gonadotropes, DA was also shown to exert indirect effects on gonadotropic function in goldfish by acting on GnRH neurons, either by blocking GnRH synthesis, or by inhibiting GnRH release from the pituitary nerve terminals (Yu and Peter 1990; Peter et al. 1991). This long series of experiments show that DA inhibition overrides the stimulatory action of GnRH and actively regulates the final steps of gametogenesis (spermiation and ovulation) in goldfish.

12.2.2 Evidence of DA Inhibition of Final Maturation in Several, but Not All Teleost Species

Following the initial discovery in goldfish, the inhibitory role of DA, superseding the stimulatory effects of GnRH on LH release, has been confirmed in various adult teleosts, including other cyprinids (Lin et al. 1988), and also silurids (De Leeuw et al. 1986), salmonids (Saligaut et al. 1999), and some percomorphs (Yaron et al. 2003; Aizen et al. 2005). Similar to goldfish, pharmacological evidence suggest that the dopaminergic influence on gonadotropes in these species is mediated through D2-like receptors. Also, DA has been shown to inhibit both basal and GnRH-stimulated LH expression and release in these species, for instance through down-regulation of GnRH receptor levels (receptor binding activity, De Leeuw et al. 1989; receptor mRNA expression, Levavi-Sivan et al. 2004) and intracellular signaling pathways of gonadotropes following binding of GnRH (for review, see Peter et al. 1986; Yaron et al. 2003).

Dopaminergic activity varies during the seasonal reproductive cycle (Senthilkumaran and Joy 1995), suggesting that gonadal factors are involved in its regulation. Accordingly, it has been shown in adult females of several teleost species that estradiol (E2) plays an important role in regulating DA activity by increasing the inhibitory tone on LH during vitellogenesis (catfish, *Heteropneustes fossilis*, Senthilkumaran and Joy 1995; rainbow trout, *Oncorhynchus mykiss*, Linard et al. 1995; Saligaut et al. 1999). In accordance with this, DA cells in the NPOav in trout express E2 receptors, providing a direct route for the increased dopaminergic tone upon E2 feedback (Linard et al. 1996). This means maximum DA inhibition of LH during vitellogenesis and towards the end of gametogenesis, while it plunges concurrently with E2 levels during induction of ovulation or spermiation. Ovulation/spermiation is in turn regulated by various endogenous and environmental factors (Weltzien et al. 2004). When investigated in teleosts, testosterone (T) had the same effect as E2 and was believed to act only following local aromatization into E2 (Trudeau et al. 1993). The effect of non-aromatizable androgens has not been tested.

The above studies involve effects of DA on LH synthesis and release, while potential effects of DA on FSH have been investigated only in rainbow trout: pharmacological data indicate that DA, again acting through D2-like receptors, inhibits also the release of FSH in late vitellogenic females (Vacher et al. 2000). Furthermore, as for LH, it seems that E2 is the main regulator of the dopaminergic system to inhibit FSH release during vitellogenesis, possibly to prevent further follicular recruitment (Vacher et al. 2002; Vetillard et al. 2003). The shortness of data on the regulation of FSH is due to a lack of analytical tools to quantify this gonadotropic hormone – homologous immunoassays have been available only for salmonids (Swanson 1991). However, the recent development of reliable in vitro expression systems for the production of FSH may open for interesting developments in this regard (Kamei et al. 2006; Levavi-Sivan et al. 2006). The development of qRT-PCR assays will also enable the quantification of FSHβ expression levels in the eel (e.g. Aroua et al. 2007).

Dopaminergic inhibition of reproduction has not been found in every teleost species studied. This was first emphasized in the Atlantic croaker, *Micropogonias undulates* (Atherinomorpha), where DA even showed a slight stimulatory effect on LH release (Copeland and Thomas 1989). Likewise, DA was not involved in the control of LH in the sea bream, *Sparus aurata* (Percomorpha), an important species for aquaculture (Zohar et al. 1995). These results show that the role of DA in reproduction cannot be directly related to phylogenetical position among teleosts since DA inhibition has been evidenced in other percomorph species (see above). Although the experimental conditions vary between different laboratories (sex, reproductive stage, pharmacological dose, experimental duration and design, assay system, etc.), there seems to be large variations in the intensity of the inhibition among the species where dopaminergic inhibition exists: DA may play a major role in some species (e.g. goldfish), and a minor one in others (e.g. rainbow trout). This may reflect the large evolutionary and biological diversity of teleosts, being the largest vertebrate group with more than 30,000 species. It appears thus that DA neuroendocrine inhibition may have been differentially conserved and expressed among teleosts in relation to the diversity of their reproductive cycles and their dependency upon various environmental factors.

12.2.3 Is DA Involved at Earlier Steps of Gametogenesis?

Although the involvement of DA in the regulation of ovulation and spermiation has been established in a certain number of adult teleost species (see previous section), less is known regarding its possible role in the earlier stages of gametogenesis. Indeed, the observation in many species that E2 increase the inhibitory tone during vitellogenesis, suggests that DA inhibition is an adult-specific control of the last steps of gametogenesis.

Accordingly, results obtained in juvenile striped bass, *Morone saxatilis* (Percomorpha), indicate that DA is not involved in the control of puberty in this species. In vivo treatment with T alone or in combination with GnRHa increased pituitary LH, while LH-release could be induced by a combination of GnRHa and T. The DA antagonist pimozide did not affect pituitary or plasma LH levels (Holland et al. 1998). Similar results were obtained in another percomorph (red sea bream, *Pagrus major*), where GnRHa alone induced precocious puberty, and no further effects were observed using a DA antagonist (Kumakura et al. 2003). The involvement of DA inhibition of puberty seems low or nonexistent also in rainbow trout where precocious puberty could be induced using a combination of GnRHa and steroids (Crim and Evans 1983).

On the other hand, a few publications indicate that DA may play an inhibitory role in the control of teleost puberty. In a preliminary study in juvenile spadefish, *Chaetodipterus faber* (Percomorpha), Marcano et al. (1995) found a decrease in dopaminergic metabolism in hypothalamus at the initiation of puberty, suggesting that a release of DA inhibition is linked with initiation of puberty in this species. Recently, Aizen et al. (2005) provided evidence that DA inhibition may be involved at the early stages of vitellogenesis in grey mullet, *Mugil cephalus* (Percomorpha). A single injection with the DA antagonist domperidone accelerated oocyte growth in vivo as compared to control or GnRHa-injected females.

12.3 Role of DA in Eel Puberty

Investigations on the pubertal development of eels have a long history. Maurice Fontaine and colleagues at MNHN in Paris showed already in the 1930s that sexual maturation could be induced in male silver eels by injections of pregnant women's urine extract (later found to contain human chorionic gonadotropin) (Fontaine 1936). Later, Fontaine also induced sexual maturation in female silver eels by administering pituitary extracts from maturing carp (Fontaine et al. 1964). These experiments clearly demonstrated that the pre-pubertal blockage resulted from a deficient gonadotropic function (for review, see Dufour et al. 2003). Later experiments have shown that although maturation can be induced by exogenous gonadotropin treatment, it is very difficult to initiate endogenous release of gonadotropins from the prepubertal eel pituitary. In the early 1980s, Dufour et al. (1983) demonstrated that LH synthesis could be stimulated by administration of sexual steroids, similar to what was also shown in other juvenile teleosts (for review, see Goos 1987). These experiments revealed the positive-feedback by gonadal steroids on LH synthesis in immature fish. However, even though steroids lead to a strong accumulation of LH in the eel pituitary, LH was not released into the circulation and so, no significant gonadal development took place (Dufour et al. 1983). This discovery gave rise to new experiments where steroid pre-treated females were treated with GnRHa over periods of up to 14 weeks. However, the combination of steroids and GnRHa was also unsuccessful in inducing LH release (Dufour et al. 1988).

12.3.1 Experimental Evidence for DA Inhibition of Puberty in the Eel

Inspired by the recent discovery in goldfish of DA as a gonadotropin release inhibiting factor during final maturation (see above), Dufour et al. (1988) investigated the possibility of DA also being involved in the initiation of puberty in the eel. Long-term E2 treatment increased the pituitary content of LH, but no secretion was observed. In fact, only a long-term treatment with E2 followed by a combined treatment with both GnRHa and pimozide was able to induce release of LH from the pituitary. These data supported the hypothesis that DA inhibits gonadotropic function as early as the pre-pubertal stage.

E2 was later shown to increase the inhibitory dopaminergic tone in some other adult teleosts (see above, Linard et al. 1995; Saligaut et al. 1999). Thus, the conclusions in Dufour et al. (1988) could have been due to an artifactual reinforcement of the DA inhibitory tone in eels receiving long-term E2-treatment. To clarify the participation of DA in the control of puberty in the European eel, we used (Vidal et al. 2004) in collaboration with the group of Yonathan Zohar (COMB, University of Maryland) a protocol identical to one previously developed for juvenile striped bass (see Holland et al. 1998). This treatment had shown that DA plays no significant role at puberty in female striped bass, injections with GnRHa being equally effective as a combined treatment with GnRHa and DA receptor antagonists. The protocol involves sustained treatments with a GnRHa, pimozide, and T (instead of the previously used E2), each compound given either alone or in combination. In the eel, in accordance with the results of Dufour et al. (1988), only the triple treatment with a sex steroid hormone (T in this experiment), GnRHa, and pimozide produced a major increase in both synthesis and release of LH (Fig. 12.5). The triple treatment also caused increased plasma levels of vitellogenin, gonadal growth, and stimulated ovarian vitellogenesis (Fig. 12.6; Vidal et al. 2004). These two studies (Dufour et al. 1988; Vidal et al. 2004) were the first to show that DA plays a key role in the control of LH release and pubertal development in a juvenile teleost. They also suggest that DA inhibition of gonadotropic function is an ancient evolutionary component in the neuroendocrine regulation of teleost reproduction.

As true also for other fish species, most of the previous works on eel puberty have focused on synthesis and release of only one of the gonadotropic hormones – LH, because FSH was only recently isolated from eel (Schmitz et al. 2005). Based on this, we have recently developed an assay for mRNA quantification of the European eel FSHβ subunit (Aroua et al. 2007), and we are currently investigating the neuroendocrine regulation of this hormone in pre-pubertal silver eels.

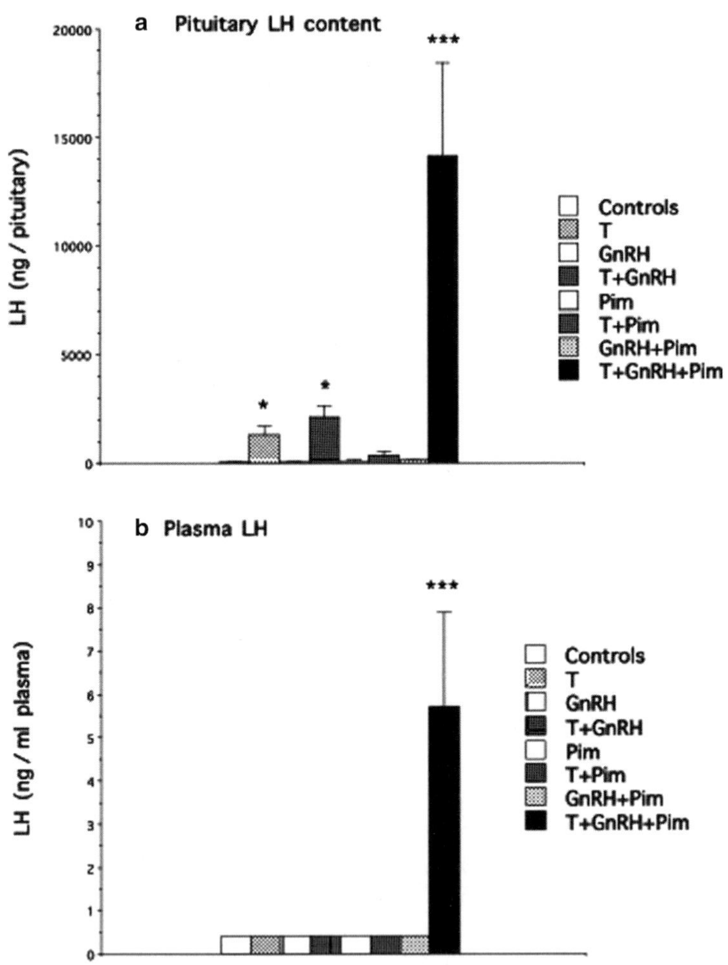

Fig. 12.5 Effects of chronic treatment with testosterone (T) GnRH, analog, and dopamine (DA) agonist (pimozidepim) on (**a**) pituitary and (**b**) plasma levels of luteinizing hormone (LH) in female silver eels. GnRH was quantified using radioimmunoassay (RIA). Eels were injected bi-weekly for 6 weeks with T, GnRH, and pimozide, alone or in combination. Control eels received vehicle only. Values are presented as mean ± SEM (n = 7). $^{*}p < 0.05$, $^{***}p < 0.001$ (From Vidal et al. 2004. With permission)

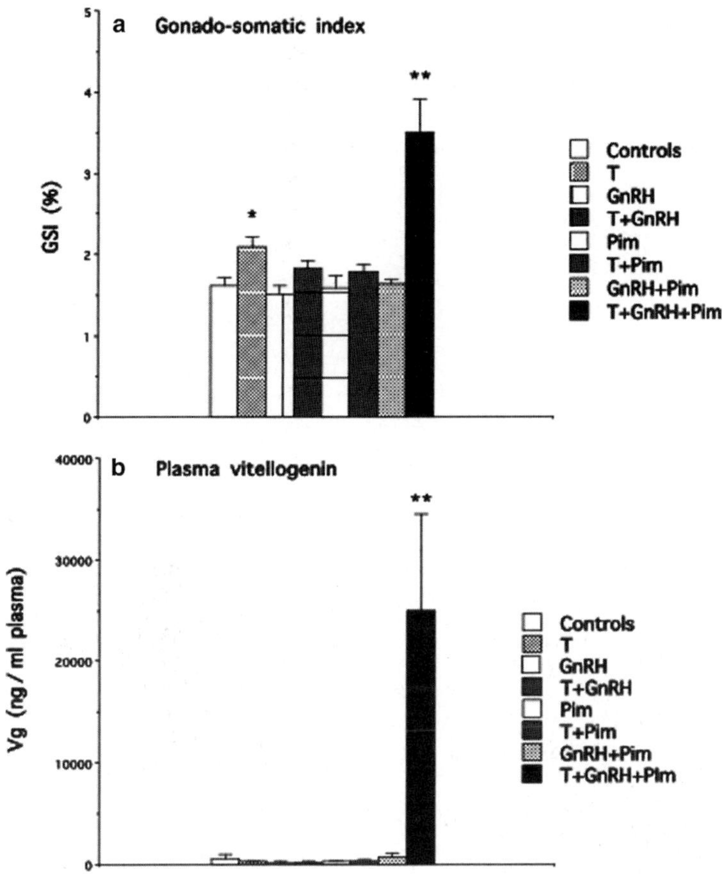

Fig. 12.6 (**A**) Effects of chronic treatment with testosterone (T), GnRH analog, and dopamine (DA) agonist (pimozidepim) on (a) gonado-somatic index and (b) plasma vitellogenin in female silver eels. Plasma vitellogenin was quantified using enzyme-linked immunosorbent assay (ELISA). Eels were injected bi-weekly for 6 weeks with T, GnRH, and pimozide, alone or in combination. Control eels received vehicle only. Values are presented as mean ± SEM (n = 7). $^{*}p < 0.05$, $^{**}p < 0.01$ (From Vidal et al. 2004. With permission). (**B**) Effects of chronic treatment with testosterone (T), GnRH analog, and dopamine (DA) agonist (pimozide) on ovarian histology of female silver eels. Eels were injected bi-weekly for 6 weeks with T, GnRH, and pimozide, alone or in combination. Control eels received vehicle only. Oocytes from control eels (a) showed small nucleoli (n) at the periphery of the nucleus (N) and contained numerous lipid vesicles (LV) in the ooplasma, a feature characteristic of the early vitellogenic stage (oil-droplet stage). Oocytes from T-treated eels (b), T- and GnRH-treated eels (c), and T- and pimozide-treated eels (d) were at the same stage. Oocytes from eels treated with T, GnRH, and pimozide (e) were enlarged and contained, in addition to large lvs, deeply stained yolk granules (black arrow) resulting from the incorporation of vitellogenin into the oocyte. A thickened zona radiata (black arrowhead) and visible follicular cells (white arrowhead) are observed in oocytes from T-, GnRH-, and pimozide-treated eels at higher magnification (f). These features (e and f) are characteristics of the yolk stage of vitellogenesis. Bar 20 μm (From Vidal et al. 2004. With permission)

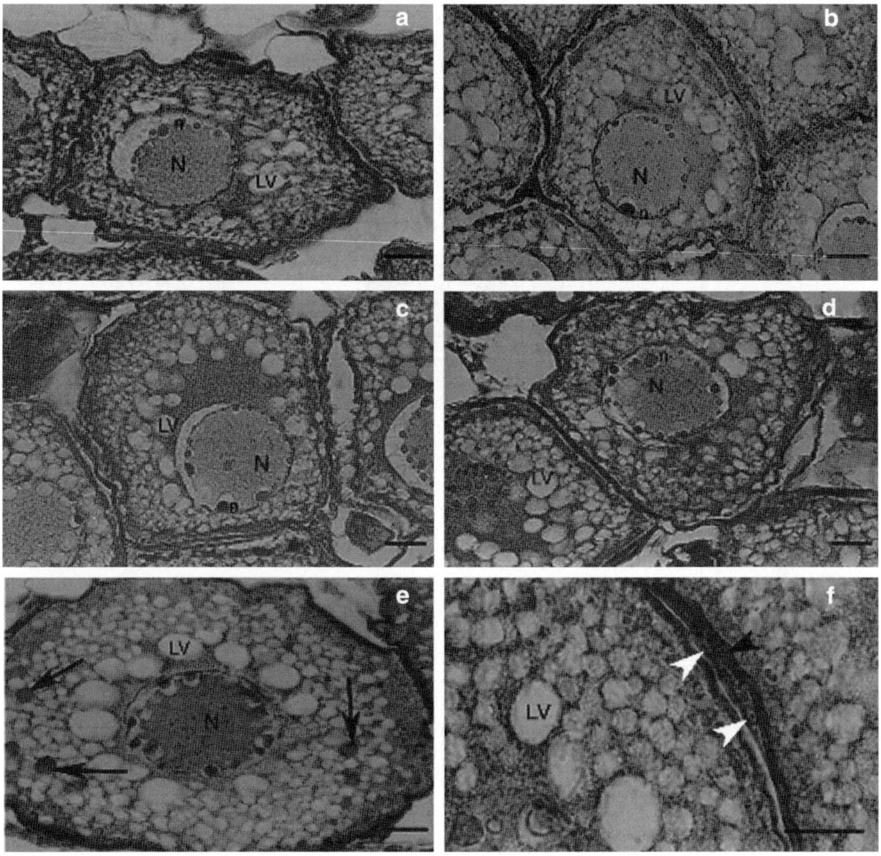

Fig. 12.6 (continued)

12.3.2 Anatomical Support for a DA Inhibitory Neuronal Pathway in the Eel

To investigate the anatomical basis for the dopaminergic inhibition of puberty in the eel, we have used different methodological approaches to study the expression of tyrosine hydroxylase (TH), the rate-limiting enzyme in the biosynthesis of DA (Fig. 12.2).

The full-length TH cDNA comprising 2,126 bp, including a coding region of 1,464 bp was isolated from the European eel by Boularand et al. (1998). Based on the isolated sequence, we have developed cRNA probes for in situ hybridization analysis of TH mRNA expression in the eel brain (Boularand et al. 1998; Weltzien et al. 2006). We have also successfully used anti-rat TH polyclonal antisera to investigate the distribution of TH cell bodies and projections in the eel brain and pituitary by immunohistochemistry (Kapsimali et al. 2000; Weltzien et al. 2006). We compared these results to those of Roberts et al. (1989), who had characterized the brain distribution of dopaminergic neurons by immunohistochemistry using a DA-specific antibody.

The hypophysiotropic nature of identified TH neurons was further established using retrograde migration of DiI (a carbocyanine lipophilic fluorescent tracer) implanted in the pituitary PPD (see Weltzien et al. 2006). All these different approaches have enabled us to establish the distribution and origin of dopaminergic neurons in the eel brain that may be involved in pubertal development and reproduction.

The highest number of TH immunoreactive cell bodies and nuclei in the eel brain are located in the olfactory bulbs, various nuclei of the telencephalon, and the hypothalamus including the preoptic area (Fig. 12.3; Kapsimali et al. 2000; Vidal et al. 2004; Weltzien et al. 2006). In the preoptic area, most TH immunoreactive cell bodies are situated in the area ventral and ventrolateral to the preoptic recess, just above the anterior portion of the optic chiasma. These cells constitute a nucleus referred to as the nucleus preopticus anteroventralis (NPOav) in the adult goldfish (Kah et al. 1987). Roberts et al. (1989) clearly identified these perikarya as being dopaminergic in the eel. They were termed nucleus preopticus parvocellularis anterior, but seem identical to the goldfish NPOav. We have therefore adopted the nomenclature from goldfish, and use the term NPOav for this preoptic nucleus (Fig. 12.7). From the NPOav, a dense tract of TH-immunoreactive axons projecting posteriorly, ventrally to the preoptic

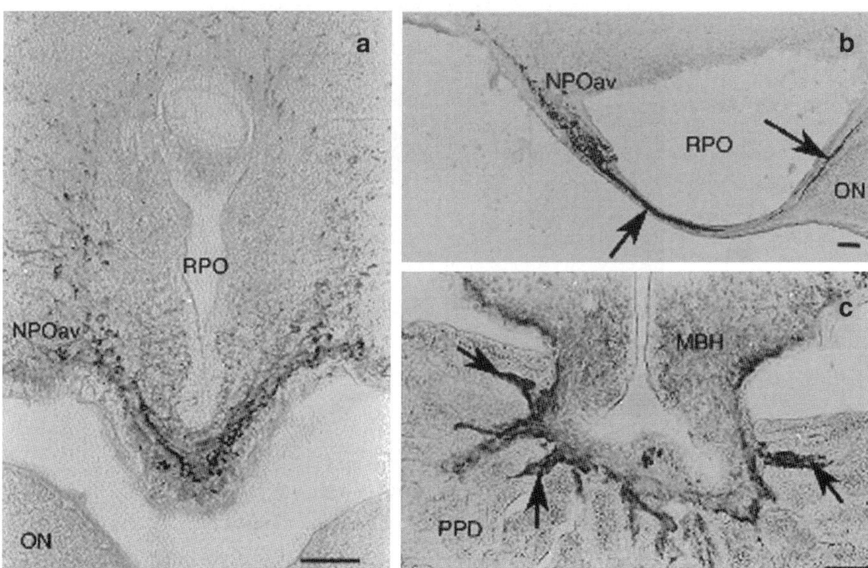

Fig. 12.7 Characterization by immunocytochemistry of the preopticohypophysial dopaminergic neuronal pathway. Immunoreactive (ir) neurons were labeled with an antibody to TH. (a) Transverse section at the level of the preoptic recess (RPO). Numerous ir cell bodies and fibers are located in the anterior preoptic area, in the nucleus preopticus anteroventralis (NPOav). (b) Longitudinal section at the level of the RPO. A dense ir axonal tract (arrow) originating from the ir neurons of the NPOav project posteriorly, ventrally to the RPO and contouring the optic nerve (ON). (c) Transverse section at the level of the medial basal hypothalamus (MBH) and PPD of the pituitary. Numerous ir axonal endings (arrow) innervate the PPD, the region where the gonadotrope cells are located. Bar 100 μm (From Vidal et al. 2004. With permission)

recess, and then turning around the optic chiasma and reaching the mediobasal hypothalamus. At the pituitary level, strongly labeled TH-immunoreactive axon terminals were observed innervating the proximal pars distalis (PPD; Fig. 12.7; Vidal et al. 2004), the area where the gonadotropes are located (Schmitz et al. 2005). This pathway precisely corresponds to that mediating the inhibition of LH and ovulation in adult teleosts. The hypophysiotropic nature of these neurons was further established utilizing retrograde migration of DiI implanted in the pituitary PPD. Following several weeks of incubation, DiI could be observed in cell bodies of hypophysiotropic neurons located in various areas of the diencephalon, including the neurons of the NPOav (Fig. 12.8; Weltzien et al. 2006). This finding further supports the hypophysiotropic role of the DA neurons in this preoptic nucleus. For a more extensive overview of brain DA systems in the eel, and methodological approaches to investigate these, see review by Sébert et al. (2008).

a

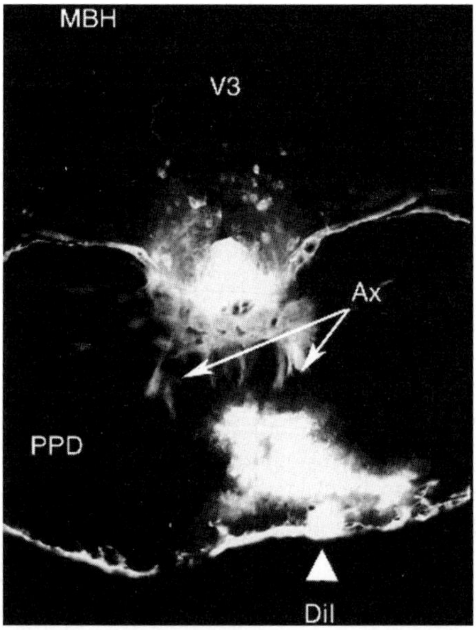

Fig. 12.8 (**a**) Transverse section showing DiI retrograde tracing from the eel pituitary. The inserted DiI microcrystal (arrowhead) is apparent in the ventromedial part of the proximal pars distalis (PPD) of the pituitary, while most of the fluorescent dye has diffused through the pars distalis and into the axonal endings (arrows, Ax) of the hypophysiotropic neurons, which directly innervate the adenohypophysis in teleosts. MBH: mediobasal hypothalamus; V3: third ventricle (From Weltzien et al. 2006. With permission). (**b**) Transverse sections of the rostral preoptic area from prepubertal female eels. Cells in the nucleus preopticus anteroventralis NPOav are hypophysiotropic, as shown by the DiI retrograde tracing (A). TH immunohistochemical analyses (B) show specific staining of cell bodies in the NPOav. V3, third ventricle; OT, optic tract. Scale bar 100 μm. (From Weltzien et al. 2006. With permission)

b

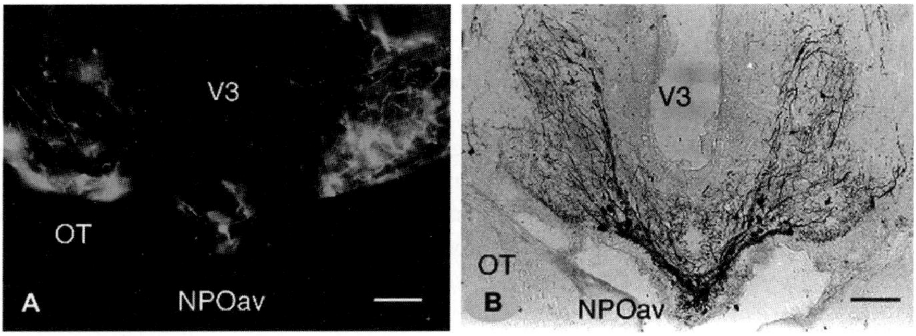

Fig. 12.8 (continued)

12.4 Regulation of Dopaminergic Inhibitory Tone on Eel Puberty

Following the unequivocal demonstration that DA is responsible for the strong inhibition of pubertal development in the female eel (Vidal et al. 2004), it is important to investigate regulatory factors that may be involved in the set-up, increase, or removal of the dopaminergic inhibition. On a general basis, we may expect that different regulatory mechanisms are involved in a juvenile fish compared to those observed in adults. Specific mechanisms related to the (at times extreme) environmental conditions encountered by the eel during its biological cycle might be implicated. Hormonal and environmental factors associated with the silvering process may be good candidates for establishing DA inhibition in the pre-pubertal silver eel. Conversely, endogenous factors and environmental cues encountered during the reproductive oceanic migration and/or at the spawning ground may be good candidates for the removal of DA inhibition and commencement of gametogenesis. An identification of specific regulatory factors will aid us in understanding how the DA block of puberty can be down-regulated, opening the way to innovative protocols to induce eel reproduction for farming and species protection.

12.4.1 Effect of Endogenous Factors

Gonadal steroids are known to influence the BPG axis through positive and negative feedback mechanisms. In the eel, previous studies have shown that gonadal steroids act on synthesis and release of both GnRH (Montero and Dufour 1996) and FSH/LH (Schmitz et al. 2005) (see also Chapter 11 this book). Prior to their reproductive migration, female silver eels undergo several morphological and physiological changes, including a slight increase in gonadosomatic index and also an increase in plasma steroid levels, both androgens and estrogens (Lokman et al. 1998; Sbaihi et al. 2001).

Also, as mentioned previously, E2 exerts a positive feedback on dopaminergic systems in adult teleosts of other species, increasing the dopaminergic tone towards final maturation (Goos 1987; Linard et al. 1995; Saligaut et al. 1999; Yaron et al. 2003).

Due to these reasons, we investigated how gonadal steroids regulate brain DA systems in the eel (Weltzien et al. 2005a, 2006). Female silver eels received weekly injections for 8 weeks of 2 mg per kg body weight T, E2, 5α-dihydrotestosterone (DHT; a non-aromatizable androgen), or vehicle control. For TH transcript analyses by qRT-PCR (Weltzien et al. 2005b), brains were dissected into six different parts (Fig. 12.4). The in vivo steroid treatment showed that sex steroids exerted regionally dependent and differential effects on brain TH mRNA levels. Specifically, androgens (T or DHT) stimulated TH mRNA expression in the olfactory bulbs and in the NPOav (Fig. 12.9). In contrast, E2 had no effect in either area, but inhibited TH mRNA expression in the diencephalon (Weltzien et al. 2006). These results are the first report of androgen-dependent modulation of TH expression in the central nervous system of a female vertebrate.

In the olfactory bulbs, the effect of androgens was localized to cells in the periglomerular cell layer (Fig. 12.10). Roberts et al. (1989) previously showed these cells to be dopaminergic in the eel. The precise function of these dopaminergic neurons is not known, but from studies in mammals, they are assumed to be essential for olfactory processing and also for odor learning and memory. This evidently has important implications for eel reproduction. Under natural conditions, plasma androgen levels in the European eel increase during silvering (Lokman et al. 1998; Sbaihi et al. 2001), and are further increased during experimental maturation in both sexes (Leloup-Hatey et al. 1988). Giorgi et al. (1994) reported an increased DA activity in eel olfactory bulbs during silvering, and Weltzien et al. (2006)

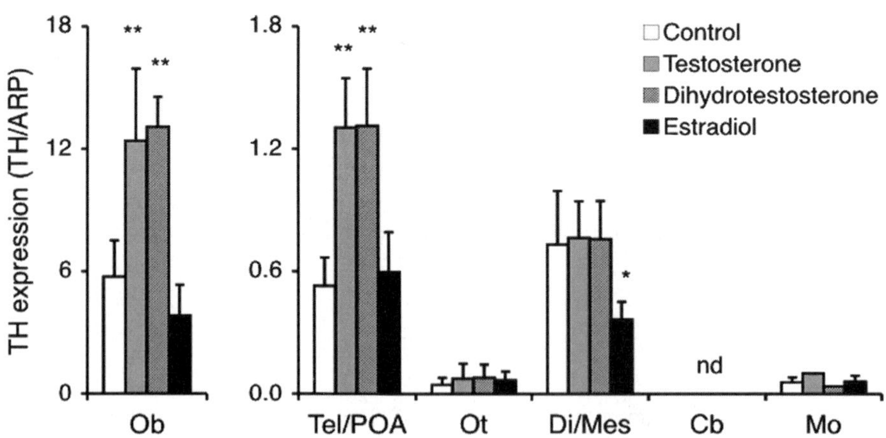

Fig. 12.9 Effect of in vivo treatment with testosterone, 17β-estradiol, or 5α-dihydrotestosterone (non-aromatizable androgen) on TH transcript levels in different brain regions of female silver eels, as quantified by qrt-PCR. Eels received eight weekly injections of steroid or saline (control). For normalization of data and for brain regions, see legend of Fig. 12.4. Means are given ± S.E.M. (n = 8). Significant differences between treated and control values from the same region: $^{*}P < 0.05$; $^{**}P < 0.01$ (Kruskal-Wallis). Nd, not detectable. (From Weltzien et al. 2006. With permission)

demonstrated that androgens stimulate DA activity in the olfactory bulbs. Thus, we may hypothesize that androgens enhance central processing of olfactory cues that may be essential for navigation during the eel catadromous migration (Westin 1990) towards the Sargasso Sea spawning grounds (Schmidt 1923).

The second area where we observed an androgen-dependent stimulation of TH mRNA levels, the NPOav (Fig. 12.11), was demonstrated in the eel to be dopaminergic (Roberts et al. 1989) and also hypophysiotropic (Vidal et al. 2004; Weltzien et al. 2006). In this same nucleus, E2 had no effect on TH mRNA expression. These results show that in the prepubertal eel, androgens, and not E2 as in adults of other teleost

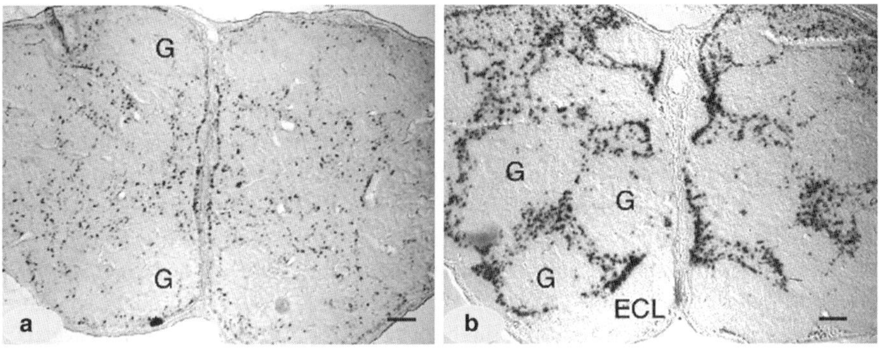

Fig. 12.10 Transverse sections of olfactory bulbs from female silver eels. *In situ* hybridization labeling of TH mRNA in control (**a**) and testosterone-treated (**b**) eels show increased TH mRNA in the periglomerular area upon testosterone treatment as compared to control. G, glomerulus; ECL, external cell layer. Scale bar 100 µm. (From Weltzien et al. 2006. With permission)

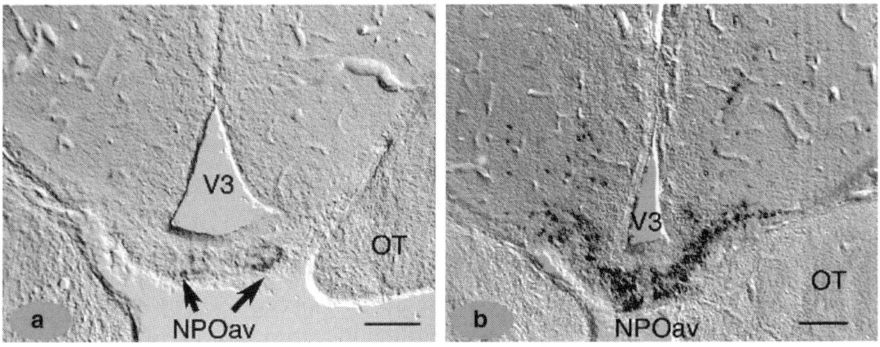

Fig. 12.11 Transverse sections of the rostral preoptic area from female silver eels. In situ hybridization labeling of TH mRNA in control (**a**) and testosterone-treated (**b**) eels show increased TH mRNA in the nucleus preopticus anteroventralis (NPOav) upon testosterone treatment as compared to control. V3, third ventricle; OT, optic tract. Scale bar 100 µm. (From Weltzien et al. 2000. With permission)

species, increase DA activity in the NPOav, indicating that androgens may contribute to the early setting of DA inhibition of pubertal development in this species until the oceanic migration can take place. Androgens recently have been shown to be responsible for various morphological and physiological changes occurring at silvering (Lokman et al. 2003; Aroua et al. 2005 and this book). The control of DA inhibition can now be included among these silvering-related androgen-specific regulations. A recent study in collaboration with the group of C-F Chang (NTOU, Taiwan) supports the importance of androgen-dependent regulations of the eel brain in that brain aromatase (enzyme complex responsible for local conversion of androgens into estrogens) activity in eels is very low compared to other species (Jeng et al. 2005).

12.4.2 Effect of Environmental Factors

We hypothesize that, under natural conditions, release of the DA inhibitory tone in silver eels and subsequent gonadal maturation must be induced by environmental cues related to the oceanic migration and/or the spawning grounds (for review, see Dufour et al. 2003; Sébert et al. 2008).

Compared to the juvenile yellow stage in rivers or estuaries, the environmental conditions change radically during the eel oceanic migration. From life in fresh or brackish water, they now enter into seawater. From living in shallow waters, eels are believed to migrate at depths between 4–600 m, and possibly as deep as 2,000 m (Robins et al. 1979; Tesch and Rohlf 2003), resulting in increased hydrostatic pressure and decreased water temperature. Their activity pattern also changes from a sedentary growth stage to a long-distance migration of 4–6,000 km. One or several of these factors probably play a role in the control of the eel BPG axis, either individually or in synergy.

Eels reproduce in seawater and the transition from freshwater/brackish water to a seawater environment may favor gonadal maturation. This is supported by the fact that cultured Japanese eels routinely are exposed to seawater prior to starting gonadotropic treatments for experimental maturation, because this is believed to be an important factor in inducing vitellogenesis (Kagawa 2003). However, Nilsson et al. (1981) found no effect of seawater exposure (nor of reduced temperature) on the activity of the BPG axis in silver eels. Specific effects of seawater on brain dopaminergic systems have not been investigated.

The depth at which eels swim during their oceanic migration, may represent a major environmental factor (i.e. high hydrostatic pressure). Early experiments in cooperation with the Oceanographic Museum of Monaco using eels immersed in cages at a depth of 450 m for 3 months, resulted in significant increases in pituitary LH content and gonadosomatic index (Dufour and Fontaine 1985; Fontaine et al. 1985). We have recently followed up on this, investigating effects of high pressure on the BPG axis, using hyperbaric chambers in cooperation with Philippe Sébert (Faculté de Médecine, Brest, France; see Sébert et al. 1990). Male and female silver eels were exposed to a constant pressure of 101 ATA for 3 (females) or 7 (males) weeks. At the pituitary level, LHß mRNA expression increased while FSHß mRNA expression decreased, showing a similar expression pattern to that observed during experimental maturation (Sébert et al. 2007a). At the gonad level, high pressure resulted in increased oocyte diameter and plasma levels of 11-kelotestosterone (11-KT), E2, and vitellogenin (Sébert et al. 2007a). It seems however, that additional internal and/or environmental factors are necessary to induce full maturation. We are currently investigating the effect of high pressure on DA activity (expression of TH). Preliminary data indicate that TH mRNA levels decreased in specific brain regions following exposure, leading to a reduction of DA inhibitory tone in eels submitted to high pressure (Sébert et al. 2007b). This is in accordance with the stimulatory effect on some parameters of the pituitary-gonadal axis (Sébert et al. 2007a), suggesting that high pressure participate in the activation of the eel BPG axis.

It is known that catadromous and anadromous teleost species go through considerable physiological and endocrine changes as a result of migratory exercise. Because European eels cover distances of more than 4,000 km to reach their spawning grounds, Guido van den Thillart and colleagues at Leiden University, the Netherlands, investigate whether exercise may induce gonadal maturation (see this book). Experimental swim-tunnel trials where silver eels swim the equivalent of more than 5,000 km have resulted in increased activity of various parameters of the BPG axis (G. van den Thillart et al. unpublished results 2004, see Van Ginneken et al. 2005). However, the effect of long-distance swimming on brain DA systems has so far not been investigated.

The above results seem to imply that one single environmental factor is insufficient to induce full sexual maturation of silver eels. Hence, synergistic effects of various factors, although experimentally challenging, should be the focus of future studies.

In addition to its inhibitory role at puberty in the eel, DA may also be involved in controlling the last steps of gametogenesis (oocyte maturation and ovulation in the female, spermiation in the male), as demonstrated in the goldfish and other adult teleosts. Future studies should analyze the pattern of DA tone during eel experimental maturation. From this point of view, the hypothesis should be raised that pheromones produced by mature males and females could participate in the final drop of DA inhibition and the induction of final gametogenesis and spawning.

Further investigations should also aim at developing new protocols, by using DA antagonist and GnRH agonist, and/or candidate pheromones, to induce spawning in experimentally matured eels (pretreated with gonadotropic hormones) in order to obtain better quality gametes.

12.5 DA Receptors – Comparative View and Investigations in the Eel

At the cellular level, the effects of DA in both teleosts and other vertebrates are mediated by two classes of DA receptors, termed D1 and D2. These receptors were initially distinguished by their ability to activate (D1) or to inhibit (D2) the enzyme adenylyl cyclase (Kebabian and Calne 1979). Molecular studies have revealed that D1 and D2 receptors are composed of several membrane proteins, each belonging to the seven-transmembrane-domain G-protein-coupled receptor (GPCR) family, and defined by their shared overall topology and common signal transduction mechanism, which triggers GDP/GTP exchange on heterotrimeric G proteins (Oliveira et al. 1994). Two D1-like receptors (D1 and D5, or D1A and D1B, respectively) and three D2-like receptor subtypes (D2, D3, and D4), each encoded by distinct genes have been isolated in mammals (Civelli et al. 1993). The receptor subtypes show overlapping brain distribution patterns, illustrating the pleiotropic activity of each receptor.

12.5.1 Cloning and Distribution of Four Distinct Eel D1 Receptors

In goldfish, D1 receptor agonists were shown to specifically inhibit GnRH release from preoptic-anterior hypothalamic slices *in vitro*, while D2 receptor antagonists inhibited GnRH release from pituitary fragments (Yu and Peter 1992). The possible effects of DA and its agonists on GnRH have not yet been investigated in the eel.

For many years and based primarily on results from mammals, it was believed that all vertebrates had two distinct D1 receptor subtypes (D1A and D1B; Civelli et al. 1993). However, additional subtypes were isolated about 10 years ago from non-mammalian vertebrates like Xenopus (D1c; Sugamori et al. 1994) and chicken (D1d; Demchyshyn et al. 1995). There were also reports of several D1-like receptor genes in teleost fish (Macrae and Brenner 1995). However, it was not entirely clear whether these genes represented truly distinct D1 receptor subtypes or if their presence was merely restricted to these particular species. Therefore, in collaboration with the group of Philippe Vernier (CNRS, Gif-sur-Yvette, France), we searched for the full repertoire of D1-like receptor diversity in the European eel. Using a combination of PCR and RACE, this resulted in the isolation of four D1 receptor subtypes, termed D1A1, D1A2, D1B, and D1C (Cardinaud et al. 1997). None of the four eel D1-like receptor genomic sequences contained introns interrupting their coding regions. This is characteristic for all known vertebrate D1 receptor sequences to date. Sequence comparisons with other members of the D1 receptor family clearly indicated that these four sequences encode fish homologues of mammalian D1 receptors, which can be subdivided into D1A, D1B, and D1C receptor subtypes (Cardinaud et al. 1997).

Pharmacological profiles and phylogenetic analyses based on deduced amino acid sequences supports the notion that the European eel possesses two distinct D1A (D1A1 and D1A2), one D1B, and one D1C receptor genes (Cardinaud et al. 1997). Other D1-like receptors have been isolated from other fish, allowing for the assessment of their evolutionary relationships. Phylogenetic analysis suggests that eel D1C, Tilapia D1C, Fugu D5-like, and Xenopus D1C are orthologues (bootstrap value 98%) and that they constitute a separate subtype of vertebrate D1 receptors.

The existence of the same paralogous D1-like receptor subtypes in teleost fish and tetrapods demonstrates that the gene duplication events at the origin of the D1-like receptor diversity arose prior to the separation of actinopterygian fishes from the other vertebrates, 420 million years ago.

All four D1 receptor-subtypes are expressed in the brain of the European eel. The specific brain localization of the different receptors was analyzed by *in situ* hybridization (Kapsimali et al. 2000). The D1A and D1B receptor transcripts exhibited largely overlapping expression patterns, with relatively abundant expression in the olfactory bulbs and the dorsal and subpallial telencephalic areas. More caudally, several preoptic and other diencephalic nuclei, optic tectum and corpus cerebellum, all contained various amounts of D1A and D1B receptor transcripts. Finally, D1A and D1B receptor mRNAs were present in several cranial nerve associated nuclei. The D1C receptor transcript was restricted to the diencephalon and cerebellum. The two D1A receptor subtypes were generally the most abundantly expressed. In the pituitary, expression of D1A1 and D1A2 were detected exclusively in the pars intermedia, while the other receptor subtypes were not detected. None of the D1 receptor transcripts were detected in the pituitary proximal pars distalis (the region where the gonadotropes are situated). There is no data on the regulation of D1 receptors in eel, but we are planning to investigate the effects of gonadal steroids in future experiments.

In summary, the four isolated eel D1-like receptors expressed molecular, pharmacological, and functional signatures characteristic of vertebrate D1A, D1B, and D1C receptors.

12.5.2 Cloning and Distribution of Two Distinct Eel D2 Receptors

The previously shown absence of D1 receptors in the eel proximal pars distalis and particularly on the gonadotrope cells, and the fact that antagonists of mammalian D2 receptors can remove the inhibition of eel puberty, prompted us to look for the presence of a D2-like receptor in the eel brain and pituitary.

Degenerate oligonucleotides were designed based on conserved nucleotide sequences of D2, D3, and D4 receptors cloned from different vertebrates species. These oligonucleotides were used in PCRs to isolate eel D2-like receptor fragments. From genomic DNA, we first determined that the eel D2-like gene structure shows a high degree of similarity with the known DA receptor genes; similar to in mammals and fugu (Japanese pufferfish), it is composed of six introns and seven coding exons. We then isolated and cloned two different cDNA sequences from eel pituitaries and olfactory bulbs, respectively. Finally, using 5′ and 3′ RACE PCR, two full-length cDNAs encoding two distinct putative proteins were obtained. These putative proteins share the typical arrangement of the seven transmembrane domain G-protein coupled receptor (GPCR) family and contain highly conserved amino acid residues thought to be important for the binding of DA. They exhibit 81.5% overall identity and a higher similarity to known D2 compared to D3 or

D4 DA receptors. The isolated transcripts were thus designated as D2A and D2B (GenBank accession number: DQ789976 and DQ789977; Pasqualini et al. 2007). For both receptors, the transmembrane domains and the functionally important amino acid residues in the intra- and extra-cellular loops, as well as those forming the primary binding pocket, the ancillary pocket, and the allosteric sodium-binding pocket (Neve et al. 2003), were well conserved in the eel compared to other vertebrate sequences (Pasqualini et al. 2007).

Phylogenetic analysis also shows that both eel sequences belong to the D2 receptor subtype, perhaps resulting from a major genome duplication event that occurred specifically in the teleost lineage (Christoffels et al. 2004; Vandepoele et al. 2004). As expected, the teleost D2 receptors form a homogeneous group of sequences, distinct from those of other vertebrates. However, within this group of fish sequences, the detailed relationships have yet to be resolved, pointing to the ancient and specific divergence of gene sequences in this very large vertebrate group.

Quantitative real-time RT-PCR analysis reveals a differential tissue distribution of D2A and D2B receptors in the eel (Pasqualini et al. 2007). Strong positive signals for both D2A and D2B are found in the brain, with differential ratios according to brain area. In the pituitary, although D2A mRNAs can be detected, the D2B gene is expressed at levels ten times higher. For the D2A gene only, mRNAs are also present in the retina, olfactory epithelium, and spinal cord, similar to what has been observed in mammals (Missale et al. 1998). No signals could be detected in various other peripheral tissues (e.g. ovaries, liver, kidney, muscle, gills).

Thus, the molecular diversity of D2 receptors is greater in the eel than in tetrapods: in contrast to mammals and birds, in which only one gene coding for D2 is present, at least two are found in the eel, as in rainbow trout (GenBank accession number: AJ347728 and AJ347729; Vacher et al. 2003) and three in the zebrafish (Boehmler et al. 2004). The few but functionally relevant differences observed among regional distribution of the two subtypes were probably an important constraint for the conservation of both genes after duplication. These results, together with those that we previously obtained for D1 receptors, support the idea that the organization of dopaminergic systems appeared early in vertebrate evolution and that the complexity of these systems may be greater in fish than in mammals.

Our ongoing studies are also aiming at characterizing the specific pharmacological profiles of the two eel D2 receptors. To that end, full-length cDNA of each receptor has been cloned. Recombinant receptors will be expressed in cell lines and tested for the ability of a variety of DA agonists and antagonists to activate/inactivate intracellular signaling pathways. From a fundamental point of view, this study will indicate whether different affinities/intracellular pathways may have participated in the selection of duplicated receptors in the eel. From an applied point of view it will allow us to select the most adequate antagonists to be used for the development of new protocols for eel maturation. We are also currently investigating the effect of sexual steroids on eel D2A and D2B receptor expression in order to decipher whether steroids may regulate DA inhibitory tone on eel reproduction, by acting both on DA production (paragraph 12.4.1) and action.

12.6 Conclusions and Future Perspectives

DA effectively inhibits pubertal development in the European eel. A triple treatment with GnRHa, pimozide and T is necessary to overcome this inhibition. The DA neuronal cell bodies responsible for this effect are located in the nucleus anteroventralis of the preoptic area (NPOav). Their axonal projections reach the pituitary proximal pars distalis (PPD), where the gonadotropes are localized. The inhibitory action of DA on the pituitary is up-regulated specifically by androgens during the silvering process, whereas estrogens have no effect. We have indications that environmental factors like high hydrostatic pressure may be involved in releasing the DA inhibition during the eel oceanic migration, but further studies are needed, particularly combining various factors to look for synergistic effects.

In all vertebrates, DA acts on its target cells via various receptor subtypes. Four D1 receptor subtypes have been characterized in the eel. However, none of them are expressed in the PPD, as revealed by *in situ* hybridization. According to pharmacological data, DA should act on gonadotropes through D2 type receptors. We have recently isolated two distinct D2 receptor subtypes in the eel, D2A and D2B, in contrast to the single subtype found in mammals. We are currently investigating the regulation of the expression and activity of these receptors, as well as their ligand specificity.

Investigations on DA in the eel, a representative species of a ancient group of teleosts, provide new insights on the evolution of DA systems in vertebrates, new information on the role of DA in the inhibition of reproductive function, and new tools for the development of innovative protocols for inducing eel reproduction.

Acknowledgements MNHN, CNRS, MENRT (Ph.D. fellowship to MES), and the European Commission (EELREP Project No. Q5RS-2001-01836 to SD; and Marie Curie Individual Fellowship No. MCFI-2002-01609 to FAW), supported this work. We thank Nadine Le Belle and Sylvie Baloche (CNRS-MNHN, Paris) for excellent technical assistance. We also acknowledge the various contributions from our long-time collaborators: Sylviane Boularand, Marika Kapsimali, Mario Wullimann, Philippe Vernier (CNRS, Gif-sur-Yvette), Philippe Chemineau (INRA, Nouzilly), Olivier Kah (CNRS-Univ Rennes 1), Philippe Sébert and coworkers (Fac Medecine, Brest), Monika Schmitz (Swedish University of Agriculture), Ching-Fong Chang and coworkers (National Taiwan Ocean University), Yonathan Zohar and coworkers (University of Maryland).

References

Aizen J, Meiri I, Tzchori I, Levavi-Sivan B, Rosenfeld H (2005). Enhancing spawning in the grey mullet (*Mugil cephalus*) by removal of dopaminergic inhibition. Gen Comp Endocrinol 142: 212–221

Aroua S, Schmitz M, Baloche S, Vidal B, Rousseau K, Dufour S (2005). Endocrine evidence that silvering, a secondary metamorphosis in the eel, is a pubertal rather then a metamorphic event. Neuroendocrinology 82: 221–232

Aroua S, Weltzien F-A, Le Belle N, Dufour S (2007). Development of real-time RT-PCR assays for eel gonadotropins and their application to the comparison of in vivo and in vitro effects of sex steroids. Gen Comp Endocrinol 153: 333–343

Boehmler W, Obrecht-Pflumio S, Canfield V, Thisse C, Thisse B, Levenson R (2004). Evolution and expression of D2 and D3 dopamine receptor genes in zebrafish. Dev Dyn 230: 481–493

Boularand S, Biguet NF, Vidal B, Veron M, Mallet J, Vincent JD, Dufour S, Vernier P (1998). Tyrosine hydroxylase in the European eel (*Anguilla anguilla*): cDNA cloning, brain distribution, and phylogenetic analysis. J Neurochem 71: 460–470

Cardinaud B, Sugamori KS, Coudouel S, Vincent JD, Niznik HB, Vernier P (1997). Early emergence of three dopamine D-1 receptor subtypes in vertebrates – molecular phylogenetic, pharmacological, and functional criteria defining D-1A, D-1B, and D-1C receptors in European eel *Anguilla anguilla*. J Biol Chem 272: 2778–2787

Chang JP, Peter RE (1983). Effects of dopamine on gonadotropin release in female goldfish, *Carassius auratus*. Neuroendocrinology 36: 351–357

Chang JP, Yu KL, Wong AO, Peter RE (1990). Differential actions of dopamine receptor subtypes on gonadotropin and growth hormone release in vitro in goldfish. Neuroendocrinology 51: 664–674

Christoffels A, Koh EGL, Chia JM, Brenner S, Aparicio S, Venkatesh B (2004). Fugu genome analysis provides evidence for a whole-genome duplication early during the evolution of ray-finned fishes. Mol Biol Evol 21: 1146–1151

Civelli O, Bunzow JR, Grandy DK (1993). Molecular diversity of the dopamine receptors. Ann Rev Pharmacol Toxicol 33: 281–307

Copeland PA, Thomas P (1989). Control of gonadotropin release in the Atlantic croaker (*Micropogonias undulatus*): evidence for lack of dopaminergic inhibition. Gen Comp Endocrinol 74: 474–483

Crim LW, Evans DM (1983). Influence of testosterone and/or luteinizing hormone releasing hormone analogue on precocious sexual development in the juvenile rainbow trout. Biol Reprod 29: 137–142

De Leeuw R, Goos HJ, van Oordt PG (1986). The dopaminergic inhibition of the gonadotropin-releasing hormone–induced gonadotropin release: an in vitro study with fragments and cell suspensions from pituitaries of the African catfish, *Clarias gariepinus* (Burchell). Gen Comp Endocrinol 63: 171–177

De Leeuw R, Habibi HR, Nahorniak CS, Peter RE (1989). Dopaminergic regulation of pituitary gonadotropin-releasing hormone receptor activity in the goldfish (*Carassius auratus*). J Endocrinol 121: 239–247

Demchyshyn LL, Sugamori KS, Lee FJS, Hamadanizadeh SA, Niznik HB (1995). The dopamine D1D receptor – cloning and characterization of three pharmacologically distinct D1-like receptors from *Gallus domesticus*. J Biol Chem 270: 4005–4012

Dufour S, Fontaine YA (1985). The reproductive migration of the European eel (*Anguilla anguilla*) – the probable importance of hydrostatic-pressure in stimulation of the gonadotropic function. Bull Soc Zool France-Evol Zool 110: 291–299

Dufour S, Delerue LeBelle N, Fontaine YA (1983). Effects of steroid-hormones on pituitary immunoreactive gonadotropin in European fresh-water eel, *Anguilla anguilla* L. Gen Comp Endocrinol 52: 190–197

Dufour S, Lopez E, Le Menn F, Le Belle N, Baloche S, Fontaine YA (1988). Stimulation of gonadotropin release and of ovarian development, by the administration of a gonadoliberin agonist and of dopamine antagonists, in female silver eel pretreated with estradiol. Gen Comp Endocrinol 70: 20–30

Dufour S, Burzawa-Gérard E, Le Belle N, Sbaihi M, Vidal B (2003). Reproductive endocrinology of the European eel, *Anguilla anguilla*. In: Aida K, Tsukamoto K, Yamauchi K (eds), Eel Biology, Springer, Tokyo, pp 373–383

Dufour S, Weltzien FA, Sébert ME, Le Belle N, Vidal B, Vernier P, Pasqualini C (2005). Dopaminergic inhibition of reproduction in teleost fishes – Ecophysiological and evolutionary implications. Ann NY Acad Sci 1040: 9–21

Fontaine M (1936). Sur la maturation complete des organes genitaux de l'anguille male et l'emission spontanee de ses produits sexuels. CR Acad Sci Paris 202: 1312–1315

Fontaine M, Bertrand E, Lopez E, Callamand O (1964). Sur la maturation des organes genitaux de l'anguille femelle (*Anguilla anguilla* L.) et l'emission spontanee des oeufs en aquarium. CR Acad Sci Paris 259: 822–824

Fontaine YA, Dufour S, Alinat J, Fontaine M (1985). A long immersion in deep sea stimulates the pituitary gonadotropic function of the female European eel (*Anguilla anguilla* L.). CR Acad Sci Paris 300: 83–87

Fremberg M, van Veen T, Hartwig HG (1977). Formaldehyde-induced fluorescence in telencephalon and diencephalon of eel (*Anguilla anguilla* L.) – Fluorescence-microscopic and microspectrofluorometric investigation with special reference to innervation of pituitary. Cell Tissue Res 176: 1–22

Giorgi O, Deiana AM, Salvadori S, Lecca D, Corda MG (1994). Developmental changes in the content of dopamine in the olfactory bulb of the European eel (*Anguilla anguilla*). Neurosci Lett 172: 35–38

Goos HJ (1987). Steroid feedback on pituitary gonadotropin secretion. In: Idler DR, Crim LW, Walsh J (eds), Proceedings of the 3rd International Symposium on Reproductive Physiology of Fish, St. John's University, St. John, Canada: Fish Symposium, pp 16–20

Holland MC, Hassin S, Zohar Y (1998). Effects of long-term testosterone, gonadotropin-releasing hormone agonist, and pimozide treatments on gonadotropin II levels and ovarian development in juvenile female striped bass (*Morone saxatilis*). Biol Reprod 59: 1153–1162

Huang YS, Rousseau K, Le Belle N, Vidal B, Burzawa-Gerard E, Marchelidon J, Dufour S (1998). Insulin-like growth factor I stimulates gonadotrophin production from eel pituitary cells: a possible metabolic signal for induction of puberty. J Endocrinol 159: 43–52

Huang YS, Rousseau K, Sbaihi M, Le Belle N, Schmitz M, Dufour S (1999). Cortisol selectively stimulates pituitary gonadotropin beta-subunit in a primitive teleost, *Anguilla anguilla*. Endocrinology 140: 1228– 1235

Jeng SR, Dufour S, Chang CF (2005). Differential expression of neural and gonadal aromatase enzymatic activities in relation to gonadal development in Japanese eel, *Anguilla japonica*. J Exp Zool 303A: 802–812

Kagawa H (2003). Artificial induction of oocyte maturation and ovulation. In:Aida K, Tsukamoto K, Yamauchi K (eds), Eel Biology, Springer, Tokyo, pp 401–414

Kah O, Chambolle P, Thibault J, Geffard M (1984). Existence of dopaminergic neurons in the preoptic region of the goldfish. Neurosci Lett 48: 293–298

Kah O, Dulka JG, Dubourg P, Thibault J, Peter RE (1987). Neuroanatomical substrate for the inhibition of gonadotrophin secretion in goldfish: existence of a dopaminergic preoptico-hypophyseal pathway. Neuroendocrinology 45: 451–458

Kamei H, Kaneko T, Aida K (2006). In vivo gonadotropic effects of recombinant Japanese eel follicle-stimulating hormone. Aquaculture 261: 771–775

Kapsimali M, Vidal B, Gonzalez A, Dufour S, Vernier P (2000). Distribution of the mRNA encoding the four dopamine D(1) receptor subtypes in the brain of the European eel (*Anguilla anguilla*): comparative approach to the function of D(1) receptors in vertebrates. J Comp Neurol 419: 320–343

Kaslin J, Panula P (2001). Comparative anatomy of the histaminergic and other aminergic systems in zebrafish (*Danio rerio*). J Comp Neurol 440: 342–377

Kebabian JW, Calne DB (1979). Multiple receptors for dopamine. Nature 277: 93–96

Kumakura N, Okuzawa K, Gen K, Kagawa H (2003). Effects of gonadotropin-releasing hormone agonist and dopamine antagonist on hypothalamus-pituitary-gonadal axis of pre-pubertal female red seabream (*Pagrus major*). Gen Comp Endocrinol 131: 264–273

Lauder GV, Liem KF (1983). Patterns of diversity and evolution in ray-finned fishes. In: Northcutt RC, David RE (eds), Fish Neurobiology, 1st edn, University of Michigan Press, Ann Arbor, pp 2–24

Lefranc G, L'Hermite A, Tusques J (1969). Mise en évidence de neurones monoaminergiques par la technique de fluorescence dans l'encéphale d'Anguille. Comptes Rendus de la Société de Biologie (Paris) 163: 1193–1196

Lefranc G, L'Hermite A, Tusques J (1970). Etude topographique et cytologique des différents noyaux monoaminergiques de l'encéphale d'*Anguilla vulgaris*. Comptes Rendus de la Société de Biologie (Paris) 164: 1629–1632

Leloup-Hatey J, Hardy A, Nahoul K, Querat B, Zohar Y (1988). Influence of a gonadotrophic treatment upon the ovarian steroidogenesis in European silver eel. In: Reproduction in Fish: Basic and Applied Aspects in Endocrinology and Genetics. INRA, Paris, pp 127–130

Levavi-Sivan B, Safarian H, Rosenfeld H, Elizur A, Avitan A (2004). Regulation of gonadotropin-releasing hormone (GnRH)–receptor gene expression in tilapia: effect of GnRH and dopamine. Biol Reprod 70: 1545–1551

Levavi-Sivan B, Biran J, Fireman E (2006). Sex steroids are involved in the regulation of gonadotropin-releasing hormone and dopamine D2 receptors in female tilapia pituitary. Biol Reprod 75: 642–650

L'Hermite A, Lefranc G (1972). Recherches sur les voies monoaminergiques de l'encéphale d'Anguilla vulgaris. Archives d'Anatomie microscopique et de morphologie experimentale 61: 139–152

Lin H-R, van der Kraak G, Zhou X-J, Liang J-Y, Peter RE, Rivier JE, Vale WW (1988). Effects of [$_D$-Arg^6, Trp^7, Leu^8, Pro^9 NEt]-luteinizing hormone–releasing hormone (sGnRH-A) and [$_D$-Ala^6, Pro^9 NEt]- luteinizing hormone–releasing hormone (LHRH-A), in combination with pimozide or domperidone, on gonadotropin release and ovulation in the Chinese loach and common carp. Gen Comp Endocrinol 69: 31–40

Linard B, Bennani S, Saligaut C (1995). Involvement of estradiol in a catecholamine inhibitory tone of gonadotropin release in rainbow trout (*Oncorhynchus mykiss*). Gen Comp Endocrinol 99: 192–196

Linard B, Anglade I, Corio M, Navas JM, Pakdel F, Saligaut C, Kah O (1996). Estrogen receptors are expressed in a subset of tyrosine hydroxylase-positive neurons of the anterior preoptic region in the rainbow trout. Neuroendocrinology 63: 156–165

Lokman PM, Verneulen GJ, Lambert JGD, Young G (1998). Gonad histology and plasma steroid profiles in wild New Zealand freshwater eels (*Anguilla dieffenbachii* and *A. australis*) before and at the onset of the natural spawning migration. I. Females. Fish Physiol Biochem 19: 325–338

Lokman PM, Rohr DH, Davie PS, Young G (2003). The physiology of silvering in anguillid eels: androgens and control of metamorphosis from the yellow to silver stage. In: Aida K, Tsukamoto K, Yamauchi K (eds), Eel Biology, Springer, Tokyo, pp 331–349

Macrae AD, Brenner S (1995). Analysis of the dopamine-receptor family in the compact genome of the puffer fish *Fugu rubripes*. Genomics 25: 436–446

Marcano D, Guerrero HY, Gago N, Cardillo E, Requena M, Ruiz L (1995). Monoamine metabolism in the hypothalamus of the juvenile teleost fish, *Chaetodipterus faber*. Proceedings of the 5th International Symposium on the Reproductive Physiology of Fish, University of Texas Press, Austin, pp 64–66

Missale C, Nash SR, Robinson SW, Jaber M, Caron MG (1998). Dopamine receptors: from structure to function. Physiol Rev 78: 189–225

Mok EY-M, Munro AD (1998). Effects of dopaminergic drugs on locomotor activity in teleost fish of the genus *Oreochromis* (Cichlidae): Involvement of the telencephalon. Physiol Behav 64: 227–234

Montero M, Dufour S (1996). Gonadotropin-releasing hormones (GnRH) in fishes: Evolutionary data on their structure, localization, regulation, and function. Zool Stud 35: 149–160

Neve KA, Durand CJ, Teeter MM (2003). Structural analysis of the mammalian D2, D3, and D4 dopamine receptors. In: Sidhu A, Laruelle M, and Vernier P (eds), Dopamine Receptors and Transporters: Function, Imaging, and Clinical Implication, Marcel Dekker, New York, pp 77–144

Nilsson L, Nyman L, Westin L, Ornhagen H (1981). Simulation of the reproductive migration of European eels (*Anguilla anguilla* (L)) through manipulation of some environmental-factors under hydrostatic compression. Spec Sci Technol 4: 475–484

Norris DO (1997). Vertebrate Endocrinology, 3rd edn, Academic, San Diego, CA

Oliveira L, Paiva ACM, Sander C, Vriend G (1994). A common step for signal-transduction in G-protein-coupled receptors. Trends Pharmacol Sci 15: 170–172

Pasqualini C, Weltzien F-A, Vidal B, Baloche S, Dufour S, Vernier P (2007). Cloning and distribution of two distinct dopamine D2 receptors in the brain and pituitary of the European eel. Proceedings of the 8th International Symposium on Reproductive Physiology of Fish. Saint-Malo, June 2007

Peter RE, Crim LW (1978). Hypothalamic lesions of goldfish: effects on gonadal recrudescence and gonadotropin secretion. Ann Biol Anim Biochim Biophys 18: 819–823

Peter RE, Paulencu CR (1980). Involvement of the preoptic region in the gonadotropin release-inhibition in the goldfish. Neuroendocrinology 31: 133–141

Peter RE, Crim LW, Goos HJTh, Crim JW (1978). Lesioning studies on the gravid female goldfish: neuroendocrine regulation of ovulation. Gen Comp Endocrinol 35: 391–401

Peter RE, Chang JP, Nahorniak CS, Omeljaniuk RJ, Sokolowska M, Shih SH, Billard R (1986). Interactions of catecholamines and GnRH in regulation of gonadotropin secretion in teleost fish. Recent Prog Horm Res 42: 513–548

Peter RE, Trudeau V, Sloley BD, Peng C, Nahorniak CS (1991). Actions of catecholamines, peptides, and sex steroids in regulation of gonadotropin-II in the goldfish. In: Reproductive Physiology of Fish. Proceedings of the 4th International Symposium on Reproductive Physiology of Fish. Fish Symposium 91, University of East Anglia Sheffield, UK, pp 30–34

Roberts BL, Meredith GE, Maslam S (1989). Immunocytochemical analysis of the dopamine system in the brain and spinal cord of the European eel, *Anguilla anguilla*. Anat Embr 180: 401– 412

Robins CR, Cohen DM, Robins CH (1979). Eels, *Anguilla* and *Histiobranchus*, photographed on the floor of the deep Atlantic in the Bahamas. Bull Mar Sci 29: 401–405

Saligaut C, Linard B, Breton B, Anglade I, Bailhache T, Kah O, Jego P (1999). Brain aminergic systems in salmonids and other teleosts in relation to steroid feedback and gonadotropin release. Aquaculture 177: 13–20

Sbaihi M, Fouchereau-Peron M, Meunier F, Elie P, Mayer I, Burzawa-Gérard E, Vidal B, Dufour S (2001). Reproductive biology of the conger eel from the south coast of Brittany, France and comparison with the European eel. J Fish Biol 59: 302–318

Schmidt J (1923). Breeding places and migration of the eel. Nature 111: 51–54

Schmitz M, Aroua S, Vidal B, Le Belle N, Elie P, Dufour S (2005). Differential feedback by sexual steroids is involved in the opposite regulation of luteinizing hormone and follicle-stimulating hormone expression during ovarian development in the European eel. Neuroendocrinology 81: 107–119

Schulz RW, Bogerd J, Goos HJTh (2000). Spermatogenesis and its endocrine regulation. In: Norberg B, Kjesbu OS, Taranger GL, Andersson E, Stefansson SO (eds), Proceedings of the 6th International Symposium on the Reproductive Physiology of Fish, John Grieg AS, Bergen, Norway, pp 225–232

Sébert M-E, Weltzien F-A, Moisan C, Pasqualini C, Dufour S (2008). Dopaminergic systems in the European eel: characterization, brain distribution and potential role in migration and reproduction. Hydrobiologia 602: 27–46

Sébert M-E, Amérand A, Vettier A, Weltzien F-A, Pasqualini C, Sébert P, Dufour S (2007a). Effects of high hydrostatic pressure on the pituitary-gonad axis in the European eel, *Anguilla anguilla* (L.). Gen Comp Endocrinol 153: 289–298

Sébert M-E, Amerand A, Vettier A, Weltzien F-A, Baloche S, Sébert P, Dufour S (2007b). Involvement of high hydrostatic pressure in the stimulation of eel sexual maturation. Proceedings of the 8th International Symposium on Reproductive Physiology of Fish. Saint-Malo, June 2007

Sébert P, Barthélémy L, Simon B (1990). Laboratory system enabling long-term exposure to hydrostatic pressure of fishes or other animals breathing water. Mar Biol 104: 165–168

Senthilkumaran B, Joy KP (1995). Changes in hypothalamic catecholamines, dopamine-beta-hydroxylase, and phenylethanolamine-n-transferase in the catfish *Heteropneustes fossilis* in relation to season, raised photoperiod, and temperature, ovariectomy, and estradiol-17ß replacement. Gen Comp Endocrinol 97: 121–134

Sugamori KS, Demchyshyn LL, Chung M, Niznik HB (1994). D1A, D1B, and D1C dopamine receptors from *Xenopus laevis*. Proc Natl Acad Sci USA 91: 10536–10540

Swanson P (1991). Salmon gonadotropins: reconciling old and new ideas. In: Scott AP, Sumpter JP, Kime DE, Rolfe MS (eds), Proceedings of the 4th International Symposium on the Reproductive Physiology of Fish. Fish-Symp91, Sheffield, UK, pp 2–7

Tesch FW, Rohlf N (2003). Migration from continental waters to the spawning grounds. In: Aida K, Tsukamoto K, Yamauchi K (eds), Eel Biology, Springer, Tokyo, pp 223–234

Trudeau VL, Sloley BD, Peter RE (1993). GABA stimulation of gonadotropin-II release in goldfish: involvement of GABA receptors, dopamine, and sex steroids. Am J Physiol 265: 348–355

Vacher C, Mananos EL, Breton B, Marmignon MH, Saligaut C (2000). Modulation of pituitary dopamine D1 or D2 receptors and secretion of follicle-stimulating hormone and luteinizing hormone during the annual reproductive cycle of female rainbow trout. J Neuroendocrinol 12: 1219–1226

Vacher C, Ferriere F, Marmignon MH, Pellegrini E, Saligaut C (2002). Dopamine D2 receptors and secretion of FSH and LH: role of sexual steroids on the pituitary of the female rainbow trout. Gen Comp Endocrinol 127: 198–206

Vacher C, Pellegrini E, Anglade I, Ferriere F, Saligaut C, Kah O (2003). Distribution of dopamine D-2 receptor mRNAs in the brain and the pituitary of female rainbow trout: An in situ hybridization study. J Comp Neurol 458: 32–45

Van Ginneken V, Antonissen E, Müller UK, Booms R, Eding E, Verreth J, van den Thillart G (2005). Eel migration to the Sargasso: remarkably high swimming efficiency and low energy costs. J Exp Biol 208: 1329–1335

Vandepoele K, de Vos W, Taylor JS, Meyer A, Van de Peer Y (2004). Major events in the genome evolution of vertebrates: paranome age and size differ considerably between ray-finned fishes and land vertebrates. Proc Natl Acad Sci USA 101: 1638–1643

Vetillard A, Atteke C, Saligaut C, Jego P, Bailhache T (2003). Differential regulation of tyrosine hydroxylase and estradiol receptor expression in the rainbow trout brain. Mol Cell Endocrinol 199: 37–47

Vidal B, Pasqualini C, Le Belle N, Holland MCH, Sbaihi M, Vernier P, Zohar Y, Dufour S (2004). Dopamine inhibits luteinizing hormone synthesis and release in the juvenile European eel: a neuroendocrine lock for the onset of puberty. Biol Reprod 71: 1491–1500

Weltzien F-A, Andersson E, Andersen Ø, Shalchian-Tabrizi K, Norberg B (2004). The brain-pituitary-gonad axis in male teleosts, with special emphasis on flatfish (Pleuronectiformes). Comp Biochem Physiol 137A: 447– 477

Weltzien F-A, Pasqualini C, Le Belle N, Vidal B, Vernier P, Dufour S (2005a). Brain expression of tyrosine hydroxylase and its regulation by steroid hormones in the European eel quantified by real-time PCR. Ann NY Acad Sci 1040: 518–520

Weltzien F-A, Pasqualini C, Vernier P, Dufour S (2005b). A quantitative real-time RT-PCR assay for European eel tyrosine hydroxylase. Gen Comp Endocrinol 142: 134–142

Weltzien F-A, Pasqualini C, Sébert M-E, Vidal B, Le Belle N, Kah O, Vernier P, Dufour S (2006). Androgen-dependent stimulation of brain dopaminergic systems in the female European eel (*Anguilla anguilla*). Endocrinology 147: 2964–2973

Westin L (1990). Orientation mechanisms in migrating European silver eel (*Anguilla anguilla*): temperature and olfaction. Mar Biol 106: 175–179

Winberg S, Nilsson GE (1993). Roles of brain monoamine neurotransmitters in agonistic behaviour and stress reactions, with particular reference to fish. Comp Biochem Physiol 106C: 597–614

Yaron Z, Gur G, Melamed P, Rosenfeld H, Elizur A, Levavi-Sivan B (2003). Regulation of fish gonadotropins. Int Rev Cytol 225: 131–185

Yu KL, Peter RE (1990). Dopaminergic regulation of brain gonadotropin-releasing hormone in male goldfish during spawning behavior. Neuroendocrinology 52: 276–283

Yu KL, Peter RE (1992). Adrenergic and dopaminergic regulation of gonadotropin-releasing hormone release from goldfish preoptic-anterior hypothalamus and pituitary in vitro. Gen Comp Endocrinol 85: 138–146

Zohar Y, Harel M, Hassin S, Tandler A (1995). Broodstock management and manipulation of spawning in the gilthead sea bream, *Sparus aurata*. In: Broodstock Management and Egg and Larval Quality, Blackwell, Oxford, pp 94–117

Chapter 13
Artificial Maturation and Reproduction of the European Eel

Arjan Palstra and Guido van den Thillart

13.1 Introduction

Mature eels cannot be obtained from the wild situation since nobody has ever observed migrating and spawning silver eels in the ocean. Our existing knowledge comes from the artificial induction of maturation by hormonal injections with carp or salmon pituitary extract (CPE/SPE; Fontaine et al. 1964) in female European eels and with human chorionic gonadotropin (hCG) in males (Fontaine 1936). In 1974, Yamamoto and his colleagues continued their investigations on artificially induced maturation in Japanese eels and were able to produce larvae but these died within a few days (Yamamoto and Yamauchi 1974; Yamauchi et al. 1976). In the 1990s, efforts resulted in fertilized eggs and non-feeding larvae.

Research on the induction of maturation and larval rearing of the Japanese eel, *Anguilla japonica*, is mainly conducted at the National Research Institute of Aquaculture (Nansei, Japan). The maturation procedure for females involves injection of salmon pituitary extract (20 mg per week) for 8–13 weeks (Kagawa et al. 2005). At 24-h after the last injection, 17, 20 β-dihydroxy-4-pregnen-3-one (DHP) is injected (2 μg g^{-1} body weight) to stimulate final maturation and induce ovulation within 14–23 h (Ohta et al. 1996). On the other hand to mature male eels, 10 to 14 weekly injections of human chorionic gonadotropin (1 IU g^{-1} body weight) are sufficient to induce spermiation (Kagawa et al. 2005). Up to 89.6% fertility and 47.6% hatchability was obtained (Ohta et al. 1996). Because of the difficulty of achieving simultaneous ovulation and spermiation, particular attention has been paid to finding an optimum diluent for cryo-preservation of spermatozoa (Ohta and Izawa 1996).

The main emphasis now is on larval rearing. Getting larvae past the first 9 days proved thus far impossible with live food (rotifers), so a complex artificial diet was developed in the form of a paste. The initial problem was to get the larvae to feed at all. Now that this has been achieved, the problem is to find an optimum diet to ensure

A. Palstra and G. van den Thillart
Integrative Zoology, Institute of Biology, Leiden University (IBL), van der Klaauw Laboratory, Kaiserstraat 63, 2311 GP Leiden, The Netherlands

G. van den Thillart et al. (eds.), *Spawning Migration of the European Eel*,

survival and optimal growth. Growth is slower than in wild larvae up to the first metamorphosis to the leptocephalus stage, thereafter it slows even more. The larvae are, much smaller than wild leptocephali, when they reach the coast of Taiwan (Tsukamoto 1992; Fricke and Tsukamoto 1998).

Jan and Inge Boetius were the first to fertilize the eggs of European eel (Boëtius and Boëtius 1980). A few years later, a Belorussian group (Bezdenezhnykh et al. 1983, 1984) obtained larvae from the European eel, but these larvae showed the same problems as with the Japanese eel, they died within a few days after hatching. The Japanese researchers used their protocol to induce ovarian development for Japanese eel also on European eels (Chiba et al. 1994). However, until now they have been unable to reproduce European eel with same success as for the Japanese eel, showing that maturation of the European eel differs to a great extent. Lokman and Young (2000) used the Ohta et al. (1996) protocol on New Zealand freshwater eels (*A. dieffenbachii* and *A. australis*). They obtained larvae of *A. australis* and kept them alive for a few days. Recently, Palstra et al. (2005) used the Ohta protocol and produced 4 day old European eel embryos (Fig. 13.9). Pedersen (2003, 2004) applied variations of the same protocol and obtained a few larvae that stayed alive for 2 days. However, those larvae showed delayed hatching and abnormal morphology. In 2006 and 2007 Tomkiewicz and colleagues (DIFRES, Denmark), were able to produce larvae, which could be repeated several times (presented at EAS meeting Istanbul 2007). They have been able to keep them alive up to 22 days, still in the pre-feeding stage. In 2007 Kurwie and her colleagues were able to produce larvae of *A. australis* and kept them alive up to 20 days (presented at website, http://www.aquarticles.com/articles/breeding/index.html). Also in 2007, Oliveira and colleagues produced 1-day old larvae of *A. rostrata* (not published). However, for all three species it has not been possible yet to initiate feeding.

The same breeding protocol with *Anguilla japonica* is currently more successful; in 2006 Japanese scientists produced feeding larvae on a weekly basis (Dr. Takao Yoshimatsu, National Institute of Aquaculture, Japan, presented at IPA workshop Bilthoven 2006). In addition Tanaka (2003) was already able to get some leptocephali larvae through metamorphosis and to obtain glass eels at approximately 250 days after hatching. Still, despite the large investments in time and energy, the question of controlled reproduction of eels is not resolved. The fact that many fertilized eggs do not develop, and most larvae die prematurely indicate that the problems may lay in the quality of eggs and possibly also in the quality of sperm.

13.2 Induction of Spermatogenesis and Sperm Production

In 1936 Fontaine and co-workers were already able to induce full spermatogenesis in male European eels by injecting urine extract from pregnant women, known later to contain large amounts of human chorionic gonadotropin – hCG (Fontaine 1936; Dufour et al. 2003). In recent studies, hCG was used to induce spermiation 3 months after a single injection (Kahn et al. 1987) or much sooner by multiple weekly injections. Eels start to spermiate after 4 (Pedersen et al. 2003) and 5 weeks (Müller

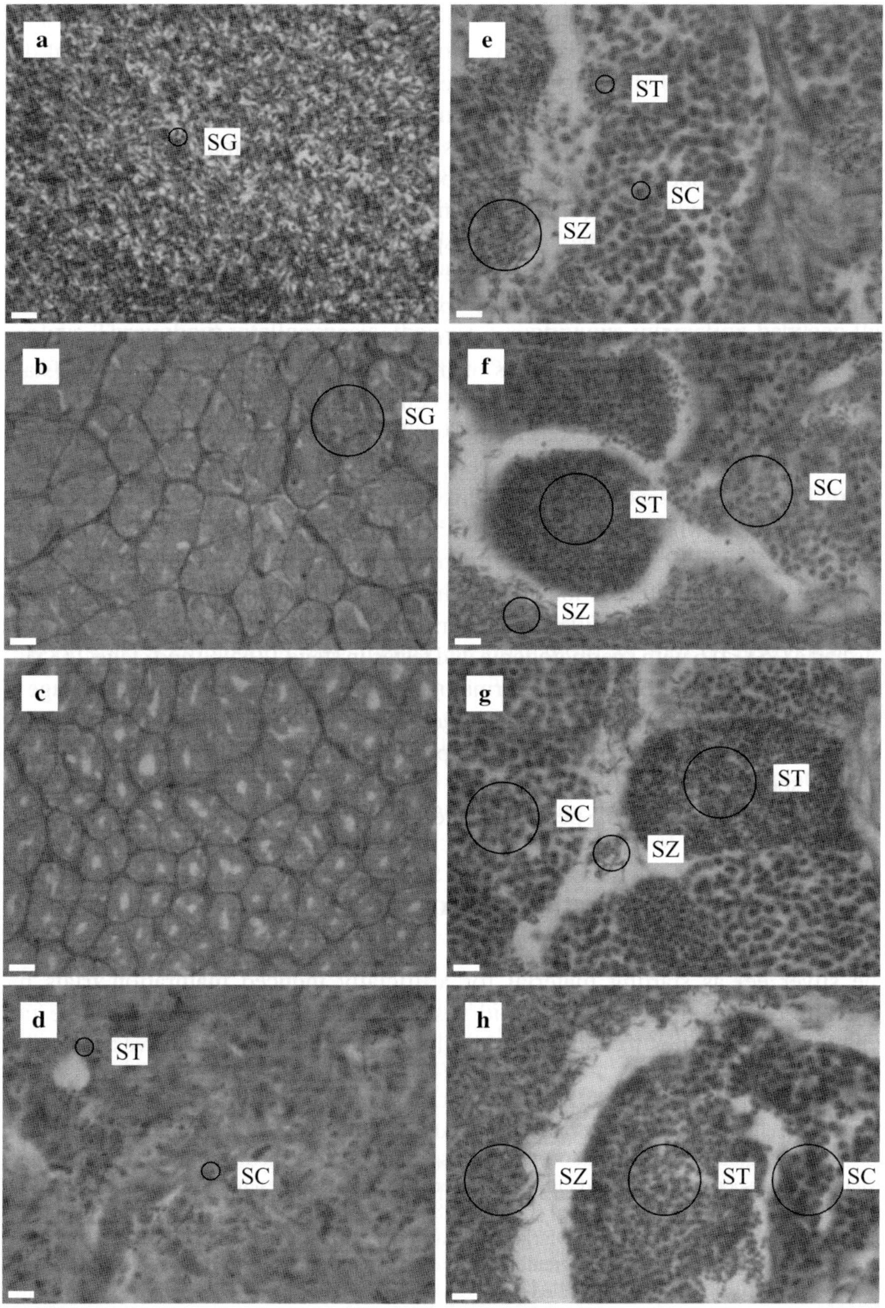

Fig. 13.1 Staging of testis maturation. Testis histology sections obtained from European eels (silver males). Twenty wild migratory eels were either untreated (n = 10) or treated with weekly hCg injections (n = 10) for 19 weeks after which testes were preserved in Bouin. Samples were embedded in paraffin and sections were stained with Mayer's Hemotoxylin-eosin (A. Palstra

these events are induced simultaneously by PE treatment it seems that the oocytes of migrating silver eels may therefore be too premature to be stimulated by PE. This phenomenon may thus be the cause of the poor gamete quality.

Unfortunately, a natural standard for gamete quality of European eel is missing since mature European eels cannot be obtained from the wild. Comparisons to the natural situation can, however, still be done in three ways: (1) by comparing with naturally induced early maturation, for instance by swimming (Chapter 10), (2) by comparing with wild silver eels from other *Anguilla* species, such as with the long-finned eels *Anguilla dieffenbachii* and short finned eels *A. australis*. These species are in a far more advanced state when they leave the continent for spawning migration (Todd 1981; Suetake et al. 2003) and (3) by comparing with a related species that matures naturally in captivity such as Conger eels, *Conger conger* (Fig. 13.2).

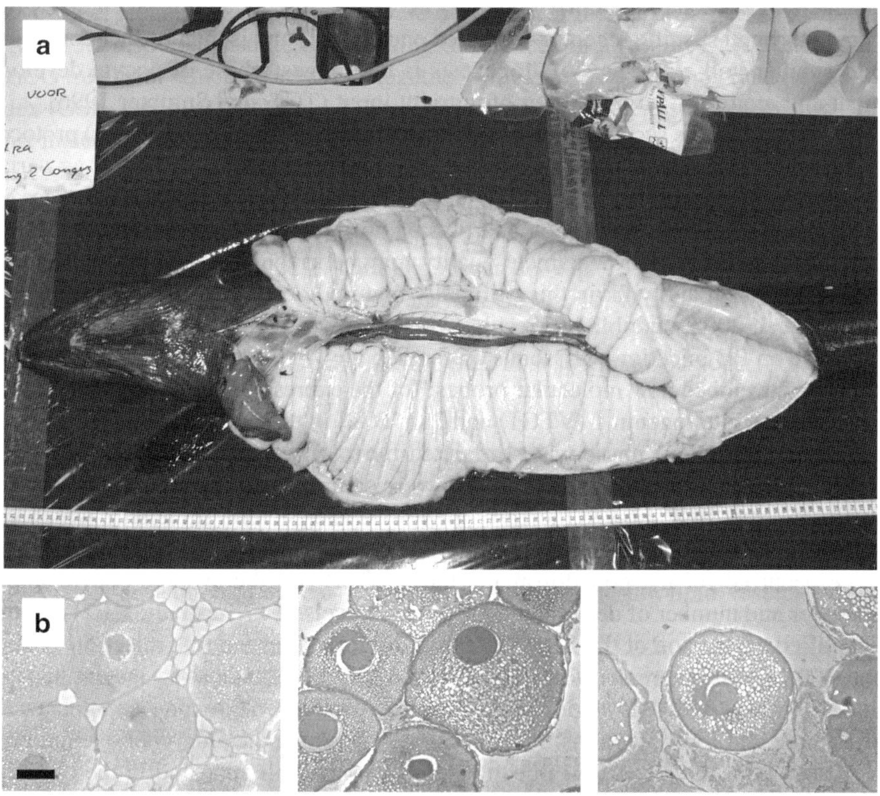

Fig. 13.2 Female conger eels (*Conger conger*) during final maturation Conger 1: 204 cm, 40.1 kg, GSI = 8.7, oocyte diameter 295 ± 21 µm. Conger 2: 185 cm, 41.0 kg, GSI = 21.2, oocyte diameter 435 ± 28 µm. Conger 3: 171 cm, 14 kg, GSI = 30.0, oocyte diameter 513 ± 50 µm. (a) Dissected Conger 3 showing the enormous ovaria. (b) Oocyte sections of respectively conger 1–3; with still large extracellular lipid stores, with larger vitellogenic oocytes, and with mature ovulated oocytes. The scale bar represents 100 µm

13.4 Follicular Maturation and Ovulation

During early maturation oocytes are non-transparent and filled with fat droplets and yolk proteins, which are products of the secondary yolk/midvitellogenic stage (Adachi et al. 2003). Fast growth and increase in transparency occurs in the tertiary yolk stage/late vitellogenic stage (Adachi et al. 2003). The latter is considered as a result of fusion of yolk globules and hydration (reviewed in Wallace and Selman 1981). Oocytes now undergo their final maturation during which the chromosomes resume meiosis and proceed to the second meiotic metaphase with the concomitant formation of the first polar body (Goetz 1983). The increase in transparency coincides with swelling of the oocytes due to hydration (up to 800–900 μm, Palstra et al. 2005). Pronounced hydration is typical for marine teleosts spawning pelagic eggs (Wallace and Selman 1981). Final maturation of eel oocytes is apart from hydration characterised by the fusion of lipid droplets and germinal vesicle migration (GVM). Figure 13.3 shows the characteristics of a ripe eel oocyte with many lipid droplets and a nucleus moving to the periphery (GVM).

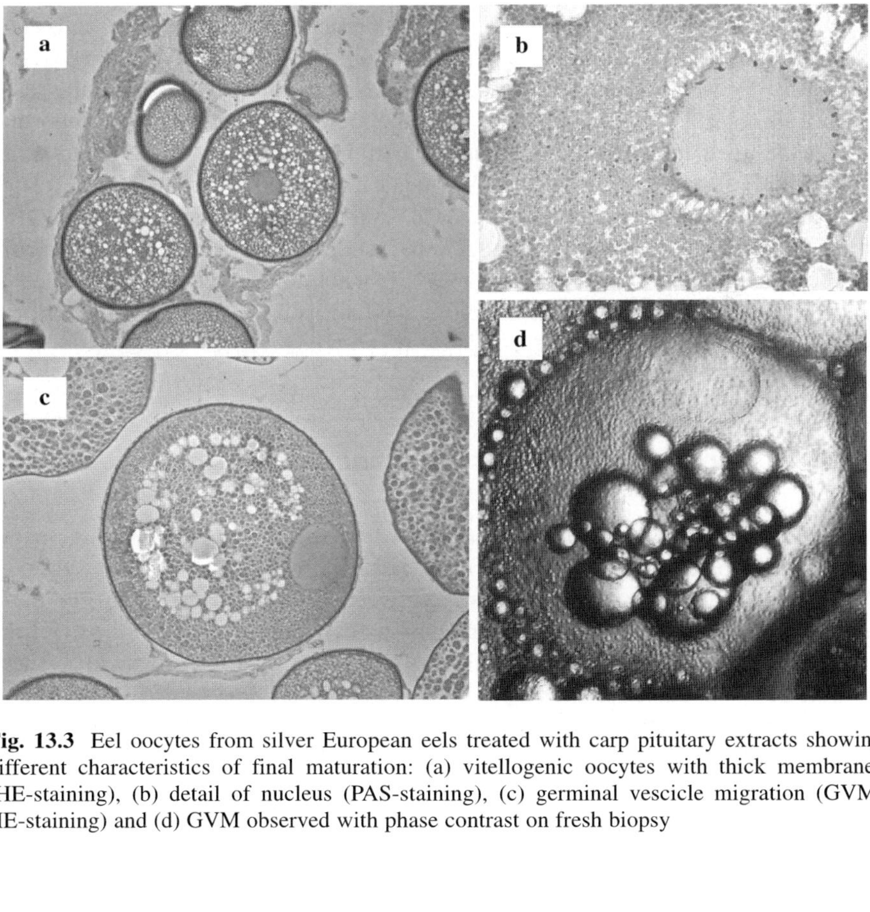

Fig. 13.3 Eel oocytes from silver European eels treated with carp pituitary extracts showing different characteristics of final maturation: (a) vitellogenic oocytes with thick membranes (HE-staining), (b) detail of nucleus (PAS-staining), (c) germinal vescicle migration (GVM; HE-staining) and (d) GVM observed with phase contrast on fresh biopsy

Shortly after migration to the periphery the germinal vesicle breaks down (GVBD) followed by the extrusion of the first polar body.

The hydration of the oocytes during final maturation results in a significant increase of the body weight, actually at ovulation the GSI can be between 20% and 70% which is the result of the simultaneous swelling of a very large number of oocytes (>1 million). Most of the hydration occurs within a couple of weeks (Fig. 13.4). This increase of the body weight during final maturation was used by Ohta et al. (1996) to indicate the moment for induction of ovulation. With the current Japanese hormonal stimulation protocol (summarized by Kagawa et al. 2005), eels do not ovulate spontaneously. Final maturation and ovulation are however successfully induced by injection of 17,20β-dihydroxy-4-pregnen-3-one (DHP; Ohta et al., 1996). The difficulty is, however, that not the eels but the experimentor decides when ovulation needs to be induced. Final maturation in the Japanese eel is regular and predictable on basis of the percentage increase in body weight (body weight index = BWI) due to the hydration response of the oocytes. In contrast, oocyte maturation in the European eel is slower, more variable and therefore less predictable (Fig. 13.4), as indicated by Pedersen (2003) and Palstra et al. (2005). The effect of regular injections with carp pituitary extracts on the body weight of the European eel (according to the Japanese protocol) is depicted in Fig. 13.4. The nine curves show a rather variable response to the hormonal stimulation, which is much different from the responses observed with the Japanese eel. The variable hydration response may partly result from the difference in responsiveness between batches of oocytes within the same animal (asynchronous development). This phenomenon seems however abnormal since (a) fertility dropped with every new batch that matured in a later stage (Ohta et al. 1996; Palstra et al. 2005), (b) fertilization success of each batch correlates positively with its initial size (Palstra et al. 2005) and (c) artificial reproduction of the Japanese eel, shows that homogenous maturation is more successful. From the above discussion it follows that the BWI is not suitable as a single indicator of the last phase of ovarian maturation of the European eel, therefore additional indicators are required.

13.4.1 Characteristical Events in Final Oocyte Maturation

In addition to the BWI, the maturation status can be determined by taking oocyte biopsies. In a recent study by Palstra et al. (2005) it was shown that transparency of the oocytes increased over the last few weeks and coincided with hydration (increase of diameter). Fat droplet diameters increased also, while at the same time the number of fat droplets decreased from about 190 in the first biopsy to a few in the last. They concluded that five processes took place in the same period: hydration, increase in transparency, fusion of fat droplets, GVM and GVBD. Hydration occurs in the first stage of final maturation, while fat fusion was observed to develop over several weeks.

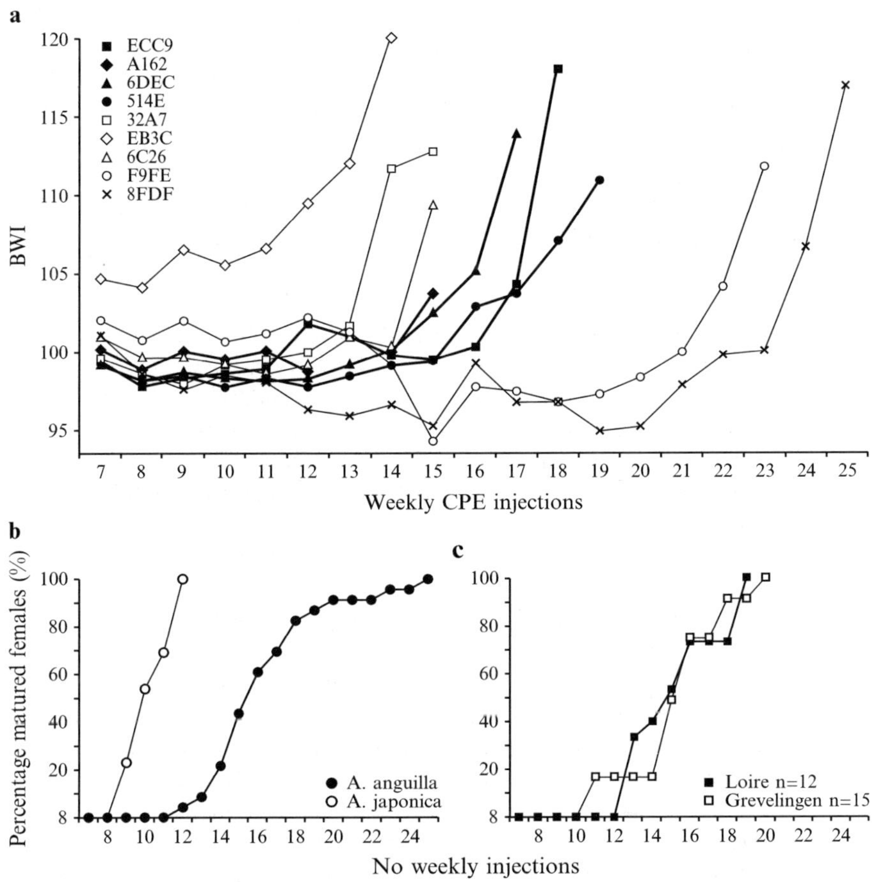

Fig. 13.4 Maturation of European female silver eels in response to CPE injections. (a) body weight index (BWI) of 9 maturing female eels over 7 to 25 weeks, (b) the percentage of matured females vs. time of *A. anguilla* vs. *A. japonica* (Ohta et al. 1996); the European eel shows a much slower response, (c) Maturation time of *A. anguilla* from Lake Grevelingen (The Netherlands) vs. eels from River Loire (France); both groups show similar timing. (Modified from Palstra et al. 2005 and Palstra 2006)

As GVM and particularly GVBD are characteristic for the last phase of oocyte maturation, the five processes were used to describe the development of the oocytes in seven different stages, thus identifying the developmental stage of the ovaria (Fig. 13.5). Stage 1 represents opaque oocytes with a centred nucleus that becomes visible. Its diameter is between 532 and 776 µm, and it still contains over 200 fat droplets. Stage 2 oocytes are fully transparent, significantly larger (723–864 µm) with 183 to 233 clustered fat droplets that are significantly larger. Stage 3 oocytes display GVM and contain 90 to 98 fat droplets that translocate in the opposite direction of GVM. Stage 4 oocytes have the nucleus at the periphery and are again

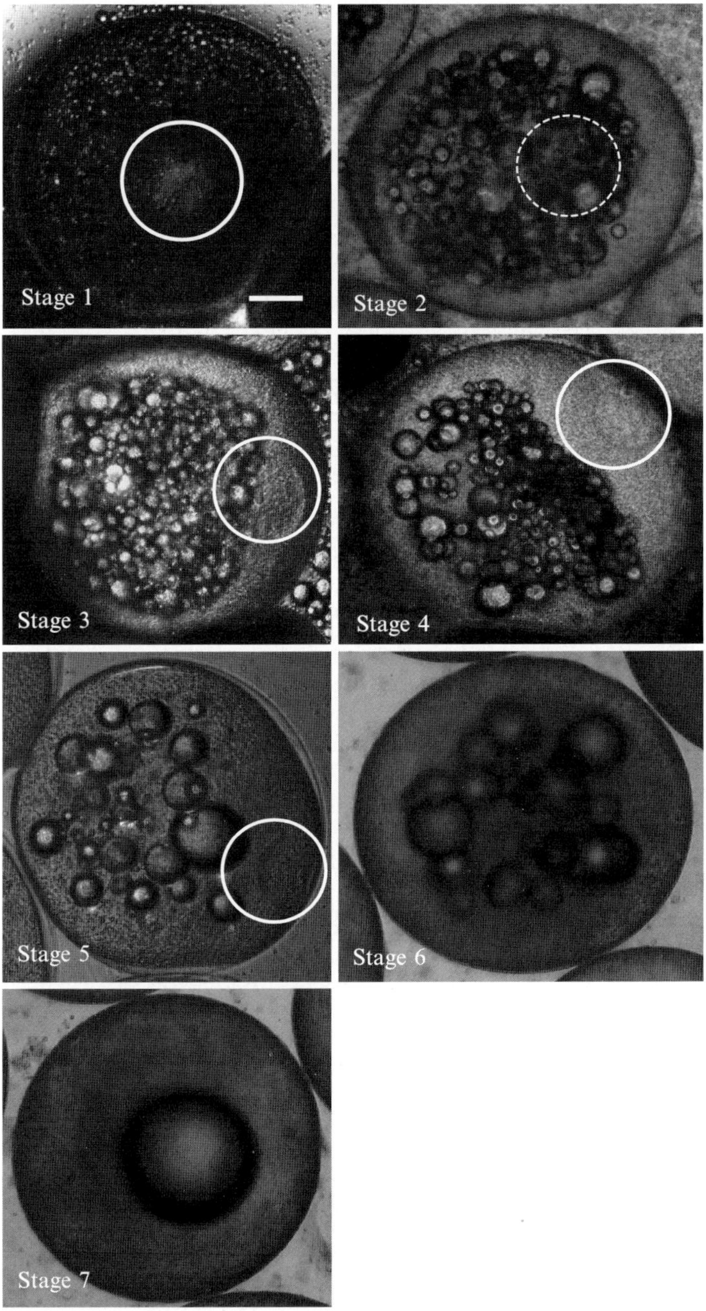

Fig. 13.5 Developmental stages of eel oocytes during final maturation. (scale bar = 100 μm). Encircled are the positions of the migrating germinal vesicle. (Stage 1–5 phase contrast microscopy, stage 6–7 bright light microscopy). Stage 1: opaque oocytes with a centred nucleus becoming visible; Stage 2: fully transparent oocyte; fat droplets clustered; Stage 3, fully transparent oocyte with GVM; Stage 4, fully transparent oocyte with nucleus at periphery; Stage 5, fully transparent oocyte with nucleus at periphery with few large fat droplets; Stage 6, fully transparent oocyte with GVBD; few fat droplets; Stage 7, fully transparent oocyte with GVBD; single fat droplet. (Palstra et al. 2005)

significantly larger (784–897 μm). Stage 5 oocytes also have the nucleus at the periphery but the number of fat droplets has reduced to 34–39. Stage 6 oocytes show GVBD and contain significantly fewer (11–25) and larger fat droplets. Stage 7 oocytes have a single fat droplet that occupies 8.3% of the oocyte volume. Many of these oocytes are overripe and deteriorate quickly.

The correct staging of maturation appears to be crucial for the induction of ovulation, as the last phase varies among animals. Thus, timing of treatment, such as hours between injection with a booster CPE and DHP, is not reliable, at least for the European eel. Therefore an additional parameter is required to determine the correct moment for hormonal stimulation. Using the average of the stages occurring at the same time in the ovary, a key emerges with a value between 0 and 7, which can be used to determine the maturation level of silver eels. Figure 13.6 explains the process. Oocyte biopsies taken from the same silver eel during the last 6 days before ovulation, show a developmental pattern where on each successive day most of the oocytes are in the next stage (one stage per day). Figure 13.7 depicts the mean stage of the oocytes during the last 6 days of final maturation for 18 different silver eels. All those eels were successful in that they produced ovulated eggs that could be fertilized. Figure 13.7 shows also under what condition the oocyte samples were taken: at the last CPE injection, at moment of DHP injection, and at ovulation. The average oocyte stage was at CPE booster injection 4.0 ± 1.2, at DHP injection 5.1 ± 1.2 and at ovulation 5.9 ± 1.2 (Fig. 13.7). Hence we conclude that fertilization thus occurred in eggs during the process of GVBD, when only few fat droplets were left. The oocyte maturation key proved to be a valuable tool in timing of administration of a CPE booster dose and the induction of ovulation by DHP in order to obtain eggs in the right state for fertilization (Palstra et al. 2005, Palstra, 2006). The mean stage of the oocyte is a very potent parameter; while the BWI indicates (indirectly) the hydration response, and the generation size (amount of oocytes responding), the biopsy method on the other hand provides the real maturation stage of the ovaries in each animal.

13.4.2 Body Girth Index (BGI) as External Indicator of Oocyte Developmental Stage

Since BWI appears to be an unreliable indicator of the oocyte hydration response, at least for the European eel, we have investigated in a recent study (A. Palstra et al. 2007 unpublished data) whether the body girth index (BGI); the girth of the body measured at the start of the dorsal fin relative to the BL, would be a better external indicator for final oocyte maturation in the European eel. To do so, we performed a pairwise Spearman correlation analysis between the BWI, BGI and oocyte stage (OS) according to the oocyte development index (Palstra et al. 2005) on the basis of 36 independent observations. Results are given in Fig. 13.8. BWI and BGI were significantly positively correlated ($P = 0.05$; $R^2 = 0.19$). BWI was significantly correlated with OS ($P = 0.009$; $R^2 = 0.27$). The correlation between BGI and OS was more significant ($P < 0.001$) and showed an almost 2-fold better fit ($R^2 = 0.47$).

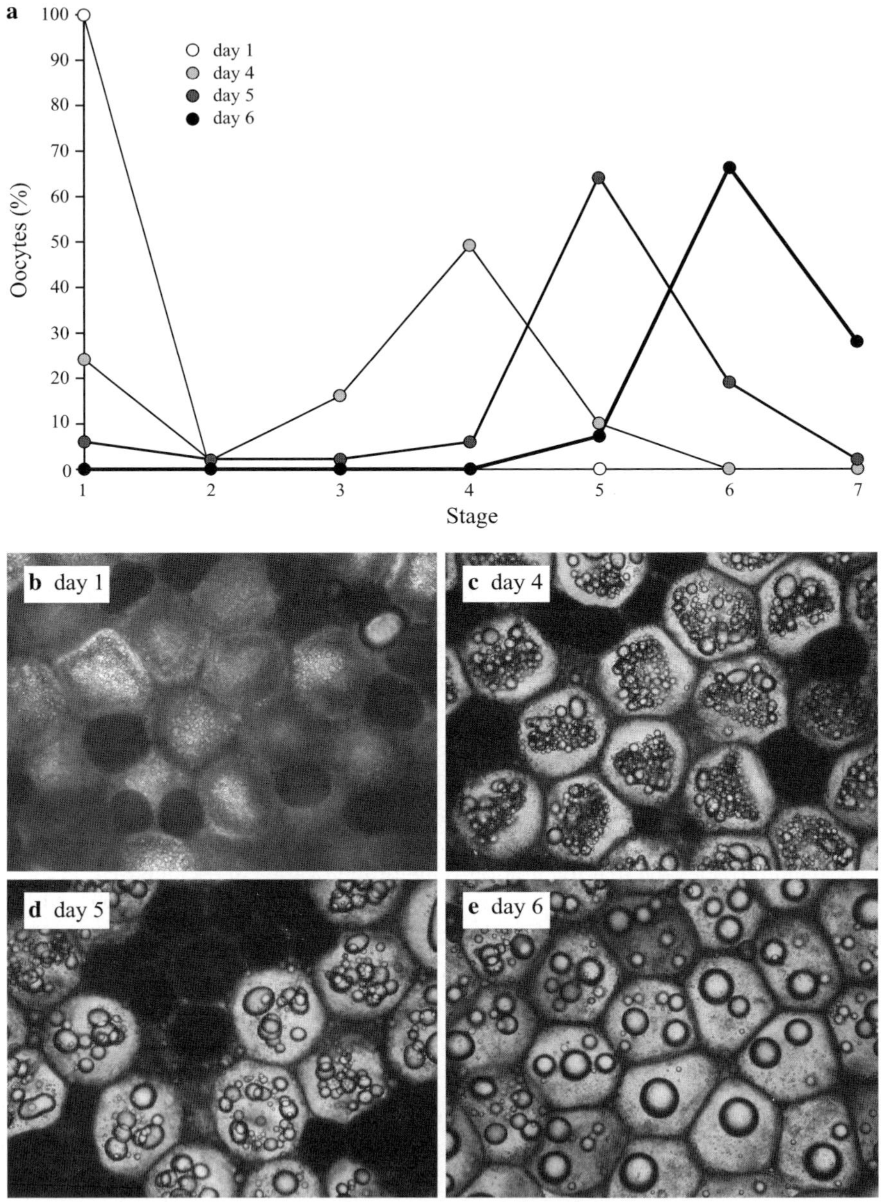

Fig. 13.6 Oocyte development during the last 6 days before ovulation from the same female. (a) The percentages of oocytes per stage in four subsequent biopsy samples at day 1, 4, 5, and 6. (b–e) The corresponding microscopic pictures: (b) stage 1 oocytes just starting to hydrate; (c) stage 3 + 4 oocytes showing germinal vescicle migration; (d): stage 5 oocytes with the nucleus at the periphery and few lipid droplets left; (e): stage 6 + 7 oocytes showing germinal vescicle breakdown and fusion to a single lipid droplet

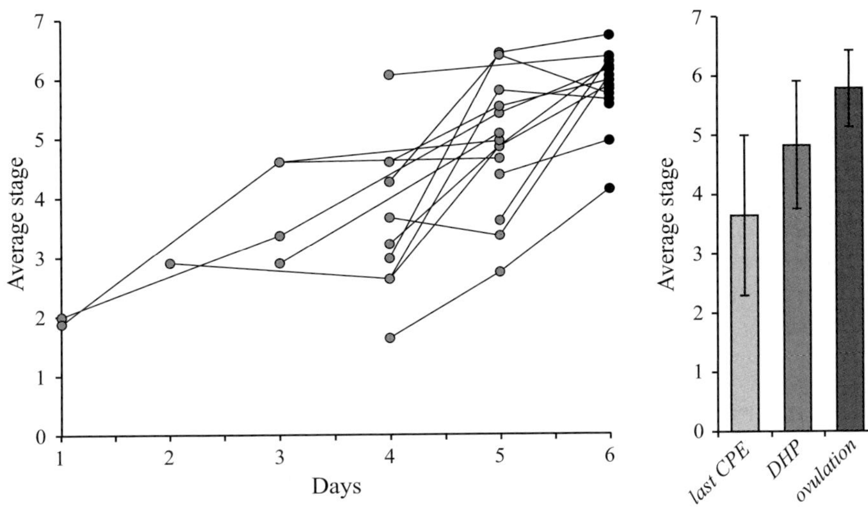

Fig. 13.7 Average oocyte stage in biopsy samples of 18 individual females of which batches were fertilised, taken during the last 6 days of final maturation. Samples were taken at four different conditions: (1) after sufficient weight increase, (2) after CPE injection, (3) after DHP injection, (4) after ovulation (day 6). The insert shows oocyte stages averaged (average ± SD) for these females at last CPE injection, DHP injection and ovulation

On basis of the BGI it is advisable to start oocyte biopsies at BGI ≥0.23. Stripping can best be done of eels with BGIs higher than 0.25. We had no records of eels with a BGI >0.30. Thus we conclude that body diameter index (BGI) is a better indicator of oocyte developmental stage than BWI.

13.4.3 DHP-Induced Ovulation

As in the Japanese eel, ovulation in the European eel can be successfully induced by injecting 17, 20 β-dihydroxy-4-pregnen-3-one (DHP). Between 10 and 24 h after DHP injection ovulation occurs (Pedersen 2003; Palstra et al. 2005). Ohta et al. (1996) and Kagawa et al. (1997) found ovulating females of the Japanese eel between 15 and 21 h after DHP injection. Palstra et al. (2005) found indications for a negative correlation between time to ovulation and developmental stage of the oocytes. Goetz and Theofan (1979) and Goetz (1983) observed the same negative correlation, although the level of synchrony between DHP as inducer of final maturation and ovulation at the used dose is uncertain. A negative correlation between time to ovulation and oocyte diameter was found by Ohta et al. (1997). *In vitro* experiments on the Japanese eel showed that oocytes between 700–800 μm are very sensitive to DHP (Ohta et al. 1997). Oocytes over 800 μm in diameter are even more sensitive for this steroid (Ohta et al. 1997). For DHP – induced ovulation of Japanese

		BGI	OS
BWI	corr. coeff.	0.287*	0.392*
	sig.	0.050	0.009
	N	34	36
BGI	corr. coeff.		0.680**
	sig.		0.000
	N		34

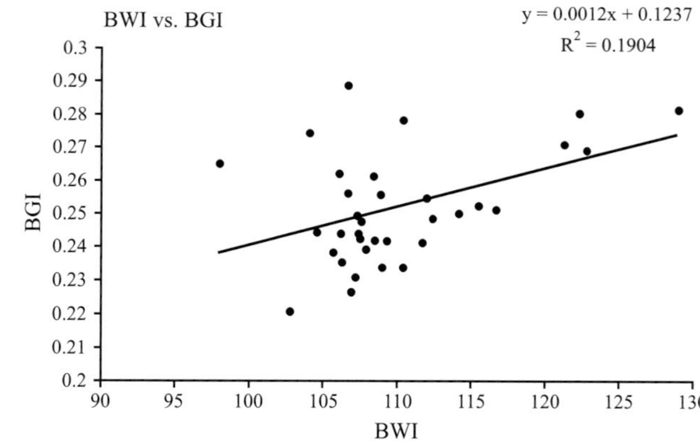

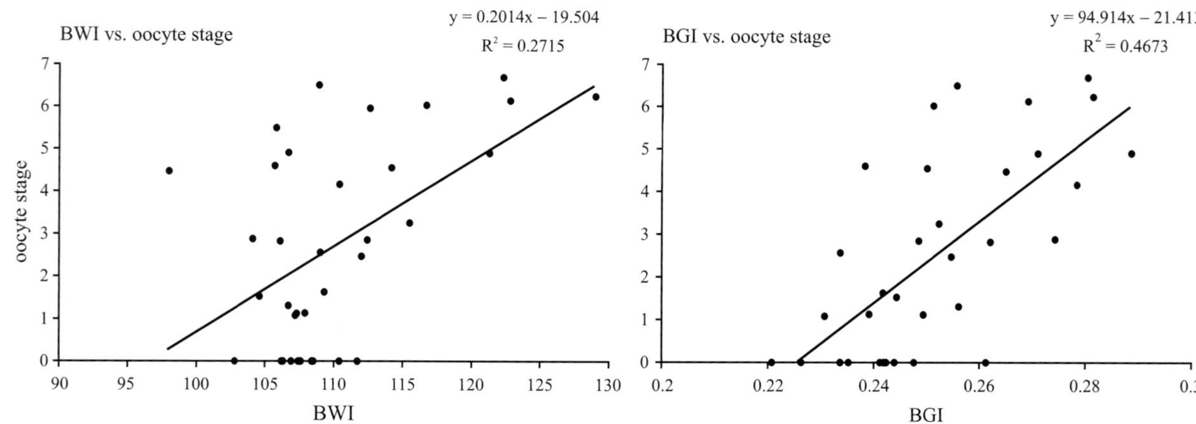

Fig. 13.8 Pearson correlations and correlation plots between body weight index (BWI; bodyweight vs. Bodyweight at the start in %), body diameter index (BGI; body grith relative to body length in %) and oocyte stage (OS)

eel, a minimum oocyte diameter of 750 μm is used as a criterion (B. Pedersen 2004, personal communication). Oocytes of the European eel in our (Palstra et al. 2005) and Pedersen's (2003) study were on average larger than oocytes from the Japanese eel (Ohta et al. 1997) at the time of DHP injection. Oocytes at the desired developmental stage must ovulate within 17 h after DHP injection. This follows from a study of Ohta et al. (1996) who found that fertility and also hatching rates decreased rapidly after this period. Also T. Kurwie and colleagues (MahurangiTechnical Institute – New Zealand, 2008, personal communication) found that the most successful batches in *A. australis* were stripped at ovulation after about 12 h, they also observed that fertility diminished rapidly thereafter.

13.5 Fertilization and Embryology

The early ontogeny of eel species is still largely unknown. Although the limited success of artificial reproduction of eel species allows attempts to investigate early developmental events, still no detailed studies exist. Fragmented reports on embryo and larval development of Japanese eels are mainly focussed on hatching (Hiroi et al. 2003) and feeding (Tanaka 2003; Kurokawa and Pedersen 2003). Pedersen (2004) reported on the chronology of embryological stages of the European eel and Prokhorcik (1986) described broadly the first 4.5 days of postembryonic development.

Experiments in 2002–2004 at Leiden University resulted in fertilization of egg batches of 18 females that were fertilized of which two showed embryonic and early larval development, although the larvae did not hatch (Fig. 13.9). During the first 3 h post fertilization (h.p.f.), most eggs in all batches showed meroblastic cleavage up to the eight cell stage. Egg batches of two females resulted in the development of about 1,600 embryos at 31–32 h.p.f.

Embryos of one female (n = 100) continued to develop and were found vigorously moving with a pigmented tail and heart beat at 58–60 h post fertilization. At this time the protein part of the yolk sac had disappeared leaving only the fat droplet. Bezdenezhnykh et al. (1983) reported complete re-absorption of yolk of 3.5 day old embryos, which coincided with the time of death. This is in contrast to larvae of the Japanese eel, which resorb the yolk in 8 days (at 21–22°C; Pedersen et al. 2003). Such early yolk re-absorption might be an artefact due to incomplete vitellogenesis during (artificial) maturation (Palstra et al. 2004). Embryonic development in the above mentioned experiments continued until 100 h post fertilization when the last embryos died. At that stage the embryos showed clear larval characteristics like pigmented tails, developed eyes and a heart beat, hatching was not observed.

Embryos of a second successful female (n = 1,500) showed serious oedema of the yolksac, a deformed head region and absence of a heartbeat. Such embryonic malformations are typical for PCB-exposed eggs and indicate negative interference with dioxin-like contaminants. Especially long-lived fatty bottom dwelling fishes, particularly those on top of the food chain like *Anguilla* spp., are vulnerable to

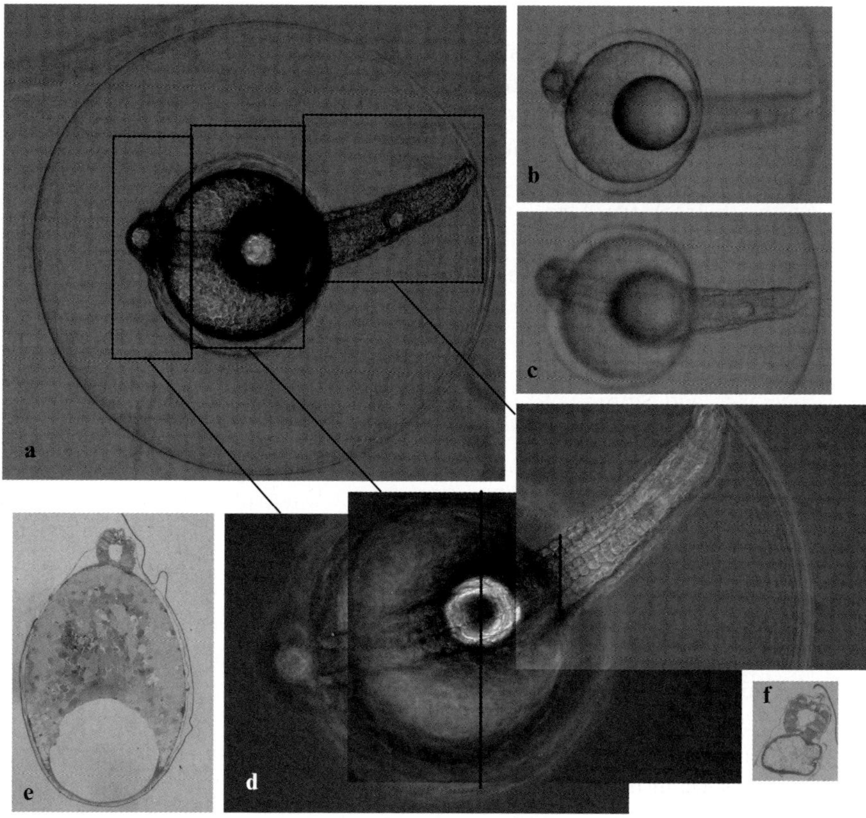

Fig. 13.9 Stretched embryo at 32 h.p.f. (Leiden University) photographed with (a) phase contrast microscopy and (b) and (c) with bright field microscopy, (d) zoomed-in views showing the developing somites. Figures 13.9e and f show cross-sections at two locations of the embryo as indicated by the vertical lines in 13.9d (embedded in epon, toluidine blue staining; source: Palstra 2006a)

accumulation of high levels of lipophilic pollutants like polychlorinated biphenyls (PCBs). These compounds have serious adverse effects on oocyte maturation, fertility and embryo development in every species studied (Stouthart et al. 1998; Gutleb et al. 1999; Pocar et al. 2003). Indeed, also for the European eel we found a strong negative correlation between PCB levels in the maternal gonads and the survival period of fertilized eggs and embryonic development (Palstra et al. 2006). Eels are rather sensitive to dioxin – like compounds since this occurred already at very low levels of 1 ng TEQ kg^{-1} gonad (TEQ expresses the total dioxin-like toxic potency) which implies that reproduction of eels from most European locations is seriously impaired. This knowledge not only introduces a significant contributor to the collapse of eel populations, it also presents a crucial criterion for selecting eels for artificial reproduction. Eels used for reproduction studies should be devoid from dioxin-like contaminants and thus purchased from very clean European habitats (i.e. Ireland) or from farms where they were fed with pollutant – free feeds.

To obtain 'normal' ovulation, embryonic development, and larval growth, many unexplored possibilities still exist. Improvement might be reached by: (a) Timing DHP injection with oocyte development, (b) Activating final maturation with pheromones (Huertas et al. 2006) and/or with natural spawning behaviour (Van Ginneken et al. 2005), (c) Applying high pressure profiles as European eels are believed to spawn at great depths.

Hiroi et al. (2003) found that high pressure treatment remarkably delayed the timing of hatching in Japanese eel up to 4 days. At high pressure, embryos may develop differently which could be a requirement for survival. Eggs of European eel show pronounced hydration which is typical for pelagically spawning marine teleosts (Wallace and Selman 1981). Van Ginneken et al. (2005) found that eggs rise in the water column with maximum speeds of 2.24 m h^{-1}. Main hatching times for European eels are between 47 and 60 h after fertilization (Bezdenezhnyk et al. 1983; Pedersen 2003, 2004). At a spawning depth of 300 m, the eggs/larvae will reach a depth of 100 m in about 4 days. It seems therefore that 4 days is a more natural period for embryonic development.

13.6 Female Responsiveness and Conditioning

When silver eels start their migration, they are still immature, being at a prepubertal stage. As published in the EELREP report (2005) and by Durif et al. (2006), the capacity of eels from different locations to mature has been evaluated by measuring short term and long term responses to hormonal treatment with carp pituitary extracts. In this study more than 250 eels from various locations were treated and analyzed. Yellow eels are in any case not responsive, indicating that silvering is a strict prerequisite for maturation. Among silver eels, the best performance was correlated with the highest silver index. In addition we observed that eels combining a high silver index, high condition factor and large body size had the best maturation response in terms of gonad yield (Table 13.1). The impact of the high condition factor reflects the requirement of high energy stores for gonadal growth. Final maturation could be induced in 50% of the tested female eels and in >90% of the

Table 13.1 Identification table for the reproduction capacity of female European silver eels. Fat content is given as percentage of wet weight. Silver stages of female eels are based on silver index scores (Durif et al. 2005).

Silver stage	I	II	III	IV	Va	Vb
Body length	–	–	–	–	<70 cm	>70 cm
Fat content	–	–	–	<13%	13–20%	>20%
Reproductive capacity	0	0	0	*	***	******

Reproduction capacity is indicated from very likely (******) to absent (0)

male eels. High hormone sensitivity and successful fertilization was found in large silver eels (>0.7 kg) and no differences were observed among locations. Recently we observed that older eels require a shorter hormonal treatment to mature (Palstra 2006), indicating age dependent sensitivity for maturation. More evidence supporting this hypothesis is provided by Palstra et al. (2007; Chapter 10) from swimming-induced maturation and by Durif et al. (2006) from positive correlations between age *vs.* vitellogenin level.

Temperature dependency of ovarian development was tested at low temperature (11°C, standardized 4 week in vivo sensitivity test). An even lower temperature (6–8°C) was also tested in some long term experiments to induce ovarian development, which showed virtually no response when compared to standard temperatures. This finding has important ecological implications, as the spawning migration is assumed to take place at great depth with temperatures well below 10°C. The results suggest that maturation of eel gonads does not progress during their oceanic journey until they reach the Sargasso Sea where temperatures are at the same depth some 10°C higher.

13.6.1 Conditioning

Especially when farmed eels are used as starting material for reproduction experiments, preconditioning of genitors is very important. It has been shown consistently that wild stocks of fish have higher gamete quality than their captive conspecifics (reviewed by Brooks et al. 1997). This gamete quality is believed to depend largely on a large number of environmental factors. Thus far at least three factors have been tested and applied for eels: diet, salinity and swimming.

Diet is believed to be one of the key determinants of egg quality (reviewed by Brooks et al. 1997), as many constituents are also transported into the egg yolk. Improving egg quality by adjusting the diet in eels is mainly focussed on altering the lipid composition e.g. n-3 fatty acids, based on results from seabass (Carrillo et al. 1995) and gilthead sea bream (Harel et al. 1994). Tomkiewicz and colleagues select the largest individuals at the farm after which these eels are subjected to an altered diet for 3 months in such a way that the lipid composition of most tissues resembles that of wild silver eels (personal communication). After these 3 months eels are transferred to salt water, after which feeding ceases.

Kagawa and colleagues investigated the effects of rearing in seawater on induced maturation in cultivated female Japanese eels (Kagawa 2003). After 3 months in seawater, the GSI had increased from less than 1 to 1.5, at the same time the oocyte diameter had increased from 160 to 200 μm (Kagawa et al. 1998). Unlike European eels, seawater exposure of Japanese eels induces vitellogenesis. This results in a gonad stage similar to that of wild migrating silver eels. The number of SPE-injections required for final maturation is significantly less in seawater vs freshwater Japanese eels (Kagawa et al. 1998).

A third factor that induces the next stage of female maturation is swimming (Chapter 10). Swimming stimulates fat deposition in the oocytes, without any deposition of vitellogenin. This process looks rather similar to that in other fish species. As discussed, oocytes of non-exercised European silver eels are probably still too premature for hormonal stimulation. During artificial induction by hormonal injections, hepatic vitellogenesis is immediately induced (Palstra et al. unpublished results). Fat and vitellogenin (VTG) incorporation occur simultaneously in artificially matured Japanese eel (Adachi et al. 2003) as well as in the European eel (Palstra et al., unpublished results), which suggests an unnatural situation. By stimulating incorporation of fats in the oocytes (Palstra et al. 2007) and inhibiting vitellogenesis (Chapter 10), swimming may result in a natural sequence of events and in a higher egg quality.

13.7 A New Technique: Induction of Maturation by Cell Implants

The procedure of weekly hormonal CPE injections has some important disadvantages with major potential implications for the poor quality of the maturation process and gametes:

1. Weekly handling causing stress
2. Weekly injections causing weekly peaking, unnatural hormone levels
3. Carp pituitaries may contain besides LH other hormones (e.g. GH) that interfere with maturation

One way to circumvent these problems is by application of a slow delivery system for large glycoproteins like the gonadotropins, for instance by means of the LU-patented hormone producing cells (Patent by Spaink et al. 2005). These can be injected only once, so weekly handling stress can be avoided. The cell implants allow a slow, more natural release of the required hormones and can be designed for the release of different hormones with different release rates. The method comprises the use of genetically manipulated embryonic zebrafish cell lines that are implanted subcutaneously into female silver eels (Schnabel et al. 2007). It was shown that even over a period of 4 weeks the implanted cells remained viable and metabolically active. Therefore the cell lines can be used for the delivery of a chosen transgenically produced secreted protein in animal tissues. Cells that stably produced the α- and β-subunits of FSH were injected in female silver eels and hormonal stimulation was shown by increased eye size, pectoral fin length (external maturation parameters) and expression levels of vitellogenin (Vtg 1 and 2), using quantitative Q-rtPCR (A. palstra et al., 2005, unpublished data). This opens the way to use cell implants for slow release production of other gonadotropins such as LH or combination of various hormones, and may enable successful breeding in captivity of eel in the near future.

13.8 Conclusions

The most successful protocol for artificial reproduction of eel has been published by Ohta, later protocols appear variations thereof. Spermatogenesis is usually induced by hCG, one injection can even be enough for full maturation and continued spermiation at which stage the GSI increases to about GSI = 10. The different stages of testis development for European eel are described in this paper, and can be used to identify the level of maturation of different stages in silvering. Induction of oogenesis is more complex and requires 12–25 weekly injections for the European eel, which is significantly more than for the Japanese eel, which matures much faster. This difference in maturation period might be related to the length of migration to the spawning site of each species, as the distance for European eel is ca 6,000-km while that for Japanese eel is ca 4,000-km. Thus it may be advantageous for the European eels to have maturation stronger suppressed than for Japanese eels.

The different end stages of oocyte development are well described in this paper, which are very useful to determine the moment for induction of ovulation with DHP. The correct timing of this step appears important for the success of fertilization. Female maturation is generally induced by pituitary extracts, which contain a mixture of hormones. With this treatment the oocytes mature, however, it appears that lipid and vitellogenin deposition occur simultaneously. This seems to be an artefact, as simultaneous deposition does not happen in other fish species. Furthermore the weekly injections with pituitary extracts are reflected as bands in the oocyte, which is another artifact. We assume therefore that the hormone mixture in the pituitary extract in combination with the large variations in hormone levels result in major artefacts responsible for the low fertility and survival rates.

We observed lipid deposition in the oocytes when silver eels were forced to swim for a long period, which can be considered as a simulation of spawning migration. During this forced activity the eels did not produce vitellogenin but did deposit lipids in the oocytes. It looks, therefore, that also in eels the two processes are separated in nature. We assume therefore that at least two steps are crucial in reaching successful reproduction: (1) separate deposition of lipids and vitellogenin during oocyte maturation, (2) stimulate maturation with low and constant hormone levels. For the first step long term swimming might be a good way to activate lipid deposition, for the second step the solution must be found in slow hormone release systems such as hormone producing cell implants.

Acknowledgements The authors wish to thank E. Cohen, M. Fekkes, E. Clavero, W. Spoor, G. Romano, J. van Rijssel, V. van Ginneken, P. Niemantsverdriet, F. Witte, P. Jouk and P. McClean for their contributions. Research of A. Palstra was subsidised by the EU contract EELREP no. Q5RS-2001-01836.

References

Abraham M, Higle V, Lison S, Tibika H (1984) The cellular envelope of oocytes in teleosts. Cell Tissue Res. 235: 403–410.

Adachi S, Ijiri S, Kazeto Y, Yamauchi K (2003) Oogenesis in the Japanese Eel, *Anguilla japonica*. In: Aida K, Tsukamoto K, Yamauchi K (Eds), Eel Biology, Springer, Heidelberg, pp. 502–518.

Bezdenezhnykh VA, Prokhorcik GA, Petrikov AM, Petukov VB, Plyuta MV (1983) Obtaining the larvae of European eel *Anguilla anguilla* L. (Pisces, Anguillidae), under experimental conditions. Dokl. Akad. Nauk SSSR 268 (5): 1264–1266 (In Russian).

Bezdenezhnykh VA, Prokhorcik GA (1984) Period of oocyte maturation and assessment of egg quality of eel, *Anguilla anguilla* (Anguillidae), under stimulation of maturation with gonadotropic hormones. Vopr. Ikhtiol. 5: 814–821 (In Russian).

Boëtius J, Boëtius L (1980) Experimental maturation of female silver eels, *Anguilla anguilla*. Estimates of fecundity and energy reserves for migration and spawning. Dana 1: 1–28.

Brooks S, Tyler CR, Sumpter JP (1997) Egg quality in fish: what makes a good egg? Rev. Fish Biol. Fisher. 7: 387–416.

Carrillo M, Zanuy S, Prat F, Cerda J, Mananos E, Bromage N, Ramos J. and Kah O (1995) Nutritional and photoperiodic effects on hormonal cycles and quality of spawning in sea bass (Dicentrarchus labrax L.). The Netherlands, J. Zool. 45: 204–209.

Chiba H, Iwatsuki K, Hayami K, Hara A, Yamauchi K (1994) Changes in serum steroid hormones and vitellogenin levels in cultured female European eels *Anguilla anguilla* during artificially induced ovarian development. J. World Aquaculture Soc. 25: 553–560.

Colombo G, Grandi G, Rossi R (1984) Gonad differentiation and body growth in *Anguilla anguilla* L. J. Fish Biol. 24: 215–228.

Cottril RA, McKinley RS, van der Kraak G, Dutil J-D, Reid KB, McGrath KJ (2001) Plasma non-esterified fatty acid profiles and 17β-oestradiol levels of juvenile immature and maturing adult American eels in the St Lawrence River. J. Fish Biol. 50: 364–379.

Couillard CM, Hodson PV, Castonquay M (1997) Correlation between pathological changes and chemical contamination in American eels, *Anguilla rostrata*, from the St Lawrence River. Can. J. Fisher. Aquat. Sci. 54: 1916–1927.

Dufour S, Burzawa-Gerard E, Le Belle N, Sbaihi M, Vidal B (2003) Reproductive endocrinology of the European eel, *Anguilla anguilla*. In: Aida K, Tsukamoto K, Yamauchi K (Eds), Eel Biology, Springer, Heidelberg, pp. 373–386.

Durif C, Dufour S, Elie P (2005) The silvering process of the eel: a new classification from the yellow resident stage to the silver migrating stage. J. Fish Biol. 66: 1–19.

Durif CMF, Dufour S, Elie P (2006) Impact of silvering stage, age, body size and condition on the reproductive potential of the European eel. Mar. Ecol. Prog. Ser. 327: 171–181.

EELREP (2005) Estimation of the reproduction capacity of European eel. Final report. http://www.fishbiology.net/eelrepsum.html

Fontaine M (1936) Sur la maturation complète des organes génitaux de l'anguille male et l'émission spontanée. C.r.hebd.Séanc. Acad. Sci. Paris 202: 132–134.

Fontaine M, Bertrand E, Lopez E, Callamand O (1964) Sur la maturation des organes genitaux de l'anguille femelle (*Anguilla anguilla* L.) et l'émission spontanée des oeufs en aquarium. Comp. Rend. Acad. Sci. 259: 2907–2910.

Fricke H, Tsukamoto K (1998) Seamounts and the mystery of eel spawning. Naturwissenschaften 85: 290–291.

Goetz FW (1983) Hormonal control of oocyte final maturation and ovulation in fishes. In: Hoar WS, Randall DJ, Donaldson EM (Eds), Fish Physiology, Volume IX, Part B, Behavior and Fertility Control, Academic, New York, pp. 117–170.

Goetz FW, Theofan G (1979) In vitro stimulation of germinal vescicle breakdown and ovulation of yellow perch (*Perca flavescens*) oocytes. Effects of 17α-hydroxy-20β-dihydroxyprogesterone and prostaglandins. Gen. Comp. Endocrinol. 37: 273–285.

Gutleb AC, Appelman J, Bronkhorst MC, van den Berg JHJ, Spenkelink A, Brouwer A, Murk AJ (1999) Delayed effects of pre- and early-life time exposure to polychlorinated biphenyls on tadpoles of two amphibian species (*Xenopus laevis* and *Rana temporaria*). Environ. Toxicol. Pharmacol. 8: 1–14.

Harel M, Tandler A, kissil GW, Gpplebaum SW (1994). The kinetics of nutrient incorporation into body tissues of gilthead seabream (Sparus Curata) females and the Subsequent effects on egg composition and egg quality. British Journal nutrition 72: 45–58.

Hiroi J, Yasumasu S, Kawazu K, Kaneko T (2003) Hatching enzymes in the Japanese eel. In: Aida K, Tsukamoto K, Yamauchi K (Eds), Eel Biology, Heidelberg, Springer pp. 445–456.

Huertas M, Scott AP, Hubbard PC, Canário AVM, Cerdà J (2006) Sexually mature European eels (Anguilla anguilla L.) stimulate gonadal development of neighbouring males: possible involvement of chemical communication. Gen. Comp. Endocrinol. 147: 304–313.

Kagawa H, Tanaka H, Ohta H, Okuzawa K, Iinuma N (1997) Induced ovulation of 17,20β-dihydroxy-4-pregnen-3-one in the artificially matured Japanese eel, with special reference to ovulation time. Fish Sci. 63, 356–367.

Kagawa H, Iinuma N, Tanaka H, Ohta H, Okuzawa K (1998) Effects of rearing period in seawater on induced maturation in female Japanese eel Anguilla japonica. Fish Sci. 64: 77–82.

Kagawa H (2003) Artificial induction of oocyte maturation and ovulation. In: Aida K, Tsukamoto K, Yamauchi K (Eds), Eel Biology, Springer, Heidelberg, pp. 401–414.

Kagawa H, Tanaka H, Ohta H, Unuma T, Nomura K (2005) The first success of glass eel production in the world: basic biology on fish reproduction advances new applied technology in aquaculture. Fish Physiol. Biochem. 31: 193–199.

Kahn IA, Lopez E, Leloup-Hatey J (1987) Induction of spermatogenesis and spermiation by a single injection of human chorionic gonadotropin in intact and hypophysectomised immature European eel (*Anguilla anguilla* L.). Gen. Comp. Endocrinol. 68: 91–103.

Kurokawa T, Pedersen B (2003) The digestive system of eel larvae. In: Aida K, Tsukamoto K, Yamauchi K (Eds), Eel Biology, Springer, Heidelberg, pp. 435–444.

Lokman PM, Young G (2000) Induced spawning and early ontogeny of New Zealand freshwater eels (*Anguilla dieffenbachi* and *A. australis*). New Zeal J. Mar. Freshw. Res. 34: 135–145.

Miura T, Miura C, Yamauchi K (2003) Spermatogenesis in the Japanese eel. In: Aida K, Tsukamoto K, Yamauchi K (Eds), Eel Biology, Springer, Heidelberg, pp. 319–330.

Müller T, Binder T, Váradi B, Horn P, Bercsényi (2002) Artificial induction of sexual maturation in the European eel males (*Anguilla anguilla* L.) (Preliminary results). Acta Agraria Kaposváriensis 6 (1): 53–57.

Ohta H, Izawa T (1996) Diluent for cool storage of the Japanese eel (*Anguilla japonica*) spermatozoa. Aquaculture 142: 107–118.

Ohta H, Kagawa H, Tanaka H, Okuzawa K, Hirose K (1996) Changes in fertilization and hatching rates with time after ovulation induced by 17,20β-dihydroxy-4-pregnen-3-one in the Japanese eel, *Anguilla japonica*. Aquaculture 139: 291–301.

Ohta H, Kagawa H, Tanaka H, Okuzawa K, Linuma N, Hirose K (1997) Artificial induction of maturation and fertilization in the japanese eel, *Anguilla japonica*. Fish Physiol. Biochem. 17: 163–169.

Ohta H, Unuma T (2003) Induction of sperm maturation. In: Aida K, Tsukamoto K, Yamauchi K (Eds), Eel Biology, Springer, Heidelberg pp. 415–424.

Palstra AP, Van Ginneken V, Van den Thillart G (2004) Artificial reproduction of the European silver eel (*Anguilla anguilla* L.). EAS Special Publication No. 34.

Palstra AP, Cohen EGH, Niemantsverdriet PRW, Van Ginneken VJT, Van den Thillart GEEJM (2005) Artificial maturation and reproduction of European silver eel: Development of oocytes during final maturation. Aquaculture 249 (1–4): 533–547

Palstra AP (2006) Energetic requirements and environmental constraints of reproductive migration and maturation of European silver eel (*Anguilla anguilla* L.). Ph.D. dissertation, University of Leiden, Leiden, The Netherlands, 183 pp.

Palstra AP, Van Ginneken VJT, Murk AJ, Van den Thillart GEEJM (2006) Are dioxin-like contaminants responsible for the eel (*Anguilla anguilla*) drama? Naturwissenschaften 93: 145–148.

Palstra A, Curiel D, Fekkes M, de Bakker M, Székely C, Van Ginneken V, Van den Thillart G (2007) Swimming stimulates oocyte development in European eel (*Anguilla anguilla* L.). Aquaculture 270: 321–332

Pedersen BH (2003) Induced sexual maturation of the European eel *Anguilla anguilla* and fertilisation of the eggs. Aquaculture 224: 323–338.

Pedersen BH, Ueberschär B, Kurokawa T (2003) Digestive response and rates of growth in pre-leptocephalus larvae of the Japanese eel *Anguilla japonica* reared on artificial diets. Aquaculture 215: 321–338.

Pedersen BH (2004) Fertilisation of eggs, rate of embryonic development and hatching following induced maturation of the European eel *Anguilla anguilla*. Aquaculture 237: 461–473.

Perez L, Asturiano JF, Tomás A, Zegrari S, Barrera R, Espinos FJ, Navarro JC, Jover M (2000) Induction of maturation and spermiation in the male European eel: assessment of sperm quality throughout treatment. J. Fish Biol. 57: 1448–1504.

Pocar P, Brevini TA, Fischer B, Gandolfi F (2003) The impact of endocrine disruptors on oocyte competence. Reproduction 125: 313–325.

Prokhorcik GA (1986) Postembryonic development of European eel, Anguilla Anguilla, under experimental conditions. J. Ichthyol. 26: 121–127.

Schnabel D, Palstra A, Van den Thillart G, Spaink HP (2007) Induced sexual maturation in eel with embryonic zebrafish cell lines that constitutive produce LH and FSH. Dev. Biol. 306 (1): 384.

Spaink HP, Van den Thillart GEEJM, Schnabel Peraza D (2005) Means and methods for improving the development and maturation of eggs and/or sperm in fish using hormones produced by transplanted cells. Patent WO2006080841.

Stouthart XJHX, Huijbregts MAJ, Balm PHM, Lock RAC, Wendelaar Bonga SE (1998) Endocrine stress response and abnormal development in carp (*Cyprinus carpio*) larvae after exposure of the embryos to PCB 126. Fish Physiol. Biochem. 18:321–329.

Suetake H, Okubo K, Yoshiura Y, Aida K (2003) GTH and GnRH molecules and their expression in the Japanese eel. In: Aida K, Tsukamoto K, Yamauchi K (Eds), Eel Biology, Springer, Heidelberg pp. 351–372

Tanaka H (2003) Techniques for larval rearing. In: Aida K, Tsukamoto K, Yamauchi K (Eds), Eel Biology, Springer, Heidelberg pp. 427–434

Todd PR (1981) Morphometric changes, gonad histology, and fecundity estimates in migrating New Zealand freshwater eels (Anguilla spp.). New Zeal. J. Mar. Freshw. Res. 15: 155–170.

Tsukamoto K (1992) Discovery of the spawning area of the Japanese eel. Nature 356: 789.

Tyler CR and Sumpter JP (1996) Oocyte growth and development in teleosts. Rev. Fish Biol. Fisher. 6: 287–318.

Van Ginneken V, Vianen G, Muusze B, Palstra A, Verschoor L, Lugten O, Onderwater M, Van Schie S, Niemantsverdriet P, Van Heeswijk R, Eding E, Van den Thillart G (2005a) Gonad development and spawning behaviour of artificially-matured European eel (*Anguilla anguilla* L.). Anim. Biol. 55 (3): 203–218.

Wallace RA, Selman K (1978) Oogenesis in fundulus heteroclitus. I. Preliminary observations on oocyte maturation in vivo an in vitro. Dev. Biol.. 62: 354–369.

Wallace RA, Selman K (1981) Cellular dynamic aspects of oocyte growth in teleosts. Am. Zool. 21: 325–343.

Yamamoto K, Yamauchi K (1974) Sexual maturation of Japanese eel and production of eel larvae in the aquarium. Nature 251: 381.

Yamauchi K, Nakamura M, Takahashi H, Takano K (1976) Cultivation of larvae of Japanese eel. Nature 263: 412

Chapter 14
Testis Development, Sperm Quality Evaluation and Cryopreservation in the European Eel

Luz Pérez, David Peñaranda, Víctor Gallego, and Juan Asturiano

14.1 Introduction

Since the European eel *Anguilla anguilla* L. cannot be bred in captivity, eel farms base their annual production on the capture of glass eels from river mouths in autumn and winter. This species has suffered a dramatic reduction in its populations, due to intensive capture of both glass eels and adults, resulting in increased cost for the annual renewal of glass eel stocks. Moreover, populations have been decreasing owing to several other factors (Feunteun 2002), such as massive exportation to other countries, the deterioration of their natural habitats and the importation of allochthonous parasites (*Anguillicola crassus* Kuwahara, Niimi and Itagaki 1974) from the Asian species (Koops and Hartmann 1989; Kennedy and Fitch 1990). Therefore, the development of methods for the reproduction of this species is necessary not only from an economical point of view, to meet the demands of fish farms, but also from an ecological point of view, to reduce the pressure on natural populations. In view of this, our group has centred its research since 1997 on trying to develop several techniques to help in the production of this species in captivity. Firstly, some experiments were carried out to develop maturation-inducing hormonal methods for males. The second step was the use of different techniques to evaluate the quality of the gametes, looking for fast and accurate results. At the same time, the development of hormonal induction protocols and the induction of spawning in the females made necessary the synchronization of gamete production. With the intention of solving this problem, we tackled sperm cryopreservation. Study of the physico-chemical characteristics of seminal plasma in good quality sperm samples was the basis for the design of cryopreservation media. Later, different factors such as the ionic composition, pH, cryoprotectants, or the presence of protective proteins, as well as different freezing-thawing methods, have been considered to try to improve spermatozoa survival post-cryopreservation.

L. Pérez, D. Peñaranda, V. Gallego, and J. Asturiano
Grupo de Acuicultura y Biodiversidad, Instituto de Ciencia y Tecnología Animal, Universidad Politécnica de Valencia, Camino de Vera, s/n 46022 Valencia, Spain

G. van den Thillart et al. (eds.), *Spawning Migration of the European Eel*,

14.2 Testis Development

The European eel is an undifferentiated gonochoristic species, characterized by a juvenile intersexual stage with a bipotential gonad, the so-called Syrski organ (Colombo et al. 1984; Colombo and Grandi 1996). The undifferentiated species are defined as those that develop first an ovary-like gonad and later a portion of the population stops this process and starts to develop a testis (Piferrer 2001; Devlin and Nagahama 2002). The gonadal development in the European eel has been extensively described by Colombo et al. (1984) and Colombo and Grandi (1996), and is mainly related to the size of the fish. Eels shorter than 20 cm have undifferentiated gonads, but some oogonia and early oocytes are present in cysts. The Syrski organ appears in eels with total lengths of between 20 and 30 cm. This organ appears as a thin lamella and contain primordial germ cells, early oocytes isolated or in nests and early spermatogonia (Grandi and Colombo 1997). While early ovaries can appear in eels measuring 22 cm, early testes can be observed only in eels longer than 30 cm (Colombo and Grandi 1996). These authors proposed a model for the sexual differentiation in the European eel, where the testes are derived later from the Syrski organ, while the ovaries derived directly from undifferentiated gonads and in some cases from Syrski organs (Fig. 14.1).

Like most teleosts, European eels have a pair of testes with a lobular structure, where the spermatogonia are found all along the lobule, not only in the blind end of the tubule as occurs in the tubular testis. The external shape is different from other fish species, with white rounded lobules that join to each other as well to the dorsal mesentery of the fish.

Fish spermatogenesis goes through three major phases (Schulz and Miura 2002): mitotic proliferation, from spermatogonial stem cells to differentiated

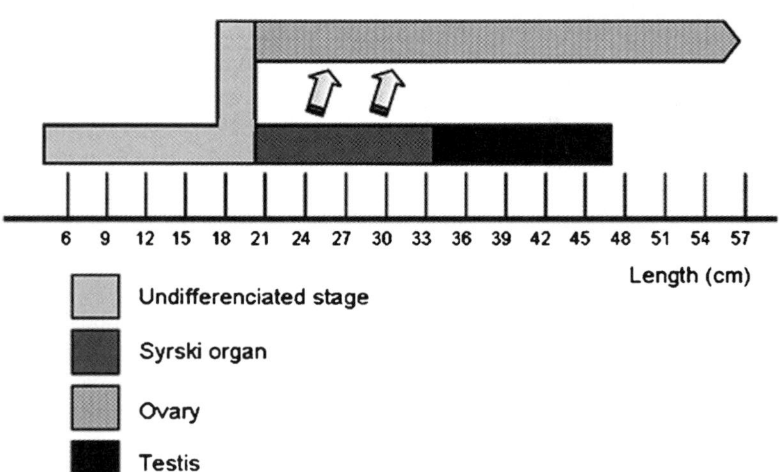

Fig. 14.1 Diagram of the development and differentiation of eel gonads in relation to body size (Modified from Colombo and Grandi 1996. With permission of Blackwell Publishing)

spermatogonia; meiosis, from spermatocytes to spermatids, and spermiogenesis, when the spermatids develop to spermatozoa. The eels used in the spermiation experiments were males grown in a fish farm (Valenciana de Acuicultura, Puzol, Valencia, Spain) from the glass eel stage to a size between 33–45 cm total length and 100–180 g weight. The age at this time is between 16 and 21 months (R. Barrera 2007, personal communication). During the first weeks of treatment (weekly injections of human chorionic gonadotropin [hCG], 1.5 IU g^{-1}) large changes can be observed in the testis structure and cellular types. Before the hCG treatment the males show spermatogonia arranged in cysts or lobules delimited by connective tissue (Fig. 14.2A), without spermatocytes or other types of germ cells. The eel

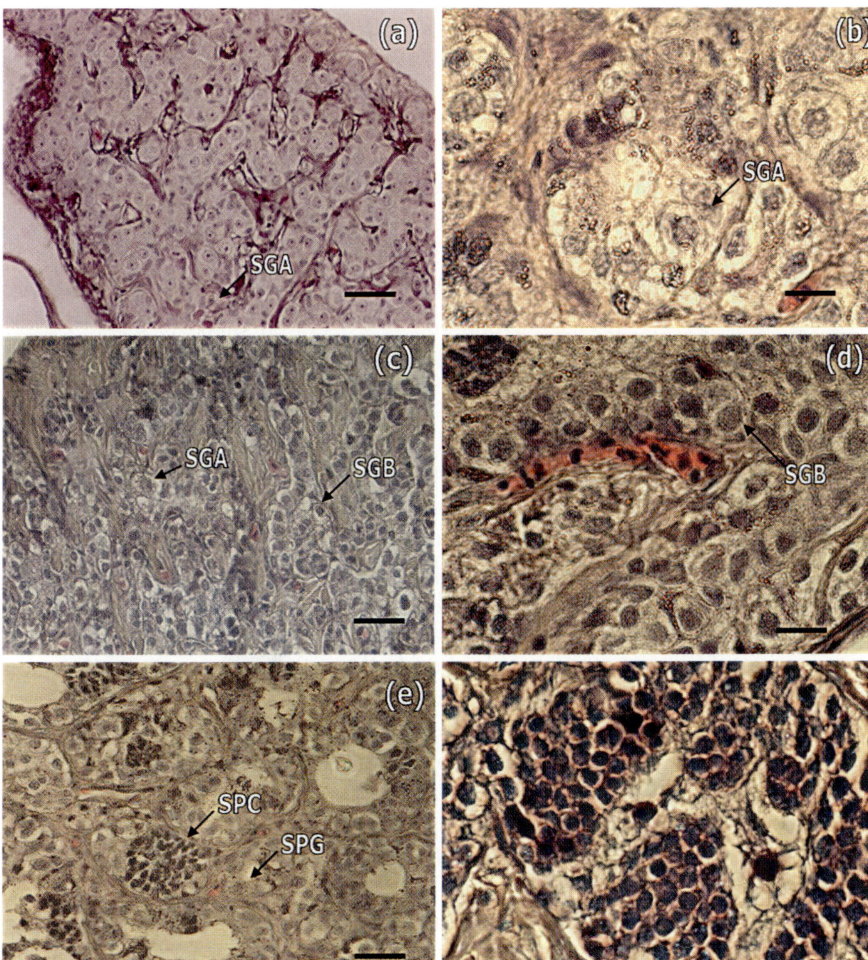

Fig. 14.2 **(a)** European eel testis before hormonal treatment. **(b–e)** Testes after one hCG injection. **(f)** Testes after 3 hCG injections. SGA: A spermatogonia; SGB: B spermatogonia; SPG: spermatogonia; SPC: spermatocyte. Scale bar: a, c, e: 25 μm; b, d, e: 10 μm

spermatogonia show a clear homogeneous nucleus with one well-defined nucleolus (Fig. 14.2A, B). There are two types of spermatogonia: A and B. The cytoplasm is also clearly coloured, wider in A than in B spermatogonia. The size of A spermatogonia is around 10 μm diameter (unpublished data), and diminishes in successive generations of spermatogonia. In the Japanese eel, *Anguilla japonica*, ten generations or mitotic cycles have been reported (Miura et al. 1991). B spermatogonia are defined as the last generations that are in rapid mitosis and are more differentiated than A spermatogonia, which divide slowly, and in some species can develop as oocytes in response to estrogens (Schulz and Miura 2002). B spermatogonia show a darker nucleus, a higher nucleus: cytoplasm ratio, and a smaller size than A spermatogonia (Fig. 14.2B, C). At this time, previous to the injections, the testes are very small, the GSI being usually lower than 0.1%. Other authors have also reported this aspect in the testis of European or Japanese eels without hormonal treatment (Colombo et al. 1987; Miura et al. 1991; Müller et al. 2005). We observed that 1 week after the first hCG injection most of males showed only small changes in the testes.

Spermatocytes have been observed in the testes of some males from the first week of treatment, 7 days after the first hCG injection (Fig. 14.2E, F). These germ cells are smaller in size than the spermatogonia (<5 μm), their nucleus is dense and the cytoplasm is quite not visible (Fig. 14.2F). After two hGC injections, several generations of spermatocytes can be observed in most of males and they are organized in cysts. At this moment, the lumen of the lobules is developing, and it appears as an acellular space filled with some kind of liquid material, or appears empty after the histological procedures (Fig. 14.3A, B). In some cases spermatozoa can be observed after two injections.

By the third week of treatment around 50% of the males showed spermatids in their testes, with a small quantity of spermatozoa in the small lumen (Fig. 14.3C). The rest of the males showed spermatocytes as the dominant germ cell. The spermatids, which originated from the second meiotic division of the spermatocytes, are very small cells, with a rounded shape (Fig. 14.3D, E). They are situated in the inner part of the cyst, near the lobule lumen and became spermatozoa after the maturation process. The eel spermatozoa have a typical head that looks like a half-moon (see following sections). In the fourth week of treatment, all the males are in spermiogenesis with abundant spermatids, or starting spermiation. During this week, small quantities of sperm were collected from males, but it is possible that some mixture between mature and immature spermatozoa, and also spermatids, has been extracted in the stripping. At this time the testes show cysts where all the germ cells can be observed (spermatogonia, spermatocytes and spermatids) and the lobular lumen, with a width between 25–140 μm, appears to be full of spermatozoa (Fig. 14.3E).

In the subsequent weeks the lobular lumen increases in width due to the accumulation of sperm. During the weeks of maximum sperm production (from the 8th to 13th) the testis is mainly composed of spermatozoa, while the other germ cells presents in the cyst wall have decreased in percentage (Fig. 14.3F). If the hormonal treatment continues after the 13th week, the sperm density decreases

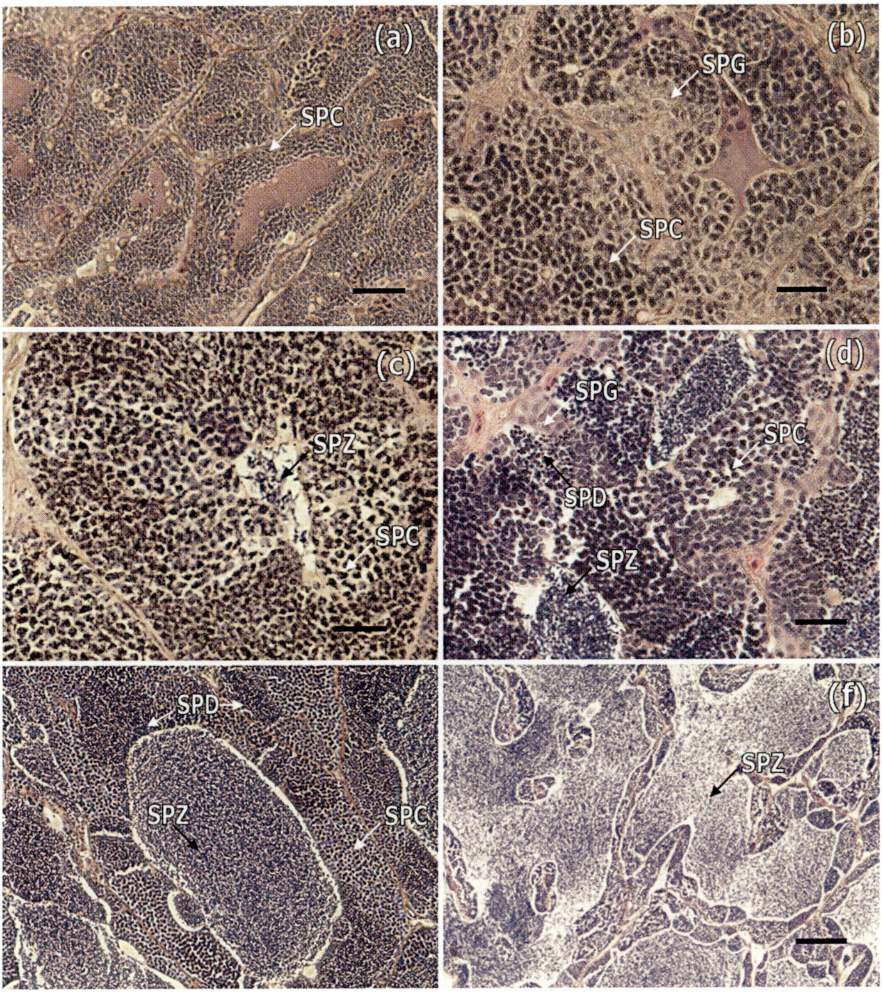

Fig. 14.3 **(a, b)** Testis after two hCG injections; acellular lumen surrounded by spermatocytes; **(c)** Testis after three injections; lobule showing spermatozoa; **(d)** Testis after four injections; **(e)** Testis after five injections; lobule filled with spermatozoa; **(f)** Testis in maximum spermiation with the lumen wider. SPG: spermatogonia; SPD: spermatid; SPC: spermatocyte; SPZ: spermatozoa. Scale bar: a, e: 50 μm; b, c, d: 25 μm; (f) 100 μm

(Pérez et al. 2000) but the sperm production (in volume) continues, being as high as in the previous weeks. From the 12th week onwards, some testes appear empty of sperm. After 15–17 hCG injections most of the testes are exhausted. The seminal lobules delimited by connective tissue appear depleted of spermatozoa and low quantities of germ cells are present in the walls of the cyst.

14.3 Development of New Methods to Evaluate Eel Sperm Quality

Fish sperm quality has traditionally been estimated by subjective evaluation of sperm motility and concentration, but during recent years an intense advance has occurred in the techniques for the objective evaluation of sperm quality. This has been achieved by the study of motility parameters with CASA systems (*computer assisted sperm analysis*), spermatozoa morphometry analysis using ASMA (*automated sperm morphometry analysis*) or by using fluorescent staining methods to evaluate membrane functionality, determining on this way the percentage of viable spermatozoa. Sperm quality is a measure of the ability of sperm to fertilise an egg successfully, but this capacity may not be reliable, as also egg quality may be variable and affect fertilisation success. The percentage of motile spermatozoa has been the most common test used to evaluate fish sperm quality (Kime et al. 2001; He and Woods 2004; Rurangwa et al. 2004). Sperm motility depends on the ATP content and the ability of mitochondria to sustain the high energy demand during motility (Christen et al. 1987). Protocols have been developed using fluorescent staining to provide rapid assessment of the mitochondrial functionality and plasma membrane integrity of fish sperm (Ogier De Baulny et al. 1997, 1999; Segovia et al. 2000; He and Woods 2004; Rurangwa et al. 2004). To assess the non-viable cells, membrane-impermeable nucleic acid stains can be used, which positively identify dead spermatozoa by penetrating into cells with damaged membranes. An intact plasma membrane will prevent these products from entering the spermatozoa and staining the nucleus. Phenanthridines, such as propidium iodide (PI; Garner et al. 1986, 1994), SYBR14 (Garner et al. 1994; Segovia et al. 2000) and bisbenzimidazole Hoechst 33258 (De Leeuw et al. 1991) have commonly been used. Mitochondrial function can be assessed using rhodamine 123 (Rh123; Segovia et al. 2000) or 5,5′, 6,6′-tetrachloro-1, 1′, 3,3′-tetraethylbenzimidazolyl-carbocyanine iodide (JC-1), allowing a distinction between spermatozoa with poorly and highly functional mitochondria (Graham 2001).

Until recently, the morphology of fish and mammalian spermatozoa has been examined using manual techniques, but these are subjective, time-consuming and expensive (Baker and Clarke 1987; Davis et al. 1995; Van Look and Kime 2003). Several reports have described the spermatozoa ultrastructure of European and Japanese eels and examined its morphology based on transmission electron microscopy (TEM; Ginsburg and Billard 1972; Çolak and Yamamoto 1974; Gibbons et al. 1983; Gwo et al. 1992) or scanning electron microscopy (SEM; Gibbons et al. 1983; Gwo et al. 1992; Okamura et al. 2000). New image analysis systems which are easier to use and enable evaluation of a great number of spermatozoa have been developed to measure several morphology parameters. Marco-Jiménez et al. (2006a) performed the first applications of ASMA in the European eel to examine the head spermatozoa morphometry, showing that it is possible to use this methodology to obtain similar results as from TEM and SEM.

14.3.1 Evaluation of Parameters of Sperm Motility by Computer Assisted Sperm Analysis (CASA)

Some specific studies have been carried out on the characteristics of the movement of the European eel spermatozoa. Woolley (1997, 1998) described different aspects of the kinematics of movement, the flagellum structure and its vibratile and rotatory bending movement. However, the first application of CASA systems on European eel sperm was reported by Asturiano et al. (2005) in an attempt to improving hormonal induction treatments. This analysis consists in the determination of the exact percentage of motile spermatozoa as well as some motility parameters: curvilinear velocity (VCL, in $\mu m\ s^{-1}$), straight line velocity (VSL, in $\mu m\ s^{-1}$), angular path velocity (VAP, in $\mu m\ s^{-1}$), and beating cross frequency (BCF, in Hz). Rurangwa et al. (2004) reviewed different studies carried out using sperm-tracking systems in African catfish, carp, goldfish, roach, Eurasian perch, trout and lake sturgeon, and concluded that the most useful parameters of velocity are the VCL (the actual velocity along the trajectory) and the VSL (the straight line distance between the start and end points of the track divided by the time of the track). In the case of European eel, data from fast and medium-velocity spermatozoa (VCL $>40\,\mu m\ s^{-1}$) have been used to compare motility parameters. Next, results were found in samples showing $49.5\% \pm 2.5\%$ motile spermatozoa: VSL: $26.1 \pm 3.3\,\mu m\ s^{-1}$, VCL: $125.5 \pm 15.6\,\mu m\ s^{-1}$, VAP: $44.6 \pm 3.7\,\mu m\ s^{-1}$, BCF: 17.0 ± 1.2 Hz (Asturiano et al. 2005).

14.3.2 Morphometry Characterisation of Spermatozoa with Assisted Sperm Morphology Analysis (ASMA) and Scanning Electron Microscopy

Previous studies reported that when spermatozoa morphology is analysed by visual methods, the intra- and inter-observer laboratory variations are usually very large (Soler et al. 2003). Spermatozoa ultrastructure and morphology of European and Japanese eels have been studied by transmission electron microscopy (TEM; Ginsburg and Billard 1972; Çolak and Yamamoto 1974; Gibbons et al. 1983; Gwo et al. 1992) or scanning electron microscopy (SEM; Gibbons et al. 1983; Okamura et al. 2000). Other techniques used to examine and analyse fish spermatozoa head are laser light-scattering spectroscopy and stroboscopic illumination (Van Look and Kime 2003). Results obtained with these techniques are subjective, time-consuming and highly variable. The search for methods of accurate, objective and repeatable assessment of sperm fertility still remains the aim of many studies. One of these computer-assisted applications is an automated system for spermatozoa head morphometry analysis (ASMA), developed and validated for mammals (Rijsselaere et al. 2004). The ASMA systems require standardisation of methods and variables, but under these conditions their repeatability and validity are much higher than any subjective morphological evaluation (Wang et al. 1991; Coetzee et al. 1998).

ASMA system has been used previously in fish, but merely to show the effect of mercuric chloride on goldfish spermatozoa morphology (Van Look and Kime 2003).

Abnormal spermatozoa head morphometry has been associated with reduced fertility in the bull, boar and stallion (Claassens et al. 1996; Van Look and Kime 2003). ASMA has increasingly been used in mammalian species, such as man (Davis et al. 1995), rat (Davis et al. 1994), rabbit (Gravance and Davis 1995), bull (Gravance et al. 1996), dog (Dahlbom et al. 1997), monkey (Gago et al. 1998) and alpaca (Buendía et al. 2002). ASMA measurements have shown toxic effects on human spermatozoa head (Davis et al. 1995) and goldfish sperm (Van Look and Kime 2003). This technique has also been used in the field of cryopreservation, in which cryoprotectants or frozen-thawed protocols are known to cause morphological damage to the spermatozoa (Billard 1983; Billard et al. 2000). Kruger et al. (1995) found that spermatozoa head morphometry, determined by ASMA, was predictive of *in vitro* fertilisation rates and its utility has also been reported in detection of fertile and subfertile stallions (Gravance et al. 1997) and rabbits (Marco-Jiménez, 2007 personal communication). ASMA has provided a series of objective parameters, which have facilitated the standardisation of morphological semen evaluation (Sancho et al. 1998). However, different problems have arisen, such as sample preparation, staining procedure and the settings of the spermatozoa morphology analyser, which must be optimised for each species (Davis and Gravance 1993; Davis et al. 1994; Ball and Mohammed 1995; Gravance and Davis 1995; Boersma et al. 1999, 2001). The computer-assisted morphometry analysis requires the standardisation of preparation, staining and sampling methods (Davis and Gravance 1993).

Until very recently, ASMA systems had never been used on eel species (Marco-Jiménez et al. 2006a, b; Asturiano et al. 2006, 2007; Garzón et al. 2008). In fact, Marco-Jiménez et al. (2006a) tried to characterise the European eel spermatozoa morphometrically, comparing the results obtained by computer-assisted spermatozoa analysis and by scanning electron microscopy. Cell morphology was analysed using ASMA software (Sperm Class Analyzer®, Morfo Version 1.1, Imagesp, Barcelona, Spain). Approximately 15,000 spermatozoa were considered, determining: head length (μm), head width (μm), head perimeter (μm) and head area (μm^2) to define the spermatozoa head morphology. This technique was validated by comparison with scanning electron microscopy. Later, the accuracy of the ASMA technique has been used to describe spermatozoa morphology changes during sperm maturation under hormonal induction (Asturiano et al. 2006), as well as evaluating the osmotic effects suffered by cells and cryoprotectant effects during the application of cryopreservation protocols (Marco-Jiménez et al. 2006b; Asturiano et al. 2007; Garzón et al. 2008). Marco-Jiménez et al. (2006a) used farmed male eels, hormonally treated as previously described by Pérez et al. (2000). The reference values of morphological characteristics of spermatozoa were obtained with scanning electron microscopy. Spermatozoa head length, width, perimeter and area were determined using a public domain ImageJ program (developed at U.S. National Institutes of Health and available at www.rsb.info.nih.gov/ij/). To examine the spermatozoa head morphometry from pictures obtained with scanning electron microscopy (SEM), one pool from ten males was used, to avoid individual differences. A total of 100 spermatozoa head measurements were taken (Fig. 14.4).

For ASMA analysis, fractions of collected sperm samples were diluted in glutaraldehyde in Dulbecco's phosphate buffered saline fixative solution, prepared with DPBS and glutaraldehyde solution (Pursel and Johnson 1974). The morphological parameters were analysed using ASMA software. To examine the spermatozoa head morphometry with ASMA system, 203 individual ejaculates, from 36 males, were analysed. Approximately 75 spermatozoa were analysed in each sample. A high number of spermatozoa were measured using ASMA system (Fig. 14.5), showing how new recently-developed image analysis systems are easy to use, allowing the evaluation of a great number of spermatozoa.

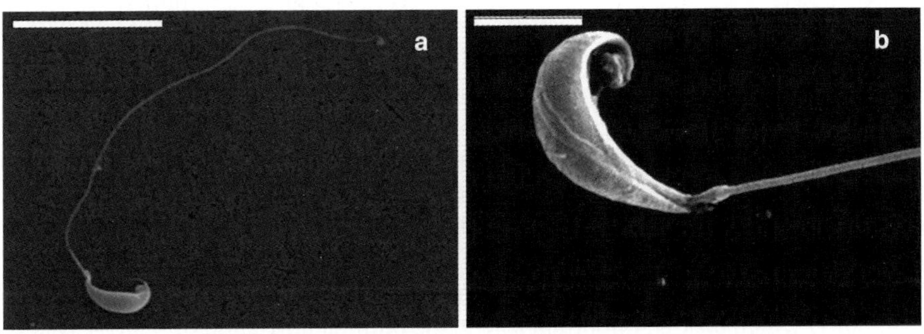

Fig. 14.4 Scanning electron microscopy of European eel spermatozoa separated from seminal plasma by centrifugation, fixed in 2% glutaraldehyde, post-fixed in 1% osmium tetroxide, dehydrated, critical point-dried in liquid CO_2 and coated with gold-palladium. Scale bars: 10 **(A)** and 2 μm **(B)** (Marco-Jiménez et al. 2006a. Copyright Elsevier 2006)

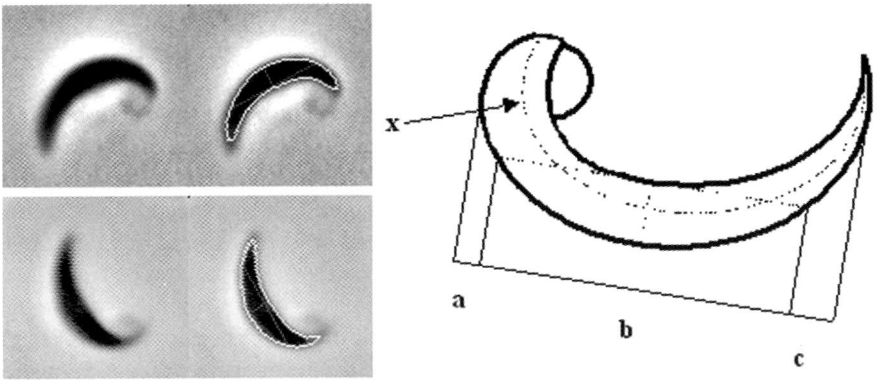

Fig. 14.5 Real pictures obtained by phase contrast optic (1000x) and schematic drawing showing how the Sperm Class Analyzer®, Morfo Version 1.1 (Imagesp, Barcelona, Spain) measures different morphology parameters of European eel spermatozoa. a = measured length by ASMA; x = real length of the cell (Marco-Jiménez et al. 2006a. Copyright Elsevier 2006)

Table 14.1 Spermatozoa morphological parameters: head length (μm), width (μm), perimeter (μm) and area (μm^2), considering spermatozoa measured by computer-assisted morphology analysis (ASMA) or scanning electron microscopy (SEM) and analysed by Image J. The mean morphometry measurements were compared by analysis of variance (General linear model, GLM). Significance level was set at $P < 0.05$. Different letters indicate significant differences between values obtained by ASMA or SEM (Marco-Jiménez et al. 2006a. Copyright Elsevier 2006)

	Length (μm)	Width (μm)	Perimeter (μm)	Area (μm^2)
ASMA	4.29 ± 0.03^b	1.15 ± 0.01^a	14.68 ± 0.13^a	5.36 ± 0.06^a
SEM	5.09 ± 0.04^a	1.12 ± 0.01^b	13.72 ± 0.19^b	5.05 ± 0.01^b
P value	0.000	0.031	0.002	0.000

The results obtained with SEM for head length and width are in good agreement with previous studies (Okamura et al. 2000; Müller et al. 2005), while eel spermatozoa head perimeter and area were measured by Marco-Jiménez et al. (2006a) for the first time. However, they showed that spermatozoa head morphology evaluated by ASMA resulted in a lower length and higher width, perimeter and area than measurements determined by SEM (Table 14.1).

Spermatozoa morphology assessment is influenced by numerous factors, such as semen preparation, magnification level of the objectives, number of evaluated spermatozoa, or the fixation and staining techniques (Gravance et al. 1997; Rijsselaere et al. 2004) which play an important role in the optimal utilisation of ASMA systems (Boersma et al. 2001). Boersma et al. (2001) suggested that stained spermatozoa heads produce better contrast in cell recognition and digitisation. However, other works indicate that the stains do not necessarily provide the appropriate grey-level contrast for accurate morphometric analysis (Verstegen et al. 2002). To stain, the samples are placed on a slide and dried in air before fixation, but it may be possible that air dried samples will shrink, flatten or collapse under these conditions (Bozzola and Russell 1991). On the other hand, fixation techniques containing aldehydes result in a low coefficient of variation and a high number of acceptable and correctly delineated heads (Sancho et al. 1998). Thus, Marco-Jiménez et al. (2006a) directly fixed the samples with glutaraldehyde solution in Dulbecco's phosphate buffered saline. The higher measurements reported by ASMA could be explained at least in part because the SEM samples were fixed and dehydrated through a graded ethanol series and finally critical point-dried in liquid CO_2: perhaps this preparation process provokes a reduction in spermatozoa head size. Moreover, Marco-Jiménez et al. (2006a) reported that Imagesp software cannot detect the curved and elongated spermatozoa head form, causing an error in the length measurements, although not in the other morphological parameters. The different spermatozoa head length measurement obtained between ASMA and SEM can be explained by this particular aspect of the head of eel spermatozoa, which are very different to those of livestock production animals. Fish sperm differs in many aspects from that of mammals (Kime et al. 2001), and the ASMA methodology used for livestock production animals is not directly applicable to fish (Van Look and Kime 2003).

The eel spermatozoa head is gently curved and elongated, with a hook-shaped upper end, which is directed inside in a crescent (Okamura et al. 2000) (Figs. 14.4 and 14.5). The shape of the head is asymmetric along the longitudinal axis. This elongation of the spermatozoa head in eels and other elopomorph fishes is a problem for the Imagesp ASMA system, causing an error in the length measurements (Fig. 14.5). Similar spermatozoa curved shapes have been reported in species such as rat (Guraya 1987), making its spermatozoa characterisation also difficult. New developments of Imagesp software are required, and different possibilities are present: (i) to obtain a skeleton of the figure by means of lines parallel to the width within the figure and drawing up to a line between the midpoints (ii) or by means of geodesic ratios, as after the digitalisation of the image the coordinates are known for each point, and it would be possible to obtain the greatest longitude between two points of the object on the inside passing through the midpoint of the straight line that defines the width (x in Fig. 14.5).

In conclusion, width, perimeter and area could be used as methods of spermatozoa morphology evaluation, whereas the length requires a new programming of the Imagesp software. This characterisation of the European eel spermatozoa morphology is one of the first applications of ASMA methodology in fish and the first one in eel species. It confirms this system as a useful tool, mainly in terms of time-saving and the reduced equipment required in comparison with electron microscopy techniques, with wide applications in future studies of fish spermatozoa membrane response under the effects of extenders, cryoprotectants, additives, etc.

14.3.3 ASMA Applications

Using sperm samples from the same experiments, Asturiano et al. (2006) used the ASMA measurements to demonstrate a significant growth of the spermatozoa during the first part of the spermiation period, specifically from the fifth to the eighth weeks of hCG treatment. Table 14.2 shows the measured parameters (head length, width, perimeter and area) from week 5 to 12 of treatment. Generally, the results show a significant spermatozoa growth from week 5 to 8, maintained until week 11 and decreasing thereafter.

However, the analysis of isolated descriptive parameters may be difficult to understand because there is a considerable variability in these parameters in each week that complicates the determination of the growth kinetic. The global size of the cells' head was calculated by applying principal component analysis (PCA), because this method establishes new components that make the interpretation of results easier, allowing a whole interpretation of the changes in the cell morphology. The Principal Component Analysis established two components (Fig. 14.6). The first can be considered as a general size component and the second one distinguishes wide and narrow cells. Our results showed an increased effect on global head size up to the eighth week of treatment, maintaining approximately these values until the 11th week, and decreasing later (component 1, Fig. 14.6A). On the other hand, the

Table 14.2 Least square means ± SEM for each of the measured parameters (head length, width, perimeter and area) from 5th to 12th weeks of hCG treatment (Asturiano et al. 2006. Copyright Elsevier 2006)

Weeks	n	Head length (μm)	Head width (μm)	Area (μm²)	Perimeter (μm)
5	471	3.99 ± 0.03[e]	1.07 ± 0.009[g]	4.90 ± 0.03[g]	13.63 ± 0.10[f]
6	1560	4.11 ± 0,01[d]	1.19 ± 0.004[b]	5.19 ± 0.02[f]	14.13 ± 0.05[d]
7	3007	4.11 ± 0.01[d]	1.21 ± 0.003[a]	5.14 ± 0.01[e]	13.94 ± 0.05[e]
8	3147	4.31 ± 0.01[b]	1.13 ± 0.003[e]	5.44 ± 0.01[b]	15.32 ± 0.04[a]
9	2357	4.28 ± 0.01[b]	1.13 ± 0.003[e]	5.38 ± 0.01[c]	15.06 ± 0.05[b]
10	3060	4.20 ± 0.01[c]	1.17 ± 0.004[c]	5.46 ± 0.01[b]	15.10 ± 0.05[b]
11	1375	4.38 ± 0.01[a]	1.10 ± 0.003[f]	5.51 ± 0.01[a]	15.09 ± 0.05[b]
12	1514	4.09 ± 0.01[d]	1.15 ± 0.004[d]	5.27 ± 0.02[d]	14.37 ± 0.06[c]

[a–g]Values in the same column with different superscripts are statistically different ($P < 0.05$).
n: Numbers of spermatozoa analyzed in each week.

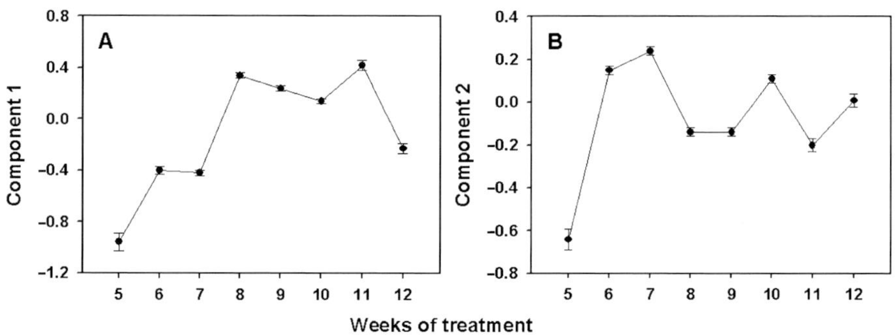

Fig. 14.6 Evolution of the different spermatozoa morphometry components, obtained by Principal Component Analysis, between 5th and 12th weeks of hCG treatment. **(A)** Spermatozoa size or component 1, **(B)** spermatozoa width or component 2. Results are shown as 95% HSD intervals over the means. One ANOVA was applied to the scores obtained after one Principal Component Analysis developed over the four morphometry parameters. Different letters indicate significant differences (Asturiano et al. 2006. Copyright Elsevier 2006)

cells' head from weeks 5 to 7 is thickened, decreasing somewhat from week 7 to 8, with the thickness staying more or less constant during the following treatment weeks. Fig. 14.6B shows a high increase in the cells width (component 2) from weeks 5 to 7 of hCG treatment. From the eighth week, changes in width were lower, but significant and without clear tendencies.

These variations in the cell's morphometry are the consequence of the hCG-induced maturation treatment. During the samplings from weeks 5 to 8, at least two cells types were extracted from testis tubules: maturing spermatids and spermatozoa, but their proportion changed during these first samplings. Spermatids are quite rounded cells that did not finish their maturation, and are especially abundant during weeks 5 and 7, while spermatozoa, the fully matured cell type, are longer, with higher perimeter and area, and form the majority of cells from week 8 onwards.

The fact of finding cells in different stages of development could be explained by the group-synchronous spawning character of this species, translated in sperm production in waves as has been previously described in other species, such as the European sea bass, *Dicentrarchus labrax*) (Asturiano et al. 2000, 2002b). The forced weekly semen collection may result in the extraction of spermatozoa from different batches or stages of development, which normally should be released in different waves of sperm production.

One of the first applications of ASMA in the European eel was the study of the effect of the cryopreservation process on spermatozoa head morphometry (Asturiano et al. 2007). They reported a significant decrease of head perimeter (7.15%) and area (12.18%) when sperm was diluted in the DMSO-freezing medium, and a further significant decrease in both parameters (12.56% and 17.90%, respectively) comparing fresh and thawed samples, indicating that cells do not recover their original size after the cryopreservation process (Table 14.3).

When the spermatozoa are exposed to the cryoprotectant, their size is reduced permanently. This change results from a water outflow from the spermatozoa to the high osmolality external medium. If the cryoprotectant is permeable, the spermatozoa initiates a progressive re-growth caused by the entry of cryoprotectant to the intracellular space, trying to reach an osmotic equilibrium. But the dehydration and re-growth depends on the cryoprotectant concentration, temperature, surface area for exchange and the permeability coefficient of the cell type.

The osmolality of the European eel seminal plasma is 320–330 mOsm kg^{-1} (Pérez et al. 2003). The basis solution used by Asturiano et al. (2007) was freezing medium isoionic with the seminal plasma, with an osmolality of 339 mOsm kg^{-1} (see Section 14.4.1). But when this media was supplemented with DMSO and L-α-phosphatidylcholine the osmolality increased to 2049 mOsm kg^{-1}. The addition of DMSO makes the freezing medium hyperosmotic and then the permeability of the cryoprotectant is a key factor to avoid osmotic differences. The results obtained by ASMA suggest that DMSO, despite being a permeable cryoprotectant, does not allow the total rehydration of the spermatozoa head. When sperm viability was assessed by Hoechst staining (see next section), a significant decrease of approximately 15% (73.09% vs. 58.26%) of living spermatozoa was registered from fresh

Table 14.3 Morphological parameters of spermatozoa (head area and perimeter) in fresh, P1-diluted and post thawing samples (n = 21 samples; 90–100 spermatozoa/sample). Numbers in parentheses are n values, indicating the total amount of spermatozoa considered in every case. Results are expressed as mean ± SEM. Different letters indicate significant differences between perimeters or areas in fresh, medium-diluted and thawed samples (Reproduced from Asturiano et al., 2007. With permission of Blackwell Verlag)

	Perimeter (μm)	Area (μm^2)
Fresh sample	14.97 ± 0.06[a] (2,025)	5.42 ± 0.01[a] (1,639)
Freezing medium	13.90 ± 0.06[b] (2,000)	4.76 ± 0.01[b] (2,033)
Post thawing	13.13 ± 0.06[c] (2,576)	4.45 ± 0.01[c] (2,576)

to thawed samples. This decrease should be considered in order to find ways of improvement of the techniques for the cryopreservation of European eel sperm.

14.3.4 Fluorescent Staining

Other attributes, such as cell viability and mitochondrial function, have also been chosen as indicators to evaluate fish sperm quality (Ogier De Baulny et al. 1997, 1999; Segovia et al. 2000; He and Woods 2004). Fluorescent staining has recently been shown to provide rapid assessment of the integrity of membranes and mitochondrial function in fish sperm (He and Woods 2004). The main benefits of these techniques are the simplicity and the high velocity of sample evaluation. On the other hand, epifluorescence microscopes must be available. Some staining products, such as propidium iodide or SYBR Green, are being assayed now in eel sperm, producing good results (D.S. Peñaranda, 2008 unpublished results). The bisBenzimide Hoechst staining has been already used to evaluate viability of eel spermatozoa, while JC-1 staining was used to check mitochondrial functionality throughout the spermiation (Asturiano et al. 2006). Spermatozoa were classified as *dead* when nuclei showed bright blue fluorescence over the sperm head, and *alive* when they did not showed this strong fluorescence. Figure 14.7A shows the sperm viability results obtained by Hoechst 33258 staining. The proportion of dead cells observed after staining with Hoechst 33258 was significantly lower during the 8th to 11th week, when the best seminal quality (higher percentage of motile cells and higher concentration) was observed. The dead cell percentage increased coinciding with the end of the spermiation at the 12th week.

One interesting aspect revealed by the use of Hoechst staining, undetected with previous techniques, is the existence of around 37% dead spermatozoa, even in the weeks when best results were registered (Asturiano et al. 2007). This result could indicate a sub-optimal effect of the hormonal maturation treatment. Similarly, Matsubara et al. (2005), studying different steroid levels caused by a weekly treatment in artificially matured female Japanese eels, concluded that weekly administrations could be inducing a sub-optimal release of steroids. If this is the case it could induce the cell mortality.

JC-1 staining has been used in European eel sperm (Asturiano et al. 2006). In the monomeric state, JC-1 fluorescence green, like rhodamine 123, it is transported into the interior of functioning mitochondria. However, as the concentration of JC-1 inside the mitochondria increases, the stain forms aggregates which fluorescence orange (Thomas et al. 1998; Garner and Thomas 1999). Spermatozoa stained with JC-1 display either green fluorescence for mitochondria with low to medium membrane potential, or yellow-orange for mitochondria with high membrane potential. Previous studies have demonstrated a positive correlation between functional mitochondria and sperm motility (Evenson et al. 1982; Auger et al. 1989). However, Asturiano et al. (2006) reported that the percentage of mitochondrial functionality determined by JC-1 staining did not show a similar pattern to that obtained with Hoechst 33258 (Fig. 14.7B). A low variability was observed throughout the weeks of treatment, varying between the best week (8th) and the worst (5th) by less than

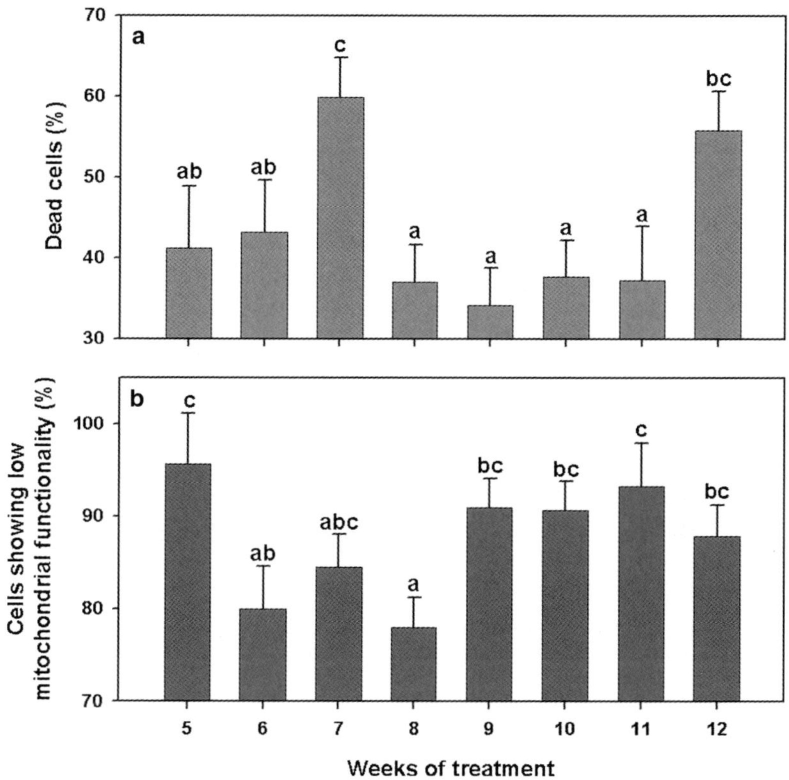

Fig. 14.7 **(A)** Sperm viability obtained by Hoechst 33258 staining, expressed as percentage of dead spermatozoa. **(B)** Mitochondrial function determined by JC-1 staining, shown as percentage of cells showing low mitochondrial functionality. The General Linear Model (GLM) was used to analyse the percentage of dead spermatozoa and mitochondrial function data. The week was included as a fixed effect in the model. Male was a blocking factor which, although having no interest in itself, was included in order to achieve a more accurate comparison between different weeks. Different letters indicate significant differences. (Asturiano et al. 2006. Copyright Elsevier 2006)

20%. It is possible that the fluorochrome stained fewer cells because the conditions were sub-optimal, since the conditions used on eel sperm (Asturiano et al. 2006) were validated in mammal studies (Thomas et al. 1998; Garner and Thomas 1999).

14.4 Sperm Cryopreservation

Though semen collection has been achieved in the Japanese eel, semen quality and quantity is highly variable and due to this fact cryopreservation has been tried for many years (Ohta and Izawa 1996; Tanaka et al. 2002a). Methods for the hormonal induction of gonad maturation in the European eel take several weeks both in males and females (Pérez et al. 2000; Asturiano et al. 2002a, 2005) making difficult to synchronize the gamete production for egg fertilization. This motivated the interest on the development of sperm cryopreservation media and methods in this species.

On the other hand, it is necessary to solve the problems in managing and assessment of sperm quality of the European eel due to its high density, with normal concentrations of 3–6 × 10^9 spermatozoa ml^{-1} (Pérez et al. 2000). Moreover, the time of sperm motility is very short (a few seconds), making sperm manipulation and its quality assessment difficult. This problem has been settled in other fish species by diluting the sperm before using it during the fertilization process. The dilution protocol has usually two phases, using first a diluting medium or extender with physico-chemical characteristics similar to the seminal plasma which does not activate the spermatozoa movement, followed by an activating solution to reach the final dilution and activate the cell movement. The use of diluting media or extenders can prolong the sperm life time and even increase the spermatozoa motility after an incubation period (Ohta et al. 1996a; Tanaka et al. 2002a). Specific extenders should be developed for the European species because trials carried out with extenders developed for Japanese eel (Ohta and Izawa 1996) gave poor results on European eel sperm (J.F. Asturiano, 2005 unpublished results).

Over the last decade, research on cryopreservation of fish spermatozoa has remained in its early stages, being especially focused on salmonids (Gwo et al. 1999) and cyprinids (Basavaraja and Hegde 2004). It is a type of technique rarely used in Mediterranean fish farming and especially not in the reproduction of the European eel. The preservation of Japanese eel *A. japonica* (Ohta and Izawa 1996; Tanaka et al. 2002a) and European eel sperm has been accomplished, using several pre-freezing and post-thawing milt dilution ratios (Asturiano et al. 2003, 2004; Müller et al. 2004; Szabó et al. 2005; Marco-Jiménez et al. 2006b; Garzón et al. 2008), cryoprotectants (Asturiano et al. 2003, 2004; Müller et al. 2004; Garzón et al. 2008), and the addition of different supplements to the cryopreservation medium (Asturiano et al. 2004; Garzón et al. 2008), but no optimal protocol has been developed at the moment, considering the relative low post-thawing survival obtained in comparison with fresh sperm samples. Sperm cryopreservation techniques will be useful for higher flexibility in broodstock management, further genetic improvement programs and preservation of genetic diversity.

14.4.1 Physico-Chemical Characteristics of Seminal Plasma and Sperm Diluents Design

In the European eel, the first studies in this regard were centered on the analysis of the seminal plasma ionic composition and physic-chemical characteristics (Pérez et al. 2003; Asturiano et al. 2004). Changes in the ionic composition of seminal plasma were correlated with changes in the sperm quality, in terms of spermatozoa motility (Pérez et al. 2003; Asturiano et al. 2004; Table 14.4). In these studies, the

Table 14.4 Ionic composition of the European eel seminal plasma (in mM) in relation to sperm motility categories, where 0 represent no motile sperm, while I <25%, II: 25–50% and III: >50% of the population were vigorously motile. Different letters show significant differences in the concentration of each iou between different motility categories (Asturiano et al. 2004)

(mM)	0	I	II	>III
Ca^{2+}	1.00 ± 0.15^{c}	0.57 ± 0.06^{bc}	0.41 ± 0.05^{ab}	0.20 ± 0.02^{a}
Mg^{2+}	5.74 ± 1.15^{b}	4.09 ± 0.59^{ab}	2.37 ± 0.31^{ab}	1.54 ± 0.28^{a}
K^{+}	27.45 ± 2.39^{b}	34.90 ± 1.83^{ab}	37.53 ± 2.95^{b}	36.16 ± 2.94^{ab}
Na^{+}	118.18 ± 6.14	116.29 ± 3.28	113.14 ± 4.15	109.34 ± 10.30

seminal plasma pH did not vary during the 14 weeks of assay, being constant at between 8.4 and 8.6. The seminal plasma osmolality showed a significant decrease during the first spermiation weeks and was maintained around 325–330 mOsm kg^{-1} during the rest of the experiment.

These results were used to design a diluting medium isoionic with the seminal plasma, called P1 medium, in mM: NaCl 125, $NaHCO_3$ 20, KCl 30, $MgCl_2$ 2.5, $CaCl_2$ 1, pH 8.5 (Asturiano et al. 2004). In the case of the Japanese eel, similar studies on seminal plasma and diluting media production have been conducted. Ohta et al. (1996a, b) managed times of conservation of diluted sperm under refrigeration longer than 3 weeks. Two diluting media (K15 and K30, Ohta et al. 1996a, b, 2001) previously used in Japanese eel sperm were tested in the European eel. Twenty-four hours after dilution, the sperm showed an important reduction in the percentage of motile spermatozoa after activation with sea water and different motility parameters (VAP, angular velocity; VCL, curvilinear velocity; VSL, straight line velocity; BCF, beating cross frequency), concluding that these media are not useful to preserve the sperm of the European eel (Asturiano et al. 2004).

To design new sperm diluting media, two different methods can be used. The first method is to employ the same ionic composition and osmolality as the seminal plasma, with the objective of stabilizing the sperm's physio-chemical conditions during storage (Villani and Catena 1991; Tan-Fermin et al. 1999). The second method consists of the use of inactivation media for fish spermatozoa, with different composition with respect to the seminal plasma (Sansone et al. 2001; Tanaka et al. 2002b; Rodina et al. 2004). In the case of the European eel, different inactivation media have been assayed, but they caused a severe motility decrease just a few hours after dilution (Peñaranda et al., unpublished results), while a medium with the same ionic composition and osmolality showed the best results, but to extend the storage time it was necessary to add one membrane protector. Taking as a basis that the bovine serum albumin (BSA) adheres rapidly to the spermatozoa membrane at the moment of dilution (Blank et al. 1976) and modifies the sperm lipid composition through lipid exchange or hydrolysis (Davis et al. 1979), and considering its good results in other works (Cabrita et al. 2005; Peñaranda et al. submitted), BSA was selected as cell protector. P1 medium plus 2% BSA was registered as the best diluting media, showing no motility significant differences with respect to fresh samples 48 h after diluting (4°C), and 30% of motile cells 1 week later (Peñaranda et al. unpublished results).

14.4.2 Cryopreservation Media, Cryoprotectants and Membrane Stabilisers

14.4.2.1 Cryopreservation Media

Cryopreservation media are based on the above solutions but must contain other components as cryoprotectants and, at least in some species, some components playing a spermatozoa membrane stabiliser role. Different sperm cryopreservation media have been tested in the European eel. Some of them were media previously used in the Japanese eel as Tanaka's medium (TNK, Tanaka et al. 2002a) or Ohta's K30 medium (Ohta et al. 2001). Some others media have been designed considering the European eel seminal plasma characteristics (Pérez et al. 2003; Asturiano et al. 2004). No significant differences were found among them and the results in terms of posthawing motility were low in all cases (Asturiano et al. 2004).

The pH and the response to the cryoprotectant are essential factors for the selection of the freezing medium, since they play an important role in the control of the sperm movement. P1 medium and TNK medium were tested as following: sperm was diluted in P1 or TNK media adjusted to different pH: 7.0, 7.5, 8.0 and 8.5 for the P1 medium, and 7.8 and 8.5 for the TNK medium (Garzón et al. 2008). Sperm samples motility was assessed immediately by diluting in sea water and no significant difference was found between them (Fig. 14.8A).

Having in mind that Tanaka et al. (2002a) suggested that DMSO is an effective cryoprotectant, Garzón et al. (2008) used P1 and TNK (at different pHs) with added DMSO (10%). Five minutes after the sperm dilution, the motility was recorded on activating with seawater (Fig. 14.8B). Samples incubated in P1 medium plus DMSO did not show sperm motility after activation with sea water, whereas samples in TNK medium plus DMSO showed a percentage of motile cells similar to that showed by the fresh pools activated with sea water. The difference between TNK and P1 media was a higher $NaHCO_3$ concentration in TNK, which coincides with the movement-inhibiting role of the bicarbonate suggested by Tanaka et al. (2002a, b). The absence of spermatozoa movement after 5 min of incubation in the presence of DMSO might be due to the cell's activation, with the consequent energy consumption, during the incubation period. In the same manner it is possible to question the mitochondria's integrity, since they are the main energy producers. If mitochondria are damaged as a consequence of osmotic stress caused by the addition of a cryoprotectant, it is possible that a decrease in ATP production occurs (Medina et al. 2005). In the same way, morphometric changes produced by DMSO addition might influence cell movement (Asturiano et al. 2006). According to these results, Garzón et al. (2008) carried out a third experiment that consisted in increasing the $NaHCO_3$ concentration (from 20 to 75 mM) in P1 medium (becoming P1 modified medium) similar to the TNK medium composition. The post-thawing motility was increased, although the final result was still quite low (22.2% ± 1.5%; Fig. 14.10B).

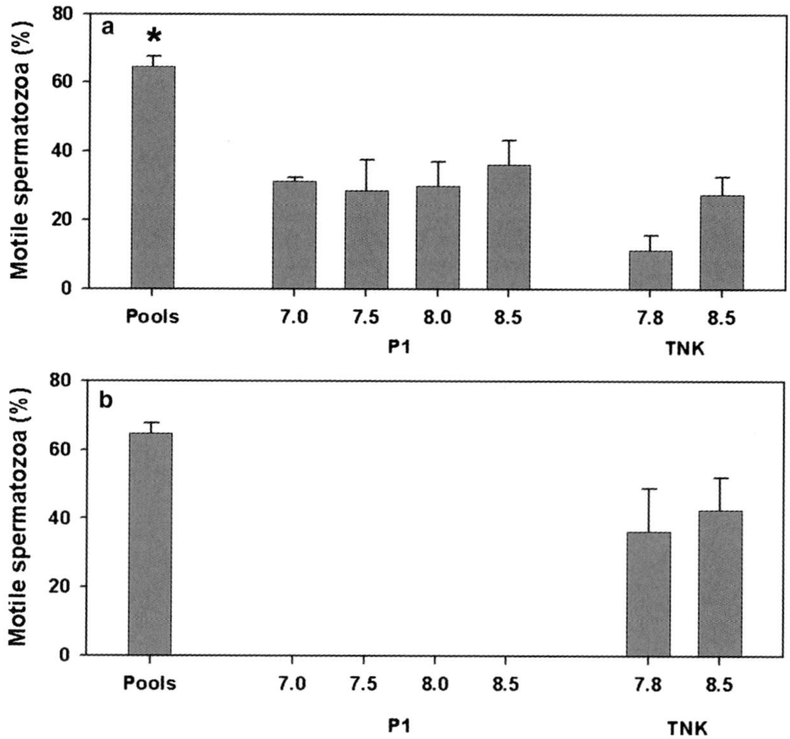

Fig. 14.8 (A) Percentage of motile (sea water-activated) spermatozoa in sperm samples just after dilution in P1 or TNK media at different pH, in comparison with fresh sperm pooled samples (n = 4). (B) Percentage of motile spermatozoa after activation with sea water in sperm samples after 5 min incubation in P1 or TNK media at different pH, in comparison with fresh pooled sperm samples (n = 4). Data are showed as means ± SEM. One ANOVA was carried out to compare mean results obtained. Multiple comparisons of means were carried out using a Student-Newman-Keuls test. Differences were considered significant at $p < 0.05$. Asterisks mean significant differences (Reproduced from Garzón et al. 2008. With permission of Blackwell Verlag)

Peñaranda et al. recently developed a final series of experiments (submitted results) trying to increase the post-thawing motility. To stop the cryoprotectant activation was considered an important factor, since this activation could partially spend the spermatozoa energy, and could be one of the causes of low post-thawing motility obtained until now. The influence of energy spent is especially important in eel sperm cryopreservation, because the time of spermatozoa motility is very short once diluted in sea water, from a few seconds to a few minutes (Gibbons et al. 1983, 1985; Wolley 1998). Different media with increasing $NaHCO_3$ concentrations (20–120 mM), and decreasing NaCl concentrations to maintain the osmolality, were used, but in this case at pH 6.5. In aqueous media $NaHCO_3$ is dissociated in several products: $CO_2 + H_2CO_3$ (free-CO_2), HCO_3^- and $CO^2{}_3^-$, and their proportion is affected by pH. If the medium is acid, most of HCO_3^- will be converted in free CO_2, and this free CO_2 will act as inhibitor factor of the motility (Tanaka et al. 2004).

This effect has been observed in studies on sperm from invertebrates to mammals (Christen et al. 1982; Johnson et al. 1983; Lee et al. 1983; Wong et al. 1981). The best media was P1 with 100 mM $NaHCO_3$ and 50 mM NaCl at pH 6.5 (called medium 5 by Peñaranda et al. submitted), arresting the cryoprotectant activation in pre-freezing conditions and obtaining a post-thawing motility around 40%, which can be considered enough for fertilization thanks to the high density of eel sperm.

14.4.2.2 Cryoprotectants

It is known that cryopreservation causes lethal damage in spermatozoa and also produces an important loss of membrane function by increasing membrane fragility in live cells (Cabrita et al. 1999). To reduce this damage, it is necessary to add cryoprotectants in the freezing medium, but they can increase the osmolality causing the spermatozoa activation. Looking for the best cryoprotectant, DMSO, acetamide, ethylene-glycol, propanol, glycerol and methanol (MeOH) were tested (Garzón et al. 2008). This effect was evaluated in terms of percentage of motile cells (activation caused by the cryoprotectants), percentage of alive cells (by Hoechst staining), and spermatozoa morphometry pre and post-cryopreservation (by ASMA).

MeOH, propanol and acetamide caused the smallest activation effect (Fig. 14.9A). Figure 14.9B shows the percentage of living cells in the fresh and in the post-thawed samples. DMSO and glycerol caused similar percentages of living cells as in the fresh pools, while the rest of cryoprotectants produced low survival values.

Finally, Fig. 14.9C and D show the results of the morphometric analyses for the spermatozoa frozen with different cryoprotectants. Cells frozen in P1 medium with DMSO added resulted in an increase of the head area of approximately 0.5 μm^2, statistically significant in comparison with the fresh pools (Fig. 14.9C). With MeOH, the spermatozoa showed an area reduction of approximately of 1 μm^2, in comparison to the fresh samples or spermatozoa frozen with other cryoprotectants. With regard to the head perimeter (Fig. 14.9D), it was observed that sperm cells frozen with DMSO did not show significant differences compared to fresh pools, while the spermatozoa frozen with the rest of cryoprotectants presented lower perimeters than fresh samples, especially in the case of cells frozen with ethylene-glycol and MeOH, which showed the major reduction in their head perimeter, 1.7 μm approximately.

The results of this experiment showed DMSO, MeOH and glycerol as the best cryoprotectants. These three cryoprotectants were chosen to make a new experiment using P1 and P1 modified as freezing media, and testing the influence of foetal bovine serum (FBS) addition (Garzón et al. 2008). All the frozen samples showed reductions of more than 50% of sperm motility compared to the fresh pools (Fig.

Fig. 14.9 (**a**) Percentage of activation effect in motile spermatozoa samples with six different cryoprotectants in comparison with fresh sperm pooled samples activated with sea water (n = 6). (**b**) Percentage of alive cells in sperm samples after cryopreservation with six different cryoprotectants in comparison with fresh sperm pooled samples activated with sea water (n = 4).

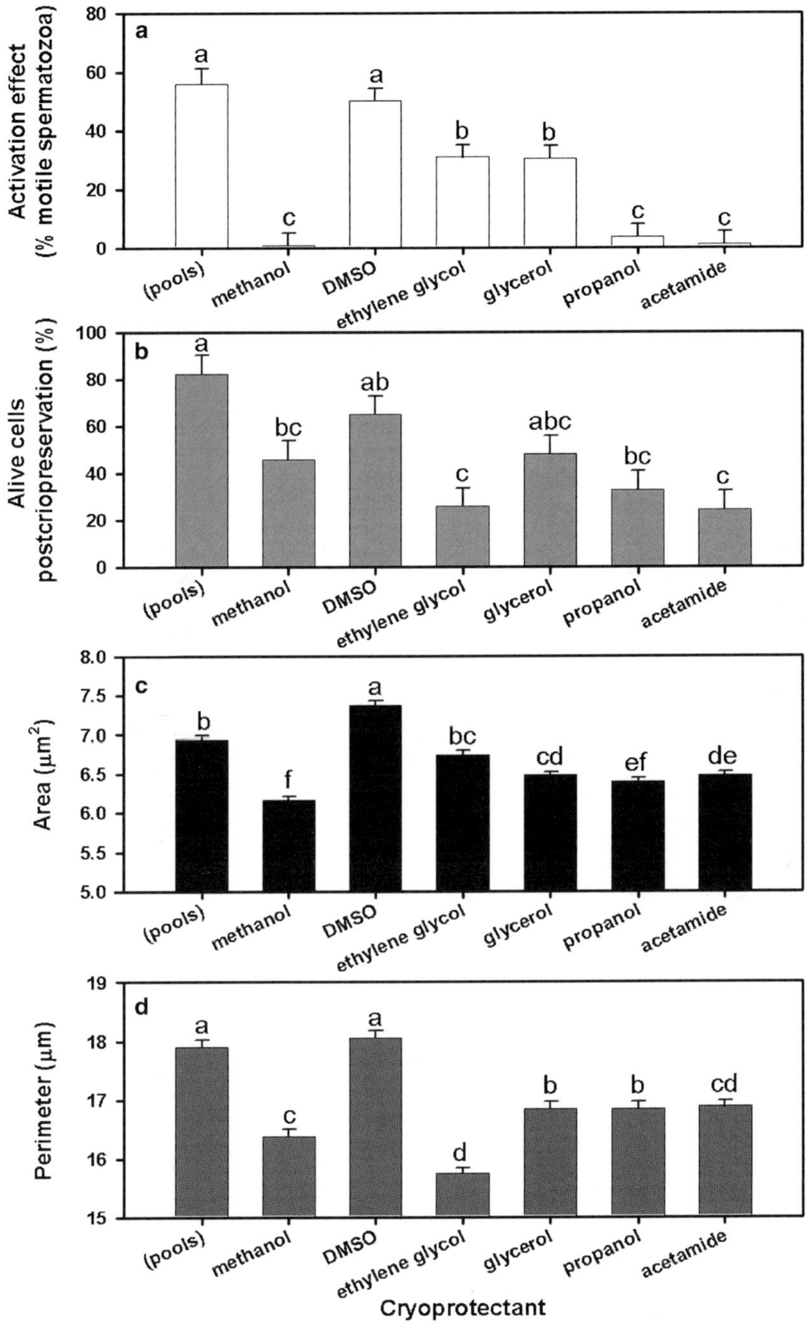

Fig. 14.9 (continued) **(c)** Effect of cryoprotectant in the spermatozoa area in μm² in comparison with fresh sperm pooled samples (n = 6). **(d)** Effect of cryoprotectant on spermatozoa perimeter (μm) in comparison with fresh pooled sperm (Reproduced from Garzón et al. 2008. With permission of Blackwell Verlag)

14.10A and B). The percentage of motile spermatozoa in the samples frozen with DMSO was significantly higher than those obtained with glycerol or MeOH. Frozen samples in P1 – modified medium with DMSO and added FBS showed a significantly higher percentage of motile spermatozoa (22.2% ± 1.5%) than in the other conditions. These results coincide with those reported by Tanaka et al. (2002a), who suggested DMSO as an effective cryoprotectant for the Japanese eel.

However, DMSO can be toxic for the sperm (He and Woods 2004), therefore the effect of different concentrations in the freezing medium (5% and 10%) was

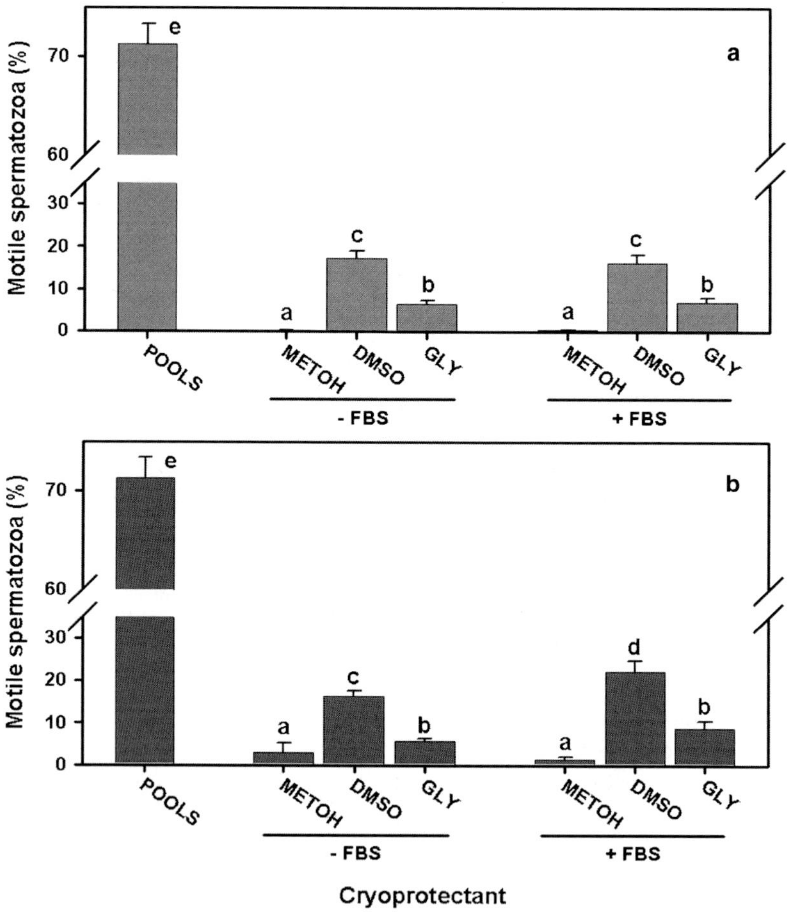

Fig. 14.10 Percentage of alive spermatozoa after cryopreservation in P1 **(a)** or P1 modified medium **(b)** with MeOH, DMSO or glycerol as cryoprotectant, and with or without FBS, in comparison with fresh sperm pooled samples activated with sea water (n = 19). Data are showed as means ± SEM. One ANOVA was carried out to compare mean results obtained. Multiple comparisons of means were carried out using a Student-Newman-Keuls test. Differences were considered significant at $p < 0.05$. Different letters indicate significant differences (Reproduced from Garzón et al. 2008. With permission of Blackwell Verlag)

studied recently. Better post-thawing motilities (38.26% ± 2.89%) were found in samples frozen with the highest level of cryoprotectant, but no differences were found in the percentage of living cells or in the cells head size (Peñaranda et al., unpublished results).

14.4.2.3 Membrane Stabilisers

It has been hypothesised that some carbohydrates, lipids and proteins can be used as membrane stabilisers (Anchordoguy et al. 1988; Crowe et al. 1990; Lahnsteiner et al. 1992; Labbe et al. 1997; He and Woods 2004). Cabrita et al. (2001) demonstrated in rainbow trout that different membrane stabilisers (bovine serum albumin, egg yolk and soya) increased cell protection against freezing injuries when compared to freezing without these agents. The positive influence of FBS (Fig. 14.10B) may be due to the serum proteins and lipids playing an important role in the cryopreservation process as a defence mechanism against instability of the plasma membrane (Watson 1995). Beesley et al. (1998) demonstrated an increase in living spermatozoa when FBS was used in conjunction with permeating cryoprotectant, enhancing its efficacy. The addition of FBS does not modify the osmotic pressure, but alters the physiological permeability of the membrane (Agca et al. 2002). The amino acids interact electrostatically with the phosphate groups in the sperm membrane phospholipids, thereby forming a layer on the spermatozoa surface that affects the permeability of DMSO (He and Woods 2003).

BSA also has been used in the European eel sperm cryopreservation media (Peñaranda et al., in press) due to its effect as buffer of the osmotic shock, because proteins can award a protection of mechanical type to the membrane, diminishing the risks of crystallization, recrystallization or ice melting during the different phases of the process of freezing and thawing (Rana 1995; Cabrita et al. 2005) but the results were significant lower than those obtained adding FBS. Maybe some proteins or lipids could exist in FBS medium that are not present in the BSA and this difference could be the responsible of improving the cell protection and motility, but further studies in this field are necessary.

Additional research is required to explain the significant interaction found between dilution ratio, cryoprotectant and FBS (Marco-Jiménez et al. 2006b) because they probably have different patterns of action in the protection of membrane viability, probably by modifying spermatozoa membrane permeability.

14.4.3 Freezing-Thawing Protocols

The freezing medium must be maintained at 4°C until sperm dilution. In our experience, dilution factors as low as 1:1 or 1:2 (sperm: freezing medium) give the best post-thawing survival and allow the maintenance of highly concentrated sperm samples in the straws. Once diluted, samples must be immediately packaged in 0.25 mL straws, sealed with modelling paste and frozen in liquid nitrogen vapour, 1.5 cm above the liquid nitrogen level for 5 min, before being plunged into the liquid nitrogen for the final freezing and storage. Thawing is carried out by immersion in a water bath at 25°C for 10–15 s.

14.5 Conclusion

This chapter review the first applications of computer assisted sperm analysis (CASA), automatic sperm morphology analysis (ASMA) and fluorescent staining in the European eel to examine spermatozoa motility, spermatozoa head morphometry and cell viability, showing that it is possible to use these methodologies to evaluate objectively a great number of cells.

The first application of these techniques in the European eel has been the evaluation of hormonal treatments. The changes observed in sperm morphometry coincide with the highest sperm quality assessed as sperm motility and concentration, as well as with the best results obtained in previous studies reporting the best sperm quality between weeks 8 to 10 of hCG treatment.

On the other hand, these techniques have been used to study the effects of the sperm dilution and the cryopreservation processes on the spermatozoa. A series of experiments showed P1 medium plus 2% BSA as the best diluting diluting solution, maintaining until 30% of motile cells (after sea water activation) for 7 days. The best freezing medium caused the highest percentage of motile spermatozoa post-freezing and the lowest spermatozoa morphometric modifications and consisted in medium 5 (Peñaranda et al. submitted) added with DMSO (10% v/v) as the best cryoprotectant. Moreover, FBS (25%) addition seems to be an important factor in protection of the integrity of the spermatozoa membrane during freezing and hereby facilitates the process of activation post-freezing.

New applications will be carried out trying to optimize the cryopreservation protocols, increasing the postfreezing survival, as well as in future studies on the physiology of the spermatozoa activation.

Acknowledgements Supported by the Generalitat Valenciana (GV04A-508, ACOMP06/193, GV/2007/202 and a grant to D.S. Peñaranda), the Spanish Ministry of Education and Science (AGL2003-05362-C02-00, including FEDER funds) and the Universidad Politécnica de Valencia (20030488).

The eel farm Valenciana de Acuicultura, S.A. (Puzol, Valencia) collaborated providing the fish used during the described experiments.

References

Agca Y, Gilmore J, Byers M, Woods EJ, Liu J, Critser JK (2002) Osmotic characteristics of mouse spermatozoa in the presence of extenders and sugars. Biol. Reprod. 67:1493–1501

Anchordoguy T, Carpenter JF, Loomis SH, Crowe JH (1988) Mechanisms of interaction of amino acids with phospholipid bilayers during freezing, Biochim. Biophys. Acta 964:299–306

Asturiano JF, Sorbera LA, Ramos J, Kime DE, Carrillo M, Zanuy S (2000) Hormonal regulation of the European sea bass (*Dicentrarchus labrax*, L.) reproductive cycle: an individualized female approach. J. Fish Biol. 56:1155–1172

Asturiano JF, Pérez L, Tomás A, Zegrari S, Espinós FJ, Jover M (2002a) Inducción hormonal de la maduración gonadal y puesta en hembras de anguila europea (*Anguilla anguilla*): primeros resultados. Boletín IEO 18:127–137 (in Spanish)

Asturiano JF, Sorbera LA, Ramos J, Kime DE, Carrillo M, Zanuy S (2002b) Group-synchronous ovarian development, spawning and spermiation in the European sea bass (*Dicentrarchus labrax* L.) could be regulated by shifts in gonadal steroidogenesis. Sci. Mar. 66(3):273–282

Asturiano JF, Pérez L, Marco-Jiménez F, Olivares L, Vicente JS, Jover M (2003) Media and methods for the cryopreservation of European eel (*Anguilla anguilla*) sperm. Fish Physiol. Biochem. 28:501–502

Asturiano JF, Pérez L, Garzón DL, Marco-Jiménez F, Peñaranda DS, Vicente JS, Jover M. (2004) Physio-chemical characteristics of seminal plasma and development of media and methods for the cryopreservation of European eel sperm. Fish Physiol. Biochem. 30:283–293

Asturiano JF, Pérez L, Garzón DL, Peñaranda DS, Marco-Jiménez F, Martínez-Llorens S, Tomás A, Jover, M (2005) Effect of different methods for the induction of spermiation on semen quality in European eel. Aquac. Res. 36:1480–1487

Asturiano JF, Marco-Jiménez F, Pérez L, Balasch S, Garzón DL, Peñaranda DS, Vicente JS, Viudes de Castro MP, Jover M (2006) Effects of hCG as spermiation inducer on European eel semen quality. Theriogenology 66:1012–1020

Asturiano JF, Marco-Jiménez F, Peñaranda DS, Garzón DL, Pérez L, Vicente JS, Jover M (2007) Effect of sperm cryopreservation on the European eel sperm viability and spermatozoa morphology. Reprod. Domest. Anim. 42:162–166

Auger J, Ronot X, Dadoune JP (1989) Human sperm mitochondrial function related to motility: a flow and image cytometric assessment. J. Androl. 10:439–448

Baker HWG, Clarke GN (1987) Sperm Morphology: Consistency of assessment of the same sperm by different observers. Clin. Reprod. Fertil. 5:37–43

Ball A, Mohammed HO (1995) Morphometry of stallion spermatozoa by computer-assisted image analysis. Theriogenology 44:367–377

Basavaraja N, Hegde SN (2004) Cryopreservation of the endangered mahseer (Tor Khudree) spermatozoa: I. Effect of extender composition, cryoprotectants, dilution ratio, and storage period on post-thaw viability. Cryobiology 49:149–156

Beesley SG, Costanzo JP, Lee Jr RE (1998) Cryopreservation of spermatozoa from freeze-tolerant and -intolerant anurans. Cryobiology 37:155–162

Billard R (1983) Ultrastructure of trout spermatozoa: changes after dilution and deep-freezing. Cell Tissue Res. 228:205–218

Billard R, Cosson J, Linhart O (2000) Changes in the flagellum morphology of intact and frozen/thawed Siberian sturgeon *Acipenser baerii* (Brandt) sperm during motility. Aquac. Res. 31:283–287

Blank M, Soo L, Britten JS (1976) Absorption of albumin on rabbit sperm membranes. J. Membr. Biol. 29:401–409

Boersma AA, Braun J, Stolla R (1999) Influence of random factors and two different staining procedures on computer-assisted sperm head morphometry in bulls. Reprod. Domest. Anim. 34:77–82

Boersma AA, Rasshofer R, Stolla R (2001) Influence of sample preparation, staining procedure and analysis conditions on bull sperm head morphometry using the morphology analyser integrated visual optical system. Reprod. Domest. Anim. 36:222–229

Bozzola JJ, Russell LD (1991) Electron Microscopy. Editorial Jones & Bartlett, Boston, MA

Buendía P, Soler C, Paolicchi F, Gago G, Urquieta B, Pérez-Sánchez F, Bustos-Obregón E (2002) Morphometric characterisation and classification of alpaca sperm heads using the Sperm-Class Analyser1 computer assisted system. Theriogenology 57:1207–1218

Cabrita E, Anel L, Herráez MP (2001) Effect of external cryoprotectants as membrane stabilizers on cryopreserved rainbow trout sperm. Theriogenology 56:623–635

Cabrita E, Robles V, Cuñado S, Wallace JC, Sarasquete C, Herráez MP (2005) Evaluation of gilthead sea bream, *Sparus aurata*, sperm quality after cryopreservation in 5 ml macrotubes. Cryobiology 50:273–284

Christen R, Schackmann RW, Shapiro BM (1982) Elevation of the intracellular pH activates respiration and motility sperm of the sea urchin, *Strongylocentrotus purpuratus*. J. Biol. Chem. 257:14881–14890

Christen R, Gatti JL, Billard R (1987) Trout sperm motility. The transient movement of trout sperm is related to changes in the concentration of ATP following the activation of the flagellar movement. Europ. J. Biochem. 166:667–671

Claassens OE, Kaskar K, Coetzee K, Lombard CJ, Franken DR, Kruger TF (1996) Comparison of motility characteristics and normal sperm morphology of human semen samples separated by percoll density gradient centrifugation. Arch. Androl. 36:127–132

Coetzee K, Kruge TF, Lombard CJ (1998) Predictive value of normal morphology: a structured literature review. Hum. Reprod. Update 4:73–82

Çolak A, Yamamoto K (1974) Ultrastructure of the Japanese eel spermatozoon. Annotationes Zool. Jpn. 47:48–54

Colombo G, Grandi G, Romeo A, Giovannini A, Pelizzola D, Catozzi L, Pifanelli A (1987) Testis cytological structure, plasma sex steroids and gonad cytosol free steroid receptors of heterologous gonadotropin (hCG)-stimulated silver eels, *Anguilla anguilla* L. Gen. Comp. Endocrinol. 65:167–178.

Colombo G, Grandi G (1996) Histological study of the development and sex differentiation of the gonad in the European eel. J. Fish Biol. 48(3):493–512

Colombo G, Grandi G, Rossi G (1984) Gonad differentiation and body growth in *Anguilla anguilla* L. J. Fish Biol. 24:215–228

Crowe JH, Carpenter JF, Crowe LM, Anchordoguy TJ (1990) Are freezing and dehydration similar stress vectors? A comparison of modes of interaction of stabilizing solutes with biomolecules. Cryobiology 27:219–231

Dahlbom M, Andersson M, Vierula M, Alanko M (1997) Morphometry of normal and teratozoospermic canine sperm heads using an image analyser: work in progress. Theriogenology 48:687–698

Davis RO, Gravance CG (1993) Standardisation of specimen preparation, staining, and sampling methods improves automated sperm-head morphometry analysis. Fertil. Steril. 59:412–417

Davis BK, Byrne R, Hungund B (1979) Studies on the mechanism of capacitation. II. Evidence for lipid transfer between plasma membrane of rat sperm and serum albumin during capacitation in vitro. Biochim. Biophys. Acta 558:257–266

Davis RO, Gravance CG, Thal DM, Miller MG (1994) Automated analysis of toxicant-induced changes in rat sperm head morphometry. Reprod. Toxicol. 8:521–529

Davis RO, Gravance CG, Overstreet JW (1995) A standardised test for visual analysis of human sperm morphology. Fertil. Steril. 63(5):1058–1063

De Leeuw AM, Den Daas JH, Woelders H (1991) The fix vital stain method. Simultaneous determination of viability and acrosomal status of bovine spermatozoa. J. Androl. 12:112–118

Devlin RH, Nagahama Y (2002) Sex determination and sex differentiation in Fish: an overview of genetic, physiological, and environmental influences. Aquaculture 208:191–364

Evenson DP, Darzynkiewicz Z, Melamed MR (1982) Simultaneous measurement by flow cytometry of sperm cell viability and mitochondrial membrane potential related to cell motility. J. Histochem. Cytochem. 30:279–280

Feunteun E (2002) Management and restoration of European eel population (*Anguilla anguilla*): An impossible bargain. Ecol. Eng. 18:575–591

Gago C, Pérez-Sánchez F, Yeung CH, Tablado L, Cooper TG, Soler C (1998) Standardisation of sampling and staining methods for the morphometric evaluation of sperm heads in the Cynomolgus monkey (*Macaca fascicularis*) using computer-assisted image analysis. Int. J. Androl. 21:169–176

Garner DL, Thomas CA (1999) Organelle-specific probe JC-1 identifies membrane potential differences in the mitochondrial function of bovine sperm. Mol. Reprod. Dev. 53:222–229

Garner DL, Pinkel D, Johnson LA, Pace MM (1986) Assessment of spermatozoal function using dual fluorescent staining and flow cytometric analyses. Biol. Reprod. 34:127–138

Garner DL, Johnson LA, Yue ST, Roth BL, Haugland RP (1994) Dual DNA staining assessment of bovine sperm viability using SYBR-14 and propidium iodide. J. Androl. 15:620–629

Garzón DL, Peñaranda DS, Pérez L, Marco-Jiménez F, Espert X, Müller T, Jover M, Asturiano JF (2008) Effects of pH, sodium bicarbonate, cryoprotectants and FBS on the cryopreservation of European eel sperm. Reprod. Domest. Anim. 43:99–105

Gibbons BH, Gibbons IR, Baccetti B (1983) Structure and motility of the 9 + 0 flagellum of eel spermatozoa. J. Submicrosc. Cytol. 15:15–20

Gibbons BH, Baccetti B, Gibbons IR (1985) Live and reactivated motility of the 9 + 0 flagellum of the eel spermatozoa. J. Submicrosc. Cytol. 5:333–350

Ginsburg AS, Billard R (1972) Ultraestructure du spermatozoïde d'Anguille. J. Microscopie 14:50–51

Graham JK (2001) Assessment of sperm quality: a flow cytometric approach. Anim. Reprod. Sci. 68:239–247

Grandi G, Colombo G (1997) Development and early differentiation of gonad in the European eel (*Anguilla anguilla* [L.], Anguilliformes, Teleostei): A cytological and ultrastructural study. J. Morphol. 231:195–216

Gravance CG, Davis RO (1995) Automated sperm morphometry analysis (ASMA) in the rabbit. J. Androl. 16:88–93

Gravance CG, Liu IK, Davis RO, Hughes JP, Casey PJ (1996) Quantification of normal head morphometry of stallion spermatozoa. J. Reprod. Fertil. 108(1):41–46

Gravance CG, Campion Z, Liu IK, Casey PJ (1997) Sperm head morphometry analysis of ejaculate and dismount stallion semen samples. Anim. Reprod. Sci. 47:149–155

Guraya SS (1987) Rat sperm head. In: Biology of Spermatogenesis and Spermatozoa in Mammals. Berlin/Heidelberg: Springer

Gwo JC, Gwo HH, Chang SL (1992) The spermatozoon of the Japanese eel, *Anguilla japonica* (Teleostei, Anguilliformes, Anguillidae). J. Submicrosc. Cytol. Pathol. 24:571–574

Gwo J-C, Ohta H, Okuzawa K, Wu HC (1999) Cryopreservation of sperm from the endangered Formosan landlocked salmon (*Oncorhynchus masou formosanus*). Theriogenology 51:569–582

He S, Woods III LC (2003) Effects of glycine and alanine on short-term storage and cryopreservation of striped bass (*Morone saxatilis*) spermatozoa. Cryobiology 46:17–25

He S, Woods III LC (2004) Effects of dimethyl sulphoxide and glycine on cryopreservation induced damage of plasma membranes and mitochondria to striped bass (*Morone saxatilis*) sperm. Cryobiology 48:254–262

Johnson CH, Clapper DL, Winkler MM, Lee HC, Epel D (1983) A volatile inhibitor immobilizes sea urchin sperm in semen by depressing the intracellular pH. Dev. Biol. 98:493–501

Kennedy CR, Fitch DJ (1990) Colonisation, larval survival, and epidemiology of the nematode *Anguillicola crassus*, parasite in the eel *Anguilla anguilla* in Britain. J. Fish Biol. 36:117–131

Kime DE, Van Look KJ, McAllister BG, Huyskens G, Rurangwa E, Ollevier F (2001) Computer-assisted sperm analysis (CASA) as tool for monitoring sperm quality in fish. Comp. Biochem. Physiol. C 130:425–433

Koops H, Hartmann F (1989) Anguillicola infestations in Germany and in German eel imports. J. Appl. Ichthyol. 1:41–45

Kruger TF, Du Toit TC, Franken DR, Menkveld R, Lombard CJ (1995) Sperm morphology: assessing the agreement between the manual method (strict criteria) and the sperm morphology analyser IVOS. Fertil. Steril. 63:134–141

Labbe C, Crowe LM, Crowe JH (1997) Stability of the lipid component of trout sperm plasma membrane during freeze-thawing. Cryobiology 34:176–182

Lahnsteiner F, Weismann T, Patzner RA (1992) Fine structural changes in spermatozoa of the grayling, *Thymallus thymallus* (Pices: Teleostei), during routine cryopreservation. Aquaculture 103:301–304

Lee HC, Johnson C, Epel D (1983) Changes in internal pH associated with initiation of motility and acrosome reaction of sea urchin sperm. Dev. Biol. 95: 31–45

Magyary I, Urbányi B, Horváth L (1996): Cryopreservation of common carp (*Cyprinus carpio* L.). J. Appl. Ichthyol. 12: 117–119

Marco-Jiménez F, Pérez L, Viudes de Castro MP, Garzón DL, Peñaranda DS, Vicente JS, Jover M, Asturiano JF (2006a) Morphometry characterisation of European eel spermatozoa with computer-assisted spermatozoa analysis and scanning electron microscopy. Theriogenology 65:1302–1310

Marco-Jiménez F, Garzón DL, Peñaranda DS, Pérez L, Viudes de Castro MP, Vicente JS, Jover M, Asturiano JF (2006b) Cryopreservation of European eels (*Anguilla anguilla*) spermatozoa: Effect of rate dilution, foetal bovine serum supplementation and cryoprotectans. Cryobiology 53:51–57

Matsubara H, Lokman PM, Kazeto Y, Adachi S, Yamauchi K (2005) Serum steroid profiles in artificially maturing female Japanese eel, *Anguilla japonica*. Aquaculture 243:393–402

Medina V, Velasco Y, Cruz P (2005) Aspectos generales de la criopreservación espermática en peces teleósteos. Revista Colombiana de Ciencias Pecuarias 18:34–48 (in Spanish)

Miura T, Yamauchi K, Nagahama Y, Takahashi H (1991) Induction of spermatogenesis in male Japanese eel, *Anguilla japonica*, by a single injection of human chorionic gonadotropin. Zool. Sci. 8:63–73

Müller T, Urbányi B, Váradi B, Binder T, Horn P, Bercsényi M, Horváth A (2004) Cryopreservation of sperm of farmed European eel *Anguilla anguilla*. J. World Aq. Soc. 35:240–246

Müller T, Baska F, Niklesz C, Horn P, Váradi B, Bercsényi M (2005) The testis histology of artificially matured European eel (*Anguilla anguilla* L.) at the end of sexual maturation, and spermatozoa ultrastructure in freshwater rearing. Acta Biol. Hung. 56(1–2):169–172

Ogier De Baulny B, Le Bern Y, Kerboeuf D, Maisse G (1997) Flow cytometric evaluation of mitochondrial activity and membrane integrity in fresh and cryopreserved rainbow trout (*Oncorhynchus mykiss*) spermatozoa. Cryobiology 34:141–149

Ogier De Baulny B, Labbe C, Maisse G (1999) Membrane integrity, mitochondrial activity, ATP content, and motility of the European catfish (*Silurus glanis*) testicular spermatozoa after freezing with different cryoprotectants. Cryobiology 39:177–184

Ohta H, Izawa T (1996) Diluent for cool storage of the Japanese eel (*Anguilla japonica*) spermatozoa. Aquaculture 142:107–118

Ohta H, Kagawa H, Tanaka H, Okuzawa K, Hirose K (1996a) Changes in fertilization and hatching rates with time after ovulation induced by 17,20 -dihydroxy-4-pregnen-3-one in the Japanese eel, *Anguilla japonica*. Aquaculture 139: 291–301

Ohta H, Kagawa H, Tanaka H, Okuzawa K, Hirose K (1996b) Milt production in the Japanese eel *Anguilla japonica* induced by repeated injections of human chorionic gonadotropin. Fish. Sci. 62:44–49

Ohta H, Kagawa H, Tanaka H, Unuma T (2001) Control by the environmental concentration of ions of the potential for motility in Japanese eel spermatozoa. Aquaculture 198:339–351

Okamura A, Zhang H, Yamada Y, Tanaka S, Horie N, Mikawa N, Utoh T, Oka HP (2000) Re-examination of the spermatozoal ultrastructure of eels: observations of the external morphology of spermatozoa in three species. J. Fish Biol. 57:161–169

Peñaranda DS, Pérez L, Fakriadis G, Mylonas CC, Asturiano JF (2008). Effects of extenders and cryoprotectant combinants on motility and morphometry of sea bass (*Dicentrarchus labrax*) spermatoza, J. Appl. Ichthyol. 24:450–455

Peñaranda DS, Pérez L, Gallego V, Jover M, Asturiano JF. Improvement of European eel sperm cryopreservation method by preventing spermatozoa movement activation caused by cryoprotectants. Cryobiology, submitted.

Pérez L, Asturiano JF, Tomás A, Zegrari S, Barrera R, Espinós FJ, Navarro JC, Jover M (2000) Induction of maturation in the male European eel (*Anguilla anguilla*): assessment of sperm quality throughout treatments. J. Fish Biol. 57: 1488–1504

Pérez L, Asturiano JF, Martínez S, Tomás A, Olivares L, Mocé E, Lavara R, Vicente JS, Jover, M (2003) Ionic composition and physio-chemical parameters of the European eel (*Anguilla anguilla*) seminal plasma. Fish Physiol. Biochem. 28:221–222

Piferrer F (2001) Endocrine sex control strategies for the feminization of teleost fish. Aquaculture 97:229–281

Pursel VG, Johnson LA (1974) Glutaraldehyde fixation of boar spermatozoa for acrosome evaluation. Theriogenology 1(2):63–68

Rana K.J., (1995) Methods in Molecular Biology: Cryopreservation and Freeze-drying Protocols: Vol 38, Cryopreservation of Fish Spermatozoa, Chapter 16.

Rijsselaere T, Soom AV, Hoflack G, Maes D, Kruif A (2004) Automated sperm morphometry and morphology analysis of canine semen by the Hamilton-Thorne analyser. Theriogenology 62:1292–1306

Rodina M, Cosson J, Gela D, Linhart O (2004) Kurokura solution as immobilizing médium for spermatozoa of Tench (*Tinca tinca* L.). Aquacult. Int. 12:119–131

Rurangwa E, Kime DE, Ollevier F, Nash JP (2004) The measurement of sperm motility and factors affecting sperm quality in cultured fish. Aquaculture 234:1–28

Sancho M, Perez-Sanchez F, Tablado L, de Monserrat JJ, Soler C (1998) Computer-assisted morphometric analysis of ram sperm heads: evaluation of different fixative techniques. Theriogenology 50:27–37

Sansone G, Fabbrocini A, Zupa A, Lavadera SL, Rispoli S, Matassino D (2001) Inactivator media of sea bass (*Dicentrarchus labrax* L.) spermatozoa motility. Aquaculture 202 257–268.Schulz RW, Miura T (2002) Spermatogenesis and its endocrine regulation. Fish Physiol. Biochem. 26:43–46

Schulz RW, Miura T (2002) Spermatogenesis and its endocrine regulation. Fish Physiol. Biochem. 26: 43–46

Segovia M, Jenkins JA, Paniagua-Chavez C, Tiersch TR (2000) Flow cytometric evaluation of antibiotic effects on viability and mitochondrial of refrigerated spermatozoa of Nile tilapia. Theriogenology 53:1489–1499

Soler C, de Monserrat JJ, Gutierrez R, Nunez J, Nunez M, Sancho M, Perez-Sanchez F, Cooper TG (2003) Use of the sperm class analyser for objective assessment of human sperm morphology. Int. J. Androl. 26:262–270

Szabó G, Müller T, Bercsényi M, Urbányi B, Kucska B, Horváth A (2005) Cryopreservation of European eel (*Anguilla anguilla*) sperm using different extenders and cryoprotectants. Acta Biol. Hung. 56:173–175

Tanaka H, Kagawa H, Ohta H (2001) Production of leptocephali of Japanese eel (*Anguilla japonica*) in captivity. Aquaculture 201:51–60

Tanaka S, Zhang H, Horie N, Yamada Y, Okamura A, Utoh T, Mikawa N, Oka HP, Kurokura H (2002a) Long-term cryopreservation of sperm of Japanese eel. J. Fish Biol. 60:139–146

Tanaka S, Zhang H, Yamada Y, Okamura A, Horie N, Utoh T, Mikawa N, Oka HP, Kurokura H (2002b) Inhibitory effect of sodium bicarbonate on the motility of sperm of Japanese eel. J. Fish Biol. 60:1134–1141

Tanaka S, Utoh T, Yamada Y, Horie N, Okamura A, Akazawa A, Mikawa N, Oka HP, Kurokura H (2004) Role of sodium bicarbonate on the initiation of sperm motility in the Japanese eel. Fish. Sci. 70:780–787

Tan-Fermin JD, Miura T, Adachi S, Yamauchi K (1999) Seminal plasma composition; sperm motility; and milt dilution in the Asian catfish *Clarias macrocephalus* (Gunther). Aquaculture 171:323–338

Thomas CA, Garner DL, DeJarnette JM, Marshall CE (1998) Effect of cryopreservation of bovine sperm organelle function and viability as determined by flow cytometry. Biol. Reprod. 58:786–793

Van Look KJW, Kime DE (2003) Automated sperm morphology analysis in fishes: the effect of mercury on goldfish sperm. J. Fish Biol. 63:1020–1033

Verstegen J, Iguer-Ouada M, Onclin K (2002) Computer-assisted semen analysers in andrology research and veterinary practice. Theriogenology 57:149–179

Villani P and Catena C (1991) Criopreservazione di gamete maschili di spigola (*D. labrax*). Riv. Ital. Acquacoltura 217–222

Wang C, Leung A, Tsoi WL, Leung J, Ng V, Lee KF, Chan SY (1991) Computer-assisted assessment of human sperm morphology; comparison with visual assessment. Fertil. Steril. 55:983–988

Watson PF (1995) Recent developments and concepts in the cryopreservation ofspermatozoa and the assessment of their post-thawing function. Reprod. Fertil. Dev. 7:871–891

Wong PYD, Lee WM, Tsang AYF (1981) The effects of extracellular sodium on acid release and motility initiation in rat caudal epididymal spermatozoa in vitro. Exp. Cell Res. 131:97–104

Woolley DM (1997) Studies on the eel sperm flagellum I. The structure of the inner dynein arm complex. J. Cell Sci. 110:85–94

Woolley DM (1998) Studies of the eel sperm flagellum. 3. Vibratile motility and rotatory bending. Cell Motil. Cytoskeleton 39:246–255

Part V
Environmental Effects

Chapter 15
The Effect of PCBs on the Spawning Migration of European Silver Eel (*Anguilla anguilla* L.)

Vincent van Ginneken, Maarten Bruijs, Tinka Murk, Arjan Palstra, and Guido van den Thillart

15.1 Introduction

Organochlorine compounds were widely used after the Second World War because they were cheap to produce and useful for many purposes, such as in agriculture for insecticides, in public health to control disease insect vectors and in industry (Pelletier et al. 2002). It is estimated that 16–30% of the 1 million tons of PCBs produced are still present in aquatic and terrestrial ecosystems (Borlakoglu and Haegele 1991). In spite of discharge restrictions, the concentrations of PCBs and chemically similar compounds in natural environments will likely remain elevated because of atmospheric transport and the internal cycling of contaminants already present in ecosystems. So, when not retrieved or destroyed, the rest of the PCBs will be released into the environment and eventually reach the oceans (Klamer et al. 1991). PCBs encompass a class of chlorinated compounds that includes up to 209 variations, or congeners, with different physical and chemical characteristics. They are ubiquitous environmental contaminants with specific modes of action (Safe 1984, 1990) and exposure to each of the congeners is associated with different levels of risk for harmful effects. Technical mixtures of PCBs, referred to by the trade names such as Aroclor, Phenoclor and Kanechlor, have been widely used for a variety of industrial purposes: hydraulic fluids in mining activities, plasticisers, fluid-filled capacitors and transformers, heat transfer fluids and paints. There are no known natural sources of PCBs. Most PCBs are oily liquids whose colour darkens and viscosity increases with rising chlorine content. PCBs with fewer chlorine atoms are more soluble, more amenable to chemical and biological degradation, and less persistent in the environment than those PCBs with more chlorine atoms bound to the biphenyl core (Safe 1984) (Fig. 15.1).

V. van Ginneken A. Palstra, and G. van den Thillart
Integrative Zoology, Institute of Biology Leiden, van der Klaauw Laboratories, P.O. Box 9516, Kaiserstraat 63, 2300 RA Leiden, The Netherlands

M. Bruijs
KEMA Technical and Operational Services. P.O. Box 9035, 6800 ET Arnhem, The Netherlands

T. Murk
Dept Toxicology Section, Wageningen University, Tuinlaan 5, POB 8000, 6700 EA, The Netherlands

G. van den Thillart et al. (eds.), *Spawning Migration of the European Eel*,

PCBs (polychlorinated biphenyls), like other lipophilic, persistent pollutants, are mostly available to aquatic organisms from contaminated sediments in lakes and water courses, which are the source of PCBs for bottom-dwelling fishes such as the eel (Tesch 1977). Fat percentages in silver eel reach up to 30% (Table 15.1), before the onset of migration (Bertin 1956), so eels are very vulnerable for accumulation of PCBs in their fat stores. PCB concentrations were highest when eels have fed on benthic macro invertebrates in the sediment (Hernandez et al., 1987). Many biomonitoring studies have produced evidence that PCBs accumulate massively in the eel (Table 15.2). Because eels have a very long juvenile phase (up to 20 years in often contaminated inland waters), they accumulate PCBs to high levels (Rahman et al. 1993; Haiber and Schöler 1994; De Boer and Hagel 1994; Hendriks et al. 1998). The total PCB concentrations in eels from rivers in northwestern Europe range from

Table 15.2 PCB levels in yellow eel from natural waters. (Data from Brusle 1991; Robinet and Feunteun 2002)

Species	Geographical position	Concentration	Source	Reference
Anguilla anguilla	North/Baltic sea	0.502–0813 $\mu g\ g^{-1}$	Whole fish	Huschenbeth 1977
	Lower Elbe river	0.058–10.93 $mg\ kg^{-1}$	Muscle	Kruse et al. 1983
	River Rhine	2.5 $mg\ kg^{-1}$	Muscle	Bimbos and Mau 1986
	River Rhine	0.48–1.6 $mg\ kg^{-1}$	Muscle	Anonymous 1987
	Doñana National Park (Spain)	0.19 $\mu g\ g^{-1}$	Muscle	Hernandez et al. 1987
	England	0.91 $\mu g\ g^{-1}$	Whole fish	Mason 1993
	Finland	33.8 $\mu g\ g^{-1}$	Whole fish	Tulonen and Vuorinen 1996
	Ireland	>0.1 $\mu g\ g^{-1}$	Muscle	Weatherley et al. 1997
	Germany	0.24×10^{-5}–$15.05 \times 10^{-5}\ \mu g\ g^{-1}$	Lipids	Wiemüller and Schlatterer 1999
Anguilla rostrata	Atlantic coast of Canada	0.44 $\mu g\ g^{-1}$	Whole fish	Sims et al. 1977
	Northwest Gulf of Mexico	15.0 $\mu g\ g^{-1}$	Whole fish	Giam et al. 1978
	Lake Ontario	4.9 $\mu g\ g^{-1}$	Fillet	Ryan et al. 1984
	St. Lawrence estuary	0.77–1.61 $\mu g\ g^{-1}$	Whole fish	Bertrand et al. 1986
	St. Lawrence River	9.81×10^{5}–$6.32 \times 10^{6}\ \mu g\ g^{-1}$	Whole fish	Castonguay et al. 1989
	New Jersey (USA)	4.86 $\mu g\ g^{-1}$	Whole fish	Kennish et al. 1992
	Kamouraska (Canada)	0.156–3.033 $\mu g\ g^{-1}$	Whole fish	Hodson et al. 1994

1.5–10 μg g^{-1} (De Boer and Hagel 1994), therefore regularly exceeding the Dutch standards for human consumption, which is 5 μg g^{-1} of total PCB for eel.

Another study by De Boer (1993) revealed that 85–90% of the toxic effect of PCBs in the yellow eel is caused by PCBs 126, 156 and 118, which are mostly used for industrial purposes. The large monitoring study on PCB contamination by De Boer and Hagel (1994) showed that the PCB levels in eels taken from the rivers Rhine and Meuse were amongst the highest reported in freshwater fishes from Europe. They reviewed other values from outside Europe, where the only higher total PCB values were reported by Sloan et al. (1983). These values were 1,500–4,000 μg g^{-1} in 1977 in different fish species from the Hudson River, New York. PCB concentrations in eel from the Rhine and Lake Ijssel have decreased substantially since the early eighties, but the levels are still above safe limits (Fig. 15.3). The yellow eel has even shown to be a very practical bio-indicator for the reflection of spatial differences and temporal trends of PCB contamination in fresh water.

In addition, in an 8-year study by De Boer et al. (1994) it was demonstrated that elimination half-lives of PCBs are in the order of years. For the higher chlorinated CBs (chlorinated biphenyls) (hexa-octa-CBs), no elimination was found at all. The combination of a long juvenile phase in contaminated inland waters and the high fat content of eels, means that eels indeed accumulate PCBs in large

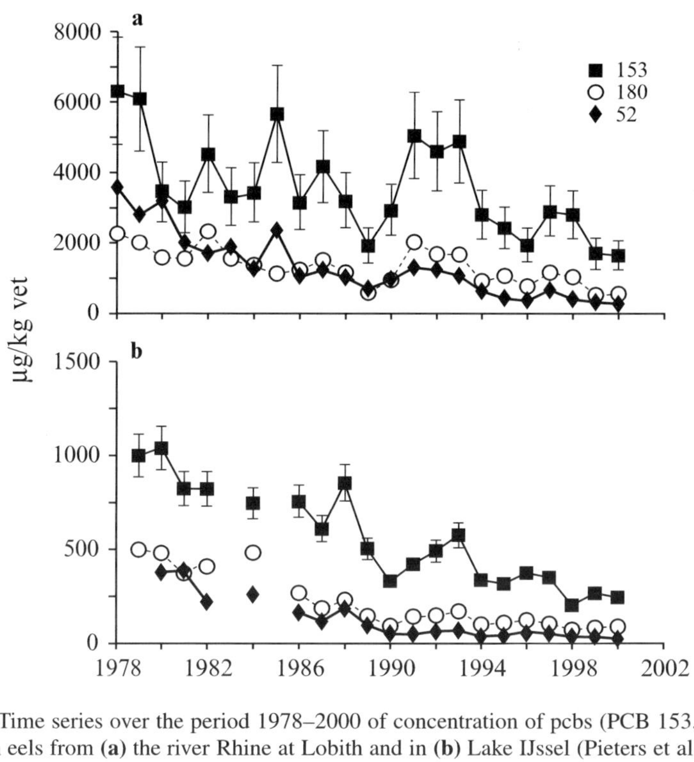

Fig. 15.3 Time series over the period 1978–2000 of concentration of pcbs (PCB 153, PCB 180, PCB 52) in eels from **(a)** the river Rhine at Lobith and in **(b)** Lake IJssel (Pieters et al. 2001)

quantities. These will be transported from the fat to the gonads and reproduction products. Also, since eels do not reproduce in inland waters, they do not lose parts of the accumulated PCBs via reproduction products.

15.3 Interference of PCBs with the Endocrine System

In the eel, thyroid hormones play a role in GtH-dependent steroidogenesis (Cyr and Eales 1988), in the final maturation of oocytes (Dickhoff et al. 1989; Weber et al., 1992) and are also shown to be correlated with the viability of larvae (Brown et al. 1989). Thyroid hormones play a role in the control of early development in fishes (Sullivan et al., 1987). T3 (3,5,3'-triiodo-L-thyronine) is the active form of thyroid hormone, which is important for stimulation of fish growth and early ovarian development (Eales and MacLatchy 1989). The systemic conversion of T3 from T4 is thought to be the primary mode of T3 production (Leatherland 1987). Furthermore, it has been reported that thyroid hormones can trigger the metamorphosis of eel lepotocephali (Vilter 1946).

PCBs and their metabolites can interfere via multiple and interactive mechanisms with the thyroid hormone system. Evidence suggests that pure congeners and mixtures of PCBs directly interfere with thyroid hormone metabolism, with enzymes such as uridine-diphosphate-glucuronyl transferases (UGTs), iodothyronine deiodinases (IDs) and sulfotransferases (SULTs) in liver and brain and with the plasma transport system of thyroid hormones (Brouwer et al. 1998). Eels have been shown to metabolize PCBs in the liver (De Boer et al. 1994), including PCB 77. The latter has been shown to be converted into hydroxylated metabolites known to mimic thyroid hormone (Murk et al. 1994).

PCBs and other toxicants have been shown to interact with several endocrine control mechanisms (Brouwer et al. 1990; Goksøyr and Förlin 1992; McKinney and Waller 1994; Barron et al. 1995). It has long been recognised in mammals that endocrine tissues are affected by PCBs (review by Birnbaum 1994), however, Hontela et al. (1992, 1995) showed that adult yellow perch, *Perca flavescens*, and northern pike, *Esox lucius*, from PCB polluted areas had an impaired stress response after capturing and handling. It is thought that PCBs interfere with neuroendocrine systems, partly by exerting estrogenic effects (Soontornchat et al. 1994). PCBs may interfere with corticosteroidogenesis and corticosteroid action in a similar manner as described for gonadosteroidogenesis (Barron et al. 1995) and thyroid hormone synthesis (McKinney and Waller 1994; Murk et al. 1994). However, cortisol, which is elevated during stress, but also after chronic exposure to organochlorine compounds (Hontela et al. 1992), affects thyroid activity as cortisol treatment decreases plasma T3 levels (Leatherland 1985; Vijayan and Leatherland 1989). Also, Redding et al. (1986) found that in chronically starved eels the in vivo conversion of T4 to T3 is decreased and reduces plasma T3 levels. In addition, Leatherland (1987) found that PCBs also inhibit T3 production in yearling coho salmon, *Oncorhynchus kisutch*. Due to the high level of structural similarity between thyroid hormones and certain PCB metabolites (Brouwer et al. 1990; Murk et al. 1998), these metabolites appear to be competitive with the thyroid hormone for the thyroxin (T4)-binding sites. This will result in elevation of free thyroxin and subsequently in almost complete elimination of thyroxin from the plasma, finally leading to decreased T3 production. More studies investigating the mechanisms of action underlying the effects of PCBs on endocrine tissues revealed that PCBs can interfere with steroidogenesis and steroid receptors (McKinney and Waller 1994).

Quabius et al. (1997) investigated whether the stress response, and in particular the HPI axis, in tilapia, *Oreochromis mossambicus*, is compromised by short-term exposure to PCB 126. Fish of both sexes were fed diets containing PCB 126 (50 μg/kg fish/day) for 5 days. The authors investigated the effect on acute response to capture. Fish were acutely stressed for periods varying between 1 and 30 min at the end of the exposure period. The PCB-fed fish showed lower cortisol levels than controls, which suggests an impaired ability to acutely activate interrenal steroidogenesis in PCB treated tilapia. Also the ACTH and cAMP-stimulated in vitro cortisol release from superfused head kidney was lower in PCB exposed tilapia. This result, together with high PCB 126 concentrations in the head kidneys of exposed fish, may indicate direct toxic effects on the interrenal cells (Quabius et al. 1997).

15.4 Immuno-Suppression by PCBs

Toxicants like PCBs have been related to reduced resistance to diseases, viruses and parasites. For example, an increased disease prevalence has been observed in benthic fishes from coastal areas containing PCB – contaminated sediments (reviewed by Vethaak and Reinhalt 1992). Several field studies have also reported evidence of a disrupted immune function in fishes from inshore areas contaminated with PCBs and other xenobiotics (Warriner et al. 1988; Weeks et al. 1990; Arkoosh et al. 1991, 1994).

We hypothesize that when eels burn fat from poorly perfused storage organs such as the adipose tissue, the released PCBs will primarily be distributed to the highly perfused organs such as the liver, gills, kidney, thymus and spleen. In this respect, it is important to notice that the most important organs in the immune function are the thymus, spleen and liver.

Fat oxidation could lead to immuno-suppression, thus to reduced resistance to diseases. This has been observed in several studies: i.e. thymus atrophy was recorded in European flounder, *Platichthys flesus*, exposed to 50 mg PCB-126/kg (Grinwis et al. 2001). It was hypothesized that this may have an impact on the specific resistance against infectious diseases (viral, bacterial, parasitical) in the field situation. With respect to the effect of PCBs on the immunologically important spleen (Taysse et al. 1998) contradictory results are found. In some studies a reduction of the spleen weight was observed (Nakata et al. 2002), while in others spleen hypertrophy was observed (Greichus et al. 1975). Spleen and head-kidney are immune organs in fishes

(Taysse et al. 1998). These authors indicated for carp that both organs also contribute to the biotransformation process, necessary for the elimination of xenobiotics.

The immuno-suppressing effect of PCB 126 on different pathogens has been reviewed by Sures and Knopf (2004). They showed that in European eel PCB 126 suppresses the humoral immune response and that it increased the incidence of infection by *Anguillicola crassus*. Coplanar congeners such as PCB 126 are especially toxic and known to affect the immune response of fishes (Regala et al. 2001). This seems to be related to a structural similarity to 2,3,7,8-tetrachlorodibenzo-p-dioxin (TCDD), the most toxic of all halogenated aromatic hydrocarbons (HAH) (Regala et al. 2001). The degree of immunotoxicity of PCBs correlates with the degree of binding affinity to the cytosolic aryl hydrocarbonreceptor (Ahr) (Kafafi et al. 1993), which is a well-described transcription factor for a variety of gene products, including cytochrome P 450 1A (Hahn and Stegeman 1994). Due to its similarity with TCDD (equivalent factor of 0.1), PCB 126 should exhibit TCDD-like toxicities but at a lower level (Regala et al. 2001; Kafafi et al. 1993). HAHs were described to show a moderate suppression of antibody response in rainbow trout (Spitzbergen et al. 1986a, b) whereas PCB 126 was not found to affect the antibody response of channel catfish, *Ictalurus punctatus*, to *Edwardsiella ictaluri* (Rice and Schlenk 1995; review Sures and Knopf 2004). In contrast, Regala et al. (2001) found suppression of the antibody response in channel catfish against *Vibrio anguillarum* following intraperitoneal injection of *I. punctatus* with 1 mg kg^{-1} PCB 126, whereas the application of 0.01 mg kg^{-1} PCB 126 did not reduce the antibody responses. Hence, the immunotoxicity is clearly related to the PCB 126 concentrations applied (reviewed: Sures and Knopf 2004).

15.5 PCB Toxicity: Interaction with the Cytochrome P450 System

A number of studies have been carried out on the biotransformation of PCBs, but a complete picture of the metabolism has not been put forward. Available data on fishes indicate that most of the PCBs are only metabolized slowly, if at all (Hutzinger et al. 1972; Melancon and Lech 1976). The rapid incorporation of certain types of organic pollutants, including PCBs in fat and liver can increase the content of enzymes collectively designated as cytochrome P-450, which are active components of the mono-oxygenase system. Some of them also increase the activity of conjugating enzymes (Jimenez and Stegeman 1990). Mono-oxygenases (highest concentrations are found in the liver) and conjugating enzymes help animals to detoxify and eliminate many pollutants from their tissues (Lu et al. 1976). Several studies have demonstrated that fish P-450-mediated enzyme activity can be stimulated by PCB mixtures (Melancon et al. 1981; Kleinow et al. 1987). This is supported by field observations of high levels of mono-oxygenase activity in fish livers, which have been reported downstream from a PCB incineration plant on the river Rhine (Monod et al. 1988). Other examples of PCB-induced mono-oxygenase activity in fish liver are given in Kleinow et al. (1987). Several factors can modify the activity of detoxification enzymes in fish, such as temperature, sex, age and nutritional status. The modulating effects of these factors have been considered in reviews by Stegeman (1982), Kleinow et al. (1987), Payne et al. (1987) and Stegeman and Kloepper-Sams (1987).

Hormonal factors can influence mono-oxygenase systems in fishes. Steroids have been shown to suppress the total cytochrome P-450 and specific mono-oxygenase activities (Stegeman et al. 1982). Reproductive hormones probably regulate seasonal changes in mono-oxygenase activity. In many fish species, mono-oxygenase activity decreases shortly before or during the spawning season (Walton et al. 1983). Despite the decline of the mono-oxygenase activity during the reproductive season, it can be increased again by xenobiotic chemicals during this period (Payne and Penrose 1975). If a depressed mono-oxygenase system is induced by a contaminant during the reproductive period, this might affect reproductive success, because steroids are endogenous substrates for certain types of mono-oxygenases. The overall effect will result in to a reduction of circulating steroid levels (Sivarajah et al. 1978).

It is shown when mono-oxygenase activity is increased, gamete viability and embryological development are decreased (Spies and Rice 1988). Starved fishes exhibit lower levels of mono-oxygenase activity than well-nourished fishes (Jimenez et al. 1988). However, as in maturing animals, pollutants increase mono-oxygenase activity in starved fishes (Jimenez and Burtis 1989). Eels are thought to starve for as long as their migration endures, which is approximately 6 months. It is likely that during this period the mono-oxygenase activity increases when PCB levels in the body increase due to fat oxidation. A recent population genetics study (Maes et al. 2005) showed a significant negative correlation between heavy metal pollution and condition, suggesting an impact of pollution on the health of yellow eels. This result can be explained in the following way. The level of bioaccumulation of toxicants is not only dependent on the level of pollution, but also on the individual's capacity for detoxification, which correlates with the individual's level of heterozygosity for these enzymes.

15.6 Effect of PCBs on Migration

Whether swimming has an influence on the PCB level and what the effect would be on the metabolism is of particular interest. To test the fate of PCBs during migration, silver eels were swum for 27 days at 0.5 body length (BL) s^{-1}, during which they swam 800 km (Van Ginneken 2006). In this study we exposed the eels to PCBs at ten times the amount acceptable in fishes for human consumption. Before the start, the eels were injected intraperitoneally with a PCB-mixture (5,000 μg kg^{-1} PCB-153 + 7 μg/kg PCB-126 + 50 μg kg^{-1} PCB-77) or with the vehicle corn oil (10 ml kg^{-1}, controls). So the PCB-mixture consisted of a di-ortho-, planar- and metabolizable PCBs in a relative and absolute amount that is environmentally relevant.

Figure 15.4 shows the oxygen consumption of swimming eels with and without PCB injection. Note that the oxygen consumption rate of the PCB – loaded animals

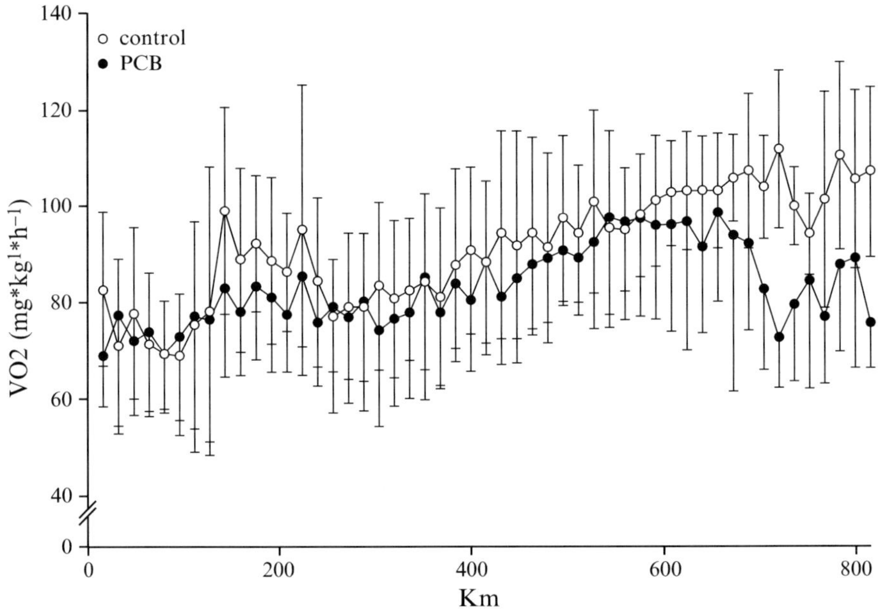

Fig. 15.4 Average oxygen consumption profiles of control and PCB-exposed eels swimming at 0.5 BL per second for 27 days (800-km). The oxygen consumption of the PCB-exposed eels is significantly depressed compared to the controls from day 18 onwards. The PCB-group was dosed intraperitoneally with an environmentally relevant mix of 5 mg PCB-153 kg^{-1}, 7 µg PCB-126 kg^{-1} and 50 µg PCB-77 kg^{-1}

was significantly lower over the last 400 km. PCB-concentrations in the muscle were 2.7 times higher in swimming compared to resting animals ($P \leq 0.0001$). The spleen somatic index was increased in the PCB-exposed swim animals ($P \leq 0.023$) but not in the PCB-exposed control animals ($P \leq 0.148$). There were no significant differences in plasma glucose and cortisol levels.

The fact that, in both treated and untreated animals, the swimmers had 2.7 times more PCBs in the muscle tissue certainly suggests that the injected PCBs accumulate faster in muscle when the animal swims. Of course during swimming the muscle is much better perfused than in resting animals. Hence, PCBs move in the first place to the better perfused organs and are not evenly distributed in the body. The PCBs in the treated group were injected shortly before the swim trial. The PCB level in the control group was about ten times lower that the treated group, but more importantly those PCBs would have been well distributed over the whole body. The PCB level in the muscle still increased 2.7 fold. This result suggests that swimming causes mobilization and redistribution of PCBs. Therefore we must conclude that, particularly during swimming, PCBs enter the circulation much faster than when the animal is resting. This must have effects on the tissues with a high level of detoxification enzymes, such as spleen and liver. We observed a doubling of the spleen/somatic index in PCB loaded swimmers compared to all other groups. Possibly the higher exposure to circulating PCBs increased the spleen weight, which contributes to a higher detoxification capacity.

The weight loss in the PCB-dosed animals was not significantly lower than in the respective controls, although the reduction (29%) corresponded with the lower oxygen consumption in the PCB treated eels. This finding is surprising because dioxin-like compounds usually result in extreme weight losses. The dioxin-related 'wasting syndrome' has also been reported in fishes and is associated with a strong suppression of protein synthesis (Kleeman et al. 1988). As the animals in our study were not fed during the experiments, the reduced weight loss in the PCB-exposed groups cannot be ascribed to an increased intake of food, and must therefore be the result of a changed intermediary metabolism. PCB-exposure significantly reduced oxygen consumption during swimming of the PCB-exposed animals from 400 km on and this effect increased with time. The Cost of Transport (COT, [mg O_2 kg^{-1} km^{-1}]) was significantly lower in PCB exposed animals from 100 km up to 800 km. In addition, the standard metabolic rate measured 2 days after the last swimming activity was significantly lower in the PCB-exposed animals (Fig. 15.5d).

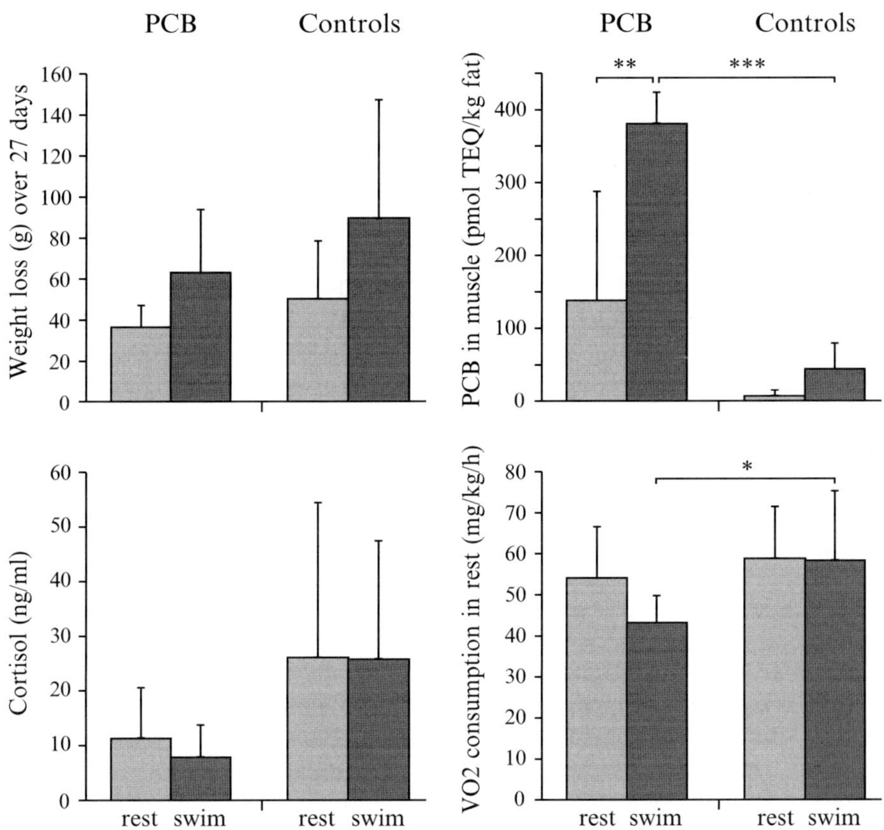

Fig. 15.5 Data from four eel groups: control and PCB-exposed eels, either resting or swimming at 0.5 BL per second for 27 days. **(a)** Weight loss, **(b)** PCB-levels in muscle, **(c)** cortisol level in plasma, **(d)** Standard Metabolic Rate (SMR). *, ** and ***: denotes a significant difference at respectively $P \leq 0.05$, $P \leq 0.01$; $P \leq 0.001$ (Van Ginneken 2006)

15.7 Effect of PCBs on Eel Reproduction

To study the relation between environmental PCB contamination and recruitment, we need to catch mature eels from the Sargasso Sea. As PCBs particularly disturb embryonic and early larval development, this can also be deduced from studying sexual maturation in eels with different PCB levels Artificial reproduction with pituitary extracts has been described for Japanese eel (Yamamoto and Yamauchi 1974; Yamauchi et al. 1976) and European eel (Van Ginneken et al. 2005a; Palstra et al. 2005; Pedersen 2003, 2004).

When the eel develop gonads, PCBs will likely be redistributed in these structures and their products as occurs in other fish species (Freeman and Idler 1975; Subramanian et al. 1987; Thomas 1988; Spies and Rice 1988; Janssen et al. 1995; Janssen 1996). The relationship between contaminant levels in spawning fishes and the passage of PCBs to their reproductive products has only briefly been studied. Some studies have reported reduced survival rates of embryos and larvae for eggs containing PCBs in the mg kg^{-1} range. Johansson (1970) reported 46–100% mortality in fertilized embryos of Atlantic salmon, *Salmo salar*, whose PCB levels ranged from 0.6 to 1.9 mg kg^{-1} on a wet weight basis and 14.4–34.0 mg kg^{-1} on a lipid basis. In addition, a significant increase in mortality with increasing PCB levels was indicated. Hogan and Brauhn (1975) reported rainbow trout eggs, which contained 0.33 μg g $lipid^{-1}$ Aroclor 1254. This resulted in 10–28% larval mortality and 60–70% of the young were deformed at 30 days post-hatch. Other studies monitored the redistribution of PCBs in the body during gonadal maturation (Guiney et al. 1979). In the pre-spawning period the elimination of PCBs was slow: it became more rapid in both sexes during the spawning season. Clearly PCBs were eliminated that were accumulated in eggs and sperm (Guiney et al. 1979).

To illustrate the potential effects of PCBs on reproduction, a study by Stouthart et al. (1998) will be discussed further in detail. This study investigated whether PCB 126 exposure during embryonic development induces oedema and endocrine stress responses in carp larvae, *Cyprinus carpio*. It was found that α-MSH, ACTH and cortisol levels were elevated in carp embryos exposed to PCBs. In animals exposed to 10^{-10} and 10^{-9} mol PCB 126 l^{-1} these parameters increased in a concentration-related manner, but the changes became evident only at 144 h post-fertilization, i.e. after resorption of the yolk-sac. Exposure to high PCB levels resulted in stable but elevated whole-body ACTH and α-MSH levels until 216 h, while the whole-body cortisol concentration gradually decreased from 168 h post fertilization and was significantly below control values at 216 h post-fertilization. Cortisol is a stress hormone, while ACTH and α-MSH are the products of one pituitary hormone precursor, proopiomelanocortin (POMC) (Alrubian et al. 2003). ACTH is involved in the control of stress and metabolism while the MSH regions in the Pars Intermedia of the pituitary are the fastest growing regions in the stages from yellow stage elvers until 12–14 cm (Grandi et al. 2003). It can be concluded that α-MSH plays a role in the prolonged aging of the yellow eel towards silvering.

PCB exposure also led to swelling of the yolk-sac and especially the pericardium (elevated wet weights of embryos/larvae), thus showing disturbed hydromineral balance (oedema). The increased ACTH and cortisol levels (at high PCB doses) indicate that PCB 126 induces a stress response in carp larvae, possibly mediated by a disturbed hydromineral balance.

Indeed, during the swim up stage when the yolk-sac had been resorbed, which is 3 days after hatching, the PCB-exposed larvae showed yolk-sac and pericardial oedema. This indicates a disturbed water balance, which in adult fishes induces a stress response, partly controlled by the Hypothalamus Pituitary Interrenal (HPI) axis via cortisol (Wendelaar Bonga and Lock 1992).

In other studies, PCB 126 has been shown to disturb water balance in developing larvae of several fish species, as indicated by severe yolk-sac and pericardial oedema (Walker and Peterson 1994; Walker et al. 1994; Zabel et al. 1995). As the HPI axis is involved in hydromineral regulation (also in larvae; Hwang and Wu 1993; Stouthart et al. 1998), oedema might be caused by inhibition of this axis by PCB exposure, but oedema itself also induces a stress response to counter the hydromineral imbalance.

In our research group at Leiden University, eggs of 13 different batches of silver eels have been fertilized. Embryonic development of healthy embryos was followed until 4 days after fertilization (Palstra et al. 2006). A negative correlation was observed between dioxin-like contaminants (>80% PCBs in eel) and embryonic survival and development (Fig. 15.6; Palstra et al. 2006). Effects occurred below the maximal allowable level in fishes for human consumption i.e. 12 ng TEQ/kg (Anonymous 2006). Embryos of PCB-contaminated eels showed serious oedema of the yolk sac, a deformed head region and absence of a heartbeat. Such embryonic malformations are typical for PCB-exposed eggs and indicate negative interference by dioxin-like

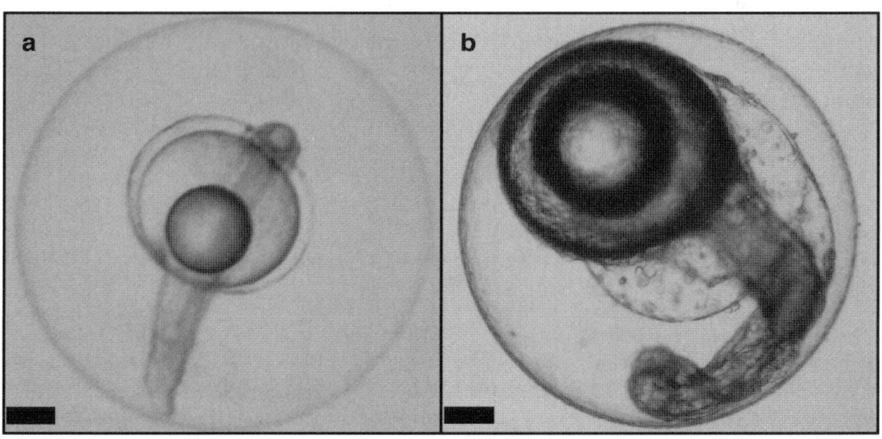

Fig. 15.6 **(a)** Healthy European eel embryo at 30–35 hpf with heartbeat and yolk sac with large fat droplet (left). **(b)** Larger embryo of an unhealthy batch at identical time of development displaying yolksac oedema, deformed head region and absence of heartbeat (Palstra et al. 2006)

contaminants (Helder 1980; Walker and Peterson 1991; Walker et al. 1994; Stouthart et al. 1998). Therefore, parental levels of dioxin-like contaminants were measured and their distribution correlated to embryonic survival and development.

15.8 Conclusions

In Fig. 15.3 the historical data series on PCB levels in eels starting in the late 1970s from the Rhine River and Lake Ijssel show a gradual decline from the start of monitoring in 1978 (Pieters et al. 2001). Figure 15.7 shows that the peak in PCB contamination is at the start of the 1970s, while the decline of eel populations started in the 1980s. This observation led Knights (1997), using a risk assessment method, to conclude that "organochlorine contamination has not been a major cause of recent declines in eel recruitment" (Knights 1997). Two remarks can be made. First, because results of the effects of migration of PCB contaminated animals were lacking at that time the conclusion of Knights (1997) seems to be premature. At this moment, only the results of an 800 km migration experiment of PCB loaded animals are available (Van Ginneken 2006) and discussion is still open. Secondly, the fact that the peak emission of PCBs in the environment preceded the decline of the European eel is not in contrast to our main conclusion of this chapter that PCBs might be a contributing factor in the decline of eel populations. The observed 10 years delay in response of the eel population to PCBs can be ascribed to the

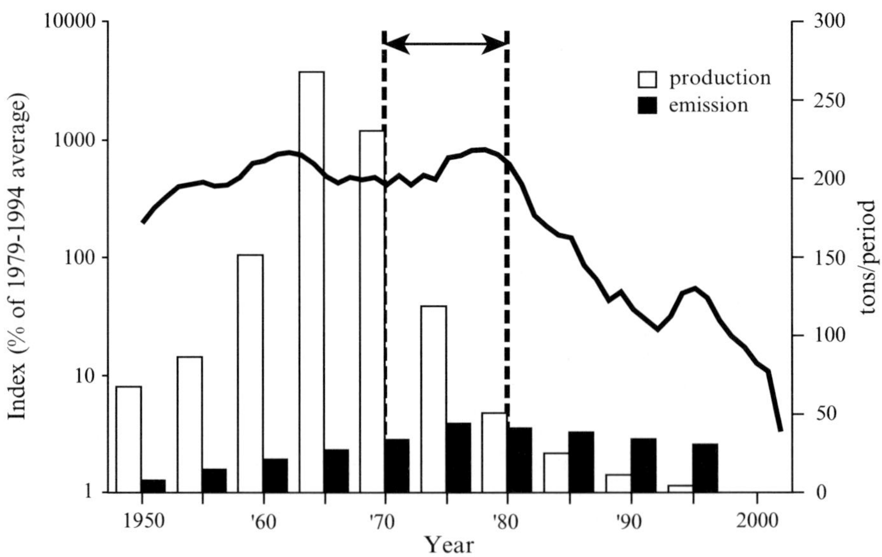

Fig. 15.7 Trends in glass eel recruitment to the European continent (Derived from Dekker 2004) and the estimated production and emissions of pcbs in OECD countries (Derived from van Leeuwen and Hermens 1995)

long juvenile phase of 10–15 years in the contaminated inland waters, where lipid reserves are stored to provide energy for the transoceanic spawning migration and gamete maturation. Therefore a correction has to be made for the generation time in order to correct for the mismatch between both trends (Robinet and Feunteun 2002).

At what concentration are pollutants like PCBs likely to reduce the breeding success in eels? De Boer et al. (1994) calculated half-lives for PCBs of between 380 and 1,450 days in their 8 years field study. For the higher chlorinated PCBs (hexa-octa-PCBs), no elimination at all was found (De Boer et al. 1994). Castonguay et al. (1994) calculated the residence of PCBs, mirex and pesticides in Lake Ontario to be 10.3 years. Because the spawning migration of European eel takes only 6 months (Van Ginneken et al. 2005a,b) it seems reasonable to assume that the contaminant level of eels after their spawning migration may be even much higher than the level at the start of their journey.

In addition, our studies indicate that PCBs have deleterious effects on the embryonic development (Palstra et al. 2006).

The disrupting effects occurred at levels below 4 ng TEQ kg gonads^{-1}: these are below the EU eel consumption standard which has been set at 12 ng TEQ kg filet^{-1} based on dioxins and dioxin like compounds (Anonymous 2006).

Furthermore, long-term swimming will increase the levels of PCBs due to specific usage of the fat stores, which will increase the PCB level in the blood and We recommend that PCB contamination in eels should be monitored in all major hydro-systems and those areas with low PCB levels in eels should be protected.

Acknowledgements This work has been supported by the European contract EELREP N°Q5RS-2001–01836.

References

Alrubian J, Sollars C, Danielson PB, Dores RM (2003) Evaluating the radiation of the POMC gene in teleosts: characterization of American eel POMC. Gen. Comp. Endocrinol. 132: 384–390.

Anonymous (1987) Résultats d'analyse de Poissons du Rhin. Agence Financière de Bassin Rhin-Meuse. Moulins lesz Metz, Rapp. No. 11686 D, 6 pp.

Anonymous (2006) COMMISSION REGULATION (EC) No 199/2006 of 3 February 2006 amending Regulation (EC) No 466/2001 setting maximum levels for certain contaminants in foodstuffs as regards dioxins and dioxin-like PCBs.

Arkoosh MR, Cassillas E, Clemons E, McCain B, Varanasi U (1991) Suppression of immunological memory in juvenile chinook salmon (Oncorhynchus tshawytscha) from an urban estuary. Fish Shellfish Immunol. 1: 261–278.

Arkoosh MR, Clemons E, Myers M, Cassillas E (1994) Suppression of B-cell immunity in juvenile chinook salmon (Oncorhynchus tschawytscha) after exposure to either polycyclic aromatic hydrocarbon or to polychlorinated biphenyls. Immunopharmacol. Immunotoxicol. 16: 293–314.

Barron MG, Galbraith H, Beltman D (1995) Comparative reproductive and developmental toxicology of PCBs in birds. Comp. Biochem. Physiol. C112: 1–14.

Bertin L (1956) Eels: A Biological Study. Cleaver-Hume Press, London.

Bertrand P, Fournier S, Vigneaulty Y (1986) Concentrations en biphenyles polychlorés et en métaux dans les sédiments de la Baie des Anglais (Québec). Rapp. Can. Sci. Halieut. Aquat. 568: 5.
Bimbos D, Mau G (1986) Scahdestoffe in hessischen Flussfischen. Situation, Bewertung und Verzehrsempfehlungen. Fischwirt 36: 69–71.
Birnbaum LS (1994) The mechanism of dioxin toxicity: relationship to risk assessment. Environ. Health Perspect. 102: 157–167.
Borlakoglu JT, Haegele KD (1991) Comparative aspects on the bioaccumulation, metabolism and toxicity of PCBs. Comp. Biochem. Physiol. C 100: 327–338.
Brouwer A, Murk AJ, Koeman JH (1990) Biochemical and physiological approaches in ecotoxicology. Funct. Ecol. 4: 275–281.
Brouwer A, Morse DC, Lans MC, Schuur AG, Murk AJ, Klasson-Wehler E, Bergman A, Visser TJ (1998) Interactions of persistent environmental organohalogens with the thyroid hormone system: mechanisms and possible consequences for animal and human health. Toxicol. Ind. Health 14(1/2) 59–84.
Brown CL, Doroshov SL, Cochran MD, Bern HA (1989) Enhanced survival in striped bass fingerlings after maternal triiodothyronine treatment. Fish. Physiol. Biochem. 7: 295–299.
Bruggeman WA (1983) Bioaccumulation of polychlorobiphenyls and related hydrophobic chemicals in fish. Rijkswaterstaat. Rijksinstituut voor Zuivering van Afvalwater; Lelystad.
Brusle J (1991) The eel (Anguilla sp) and organic chemical pollutants. Sci. Total Environ. 102: 1–19.
Castonguay M, Dutil J-D, Desjardins C (1989) Distinction between American eels (Anguilla rostrata) of different geographic origins on the basis of their organochlorine contaminant levels. Can. J. Fish. Aquat. Sci. 46: 836–843.
Castonguay M, Hodson PV, Couillard CM, Eckersley MJ, Dutil J-D, Verreault G (1994) Why is recruitment of the American eel, Anguilla rostrata, declining in the St Lawrence River and Gulf? Can. J. Fish. Aquat. Sci. 51: 479–488.
Cyr DG, Eales JG (1988) In vitro effects of thyroid hormones on gonadotropin-induces estradiol-17β, secretion by ovarian follicles of rainbow trout, Salmo gairdneri. Gen. Comp. Endocrinol. 69: 80–87.
De Boer J, Stronck CJN, Traag WA, van der Meer J (1993) Non-Ortho and Mono-Ortho substituted Chlorobiphenyls and Chlorinated Dibenzo-p-Dioxis and Dinzofurans in Marine and Freshwater Fish and Shellfish from the Netherlands. Chemoshphere CMSHAF, 26: 1823–1842.
De Boer J, Van der Valk F, Kerkhoff MAT, Hagel P, Brinkman UATh (1994) 8-Year study on the elimination of PCBs and other organochlorine compounds from eel (Anguilla anguilla) under natural conditions. Environ. Sci. Technol. 28(13): 2242–2248.
De Boer J, Hagel P (1994) Spatial differences and temporal trends of chlorobiphenyls in yellow eel (Anguilla anguilla) from inland waters of the Netherlands. Sci. Total Environ. 141: 155–174.
Dekker W (2004) Slipping through our hands. Population dynamics of the European eel, Ph.D. thesis, University of Amsterdam, 186 pp. ISBN 90–74549–10–1.
De Voogt WP (1990) QSARs for the environmental behavior of polynuclear (hetero) aromatic compounds, studies in aqueous systems. Ph.D. thesis, VU Amsterdam. Deusichem; de Betuwe B.V. 235 pp. ISBN: 90–5549–270–1.
Dickhoff WW, Yan L, Plisetskaya EM, Sullivan CV, Swanson P, Hara A, Bernard MG (1989) Relationship between metabolic and reproductive hormones in salmonid fish. Fish. Physiol. Biochem. 7: 147–155.
Eales JG, MacLatchy DL (1989) The relationship between T3 production and energy balance in salmonids and other teleosts. Fish Physiol. Biochem. 7(1–4): 289–293.
Freeman HC, Idler DR (1975) The effect of polychlorinated biphenyls on steroidogenesis and reproduction in the brook trout (Salvelinus fontinalis). Can. J. Biochem. 3: 666–670.
Fry DM, Toone CK (1981) DDT-induced feminization of gull embryos. Science 213: 922–924.
Giam CS, Chan HS, Neff GS (1978) Phthalate ester plasticizers, DDT, DDE and polychlorinated biphenyls in biota from the Gulf of Mexico. Mar. Pollut. Bull. 9: 249–251.

Giesy JP, Ludwig JP, Tillitt DE (1994) Deformities in birds of the Great Lakes Region. Environ. Sci. Technol. 28: 128A–135A.

Grandi G, Colombo G, Chicca M (2003) Immunocytochemical studies on the pituitary gland of Anguilla anguilla L., in relation to early growth stages and diet-induced sex differentiation. Gen. Comp. Endocrinol. 113: 413–428.

Goksøyr A, Förlin L (1992) The cytochrome P450 system in fish, aquatic toxicology and environmental monitoring. Aquat. Toxicol. 22: 287–312.

Greichus YA, Call DJ, Ammann BM (1975) Physiological effects of polychlorinated biphenyls or a combination of DDT, DDD, and DDE in penned white pelicans. Arch. Environ. Contam. Toxicol. 3: 330–343.

Grinwis GC, Van den Brandhof EJ, Engelsma MY, Kuiper RV, Vaal MA, Vethaak AD, Wester PW, Vos JG (2001) Toxicity of PCB-126 in European Flounder (Platichthys flesus) with emphasis on histopathology and cytochrome P4501A induction in several organ systems. Arch. Toxicol. 75: 80–87.

Guiney PD, Melancon MJ Jr, Lech JJ, Peterson RE (1979) Effects of egg and sperm maturation and spawning on the distribution and elimination of a polychlorinated biphenyl in rainbow trout (Salmo gairdneri). Toxicol. Appl. Pharmacol. 47(2): 261–272.

Hahn ME, Stegeman JJ (1994) Regulation of cytochrome P450 A1 in teleost: sustained induction of CYP1A1 mRNA, protein and catalytic activity by 2,3,7,8-tetrachlorodibenzo-p-dioxin in the marine fish Stenotomous chrysops. Toxicol. Appl. Pharmacol. 127: 187–198.

Haiber G, Schöler HF (1994) Identification of di-o,o'-cl-,mono-o-cl-and non-o-cl-substituted PCB congeners in neckar river fish. Chemosphere 28: 1913–1919.

Helder T (1980) Effects of 2,3,7,8-tetratchlorodibenzo-p-dioxin (TCDD) on early life stages of the pike (Esox lucius L.) Sci. Total Environ. 14: 255–264.

Hendriks AJ, Pieters H, De Boer J (1998) Accumulation of Metals, Polycyclic (Halogenated) Aromatic Hydrocarbons, and Biocides in Zebra Mussel and Eel from the Rhine and Meuse Rivers. Environ. Toxicol. Chem. 17: 1885–1898

Hernandez LM, Rico MC, Gonzalez MJ, Montero MC, Fernandez MA (1987) Residues of organochlorine chemicals and concentrations of heavy metals in ciconiform eggs in relation to diet and habitat. J. Environ. Health B22(2): 245–258.

Hodson PV, Castonguay M, Couillard CM, Desjardins C, Pelletier E, McLeod R (1994) Spatial and temporal variations in chemical contamination of American eels, Anguilla rostrata, captured in the estuary of the St Lawrence River. Can. J. Fish. Aquat. Sci. 51: 464–478.

Hogan JW, Brauhn JL (1975) Abnormal rainbow trout fry from eggs containing high residues of a PCB (Aroclor 1242). Prog. Fish-Cult. 37: 229–230.

Holden AV, Marsden K (1967) Organochlorine pesticides in seals and porpoises. Nature, Lond. 216: 1274–1276.

Holmes DC, Simmons JH, Tatton JOG (1967) Chlorinated hydrocarbons in British wildlife. Nature, Lond. 216: 227–229.

Hontela A, Rasmussen JB, Audet C, Chevalier G (1992) Impaired cortisol stress response in fish from environments polluted by PAHs, PCBs and mercury, Arch. Environ. Contam. Toxicol. 22: 278–283.

Hontela A, Dumont P, Duclos D, Fortin R (1995) Endocrine and metabolic dysfunction in yellow perch (Perca flavescens) exposed to organic contaminants and heavy metals in the St. Lawrence river. Environ. Toxicol. Chem. 14: 725–731.

Huschenbeth E (1977) Uberwachung der Speicherung von chlorierten Kohlenwasserstoffen im Fisch. Fischereiwiss. 28: 173–186.

Hutzinger O, Nash DM, Safe S, Defritas ASW, Norstrum RJ, Wildish DJ, Zitko V (1972) Polychlorinated biphenyls: metabolic behavior of pure isomers in pigeons, rats and brook trout. Science 178: 312–314.

Hwang PP, Wu SM (1993) Role of cortisol in hypo-osmoregulation in larvae of tilapia (Oreochromis mossambicus). Gen. Comp. Endocrinol. 92: 318–324.

Janssen PAH, Lambert JGD, Goos HJTh (1995) The annual ovarian cycle and the influence of pollution on vitellogenesis in the flounder, Pleuronectus flesus (L.) J. Fish Biol. 47: 509–523.

Janssen PAH (1996) Reproduction of the flounder, Platichthys flesus (L.), in relation to environmental pollution. Steroids and vitellogenesis. Ph.D. thesis Universiteit Utrecht, 174 pp.

Jensen S (1966). Report of a new chemical hazard. New Sci. 32: 612.

Jensen S, Johanson N, Olsson M (1970) DDT and PCB in marine animals from Swedish waters, Nature, Lond. 224: 247–250.

Jimenez BD, Burtis LS (1989) Influence of environmental variables on the hepatic mixed function oxidase system in the bluegill sunfish (Lepomis macrochirus). Comp. Biochem. Physiol. 93: 11–21.

Jimenez BD, Burtis LS, Ezell GH, Egan BZ, Lee NE, Beauchamp JJ, McCarthy JF (1988) The mixed function oxidase system of bluegill sunfish Lepomis macrochirus: correlation of activities in experimental and wild fish. Environ. Toxicol. Chem. 7: 623–634.

Jimenez BD, Stegeman JJ (1990) Detoxification enzymes as indicators of environmental stress on fish. Am. Fisher. Soc. Symp. 8: 67–79.

Johansson N (1970) PCB – indications of effects on fish, pp. 58–68. Proceedings of the PCB Conference, Stockholm, Sweden, September 29, 1970.

Kafafi SA, Afeefy HY, Ali AH, Said HH, Kafafi G (1993) Binding of polychlorinated biphenyls to the aryl hydrocarbon receptor. Environ. Health Persp. 101: 422–428.

Kennish MJ, Belton TJ, Hauge P, Lockwood K, Ruppel BE (1992) Polychlorinated biphenyls in estuarine and coastal marine waters of New Jersey: a review of contamination problems. Rev. Sci. Aquat. 6: 275–293.

Klamer J, Laane RWPM, Marquenie JM (1991) Sources and fate of PCBs in the North Sea: a review of available data. Wat. Sci. Tech. 24:77–85.

Kleeman JM, Olson JR, Chen SM, Peterson RE (1988) Species differences in 2,3,7,8-tetrachlorodibenzo-p-dioxin toxicity and biotransformation in fish. Toxicol. Appl. Pharmacol. 83: 391–401.

Kleinow KM, Melancon MJ, Lech JJ (1987) Biotransformation and induction: implications for toxicity, bioaccumulation and monitoring of environmental xenobiotics in fish. Environ. Health Persp. 71: 105–119.

Knights B (1997) Risk assessment and management of contamination of eels (Anguilla spp.) by persistent xenobiotic organochlorine compounds. Chem. Ecol. 13: 171–12.

Kruse R, Boek K,Wolf M (1983) Der Gehalt an Organchlor Pestiziden and Polychlorierten Biphenylen in Elbaalen. Arch. Lebensmitelhyg. 34: 81–86.

Leatherland JF (1985) Studies of the correlation between stress-response, osmoregulation and thyroid physiology in rainbow trout, Salmo gairdnerii (Richardson). Comp. Biochem. Physiol. 80A(4): 523–531.

Leatherland JF (1987) Lowering of serum thyroxine and triiodothyronine levels in yearling coho salmon, Oncorhynchus kisutch, by dietary mirex and PCBs. J. Fish. Res. Can. 35(10): 1285–1289.

Lu AYH, Kuntzman R, Conney AH (1976) The liver microsomalhydroxylation enzyme system. In: Frontiers of Gastrointestinal Research (van der Reis, L. ed.), vol. 2, pp1-31, Karger, Basel.

Maes GE, Raeymakers JAM, Pampoulie C, Seynaeve A, Goemans G, Belpaire C, Volckaert FAM (2005) The catadromous European eel Anguilla anguilla (L.) as a model for freshwater evolutionary ecotoxicology: Relationship between heavy metal bioaccumulation, condition and genetic variability. Aquat. Toxicol. 73: 99–114.

Mason CF (1993) Organochlorine pesticide residues and PCBs in eels Anguilla anguilla from some British reedbeds. Chemosphere 26:2289–2292.

McKinney JD, Waller CL (1994) Polychlorinated biphenyls as hormonally active structural analogues. Environ. Health Persp. 102: 290–297.

Melancon MJ Jr, Lech JJ (1976) Isolation and identification of a polar metabolite of tetrachlorobiphenyl from bile of rainbow trout exposed to 14C-tetrachlorobiphenyl. Bull. Environ. Contam. Toxicol. 15: 181–188.

Melancon MJ, Elcombe CR, Vodicnik MJ, Lech JJ (1981) Induction of cytochromes P-450 and mixed-function oxidase activity by polychlorinated biphenyls and á-naphthoflavone in carp (Cyprinus carpio). Comp. Biochem. Physiol. 69C: 219–226.

Monod G, Gevaux A, Riviere JL (1988) Effects of chemical pollution on the activities of hepatic xenobiotic metabolising enzymes in fish from the river Rhone France. Sci. Total Environ. 73: 189–202.

Murk AJ, Van den Berg JHJ, Fellinger M, Rozemeijer MJC, Swennen C, Duiven P, Boon JP, Brouwer A, Koeman JH (1994) Toxic and biochemical effects of 3,3′,4,4′-tetrachlorobiphenyl (CB 77) and Cophen A50 on eider ducklings (Somateria mollissima) in a semi-field experiment. Environ. Pollut. 86: 21–30.

Murk AJ, Leonards PEG, Hattum B, van Luit R, Van der Weiden MEJ, Smit M (1998) Application of biomarkers for exposure and effect of polyhalogenated aromatic hydrocarbons in naturally exposed European otters (Lutra lutra). Environ., Toxicol. Pharmacol. 6: 91–102.

Nakata H, Sakakibara A, Kanoh M, Kudo S, Watanabe H, Miyazaki N, Asano Y, Tanabe S (2002) Evaluation of mitogen-induced responses in marine mammals and human lymphocytes by in-vitro exposure to butyltins and non-ortho coplanar PCBs. Environ. Pollut. 120: 245–253.

Palstra AP, Cohen EGH, Niemantsverdriet PRW, Van Ginneken VJT, Van den Thillart GEEJM (2005) Artificial maturation and reproduction of European silver eel: development of oocytes during final maturation. Aquaculture 249: 533–547.

Palstra AP, van Ginneken VJT, Murk AJ, Van den Thillart GEEJM (2006) Are dioxin-like contaminants responsible for the eel (Anguilla anguilla) drama? Naturwissenschaften 93(3): 145–148.

Payne JF, Penrose WR (1975) Induction of aryl hydrocarbon benzo(a)pyrene hydroxylase in fish by petroleum. Bull. Environ. Contam. Toxicol. 14: 112–116.

Payne JF, Fancey LL, Rahimtula AD, Porter EL (1987) Review and perspective on the use of mixed-function oxygenase enzymes in biological monitoring. Comparative Biochemistry and Physiology C, Comp. Pharmacol. Toxicol. 86: 233–245.

Pedersen BH (2003) Induced sexual maturation of the European eel Anguilla anguilla and fertilization of the eggs. Aquaculture 224: 323–338.

Pedersen BH (2004) Fertilization of eggs, rate of embryonic development and hatching following induced maturation of the European eel Anguilla anguilla. Aquaculture 237: 461–473.

Pieters H, Van Leeuwen SPJ, De Boer J (2001) Contamination of eel and pike-perch: monitoring program for the Dutch anglers 2000. RIVO Report C064/01 Wageningen University, 17 pp.

Pelletier C, Doucet E, Imbeault P, Tremblay A (2002) Associations between weight loss-induced changes in plasma Organochlorine Concentrations, Serum T3 concentrations, and resting metabolic rate. Toxicol. Sci. 67: 46–51.

Quabius ES, Balm PHM, Wendelaar Bonga SE (1997) Interrenal stress responsiveness of tilapia (Oreochromis mossambicus) is impaired by dietary exposure to PCB 126. Gen. Comp. Endocrinol. 108: 472–482.

Rahman MS, Bowadt S, Larsen B (1993) Dual-column GC analysis of Mediterranean fish for ten organochlorine pesticides and sixty two chlorobiphenyls. J. High Res. Chrom. 16: 731–735.

Redding JM, DeLuze A, LeLoup-Hatey J, LeLoup J (1986) Suppression of plasma thyroid concentrations by cortisol in the European eel Anguilla anguilla. Comp. Biochem. Physiol. 83A: 409–413.

Regala RP, Rice CD, Schwedler TE, Dorociak IR (2001) The effects of tributyltin (TBT) and 3,3϶4,4϶5-pentachlorobiphenyl (PCB-126) mixtures on antibody responses and phagocyte oxidative burst activity in channel catfish, Ictalurus punctatus. Arch. Environ. Con. Toxicol. 40: 386–391.

Rice CD, Schlenk D (1995) Immune function and cytochrome P4501A activity after acute exposure to 3,3϶4϶,5-Pentachlorobiphenyl (PCB 126) in channel catfish. J. Aquat. Anim. Health 7: 195–204.

Risebrough RW, Reiche P, Peakall DB, Herman SG, Kerven MN (1968) Polychlorinated biphenyls in the global system. Nature 220: 1098–1102.

Robinet T, Feunteun E (2002) Sublethal effects of exposure to chemical compounds: a cause for the decline in Atlantic eels ? Ecotoxicology 265–277.

Ryan JJ, Lau PY, Pilon JC, Lewis D, McLeod HA, Gervals A (1984) Incidence and levels of 2,3,7,8-tetrachlorodibenzo-p-dioxin in Lake Ontario commercial fish. Environ. Sci. Technol. 18: 719–721.

Safe S (1984) Polychlorinated biphenyls (PCBs) and polybrominated biphenyls (PBBs): biochemistry, toxicology and mechanism of action. Crit. Rev. Toxicol. 13: 319–395.

Safe S (1990) Polychlorinated biphenyls (PCBs), dibenze-p-dioxins (PCDDs), dibenzofurans (PCDFs), and related compounds: Environmental and mechanistic considerations which support the development of toxic equivalence factors (TEFs). Crit. Rev. Toxicol. 21(10): 51–88.

Sims GG, Campell JR, Zemlyak F, Graham JM (1977) Organochlorine residues in fish and fishery products from the Northwest Atlantic. Bull. Environ. Contam. Toxicol. 18: 697–705.

Sivarajah K, Franklin CS, Williams WP (1978) The effects of polychlorinated biphenyls on plasma steroid levels and hepatic microsomal enzymes in fish. J. Fish. Biol. 13: 401–409.

Sloan RJ, Simpson KW, Schroeder RA, Barnes CR (1983) Temporal trends toward stability of Hudson River PCB contamination. Bull. Environ. Contam. Toxicol. 31: 377–385.

Soontornchat S, Li M-H, Cooke PS, Hansen LG (1994) Toxicokinetic and toxicodynamic influences on endocrine disruption by polychlorinated biphenyls. Environ. Health Persp. 102: 568–571.

Spies RB, Rice SW Jr. (1988) Effects of organic contaminants on reproduction of the starry flounder Platichthys stellatus in San Francisco Bay: 1. Hepatic contamination and mixed-function oxidases (MFO) activity during the reproductive season. Mar. Biol. 98: 191–200.

Spitzbergen JM, Schat KA, Kleeman JM, Peterson R (1986b) Effects of 2,3,7,8-tetrachlorodibenzo-p-dioxin (TCDD) or Arochlor 1254 on the resistance of rainbow trout, Salmo gairdneri T, to infectious haematopoietic necrosis virus. J. Fish Dis. 11: 73–83.

Spitzbergen JM, Schat KA, Kleeman JM, Peterson R (1986a) Interactions of 2,3,7,8-tetrachlorodibenzo-p-dioxin (TCDD) with immune responses of rainbow trout. Vet. Immunol. Immunopathol. 12: 263–280.

Spitzbergen JM, Walker MK, Olson JR, Peterson RE (1991) Pathologic alterations in early life stages of lake trout, Salvelinus namaycush, exposed to 2,3,7,8-tetrachlorodibenzo-p-dioxin as fertilized eggs. Aquat. Toxicol. 19: 41–72.

Stegeman JJ, Pajor AM, Thomas P (1982) Influence of estradiol and testosterone on cytochrome P-450 and monooxygenase activity in immature brook trout, Salvelinus fontinalis. Biochem. Pharmacol. 31: 3979–3989.

Stegeman JJ, Kloepper-Sams PJ (1987) Cytochrome P-450 and monooxygenase activity in aquatic animals. Environ. Health Persp. 71: 87–95.

Stouthart XJHX, Huijbregts MAJ, Balm PHM, Lock RAC, Wendelaar Bonga, SE (1998) Endocrine stress response and abnormal development in carp (Cyprinus carpio) larvae after exposure of the embryos to PCB 126. Fish Physiol. Biochem. 18: 321–329.

Subramanian AN, Tanabe S, Tatsukawa R, Saito S, Miykazaki N (1987) Reduction in the testosterone levels by PCBs and DDE in Dall's porpoises of North-western Pacific. Mar. Pollut. Bull. 18: 643–646.

Sullivan CV, Iwamoto RN, Dickhoff WW (1987) Thyroid hormones in blood plasma of developing salmon embryos. Gen. Comp. Endocrin. 65: 337–345.

Sures B, Knopf K (2004) Individual and combined effects of cadmium and 3,3',4,4'-5-pentachlorobiphenyl (PCB 126) on the humoral immune response in European eel (Anguilla anguilla) experimentally infected with larvae of Anguillicola crassus. Parasitology 128:445–454.

Taysse L, Chambras C, Marionnet D, Bosgiraud C, Deschaux P (1998) Basal level and induction of Cytochrome P450, EROD, UDPGT, and GST activities in Carp (Cyprinus carpio) immune organs (Spleen and Head Kidney). Bull. Environ. Contam. Toxicol. 60:300–305.

Tesch WW (1977) "The eel", Biology and Management of Anguillid Eels, Chapman & Hall, London, 434 pp.

Thomas P (1988) Reproductive endocrine function in female Atlantic croaker exposed to pollutants. Mar. Environ. Res. 35: 47–95.

Tulonen J, Vuorinen PK (1996) Concentrations of PCBs and other organochlorine compounds in eels (Anguilla anguilla L.) of the Vanajavesi watercourse in southern Finland, 1990–1993. Sci. Total. Environ. 187: 11–18.

USFWS (1981) Chlorinated hydrocarbons as a factor in the reproduction and survival of lake trout (Salvelinus namaycush) in Lake Michigan. U.S. Fish and Wildlife Service Tech. Paper, 105, 42 pp.

Van der Weiden MEJ (1993) Cytochrome P450 1A induction in carp as a biological indicator for the aquatic contamination of chlorinated polyaromatics. Ph.D. thesis, Utrecht University, Research Institute Toxicology, ISBN 90–393–0376–2.

Van Ginneken V, Vianen G, Muusze B, Verschoor L, Lugten O, Onderwater M, Van Schie S, Niemantsverdriet P, Van Heeswijk R, Eding E, Van den Thillart G (2005a) Gonad development and spawning behaviour of artificially matured European eel (Anguilla anguilla L.). Anim. Biol. 55: 203–218.

Van Ginneken V, Antonissen E, Müller UK, Booms R, Eding E, Verreth J, Van den Thillart G (2005b) Eel migration to the Sargasso: remarkably high swimming efficiency and low energy costs. J. Exp. Biol. 208: 1329–1335.

Van Ginneken V (2006) Simulated migration of European eel (Anguilla anguilla, Linnaeus 1758). Ph.D. thesis, Wageningen University, The Netherlands, 309 pp. ISBN: 90–8504–456–1.

Van Ginneken V, Dufour S, Sbaihi M, Balm P, Noorlander K, de Bakker M, Doornbos J, Palstra A, Antonissen E, Mayer, I. van den Thillart G (2007) Does 5500-km swim trial stimulate early sexual maturation in the European eel (*Anguilla anguilla* L.)? Comp. Biochem. Physiol. A 147: 1095–1103.

Van Leeuwen CJ, Hermens JLM (1995) eds. Risk Assessment of Chemicals: An Introduction. 374 pp. Kluwer, Dordrecht, ISBN: 0–7923–3740–9.

Van Straalen NM, van Verkleij JAC (red.) (1993) Teaching Book Oecotoxicology, IVM-publication, R-91/16, VU-publishers, Amsterdam; ISBN 90-5383-083-9, 423 pp.

Vethaak AD, Reinhalt T (1992) Fish disease as a monitor for marine pollution: the case of the North Sea. Rev. Fish Biol. Fisher. 2: 1–32.

Vijayan MM, Leatherland JF (1989) Cortisol-induced changes in plasma glucose, protein, and thyroid hormone levels, and liver glycogen content of coho salmon (Oncorhynchus kisutch Walbaum). Can. J. Zool. 67: 2746–2750.

Vilter V (1946) Action de la thyroxine sur la metamorphose larvaire de l'anguille. CR Soc. Biol. Paris 140: 783–785.

Walker MK, Peterson RE (1991) Potencies of polychlorinated dibenzo-p-dioxin, dibenzofuran and biphenyl congeners, relative to 2,3,7,8-tertachlorodibenzo-p-dioxin, for producing early life stage mortality in rainbow trout (Oncorhynchus mykiss), Aquat. Toxicol. 21: 219–238.

Walker MK, Cook PM, Batterman AR, Butterworth BC, Berini C, Libal JJ, Hufnagle LC Peterson RE (1994) Translocations of 2,3,7,8-tertachlorodibenzo-p-dioxin from adult female Lake Trout (Salvelinus namacycush) to oocytes: effects on early life stage development and sac fry survival. Can. J. Fisher. Aquat. Sci. 51: 1410–1419.

Walker MK, Peterson RE (1994) Toxicity of 2,3,7,8- tetrachlorodibenzo-p-dioxin to brook trout (Salvelinus fontinalis) during early development. Environ. Toxicol. Chem. 13: 817–820.

Walton DG, Fancey LL, Green JM, Kiceniuk JW, Penrose WR (1983) Seasonal changes in aryl hydrocarbon hydroxylase activity of a marine fish Tautogolabrus adspersus (Walbaum) with and without petroleum exposure. Comparative Biochemistry and Physiology C, Comp. Pharmacol. Toxicol. 76: 247–253.

Warriner JE, Mathews ES, Weeks BA (1988) Preliminary investigations of the chemiluminescent respons in normal and pollutant-exposed fish. Mar. Environ. Res. 24: 281–284.

Weatherley NS, Davies GL, Ellerly S (1997) Polychlorinated biphenyls and organochlorine pesticides in eels (Anguilla anguilla L.) from Welsh Rivers. Environ. Poll. 95: 127–134.

Weber GW, Okimoto DK, Richman NH, Grau EG (1992) Patterns of thyroxine and triiodothyronine in serum and follicle-bound oocytes of the tilapia, Oreochromis mossambicus during oogenesis. Gen. Comp. Endocrinol. 85: 392–404.

Weeks BA, Huggett RJ, Warriner JE, Methews ES (1990) Macrophage responses of estuarine fish as bioindicators of toxic contamination. In Biomarkers of Environmental Contamination (McCarthy JF, Shugart LR, eds.), pp. 193–201. Boca Raton, FL: Lewis Publishers.

Wendelaar Bonga SE, Lock RAC (1992) Toxicants and osmoregulation in fish. The Netherlands. J. Zool. 42: 478–493.

Wiesmüller T, Schlatterer B (1999) PCDDs/PCDFs and coplanar PCBs in eels (Anguilla anguilla) from different areas of the rivers Havel and Oder in the state of Brandenbourg. Chemosphere 38: 325–334.

Wren CD (1991) Cause-effect linkages between chemicals and populations of mink (Mustela vison) and otter (Lutra canadensis) in the Great Lakes basin. J. Toxicol. Environ. Health 33: 549–585.

Yamamoto K, Yamauchi K (1974) Sexual maturation of Japanese eel and production of eel larvae in the aquarium. Nature 251: 220–221.

Yamauchi K, Nakamura M, Takahashi H, Takano, K. (1976) Cultivation of larvae of Japanese eel. Nature 263: 412.

Zabel EW, Cook PM, Peterson RE (1995) Potency of 3,3',4,4',5-pentachlorobiphenyl (PCB 126), alone and in combination with 2,3,7,8- tetrachlorodibenzo-p-dioxin (TCDD), to produce late trout early life mortality. Environ. Toxicol. Chem. 14: 2175–2179.

Chapter 16
Impact of Eel Viruses on Recruitment of European Eel

Olga Haenen, Vincent van Ginneken, Marc Engelsma, and Guido van den Thillart

16.1 Introduction

Eels have an uncommon catadromic life cycle with exceptional migratory patterns to their spawning grounds several thousand kilometres away: the European eel (*Anguilla anguilla*) travels over 5,500 km to the Sargasso Sea (Schmidt 1923; McCleave and Kleckner 1987; Tesch 1982; Tesch and Wegner 1990); the American eel (*A. rostrata*) migrates over 4,000 km also to the Sargasso Sea (Castonguay and McCleave 1987; McCleave and Kleckner 1987; Tesch and Wegner 1990); the Australian eel (*A. australis*) travels over 5,000 km into the Pacific Ocean to spawn (Jellyman 1987); and the Japanese eel (*A. japonica*) travels over 4,000 km to an area near the Marianna Islands in the Philippines to spawn (Tsukamoto 1992). Evidently such long distance swimming will place those fishes under extra stress caused by the long starvation period, the high energy cost of the journey, and the many changes in the environment such as salt water, darkness, high pressure, and low temperatures, among other stress factors. Stress is often a basis for disease in eel, especially in intensive eel culture (Haenen and Engelsma, 2005 unpublished finding).

Nowadays, global transport of live fishes for aquaculture has facilitated the global spread of pathogens from diseased to healthy stocks. Within the last few decades, aquaculture has become an important production branch in our society. Its global production has more than doubled between 1986 and 1996 in tonnage and value, and over one quarter of human fish consumption at world scale is now produced in aquaculture (Naylor et al. 2000). The Netherlands is one of the leading eel producing & trading countries (Heinsbroek and Kamstra 1995). Blanc (1997) showed that nearly 100 pathogens have been introduced into European water bodies since the introduction of aquaculture. Worldwide many diseases are known in both wild and cultured eel. Parasites, for example trematodes, *Anguillicola*

O. Haenen and M. Engelsma
National Reference Laboratory for Fish & Shellfish Diseases, CVI, POB 65, 8200 AB Lelystad, The Netherlands

V. van Ginneken
Leiden University, Institute Biology, POB 9316, 2300 RA Leiden, The Netherlands

G. van den Thillart et al. (eds.), *Spawning Migration of the European Eel*,

crassus (nematode), and *Myxidium giardi (myxosporean)* occur naturally in wild eel populations, mostly in low numbers, without causing mortality (Køie 1988; Van Banning and Haenen 1990; Borgsteede et al. 1999). However, under culture conditions, with eels kept in high densities, they may be harmful. Eel pathogenic bacteria like *Vibrio vulnificus, Vibrio anguillarum, Pseudomonas anguilliseptica* and *Edwardsiella tarda* may also cause disease, especially when a stress factor is involved or when the eel is injured (Veenstra et al. 1993; Austin and Austin 1999; Haenen and Davidse 2001). As far as we know, the clinical signs are often more severe under culture conditions compared to in the wild.

Several eel viruses are known to cause disease in eels. Three of them are found regularly in Europe: Eel Virus European (EVE), Eel Virus European X (EVEX), and Herpesvirus anguillae (HVA). They are treated, one by one, below. Most of the studies on eel viruses are related to cultured eel, and data on wild eel viral infections are

Table 16.1 Eel viruses isolated from preadult and adult eels (unless elvers are mentioned), several species, of different geographic locations with or without clinical signs. EVE: European Virus of Eel; EVEX: Eel-Virus European X, HVA: Herpesvirus anguillae, IPNV: Infectious Pancreatic Necrosis Virus; (references in the table)

Country	Location	Species	Clinical signs	Virus	Reference
Denmark	Wild eel	*A.anguilla* (elvers & eels)	–&+	EVEX	Jørgensen et al. 1994
	Wild eel	*A.anguilla* (elvers & eels)	–&+	IPNV(EVE)	Jørgensen et al. 1994
France	Wild eel	*A.anguilla* (elvers & eels)	–&+	EVEX	Jørgensen et al.1994
	French W-coast	*A.anguilla* elvers	–	EVEX-like	Castric and Chastel 1980
	Eel farm	*A.anguilla*	+	herpes-like	Jørgensen et al. 1994
	Wild eel	*A.anguilla*	+	herpes-like	Jørgensen et al. 1994
	Wild eel	*A.anguilla* (elvers & eels)	–&+	IPNV(EVE)	Jørgensen et al. 1994
Germany	River Mosel	*A.anguilla*	+	HVA	Lehmann et al. 2005
	Eel farm	*A.anguilla*	+	HVA	Bergmann unpublished 2006
	Wild eel	*A.anguilla* elvers	–	EVEX	Shchelkunov et al. 1989
Italy	Eel farm	*A.anguilla*	–	EVEX & EVE	Van Ginneken et al. 2004
	Eel farm	*A.anguilla*	+	EVEX	Van Ginneken et al. 2004
Japan	Tokyo (source: France)	*A.anguilla* elvers	–	EVEX	Sano et al. 1977

(continued)

Table 16.1 (continued)

Country	Location	Species	Clinical signs	Virus	Reference
	Cage culture	*A.japonica*	+	EVEX-like	Kobayashi and Miyazaki 1996
	Eel farm	*A.anguilla & A.japonica*	+	HVA	Sano et al. 1990
	Eel farm	*A.japonica*	+	HVA	Kobayashi and Miyazaki 1997
	Eel farm	*A.japonica*	+	HVA	Lee et al. 1999
	Eel pond farm	*A.japonica*	+	EVE	Egusa 1970
	Eel pond farm	*A.anguilla*	+	EVE	Sano 1976
Morocco	Sebou	*A.anguilla*	–	EVEX	Van Ginneken et al. 2004
The Netherlands	Lake Grevelingen	*A.anguilla*	–	EVEX	Van Ginneken et al. 2004
	Lake Grevelingen	*A.anguilla*	+	EVEX	Van Ginneken et al. 2005
	Eel indoor farms	*A.anguilla*	+	EVEX	Haenen et al. unpublished 2001
	Lake Grevelingen	*A.anguilla*	+	HVA	Van Ginneken et al. 2004
	Lake Lauwers	*A.anguilla*	+	HVA	Van Ginneken et al. 2004
	River Rhine	*A.anguilla*	+& –	HVA	Haenen et al. unpublished 2008
	Eel indoor farms	*A.anguilla* elvers	+	HVA	Davidse et al. 1999; Van Ginneken et al. 2004
	Eel indoor farms	*A.anguilla*	+	HVA	Davidse et al. 1999
	River Merwede	*A.anguilla*	+	HVA	Haenen et al. unpublished 2008
	Eel indoor farms	*A.anguilla*	+	EVE	Haenen et al. 2001
	Eel indoor farms	*A.anguilla*	+	REOvirus	Haenen et al. unpublished 2001
	Lake Grevelingen	*A.anguilla*	–	picorna-like	Van Ginneken et al. 2004
	Eel indoor farms	*A.anguilla*	+	orthomyxo	Haenen et al. unpublished 2001
New Zealand	Tekawata	*A.dieffenbachi*	–	picorna-like	Van Ginneken et al. 2004
	Tekawata	*A.dieffenbachi*	–	EVEX	Van Ginneken et al. 2004
Sweden	Wild eel	*A.anguilla* (elvers & eels)	–&+	EVEX	Jørgensen et al., 1994
	Wild eel	*A.anguilla* (elvers & eels)	–&+	IPNV(EVE)	Jørgensen et al., 1994
United Kingdom	Wild eel	*A.anguilla* (elvers & eels)	–&+	EVEX	Jørgensen et al., 1994
	Wild eel	*A.anguilla* (elvers & eels)	–&+	IPNV(EVE)	Jørgensen et al., 1994

scarce. In Table 16.1, results from various authors on the occurrence of eel viruses in eel species from several geographical regions are presented.

16.2 Eel Virus European (EVE)

Eel Virus European (EVE) is a double stranded RNA (biRNA) virus, with an icosahedric form, a size of 68–77 nm, and serologically closely related to infectious pancreatic necrosis virus (IPNV) (Sano 1976; Sano et al. 1981; Okamoto et al. 1983). EVE was first described by Egusa (1970) in pond – cultured Japanese eel *(Anguilla japonica)*. It caused branchionephritis, with mortalities up to 50%. In 1976, EVE was isolated for the first time from European eel *(A. anguilla)* in Japan with a renal pathology (Sano 1976). The virus was isolated from European elvers, which had been imported into Japan. The infection spread to Japanese eels and this resulted in mass mortalities, with swollen and congested gills and glomerulonephritis (Egusa 1976; Sano 1976). EVE was suspected to cause disease in natural eel populations by Wolf (1988), but this has not yet been confirmed.

In Europe, Jørgensen et al. (1994) tested 2,092 pools of healthy wild elvers and cultured juvenile European eels, from Denmark, UK, France and Sweden for the presence of virus. Of the 91 virus isolates they typed 38 viral strains from all four countries to IPNV. In The Netherlands, EVE was isolated several times from diseased juvenile cultured *A. anguilla* with haemorrhages in the skin, skin lesions and a congested and red head, with mortalities up to 50% (Haenen et al. 2000 unpublished findings). Van Ginneken et al. (2004) isolated EVE from clinically healthy adult farmed European eel from Italy, which had a double virus infection of EVE and EVEX (Eel Virus European X, a rhabdovirus, see below). At present there are no publications available on EVE isolated from wild silver eels.

16.3 Eel Virus European X (EVEX)

The first rhabdovirus from eel was isolated by Sano et al. (1976) from European elvers imported into Japan from Cuba with a haemorrhagic disease. The virus was named Eel Virus American (EVA). Sano et al. (1977) described the isolation of a second rhabdovirus, from European eels imported from France, and named it Eel Virus European X (EVEX). EVEX is a RNA-virus, and has a bullet-like form and a size of 170–175 × 90–95 nm (Sano et al. 1977). Hill et al. (1980) showed that EVA and EVEX were closely related. Infected elvers in Japan showed vascular congestion of the abdominal surfaces and pectoral and anal fins, and histology revealed extensive haemorrhaging and necrosis of kidney, muscle, pancreas and liver (Sano 1976). One of the rhabdoviruses isolated by Castric and Chastel (1980) from apparently healthy elvers from French coastal waters showed similarities to EVA and EVEX. In infection experiments, this virus isolate was shown to be a-virulent for elvers (Castric and Chastel 1980; Nishimura et al. 1981). EVEX was also isolated from healthy elvers imported to the USSR from Western Germany (Shchelkunov et al. 1989), and it was shown to be pathogenic to 4-year old eel after i.p. injection. Jørgensen et al. (1994) tested 2,092 pools of healthy wild elvers and cultured eels *A. anguilla*, from Denmark, UK, France and Sweden. From 91 virus isolates, 47 were eel typed as rhabdovirus (EVEX/EVA), isolated from all four countries (Castric et al. 1984).

Kobayashi and Miyazaki (1996) found EVEX/EVA-like virus in Japanese eels from cage culture showing cutaneous erosion and ulceration at 15°C. The disease has sometimes occurred since then and caused mass mortalities in Japanese eel during stocking before shipping. In experimentally induced infections cutaneous lesions developed in eel, accompanied with necrosis of the dermal fibrocytes, haemorrhage and inflammatory cellular infiltration, and diffuse necrosis of the haematopoietic tissue, renal tubules, splenic pulps and hepatic parenchyma (Kobayashi and Miyazaki 1996). They proposed to name the disease *rhabdoviral dermatitis* in Japanese eel. Haenen et al. (unpublished findings) found clinical signs of EVEX in cultured European eel: lethargy, haemorrhages, haemorrhagic skin with tiger-like patterns, skin lesions, a red head, and severe anorexia, with mortalities below 20% at eel farms, or higher, depending on stress factors. They only rarely isolated EVEX. More recently, clinical signs of viral infection have been found in European eel from Lake Grevelingen, The Netherlands, used for a swim tunnel experiment (Van Ginneken et al. 2005). The eels showed some haemorrhages and anaemia, and died after 1,000–1,500 km swimming in the tunnel (Van Ginneken et al. 2005). EVEX was isolated from these eels. The stress factor of the swim tunnels might have induced virus replication, and disease. Van Ginneken et al. (2004) isolated EVEX from Dutch adult wild European eel, farmed Italian European eel, wild adult European eel from Morocco, and wild adult *Anguilla dieffenbachi* from New Zealand, but apart from the European eel from Italy, which showed haemorrhages and red patches on the skin, they showed no clinical signs of disease. It was suggested, that eel virus infections could be a contributing factor in the worldwide decline of the eel stocks (Van Ginneken et al. 2005). However, due to the fact that no EVEX was found in a disease screening of 80 wild silver eel originating from the River Rhine (The Netherlands; Haenen et al. 2008, unpublished findings), there seems to be a very low prevalence of EVEX in at least Dutch silver eels.

16.4 Herpesvirus anguillae (HVA)

Herpesvirus anguillae (HVA) was isolated in 1985 from diseased Japanese eels and European eel in Japan (Sano et al. 1990). It is an enveloped double-stranded DNA-virus, with an icosahedric form, and a size of the nucleocapsid of 110 nm (Davidse et al. 1999). Infected eels showed reddening of the ventral skin, especially below the mouth (Fig. 16.1), and a swollen and haemorrhagic kidney and spleen. Furthermore, necrosis was found in skin, gills, liver and spleen. Kobayashi and Miyazaki (1997) isolated HVA from Japanese eels with skin lesions, and tested the pathogenicity of HVA from cutaneous lesions of Japanese eel. Lee et al. (1999) described gill filament necrosis in farmed Japanese eels infected with HVA.

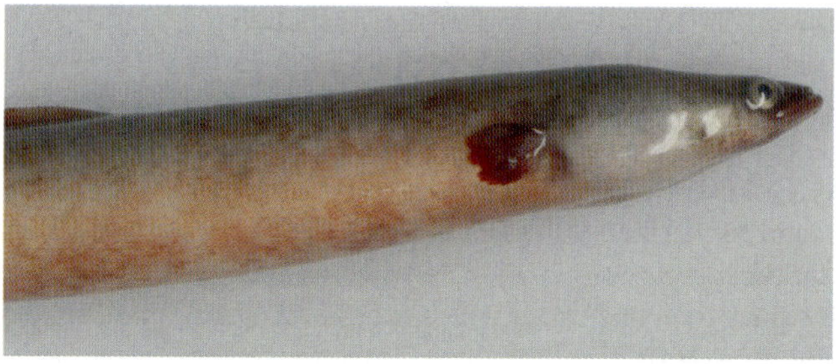

Fig. 16.1 Diseased cultured European eel *Anguilla anguilla* with a HVA-infection. The eel shows a patchy pattern of haemorrhages in the skin, and severe haemorrhagic fins (Haenen et al. 2002, with permission)

In Europe, Békési et al. (1986) found herpes-like particles in skin lesions of European eel. From 1977–1992 Jørgensen et al. (1994) found herpes-like virus in three pools of farmed and two pools of wild adult healthy European eels from France, by electron microscopy and syncytia-induction in cell culture. By use of EK-1 (eel kidney cells, Chen et al. 1982) Davidse et al. (1999) isolated, replicated, and typed the virus for the first time in Europe from diseased cultured European eel. Optimal growth temperature was 20–25°C (Smail and Munro 2001) and 26°C (Davidse et al. 1999). Clinical signs of HVA outbreaks in farmed European eel were variable: predominantly lethargy, anemia, haemorrhages in skin and fins with a tiger-like pattern, ulcerative skin and fin lesions, haemorrhages in the gills and fins, a red head, a slightly congested head, anorexia, an enlarged spleen, a pale spleen, kidney, and liver with petechial haemorrhages in the liver, pinkfat by small diffuse haemorrhages, haemorrhages in the mesenteria and muscle, and ascites with blood, and congested gill epithelium (Davidse et al. 1999; Haenen et al. 2002).

Van Nieuwstadt et al. (2001) infected clinically healthy farmed juvenile European eel with HVA. The eels showed a response in specific antibodies against the virus, but no disease. It was concluded, that they had probably prior contact with HVA, and were protected by their specific antibodies. HVA was isolated from the eels, and treatment with the stress hormone dexamethasone provoked virus release. The recrudescence of HVA by a stress hormone suggested, that HVA can establish a latent infection in eel (Van Nieuwstadt et al. 2001). Infection of eel with HVA resulted in mortalities from 1% to 6.8% in Japan (Sano et al. 1990), and 10% or higher in cultured European eel in Europe, depending on stress factors like a suboptimal water quality (Haenen et al. 2002).

Van Ginneken et al. (2004) isolated HVA from diseased adult wild and farmed European eels in the Netherlands. The wild eels originated from Lake Grevelingen (one of ten fish positive) and Lake Lauwers (ten of ten fish positive). In both cases the eel showed typical symptoms like reddening with petechial haemorrhages (Van Ginneken et al. 2004). The virus has also been detected in diseased silver eels of the River Merwede in autumn 2003 (The Netherlands; Haenen et al. 2008, unpublished results). In autumn 2004, one out of 6 adult wild European eel originating from Kornwerderzand (The Netherlands) was found HVA positive without clinical signs. In 2005, 80 silver eels from the River Rhine were investigated for the presence of pathogens, including eel viruses (Haenen et al. 2008, unpublished findings). HVA was detected in 2005 in 40% of the eels, of which most eels showed some fin haemorrhages but no internal signs of viral disease. However, the clinical observation of fin haemorrhages was a-specific, as they were also seen in virus negative eels. From these observations we suggest that HVA is widespread in wild European eel, supporting the hypothesis of Van Nieuwstadt et al. (2001) and Haenen et al. (2002).

16.5 Other Viruses and Double Infections

Other viruses have been isolated from eels in lower frequencies. For example, in The Netherlands, between 1986–2001 a reovirus was isolated four times from diseased cultured European eel, showing a mortality of less than 5%, (Haenen et al. unpublished findings). Furthermore, two orthomyxo virus strains with a mortality of less than 25%, and one non-EVE biRNA virus with a mortality of 100% were isolated from diseased cultured sub-adult European eel in The Netherlands (Haenen et al. unpublished findings). As these viruses occur only rarely, the impact on the eel population in total will most probably be limited. In cultured eel double infections with two viruses can be observed: herpesvirus and reovirus, EVEX and reovirus, HVA and EVE, and EVEX and EVE are known combinations related to diseased cultured European eel (Haenen et al. 2003, unpublished findings). Thus far, no double infections were found in wild eels.

16.6 Discussion

16.6.1 Viral Infections in Wild Eel

As described above EVE, and EVEX both have been isolated from elvers. This implicates either a vertical transmission of the disease, from parent to offspring, or acquisition of the infection at an early stage in life. For IPNV, which is strongly related to EVE, vertical transmission has been shown in salmonids (Bullock et al. 1976; Mulcahy and Pascho 1984), via eggs and sperm of fish, but this is not yet proven for eels. In juvenile eels, all three viruses EVE, EVEX, and HVA have been found. In wild silver eels, only EVEX and HVA were found so far. In general the viruses seem to be wide spread in the wild European eel population with limited impact. This suggests EVE, EVEX and HVA to be endemic in Western Europe and the wild eel to have some resistance against these viruses.

16.6.2 Stress and Immunosuppression during Migration

Stress can be an important factor in inducing disease in eel, in its severity determining the seriousness of the disease. The stress source may either be physical (high water temperature), chemical (bad water quality, pollution), or biological (lack of feeding, predators, etc.). Eels prepare their long-distance migration by increasing energy reserves as stored fat, which can reach levels up to 30% of their body weight (Bruggeman 1983). Stored lipid reserves in wild eels are often contaminated with PCBs, which have an immuno-suppressive effect (see Chapter 15). For other pathogens like parasites and bacteria the immunosuppressive action of PCBs has been demonstrated in response to exposure to different PCBs. Sures and Knopf (2004) showed that PCB 126 suppressed humoral immune response in European eel: The exposure to PCB 126 increased the chance of infection of eel for the nematode *Anguillicola crassus*.

Also the migration and spawning itself could stress the migrating eel as a biological and physiological stress factor, and thereby suppress the resistance to pathogens. In general physiological stress, when associated with migration is a well known risk factor resulting in increased susceptibility of animals for infectious diseases. For bird migration it is known that migration stress often leads to reactivation of otherwise latent virus infections (Gylfe et al. 2000).

16.6.3 Eel Viruses and Migration

In a recent paper it was shown that a group of eels swimming continuously in swim tunnels at a speed of 0.5 BL s^{-1} became seriously ill (Van Ginneken et al. 2005). Those eels were found to be infected with the rhabdovirus EVEX (Eel Virus European X)-virus, and the other eels from the same batch did not develop this disease. The clinical signs were haemorrhages and extreme anemia in actively swimming eels. Possibly, the anemia was the result of internal blood loss, since the symptoms included blood in the abdominal fluid and haemorrhages all over the body. The observed changes in plasma lactate dehydrogenase (LDH), total protein and aspartate aminotransferase (AAT) were indicative of a serious viral infection. All experimental eels stopped swimming between 1,000 and 1,500 km during simulated migration, and died soon after that (Fig. 16.2). In contrast, non-infected animals from another batch swam 5,500 km, the estimated distance to the spawning ground of the European eel in the Sargasso Sea. Virus-positive eels showed a decline in hematocrit, which was related to the swim distance, while the healthy eels showed even a slightly increased hematocrit. Based on these observations, we concluded, that eel viral infections may adversely affect the spawning migration of eels, which might be a contributing factor to the worldwide decline of eel.

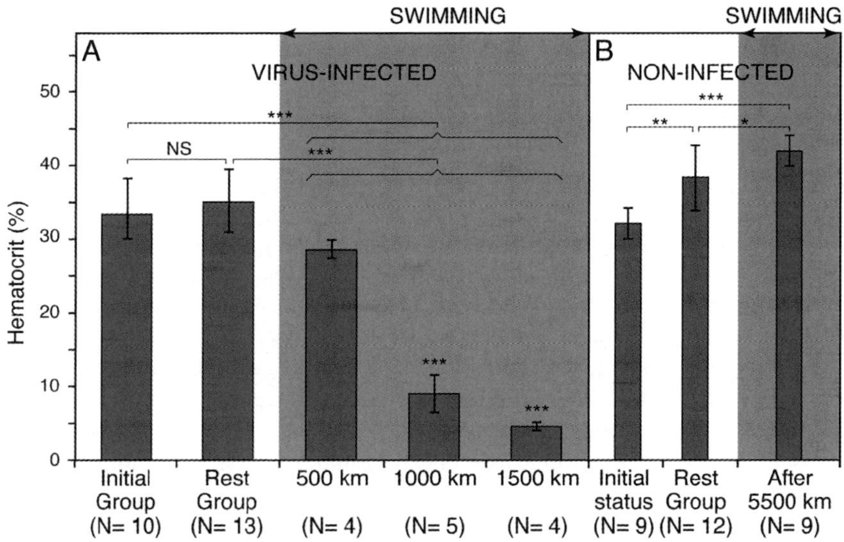

Fig. 16.2 Development of anaemia in swimming eels which were tested and were shown to be infected with EVEX. **A:** Haematocrit of European eels infected with EVEX-virus. Eels swam for up to 1 month during September 2000 in large 127-l Blazka swim tunnels. The decrease of red blood cells is negatively correlated with the distance covered. The rest group did not develop clinical signs of the disease. **B:** Haematocrit of virus-negative eels which swam 5,500-km during the period March-September at a continuous speed of 0.5 Body-length (BL) per second in the same set up. $^{*}P \leq 0.05$, $^{**}P \leq 0.01$, $^{***}P \leq 0.001$ (Student t-test) (Van Ginneken et al. 2005. With permission)

16.6.4 Eel Migration and Water Temperature

The temperature range in which wild eels migrate has been studied. Silver eels migrate downstream in late summer and autumn, at temperatures around 15°C, while at lower temperatures (<10°C) they stop their migration behavior (Tesch 2003). In the autumn the North Sea remains at 12–15°C, the same applies for the upper 200 m of a large part of the Atlantic Ocean. From telemetry studies it was observed that silver eels, which were tracked when they left the continental slope off the Bay of Biscay and west of Spain, swam to depths of at least 400 m in the daytime, while they selected shallower water (50 to 215 m) at night (Tesch 1978). Studies on tracking eels in the western Mediterranean Sea provided information on thermal preference. Eels tended to swim in the 13°C hypolimnion, preferred 344 m, at day, but at night regularly crossed the thermocline during vertical migrations to 196 m depth into surface waters as warm as 18°C, according to a diurnal rhythm (Tesch 1989).

The distance to the continental shelf starting in The Netherlands is 1,000–1,500 km, and till then eels have to travel in this upper layer, which certainly remains above 10°C. As eels migrate further into the Atlantic, they are assumed to swim at depths below the 500 m zone. Below 500 m the temperature drops rather fast from around 10°C at 500 m to 4°C at 2,000 m depth. This means, that it is likely that eels swim for most of their journey (4,000 km) at temperatures between 10°C and 4°C. The temperature profiles of the water of the Atlantic ocean below 750 m are rather stable at around 8°C (20–40° N), with the isotherm of 5°C between 1,000 and 1,500 m. Steep temperature profiles can be found in the upper water layers.

In the Sargasso Sea and along the Gulf Stream warm water reaches rather deep, mainly due to the higher salinity. The Sargasso Sea is a massive sinking gyre with a diameter of some 2,000-km. The circulating water mass sinks due to the evaporation at the surface, which increases the salinity and therefore its density. At the surface the temperatures are above 30°C, falling rather slow with depth to about 16°C at 500 m. A very stable water layer of virtually the same salinity and temperature (18–19°C) occurs between 100 and 400 m over an enormous surface area (Worthington 1959). At deeper levels the temperature drops again quickly: at 500 m 16°C and at 750 m already 11°C. So, for migrating silver eels – assuming they are swimming at depths below 750 m – it will be a rather sudden change in temperature when they arrive at the Sargasso Sea; the temperature will rise from 5–8°C to 11–12°C. This rise with 6°C will take place over a distance of a few hundred kilometres, which they cover easily in a few days. As silver eels cannot be stimulated to maturation at 12°C (see Chapter 13), it is likely that they will go to shallower waters after arriving at the Sargasso Sea. The vertical migration of the leptocephali occurs actually in the upper layers between 140 m at night and at 350 m daytime (Kleckner and McCleave 1982). These water layers have a year round constant temperature of 18–19°C (Worthington 1959). As the eggs float, going up by 100–200 m over 2 days before hatching, spawning likely takes place at a greater depth. However, as maturation is temperature dependent this will not be much more than 400–500 m. In a laboratory experiment the final preferred temperatures (FPT, the temperature an animal ultimately selects) of pre-migratory and migratory American eels were determined using chronic tests in a horizontal thermal gradient. Results indicated that both silver and yellow *Anguilla rostrata* in saltwater had mean FPTs of 17.5°C, which indicates that American silver eels have a preference for relatively high temperatures (Haro 1991).

16.6.5 Eel Viral Disease Dependent on Water Temperature

Temperature is an important factor involved in the development of viral disease in fish: Each virus has its appropriate optimal temperature for virus replication. For EVE this is between 15°C and 23°C (Haenen and Davidse 2001 unpublished findings), for EVEX between 10°C and 15°C (Shchelkunov et al. 1989), and for HVA between 20°C and 26°C (Smail and Munro 2001; Davidse et al. 1999). Outside these temperature ranges the viruses may persist but disease does not develop in the host. In this light, (latent) viral diseases of eel depend strongly on the environmental temperature. Given the fact, that, like discussed above, the eels are thought to swim for a long time at temperatures between 4°C and 10°C, and in the Sargasso Sea they are supposed to swim in water of 10–20°C, the eels experience temperatures, which, dependent on the optimum per eel virus, may contribute to the reactivation of latent virus in the eel, and replication of eel virus, causing the development of viral disease in eel.

As presented in Table 16.1, many virus infections are found in farmed and wild eel populations in The Netherlands, which could be a threat for the global eel population. HVA and EVEX have been found in eels from several countries worldwide: EVEX now and than in the European eel, *Anguilla anguilla*, in the Netherlands, Italy, and Morocco, but also in another eel species: The New Zealand eel, *Anguilla dieffenbachi*, in this latter without clinical signs though. Moreover, Herpesvirus anguillae (HVA) has been isolated and identified in eel populations in different parts of the world: In cultured eels in Taiwan (Ueno et al. 1992, 1996; Chang et al. 2002), in cultured and wild eels in The Netherlands (Van Ginneken et al. 2004; Van Nieuwstadt et al. 2001; Davidse et al. 1999; O. Haenen et al. 2008, unpublished) and in cultured eels (S. Bergmann 2006, unpublished) and wild silver eels in Germany (Lehmann et al. 2005) EVE was never found in silver eel yet, so the disease risk of this virus in migrating eel is supposed to be small.

16.7 Conclusions

This study shows that EVEX, EVE, and HVA are found at several geographic places in wild and farmed European eel. Transport of clinically healthy but virus-infected elvers and eels may cause introduction of eel viruses into the virus uninfected waters with eels, both fisheries and aquaculture. No serious mortalities are known in wild eel populations due to eel viruses, but wild diseased eels are difficult to trace, particularly migrating silver eels.

Related to the fact that migrating silver eels are stressed during their spawning migration, and as a result immunosuppressed, with the assumption, that they swim for some weeks in ambient water temperatures for virus infections, the eels might get viral disease in these water bodies. To what extent this disease would occur and threaten the wild eel population and recruitment depends on the immune status of the eels, the water temperature, and the pathogenicity of the particular virus strain. Maturing eels have a depressed immune system, so, they do run a risk of developing the disease after arriving in the Sargasso Sea, the extend of it is still difficult to estimate.

References

Austin B, Austin DA (1999) In: Bacterial Fish Pathogens: Disease of farmed and wild fish, 3rd (revised) edition. Praxis Publishing, Chichester, UK, 457 pp.

Békési L, Horváth I, Kovács-Gayer E, Csaba G. (1986) Demonstration of herpesvirus like particles in skin lesions of European eel (*Anguilla anguilla*). J. Appl. Ichthyol. 4, 190–192 (short communication)

Blanc G (1997) L'introduction des agents pathogènes dans les écosystèmes aquatiques: aspects théoriques et réalités. Bulletin Français de la pêche et de la piscine 344/345, 489–513.

Borgsteede FHM, Haenen OLM, De Bree J, Lisitsina, OI (1999) Parasitic infections of European eel (*Anguilla anguilla* L.) in The Netherlands. Helminthologia 36(4), 251–260.

Bruggeman WA (1983) *Bioaccumulation of polychlorobiphenyls and related hydrophobic chemicals in fish*. Rijkswaterstaat. Rijksinstituut voor Zuivering Van Afvalwater, Lelystad.

Bullock G, Rucker RR, Amend D, Wolf K, Stuckey HM (1976) Infectious pancreatic necrosis: transmission with iodine-treated and nontreated eggs of brook trout *(Salvenius fontinalis)*. J. Res. Board Can. 33, 1197–1198.

Castonguay M, McCleave JD (1987) Vertical distributions, diel and ontogenetic vertical migrations and net avoidance of leptocephali of *Anguilla* and other common species in the Sargasso Sea. J. Plankton Res. 9, 195–214.

Castric J, Chastel C (1980) Isolation and characterization attempts of three viruses from European eel, *Anguilla anguilla*: preliminary results. Ann. Virol.. 131E, 435–448.

Castric J, Rasschaert D, Bernard J (1984) Evidence for Lyssaviruses among rhabdovirus isolated from European eel, *Anguilla anguilla*. Ann. Virol. 135E, 33–35.

Chang PH, Pan YH, Wu CM, Kuo ST, Chung HY (2002) Isolation and molecular characterization of herpesvirus from cultured European eels *Anguilla anguilla* in Taiwan. Dis. Aquat. Organ. 50, 111–118.

Chen SN, Ueno Y, Kou G-H (1982) A cell line derived from Japanese eel (*Anguilla japonica*) kidney. Proc. Natl. Sci. Counc. R.O.C. 6, 93–100.

Davidse A, Haenen OLM, Dijkstra SG, Van Nieuwstadt AP, Van der Vorst TJK, Wagenaar F, Wellenberg GJ (1999) First Isolation of Herpesvirus of Eel (*Herpesvirus anguillae*) in diseased European eel (*Anguilla anguilla* L.) in Europe. Bull. Eur. Ass. Fish. Pathol. 19, 137–141.

Egusa S (1970) Branchionephritis prevailed among eel populations in farm-pond in the warm water 1969–1970. Fish Pathol. 5, 51–56 (in Japanese).

Egusa S (1976) Some bacterial diseases of freshwater fishes in Japan. Fish Pathol. 10, 103–114.

Gylfe A, Bergstrom S, Lundstrom J, Olsen B (2000) Reactivation of *Borellia* infection in birds. Nature 403, 724–725.

Haenen OLM, Dijkstra SG, Van Tulden PW, Davidse A, Van Nieuwstadt AP, Wagenaar F, Wellenberg GJ (2002) *Herpesvirus anguillae* (HVA) isolations from disease outbreaks in cultured European eel, *Anguilla anguilla* in The Netherlands since 1996. Bull. Euro. Assoc. Fish Pathol. 22(4), 247–257.

Haenen OLM, Davidse A (2001) First isolation and pathogenicity studies with *Pseudomonas anguilliseptica* from diseased European eel *Anguilla anguilla* (L.) in The Netherlands. Aquaculture 196: 27–36.

Haro AJ (1991) Thermal preferenda and behaviour of Atlantic eels (genus *Anguilla*) in relation to their spawning migration. Environ. Biol. Fishes 31, 171–184.

Heinsbroek LTN, Kamstra A (1995) The River Eels, Chapter 6: In: Nash CE, Novotny AJ (eds), Production of Aquatic Animals. Elsevier, Amsterdam, pp. 109–131.

Hill BJ, Williams RF, Smale CJ, Underwood BO, Brown F (1980) Physico-chemical and serological characterization of two rhabdoviruses isolated from eels. Intervirology 14, 208–212.

Jellyman D (1987) Review of the marine life history of Australasian temperate species of *Anguilla*. Am. Fish. Soc. Symp. 1, 276–285.

Jørgensen PEV, Castric J, Hill B, Ljungberg O, De Kinkelin P (1994) The occurrence of virus infections in elvers and eels *(Anguilla anguilla)* in Europe with particular reference to VHSV and IHNV. Aquaculture. 123, 11–19.

Kleckner RC, McCleave JD (1982) Entry of migrating American eel leptocephali into the gulf stream system. Helgolander Meeresuntersuchungen 35, 329–339.

Kobayashi T, Miyazaki T (1997) Characterization and pathogenicity of a herpesvirus isolated from cutaneous lesion in Japanese eel, *Anguilla japonica*. Fish Pathol. 32(2), 6, 89–95.

Kobayashi T, Miyazaki T (1996). Rhabdoviral dermatitis in Japanese eel, *Anguilla japonica*. Fish Pathol. 31, 183–190.

Køie M (1988) Parasites in eels, *Anguilla anguilla* (L.) from eutrophic Lake Estrum (Denmark). Acta Parasitol. Pol. 33, 89–100.

Lee N-S, Kobayashi J, Miyazaki T (1999) Gill filament necrosis in farmed Japanese eels, *Anguilla japonica* (Temminck and Schlegel), infected with *Herpesvirus anguillae*. J. Fish Dis. 22, 457–463.

Lehmann J, Stürenberg F-J, Kullmann Y, Kilwinski J (2005) Umwelt- und Krankheitsbelsatungen der Aale in Nordrhein-Westfalen. LÖBF-Mitteilungen 2/05, 35–40.

McCleave JD, Kleckner RC (1987) Distribution of leptocephali of the catadromous *Anguilla* species in the western Sargasso Sea in relation to water circulation and migration. Bull. Mar. Sci. 41, 789–806.

Mulcahy D, Pascho RJ (1984) Adsorption to fish sperm of vertically transmitted fish viruses. Science 225, 333–335.

Naylor RL, Goldburg RJ, Primavera JH, Kautsky N, Beveridge MCC, Clay J, Folke C, Lubchenco J, Mooney H, Troell M (2000) Effect of aquaculture on world fish supplies. Nature 405, 1017–1024

Nishimura T, Toba M, Ban F, Okamoto N, Sano T (1981) Eel rhabdovirus, EVA, EVEX and their infectivity of fishes. Fish Pathol. 15, 173–184.

Okamoto N, Sano T, Hedrick RP, Fryer JL (1983) Antigenic relationships of selected strains of infectious pancreatic necrosis and European eel virus. J. Fish Dis. 6, 19–25.

Sano M, Fukuda H, Sano T (1990) Isolation and characterization of a new herpesvirus from eel. In: Perkins TO, Cheng TC (eds), Pathology in Marine Science. Academic, New York, pp. 15–31.

Sano T, Nishimura N, Okamoto N, Fukuda H (1977) Studies on viral diseases of Japanese fishes. VII. A rhabdovirus isolated from European eel, *Anguilla anguilla*. Bull. Jpn. Soc. Sci. Fisher. 43, 491–495.

Sano T, Nishimura T, Okamoto N, Fukuda H (1976) Isolation of rhabdovirus from European eels (*Anguilla anguilla*) a Japanese port of entry. Fish Health News 5, 5–6.

Sano T, Okamoto N, Nishimura T (1981) A new viral epizootic of *Anguilla japonica* Temminck and Schlegel. J. Fish Dis. 4, 127–139.

Sano T (1976) Viral diseases of cultured fishes in Japan. Fish Pathol. 19, 221–226.

Schmidt, J. (1923) Breeding places and migration of the eel. Nature 111, 51–54.

Shchelkunov IS, Skurat EK, Sivolotskaya VA, Sapotko KV, Shimko VV, Linnik VYa (1989) Rhabdovirus anguilla in eel in the USSR and its pathogenicity for fish. Vopr. Virusisol. 34(1), 81–84. In Russian with English abstract.

Smail DA, Munro ALS (2001) The virology of teleosts. In: Roberts RJ (ed), Fish Pathology, 3rd edition, WB Saunders, London, 472 p, 205.

Sures, B.; Knopf, K. (2004) Individual and combined effects of cadmium and 3,3′,4,4′-5-pentachlorobiphenyl (PCB 126) on the humoral immune response in European eel (*Anguilla anguilla*) experimentally infected with larvae of *Anguillicola crassus*. Parasitology 128, 445–454.

Tesch, FW (1978) Telemetric observations on the spawning migration of the eel (*Anguilla anguilla*) west of the European continental shelf. Environ. Biol. Fish. 3, 203–209.

Tesch, FW (1982) The Sargasso Sea Eel Expedition 1979. Helgoländer Meeresunsters. 35, 263–277.

Tesch, FW (1989) Changes in swimming depth and direction of silver eels (*Anguilla anguilla* L.) from the continental shelf to the deep sea. Aquat. Living Resour. 2, 9–20.

Tesch, FW (2003) The Eel. Blackwell Science Ltd, Oxford, UK 5th Ed. 408p, 119–212.

Tesch, FW, Wegner G (1990) The distribution of small larvae of *Anguilla* sp. related to hydrographic conditions 1981 between Bermuda and Puerto Rico. International Revue der gesamten Hydrobiologie 75, 845–858.

Tsukamoto K (1992) Discovery of the spawning area for Japanese eel. Nature 356, 789–791.

Ueno Y, Shi J-W, Yoshida T, Kitao T, Sakai M, Chen S-N, Kou GH (1996) Biological and serological comparisons of eel herpesvirus in Formosa (EHVF) and herpevirus anguillae (HVA). J. Appl. Ichthyol. 12, 49–51.

Ueno Y, Kitao T, Chen S-N, Aoki T, Kou GH (1992) Characterization of a Herpes-like Virus isolated from cultured Japanese eels in Taiwan. Gyohyo Kenkyu 27, 7–17.

Van Banning P, Haenen OLM (1990) Effects of the swimbladder nematode *Anguillicola crassus* in wild and farmed eel *Anguilla anguilla*. In: Perkins FO, Cheng TC (eds), Pathology in Marine Science. Academic, New York. Proceedings of the 3rd International Colloquium on Pathology in Marine Aquaculture, Gloucester Point, Virginia, October 2–6 1988, pp. 317–330.

Van Ginneken V, Van Ballieux B, Willemze R, Coldenhoff K, Lentjes E, Antonissen E, Haenen O, Van den Thillart G (2005) Hematology patterns of migrating European eels and the role of EVEX virus. Comp. Biochem. Physiol., part C 140, 97–102.

Van Ginneken V, Haenen O, Coldenhoff K, Willemze R, Antonissen E, Van Tulden P, Dijkstra S, Wagenaar F, Van den Thillart G (2004) Presence of virus infections in eel species from various geographic regions. Bull. Euro. Ass. Fish Pathol. 24(5), 268–272 (short communication).

Van Nieuwstadt AP, Dijkstra SG, Haenen OLM (2001) Persistence of herpesvirus of eel *Herpesvirus anguillae* in farmed European eel *Anguilla anguilla*. Dis. Aquat. Organ. 45, 103–107.

Veenstra J, Rietra PJGM, Coster JM, Stoutenbeek CP, ter Laak EA, Haenen OLM, De Gier HHW, Dirks-Go S (1993) Human *Vibrio vulnificus* infections and environmental isolates in the Netherlands. Aquac. Fish. Manage. 24, 119–122.

Wolf K (1988) Fish Viruses and Fish Viral Diseases. Cornell University Press, Ithaca, USA. Comstock Publishing Association, New York, 476 pp.

Worthington LV (1959) The 18°C water in the Sargasso Sea. Deep Sea Res. 5, 297–305.

Chapter 17
Evolutionary Consequences of Eel Migration

Gregory Maes, Marti Pujolar, and Filip Volckaert

17.1 Introduction

Fish migration is a regularly occurring habitat transition between the spawning area and growth habitat of a species. A conceptual model that superimposes the life cycle of migratory fishes onto a closed route of migration connecting the spawning area and growth habitat has been defined as a "migration loop" (McDowall 1992; Tsukamoto et al. 2002). In principle, each fish species has a migration loop specific to its life history and geographic distribution. Therefore, the differentiation of a new migration loop has the potential to cause reproductive isolation and hence speciation. This concept helps for an understanding of the evolutionary processes of fish migration as well as the migratory behavior and life cycle of fishes.

Diadromy is a migratory strategy in fishes that involves a regular migration pattern between fresh and salt water (McDowall 1992). It is a relatively rare behavioral trait, occurring in perhaps 250 out of the some 25,000 known fish species (McDowall 1993). Diadromous migrations have been observed in several taxa of fishes, suggesting that diadromy originated independently throughout fish evolution (McDowall 1992, 1993). Therefore, each migratory fish species might have experienced different selection pressures that resulted in variations in its life history. Diadromous fishes undertake two major habitat shifts in every generation: a migration from fresh water to the ocean, and another migration in the opposite direction. Some diadromous fish are semelparous (one single reproductive migration per generation), others are iteroparous (two to several reproductive events per generation). Anadromy, catadromy and amphidromy are all variants of diadromy. Anadromy refers to the migration patterns of fish, such as salmonids, that live in the ocean but return to fresh water to spawn. In amphidromy, such as observed in gobiids, the migrations are not directly tied to spawning, but to some other activity,

G. Maes, M. Pujolar, and F. Volckaert
Laboratory for Animal Diversity and Systematics, Katholieke Universiteit Leuven, Ch. Deberiotstraat, 32, B-3000, Leuven, Belgium

M. Pujolar
Dipartimento di Biologia-Università di Padova, Via G. Colombo 3, I-35131 Padova, Italy.

G. van den Thillart et al. (eds.), *Spawning Migration of the European Eel,*

such as feeding. Catadromy refers to the migration patterns of fishes, such as the European eel, that live in fresh water but return to the ocean to spawn. The evolution of these life-strategies is triggered by several abiotic and biotic factors, such as glaciation, continental drift, habitat suitability, food availability and selection. For example, the ancestor of the Anguillidae might have been a tropical marine species with a migration loop extending to coastal waters. From there they incidentally visited estuaries and eventually obtained a reproductive advantage because of the greater amount of food available in estuaries and by extension in fresh water (Tsukamoto et al. 2002). Thus the ancestor probably developed an adaptive behavior of regularly migrating upstream as a result of a gradient in food abundance between the ocean and fresh water in the tropics. In order for this process to occur, first there must have been a euryhaline marine species and then an oceanic amphidromous adaptation before the appearance of catadromous eels (Tsukamoto et al. 2002) (Fig. 17.1). A similar scenario applies to anadromous fishes.

The evolutionary consequences of a catadromous life-strategy in fishes are poorly known, mainly because of the paucity of information on their genetic structuring, general biology and ecology (McDowall 1992; Maes & Volckaert 2007). Catadromy has important repercussions on the formation of spatial genetic heterogeneity, because the potential for gene flow among geographical locations is greater than that of strictly freshwater or anadromous fishes. The few studies on genetic structuring in catadromous fishes have shown that the extent of differentiation among populations varies and is closely linked to life history attributes, in particular the location of spawning sites and the subsequent dispersal of larvae. In barramundi (*Lates calcarifer*) and Australian bass (*Macquaria novemaculeata*), which spawn at

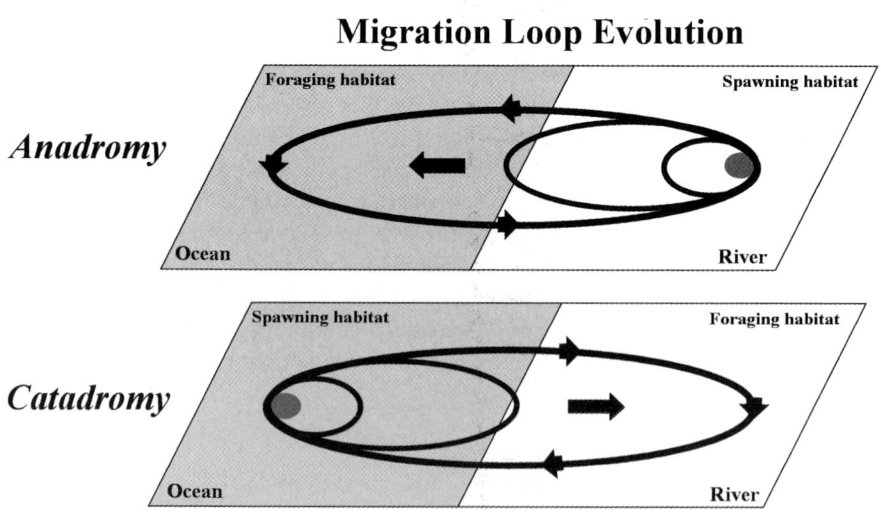

Fig. 17.1 Diagram of changing migration loops that result in the evolution of the diadromous life histories of salmon (top) and eel (bottom) (Modified from Tsukamoto et al. 2002)

the mouth of coastal drainages, disjunct riverine populations are genetically heterogeneous across Australia (Chenoweth et al. 1998; Jerry and Baverstock 1998). The genetic differentiation in these two taxa demonstrates the need for additional information from other catadromous fishes to understand more on how this particular life-strategy influences spatio-temporal genetic population structuring. Therefore, knowledge of the marine phase is crucial to understand the continental distribution and genetic patterns.

17.2 Genetics of Marine Organisms

Marine organisms experience a wide range of intrinsic and extrinsic influences during their life cycle, which considerably impact their population dynamics and genetic structure (Waples 1998). Due to their biological characteristics, marine species are strongly influenced by ocean currents and food abundance, leading to variable reproductive success and high larval mortality. Subtle genetic differences reflect the continuity of the marine environment, but also pose serious challenges to define management units (Kenchington et al. 2003). Marine fishes are expected to exhibit a high genetic variability, a high exchange of migrants between populations (gene flow) inducing a low genetic differentiation (low genetic signal/noise ratio) and a high effective (genetic) population size (Waples 1998). In diadromous species, selection pressure during larval migration and freshwater/marine residence may additionally affect the genetic pattern of populations (Mitton 1997).

Widely distributed marine species are less likely to be fully panmictic, but may be divided into subgroups in a pattern described by one of the classical population models, such as the island, stepping-stone or Isolation-By-Distance (IBD) model (Rousset 1997). The genetic architecture of natural populations integrates factors such as population size, individual dispersal, behaviour, assortative mating, reproductive success and survival (Rousset 1997; Avise 2004). In order for populations to diverge, strong temporally stable barriers to gene flow are required to restrict the dispersal of migrants. If separation is incomplete, some level of gene flow may occur between spatially adjacent populations, creating a pattern of IBD (Rousset 1997). In populations composed of a mixture of individuals reproducing at different times within a reproductive season, temporal differentiation may accompany geographical partitioning. Reproductive times, thought to be highly heritable rather than individually plastic, are thus influenced by as well genetic (the breeding value) as environmental factors (Fillatre et al. 2003; Hendry et al. 2004; Hendry and Day 2005). Under these conditions, gene flow is expected to be limited between early and late reproducers. A temporal –instead of spatial-restriction on gene flow would hence create a pattern of Isolation-By-Time (IBT) (Hendry and Day 2005).

In the case of marine organisms, IBD has been described repeatedly (Pogson et al. 2001; Wirth and Bernatchez 2001; Maes and Volckaert 2002; O'Reilly et al. 2004) and several cases of the temporal segregation of spawning cohorts have been documented (Ruzzante et al. 1996, 1999; Beacham et al. 2002; Chapman et al. 2002; Hoarau et al. 2002; McPherson et al. 2003; Jørgensen et al. 2005). Many marine species divide their reproductive effort between several events during a protracted spawning season, potentially generating genetic differentiation between consecutive spawning groups. The degree of temporal overlap between spawning groups will determine the level of differentiation between populations. If temporal overlap is small, a stable pattern of IBT may arise, its stability depending on the heritability of spawning time, the level of gene flow after dispersal and environmental conditions (Hendry and Day 2005).

Additionally, heterogeneous larval cohorts characterize many marine organisms including sea urchins (Flowers et al. 2002), oysters (Li and Hedgecock 1998) and fishes (Ruzzante et al. 1996; Planes and Lenfant 2002). Temporal heterogeneity in the genetic composition of recruits is likely to result from a large variance in parental reproductive success driven by the unpredictability of the marine environment (Waples 1998). Under the hypothesis of "sweepstakes reproductive success" (Hedgecock 1994), chance determines which adults are successful during each spawning event. Hedgecock (1994) attributed the variation in reproductive success of adults more to spatio-temporal variation in oceanographic conditions, occurring within and among seasons than to differential levels of adult maturity. Additionally, fluctuations in recruitment success are highly dependent on an ecosystem-wide match between phytoplankton production and spawning timing (match-mismatch hypothesis; Cushing 1990) and/or physical larval retention (member-vagrant hypothesis; Sinclair 1988). A genetic consequence of a high variance in reproductive success is a stochastic genetic composition of recruits (genetic patchiness; Hedgecock 1994), possibly counteracting spatio-temporal genetic differentiation, especially if patchiness surpasses forces restricting dispersal in time and space. Marine species typically divide their reproductive effort between several events during a protracted spawning season, supposedly to maximize their reproductive success (Hutchings and Myers 1993; Maes et al. 2006a).

For decades, marine research has focused on the population dynamics of marine organisms. The main source of information on population composition, abundances, stability and fitness of commercial species remains the fisheries data. From the late 1980s onwards it became clear that an increasing number of marine fish stocks were fully exploited or even overexploited, leading to a strong decline or even collapse in population size (Myers and Worm 2003; Mullon et al. 2005). A gap became apparent between the population counts (census population size; *Nc*) and the genetically determined population sizes (effective population size; *Ne*). The *Nc/Ne* ratio turned out to be several orders of magnitude lower than expected (Hauser et al. 2002; Turner et al. 2002). This has affected the interpretation of the so-thought vast numbers of marine organisms. The future of commercial fish biomass has been further compromised by the synergy between climate change and anthropogenic influences, such as heavy fisheries and habitat degradation, which thus plays a major role in the decline of commercial species (Dulvy et al. 2003). Apart from Ne, genetics has proven invaluable to discriminate independently evolving populations (Park and Moran 1994; Ward 2000) and to provide indirect estimates of dispersal, population size, demography, stock sustainability and adaptive divergence (Palumbi 1994; Waples 1998; Avise 2004; Conover et al. 2006) using various classes of markers (see Box 17.1 and Fig. 17.2).

Box 17.1 Genetic markers in marine evolutionary research

The choice of molecular markers in population biology is highly dependant on the organism of study, the quality of the tissue, the availability of molecular markers, the sensitivity required and the hierarchical level of the study (species, population, individual) (Park and Moran 1994). A short description of the genetic markers used is provided here, while Fig. 17.2 shows the level of variability and applicability of the different classes of genetic markers used in eel research.

Allozymes are identifiable polymorphic gene loci with Mendelian inheritance. Screening has several advantages; the procedure is rather simple and cheap and there is no need for specific development of markers. Limitations include the requirement for about 1 g of fresh or frozen tissue samples. In addition, the sampling of tissues from different organs often requires lethal sampling; in some cases tissue biopsies or blood may be used. In addition, some changes in DNA sequence are masked at the protein level, reducing the level of variation detected. Some changes in nucleotide sequence do not change the encoded polypeptide (silent substitutions), and some polypeptide changes do not alter the mobility of the protein in an electrophoretic gel (synonymous substitutions) (Ryman and Utter 1987). Selective neutrality is not always complete, as several studies showed the balancing or directional pattern of allozymatic variation (Bossart and Prowell 1998). In a few cases, correlations exist between selected allozyme markers and performance traits (McGoldrick and Hedgecock 1997). Individuals are screened for allozymes alleles by various kinds of gel electrophoresis; alleles are made visible with enzyme-specific staining reactions (Richardson et al. 1986).

Microsatellites consist of multiple copies of tandemly arranged simple sequence repeats (SSRs) that range in size from 1 to 6 base pairs (e.g., ACA or GATA) (Chistiakov et al. 2006). They vary in the number of tandem repeats, so that they belong to the so called "VNTRs" or Variable Number of Tandem Repeats, and correspond to non-coding regions of the nuclear DNA. Abundant in all species, microsatellites have been estimated to occur as often as once every 10 kb in fishes. Most microsatellite loci are relatively small, ranging from a few to a few hundred repeats, which is important for PCR-facilitated genotyping. Microsatellite polymorphism is based on size differences due to varying numbers of repeat units contained by alleles at a given locus. Advantages of microsatellites include: a Mendelian fashion of inheritance as codominant markers, high abundance, even genomic distribution, small locus size, and high polymorphism. Microsatellites have become an extremely popular marker in a wide range of fields, such as forensics, kinship and parentage analysis, population genetics and fisheries research (reviewed in O'Connell and Wright 1997). This marker class has a stronger discriminative potential for subtle genetic differentiation, considering only drift and mutation as

(continued)

Box 17.1 (continued)

evolutionary forces acting on their evolution (no selection). Moreover, primers developed for one species can usually be used in other related species. One disadvantage of microsatellites is the presence of null alleles that originate through a mutation in the flanking region, and which can be a serious problem in population genetic studies since they induce a strong heterozygote deficit as only one of the two alleles is amplified. Microsatellite alleles are detected as DNA fragments through polymer gel electrophoresis, typically on a sequencing unit.

Mitochondrial DNA (mtDNA) is the small (15–18 kb) non-recombining circular DNA molecule (genome) of the mitochondrion, a cytoplasmic organelle. Because of haploidy and maternal inheritance, its effective population size is four times smaller than nuclear loci and is more influenced by genetic drift. One characteristic is its rapid evolution, 5–10 times faster than nuclear DNA. mtDNA markers are excellent tools for population genetic analyses (Avise 2004). Advantages of mtDNA are the ease of amplification in a variety of taxa, the recovery of the pattern and tempo of recent historical events, and its genealogical history. Limitations are that it is a marker for historical processes in females only, and mtDNA sequences may be integrated in the nucleus. mtDNA polymorphisms are detected through sequencing of PCR amplified fragments. A common and cheaper strategy is Restriction Fragment Length Polymorphism (RFLP), where restriction enzymes cut the PCR amplified fragment, exposing polymorphism at target sequences.

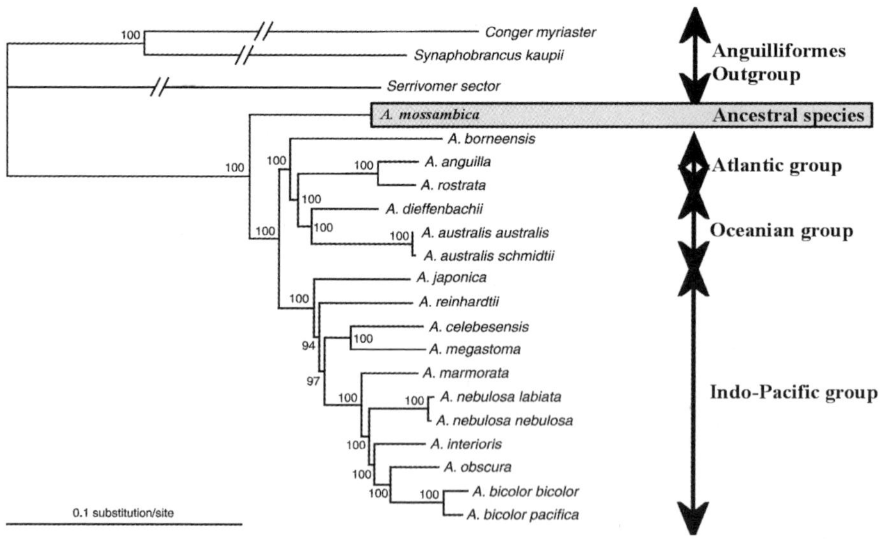

Fig. 17.2 Most recent phylogenetic tree of the genus *Anguilla* based on the complete mitochondrial genome from all 15 species and 3 subspecies (Modified from Minegishi et al. 2005)

17.3 Evolutionary Status of North Atlantic Eels

The life history of the catadromous European eel (*Anguilla anguilla* L.) hinges on oceanic conditions. Maturation, migration, spawning, larval transport and recruitment dynamics, all take place in the marine environment (Tesch 2003; Knights 2003; Kettle and Haines 2006). Its life cycle (migration-loop) typically tracks the North-Atlantic gyral system (Tsukamoto et al. 2002). Despite the key biological importance of the marine phase (Knights 2003), most research has focused on the freshwater phase. This has proven to be detrimental for management. Now that recruitment has dropped to just 1% of the 1960 level (Dekker 2003), critical biological and genetic questions are being raised. Several causes have been proposed for the decline ranging from pollution, overfishing, migration barriers, habitat destruction, parasites and diseases to global oceanic and climatic changes (Dekker 2003; Knights 2003). Synergy among all these factors seems the most likely cause (Wirth and Bernatchez 2003). Genetic data may help assessing the species' integrity within the North-Atlantic Ocean, the number of genetic stocks within the European eel, the spatio-temporal stability of the genetic structure, the influences of oceanic conditions on genetic variability, the effect of a population decline on the genetic variability and the overall fitness of eels.

Recently the European Commission has released an action plan for the European eel, which aims at strengthening the return rate of adult eels to the Sargasso Sea and includes the development of national management plans (CEC 2005). To do so, a detailed review of the genetics of European eel is required, as well as the delineation of future research areas to improve our understanding. This is of importance to maintain intraspecific genetic diversity, to develop sound restocking programs for brood stock enhancement and to help realizing profitable artificial breeding. The aim of this chapter is to review knowledge on the evolution of eels, to document evidence of the importance of genetic diversity and fitness in marine species, to synthesize the most recent population genetic knowledge on European eel and to provide an overview of the more optimal use of genetics in future management decisions in this strongly declining species. All this contributes to a comprehensive biological review of the migration potential of European eel.

17.3.1 Phylogeny of the Genus Anguilla

The Anguillidae, including the genus *Anguilla*, belong together with the Moringuidae to the suborder Anguilloidei. Based on a combination of morphological and molecular traits, 15 species have been officially recognized within the genus *Anguilla* (Aoyama et al. 2001; Minegishi et al. 2005). Recent phylogenetic analyses confirm the monophyletic status of the genus *Anguilla*. The most likely ancestor is *A. mossambica*, which might have originated close to the Indonesian archipelago during the Eocene (60 MY bp) at the latest (Fig. 17.3). Complex dispersal and subsequent speciation patterns lead to the present worldwide distribution of eel (Aoyama et al. 2001; Lin et al. 2001; Minegishi et al. 2005) (Fig. 17.4).

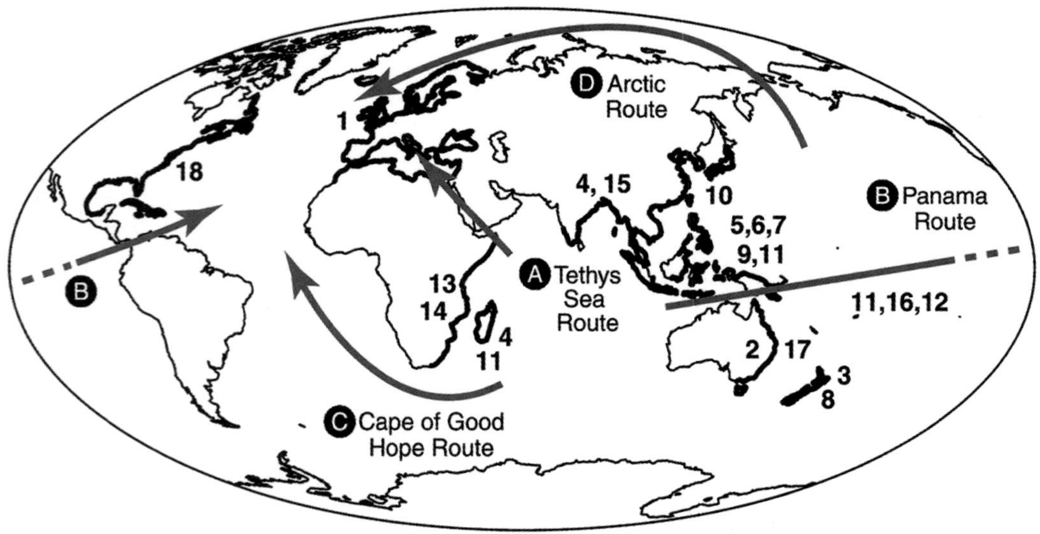

1	*A. anguilla*
2	*A. australis australis*
3	*A. australis schmidtii*
4	*A. bicolor bicolor*
5	*A. bicolor pacifica*
6	*A. borneensis*
7	*A. celebesensis*
8	*A. dieffenbachii*
9	*A. interioris*
10	*A. japonica*
11	*A. marmorata*
12	*A. megastoma*
13	*A. mossambica*
14	*A. nebulosa labiata*
15	*A. nebulosa nebulosa*
16	*A. obscura*
17	*A. reinhardtii*
18	*A. rostrata*

Fig. 17.3 Geographic distribution of the genus *Anguilla* (bold lines), and possible dispersal routes of the ancestral eels into the Atlantic Ocean (gray arrows). The centre of geographical distribution are indicated by numbers on the right side (Modified from Minegishi et al. 2005)

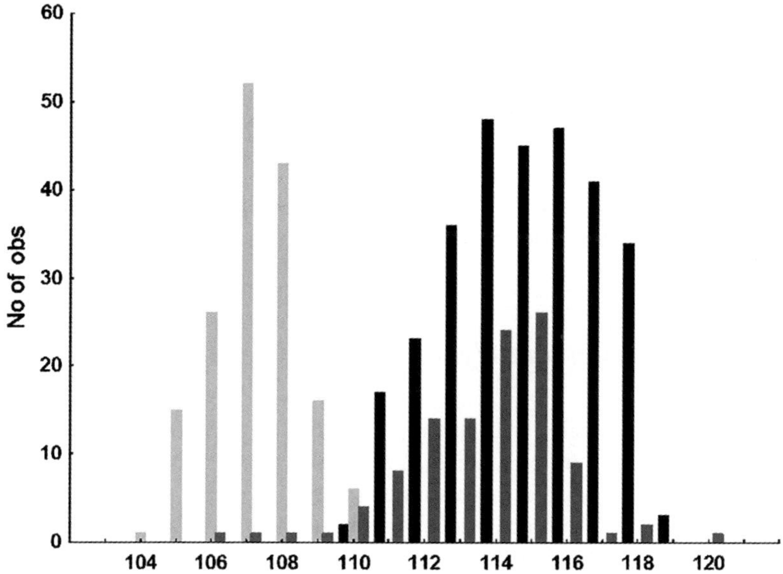

Fig. 17.4 Distribution of vertebral counts in North-Atlantic eels. Light Grey: *A. Rostrata*; Black: *A. Anguilla*. Within the Icelandic "populations" (Dark Grey), there is a fraction of eels exhibiting a number of vertebrae characteristic for American eels (Maes 2005)

Only two species inhabit the North-Atlantic Ocean, the European (*A. anguilla*) and the American eel (*A. rostrata*). Several scenarios have been proposed for their origin, based on evidence from fossil records, plate tectonics, paleocurrents and a standard fish molecular clock. A first scenario is the dispersal of ancestral organisms through the Tethys Sea (Fig. 17.4) that separated 70 MY bp Laurasia (North-America and Eurasia) from Gondwana (South America, Australia, Africa and India). Along this sea, dispersal was possible through westerly paleocircum-global equatorial currents (Aoyama and Tsukamoto 1997; Aoyama et al. 2001). Aoyama et al. (2001) suggest that *Anguilla* speciation started 43.5 MY bp and that the North-Atlantic eels speciated some 10 MY bp. Although such results were partially confirmed by Bastrop et al. (2000), Lin et al. (2001), using a much larger fragment of the mitochondrial genome (cytochrome *b* and 12sRNA), proposed that the genus *Anguilla* speciated much more recently, some 20 MY bp. They hypothesized that the Atlantic eels colonized the North Atlantic through the Central American Isthmus (Panama) and speciated some 3 MY bp (Fig. 17.4). Although a longer fragment was used and the speciation estimates were much more congruent with the accepted molecular clock, some incongruence remained. The absence of any eel species on the west coast of North America or South America and the large phylogenetic distance from *A. japonica*, the presumed ancestor of the North Atlantic eels, suggests that the radiation events are much more complicated than expected using contemporary hydrodynamic and tectonic knowledge. A recent study analyzing the complete mitochondrial genome provided additional support for the first hypothesis' dispersal route, but for the second hypothesis' speciation time (Minegishi et al. 2005). Speciation started 20 MY bp and formed two main clades, The Atlanto-Oceanian group and the Indo-Pacific group. The geographical distribution does not seem to follow phylogenetic relationships in the former, but does so in the latter group (Minegishi et al. 2005). Data from the nuclear genome might clarify these ambiguities (see also later). These results also confirm the unreliable nature of morphological characters to discriminate the evolutionary relationships between *Anguilla* species, even after a thorough revision (Ege 1939; Watanabe et al. 2004).

17.3.2 Evolution and Hybridization in North Atlantic Eels

The only two eel species with incomplete reproductive isolation are the European (*A. anguilla*) and the American eel (*A. rostrata*). Based on the number of vertebrae, the American eel (vertebral counts ranging from 103–110, mean 107.1) can be distinguished from the European eel (vertebral counts ranging from 110–119, mean 114.7) (Boëtius 1980) (Fig. 17.5).

Based on the geographic distribution of leptocephalus larvae, the spawning area of the two eel species is thought to geographically overlap in the Sargasso Sea (Schmidt 1935; Ohno et al. 1973; Comparini and Rodinò 1980; McCleave and Kleckner 1987; Tesch and Wegner 1990). There is also evidence for some temporal overlap of spawning, but judging from the distinct geographical range of the adults

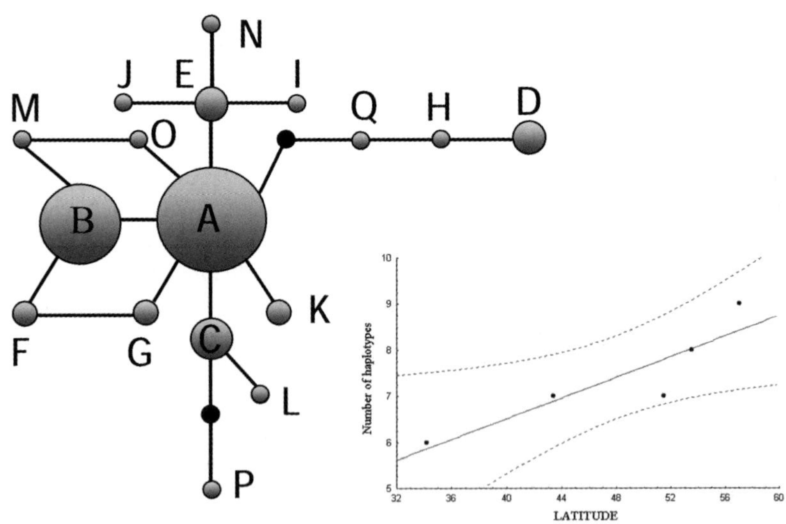

Fig. 17.5 Median network of all cytochrome *b* mitochondrial haplotypes detected in the European eel **(a)** and North-South clinal variation in haplotype diversity **(b)** (Modified from Daemen et al. 2001)

and the unique presence of hybrids in Iceland (see later), there has to be some segregation between the spawning adults of each species.

The divergence between the two North-Atlantic species has been under discussion for decades. Tucker (1959) claimed that differentiating meristic characters (number of vertebrae) were under ecophenotypic selection during the transoceanic migration. Hence the European eel would represent offspring of the American eel. Tucker (1959) suggested that European eels do not participate in reproduction, because of the distance to the remote Sargasso Sea is too great. Later work, based on genetic variation of the proteins hemoglobin, transferrins and a wide range of allozymes did confirm the two species status (Fine et al. 1967; Drilhon et al. 1966, 1967; Drilhon and Fine 1968; De Ligny and Pantelouris 1973; Comparini and Rodinò 1980; Comparini and Schoth 1982). The allozyme locus *MDH-2** exhibits a nearly fixed difference between the two species, although Williams and Koehn (1984) questioned its taxonomic reliability based on a single enzymatic locus possibly under selective pressure. A mitochondrial DNA Restriction Fragment Length Polymorphism (RFLP) study showed conclusive results, separating the two species with significant confidence at 11 out of 14 restriction endonucleases, although the two North-Atlantic species exhibited the lowest genetic distance reported between any *Anguilla* species (Avise et al. 1986; Tagliavini et al. 1995; Aoyama and Tsukamoto 1997; Ishikawa et al. 2004). Another study assessed the speciation process with jointly distributed parasites (Marcogliese and Cone 1993). They reviewed the "oceanic" and the "vicariance" hypothesis of Avise et al. (1990), suggesting that the two species diverged either in sympatry through separate currents or through the influence of the ice sheets during the Pleistocene, respectively. In the first hypothesis, eels were supposed to live along a single coast (America or Europe) and disperse through changing currents to the opposite side of the Atlantic Ocean, with subsequent assortative mating. The second hypothesis states that the ancestor had a broad continuous distribution, but split into two groups distributed on each side of the Atlantic under the influence of southward Pleistocene glaciations. The vicariance hypothesis seems to be the most likely to explain the present transcontinental distribution of the parasites, which can only be transmitted horizontally by continental residents living in freshwater (Marcogliese and Cone 1993). Probably, distinct dispersal patterns during spawning and/or unique spawning grounds provide the basis for the current division between the two species. Larvae may heritably and actively choose the currents that ultimately bring them at either side of the Atlantic Ocean (Kleckner and McCleave 1985). Another possibility is a strict genetically-determined period of metamorphosis (McCleave 1993; Cheng and Tzeng 1996), which ultimately brings the larvae into currents directing them to the American or European continent.

Even though reproductive isolation is strong, hybrids between European and American eel have been detected. Williams and Koehn (1984) compared the *MDH-2** genotypes with the number of vertebrae and concluded that there must be a significant amount of gene flow between the two species. Avise et al. (1990) evaluated mitochondrial DNA in addition to nuclear and meristic markers in Icelandic individuals. The data reflected cytonuclear disequilibria, most likely due to ongoing gene flow between the two species. Besides pure individuals of the two species hybrids were detected and American eel material in Iceland was quantified at 2–4%. Recently, Mank and Avise (2003) reassessed these conclusions with highly polymorphic microsatellites markers, but surprisingly failed. Despite the high resolution and power of microsatellite markers (Manel et al. 2002; Anderson and Thompson 2002), most likely homoplasy generated a low discriminative power between the two eel species (Mank and Avise 2004). This result prompted further investigations on the paradigm of complete isolation of European and American eels, and reopened the debate of the existence and maintenance of a hybrid zone at more than 6,000 km from the spawning site. Very recently two studies reassessed the hybridization dynamics using AFLP, moderately variable microsatellites and the number of vertebrae. Both studies found similar proportions of American eel introgressed in the mainly European Icelandic populations. There was even a suggestion for backcrosses, which points to a high survival rate of hybrids (Maes 2005; Albert et al. 2006).

17.4 European Eel: The Panmixis Hypothesis Revisited

17.4.1 Phylogeography

The phylogeography of the European eel has not been studied that intensively (Avise et al. 1986; Lintas et al. 1998; Daemen et al. 2001; but see Patarnello et al. 2007). Mitochondrial DNA was not very helpful with the geographical partitioning of genetic variability in the European eel, mainly because the high haplotype diversity complicated a quantitative analysis (Lintas et al. 1998). Bastrop et al. (2000) confirmed this result based on nuclear 16sRNA sequences. Although the European eel population is genetically more diverse than the American eel population (Avise et al. 1986; Bastrop et al. 2000) and the genetic homogeneity of the European eel seemed beyond dispute (Lintas et al. 1998; Bastrop et al. 2000), the possibility remained of multiple spawning areas. In a later study, new indications of non-random distribution of haplotypes were reported using a less variable mtDNA marker (cytochrome *b*) (Daemen et al. 2001, Fig. 17.6). Genetic variation at the cytochrome *b* locus was moderately high (17 haplotypes in 107 eels), with two common ancestral haplotypes in the haplotype network (A and B) and a significant latitudinal clinal pattern of haplotypes fitting an IBD model.

17.4.2 Population Genetics

The hypothesis that European eels migrate to the Sargasso Sea to reproduce as a single randomly mating population, the so-called panmixia theory, is generally accepted. However, molecular genetic data have provided contradictory evidence.

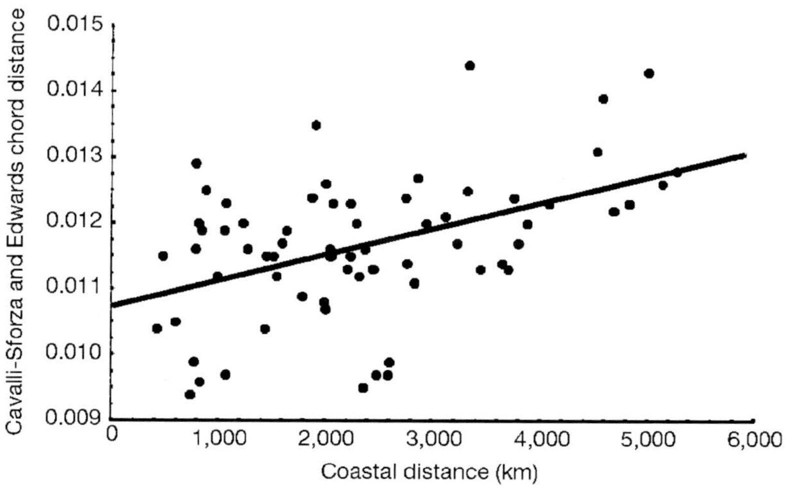

Fig. 17.6 Correlation between genetic distance (D_{CE}) and coastal geographical distance in the European eel (Reprinted with permission from Wirth and Bernatchez 2001)

We review chronologically the development of ideas on European eel population structure, fuelled by continuous advances in molecular evolutionary genetics and data analysis.

Early population genetic studies, based on differences in the proteins transferrin and liver esterase, claimed that European eel populations differed between several continental European locations (Drilhon et al. 1966, 1967; Drilhon and Fine 1968; Pantelouris et al. 1970). It was even suggested that eels in the south-eastern part of the Mediterranean Sea formed a separate group and reproduced locally. Now and then this unsupported claim keeps on surfacing in the popular literature. Although differential selection was also proposed as an explanation (Pantelouris et al. 1970, 1971), the conclusions of most allozyme-based studies from the 1960s were ultimately rejected on methodological grounds (Koehn 1972). Later on allozyme studies failed to confirm spatial genetic differentiation (de Ligny and Pantelouris 1973; Comparini et al. 1977; Comparini and Rodinò 1980; Yahyaoui et al. 1983).

Panmixia in the European eel thus became widely accepted until three independent genetic studies reported evidence for weak but significant population structure (Daemen et al. 2001; Wirth and Bernatchez 2001; Maes and Volckaert 2002). Daemen et al. (2001) detected weak but significant genetic differentiation among the British/Irish, Atlantic, Moroccan, Italian and Swedish Baltic populations, respectively, using five nuclear microsatellite loci. Later on, Wirth and Bernatchez (2001) also identified weak but highly significant genetic structure in the European eel population among 13 samples, based on seven microsatellite loci, and reported evidence for IBD (Fig. 17.7).

Maes and Volckaert (2002) reported a clinal genetic structure and IBD in the European eel population using 15 allozyme loci. They identified three

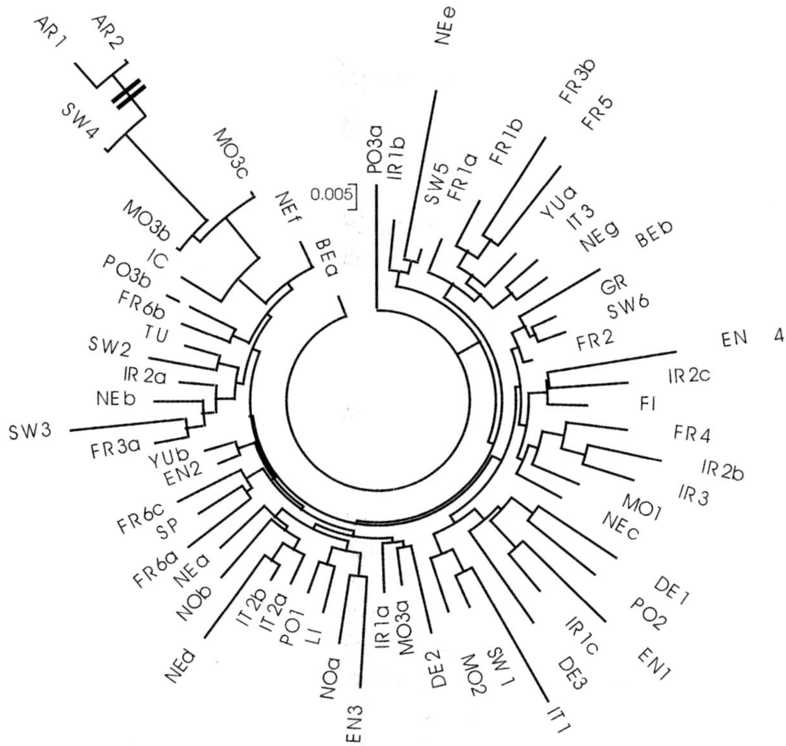

Fig. 17.7 Genetic evidence based on microsatellites combining geographical and temporal samples across Europe favour geographical homogeneity. AR: *Anguilla rostrata* (Modified from Dannewitz et al. 2005)

groups: Northern Europe, Western Europe and the Mediterranean Sea. Results from the former genetic studies pointed to a genetic mosaic, consisting of several isolated spawning groups. According to Wirth and Bernatchez (2001) and Maes and Volckaert (2002), three scenarios could explain the rejection of the panmixia hypothesis: (i) there is one common spawning area, but there is a temporal delay between the arrival of adult eels originating from different latitudes, (ii) there is one reproductive area shared by several populations where distinct sea currents carry the leptocephali back to their parent's original freshwater habitat and (iii) there is a single shared spawning area where assortative mating occurs and larval homing to the parents' habitat takes place through an unspecified mechanism.

However, a more recent and extensive genetic study on European eel expanded significantly geographical coverage (42 sites) and included crucial temporal replicates (at 12 sites) into their analyses to check for consistency in the observed spatial pattern (Dannewitz et al. 2005). Surprisingly, the signal for spatial genetic structuring was lost, while temporal variance in allele frequency exceeded the geographical component (Fig. 17.8). Sampling bias due to the mixing of life

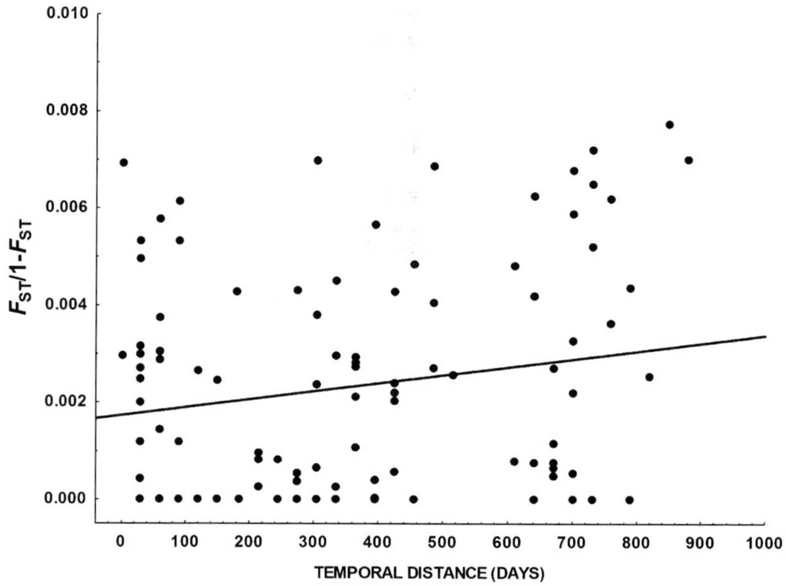

Fig. 17.8 Evidence for Isolation-By-Time in European eel (Reprinted with permission from Maes et al. 2006a)

stage and a lower effective population size than expected explain the seemingly conflicting results.

Finally, two studies looked for the contribution of temporal genetic variance in the European eel. Maes et al. (2006a) discriminated between the various forces shaping the genetic composition of recruiting juveniles of European eel (Fig. 17.9). Based on 12 polymorphic allozyme and six variable microsatellite loci, and by controlling for geographical variation, genetic differentiation was low but significant among temporal samples. Regression analysis between genetic and temporal distance was consistent with a subtle inter-annual pattern of IBT. They hypothesized that the population dynamics of the European eel may be governed by a double pattern of temporal variance in genetic composition: (i) large-scale IBT of spawning cohorts, possibly as a consequence of the large migration loop in Anguillids and strong variance in annual adult reproductive contribution; and (ii) a small-scale variance in reproductive success (genetic patchiness) within cohorts among seasonally separated spawning groups, most likely originating from fluctuating oceanic and climatic variables.

Pujolar et al. (2006) quantified the small-scale temporal genetic variation in glass eel recruitment waves and showed genetic variation over two time scales, namely between interannual samples (cohorts) and intra-annual samples within cohorts ('arrival waves'). Ten allozyme and six microsatellite loci detected highly significant genetic differentiation among arrival waves. Genetic composition was heterogeneous both among cohorts and among samples within cohorts, while

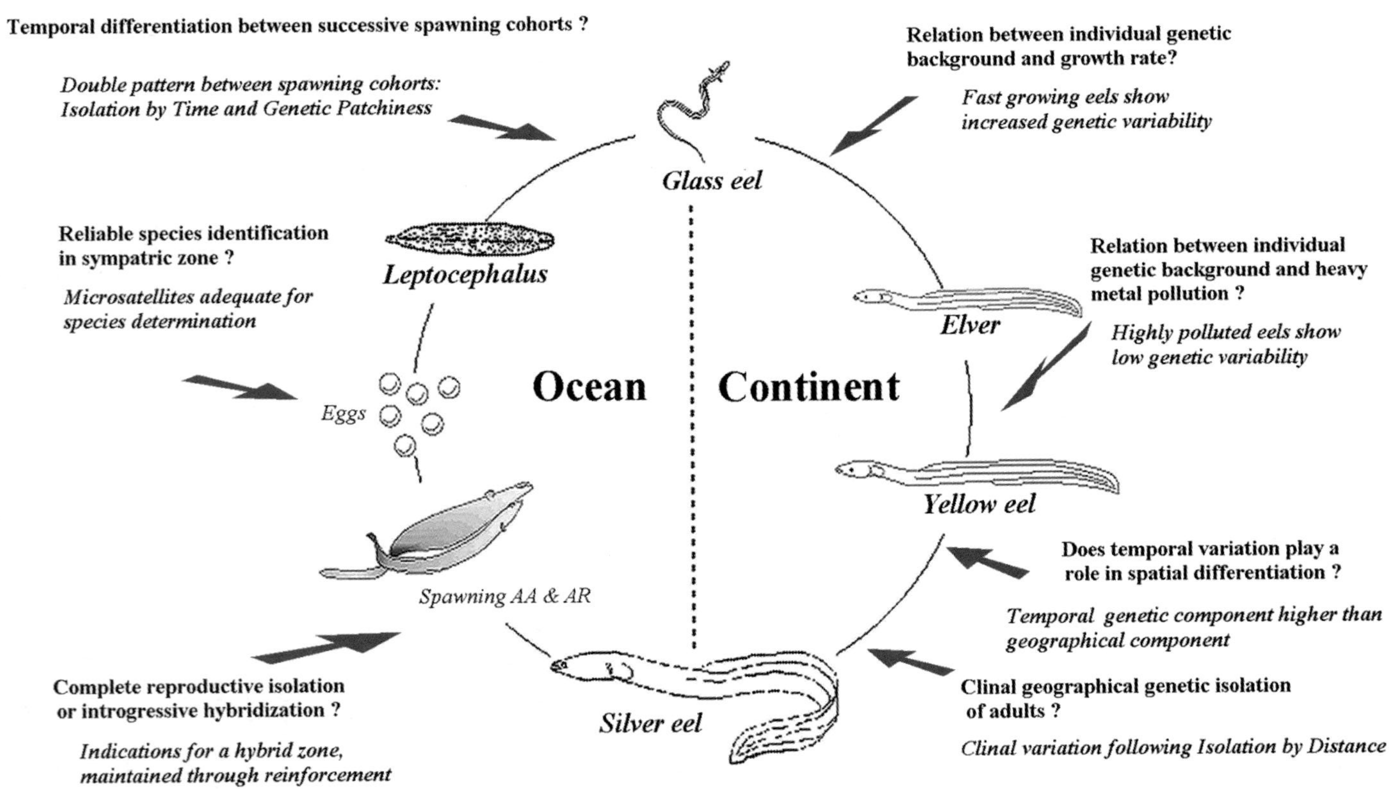

Fig. 17.9 Scheme of the life cycle of the European eel with a summary of the principal evolutionary questions and their answers (Reprinted with permission from Maes and Volckaert 2007)

genetic differentiation partitioned within cohorts reached more than tenfold the difference among cohorts. Such genetic heterogeneity may be the result of a large variance in the contribution of individuals to each cohort due to genetic drift (Hedgecock 1994). If only a subset of the adults contribute to the new recruits, effective population size in European eel should be much lower than the census size. They suggested that a low effective population size combined with fluctuating oceanic conditions might have contributed to the current dramatic decline in abundance of European eel.

In summary, a subtly heterogeneous population of European eel occupies Europe, with minimal geographical variation across the continent, but with most genetic variation present between temporally separated populations or recruitment waves. Such results reflect the high variance in reproductive success in marine species in general, inducing small and large-scale temporal changes in genetic composition between cohorts (Dannewitz et al. 2005; Maes et al. 2006a; Pujolar et al. 2006). The relationship between life-history and evolution is summarized in Fig. 17.10.

17.5 Heterozygosity-Fitness Correlations in the European Eel

Genetic diversity is of key importance to evolve in a changing environment or ouring events of strong selective pressure. Heterozygosity-Fitness Correlations (HFCs), which is the correlation between heterozygosity observed at marker loci and fitness-related traits such as growth, survival, fecundity or developmental stability, have been used as a proxy for adaptive potential in populations of many species for decades. Recently they have become integrated in ecological genomic approaches (Feder and Mitchell-Olds 2003). Positive HFCs have been reported in organisms as diverse as plants, marine bivalves, crustaceans, amphibians, salmonids and mammals (David 1998) but understandably null results are underrepresented in the literature (Hansson and Westerberg 2002). The first HFCs were observed in studies using allozyme markers, which led to the hypothesis of a direct heterozygote advantage at allozyme loci compared to the corresponding homozygotes (direct overdominance). Intrinsically higher heterozygote fitness implicated greater metabolic efficiency (Mitton 1997). The observation of positive HFCs with putative neutral DNA markers proves that at least some correlations result from the genetic association between the neutral markers and fitness genes (associative overdominance), due to either linkage disequilibrium or some degree of relatedness (David et al. 1995).

Understanding the significance of genetic variability for the fitness of eels and any organism is of importance to assess the risk related to the diversity loss due to overexploitation and environmental impact. Given the broad distribution range of European eel and the multiple environmental and anthropogenic factors influencing continental populations, the maintenance of a high level of genetic diversity is expected to be crucial to safeguard its evolutionary potential.

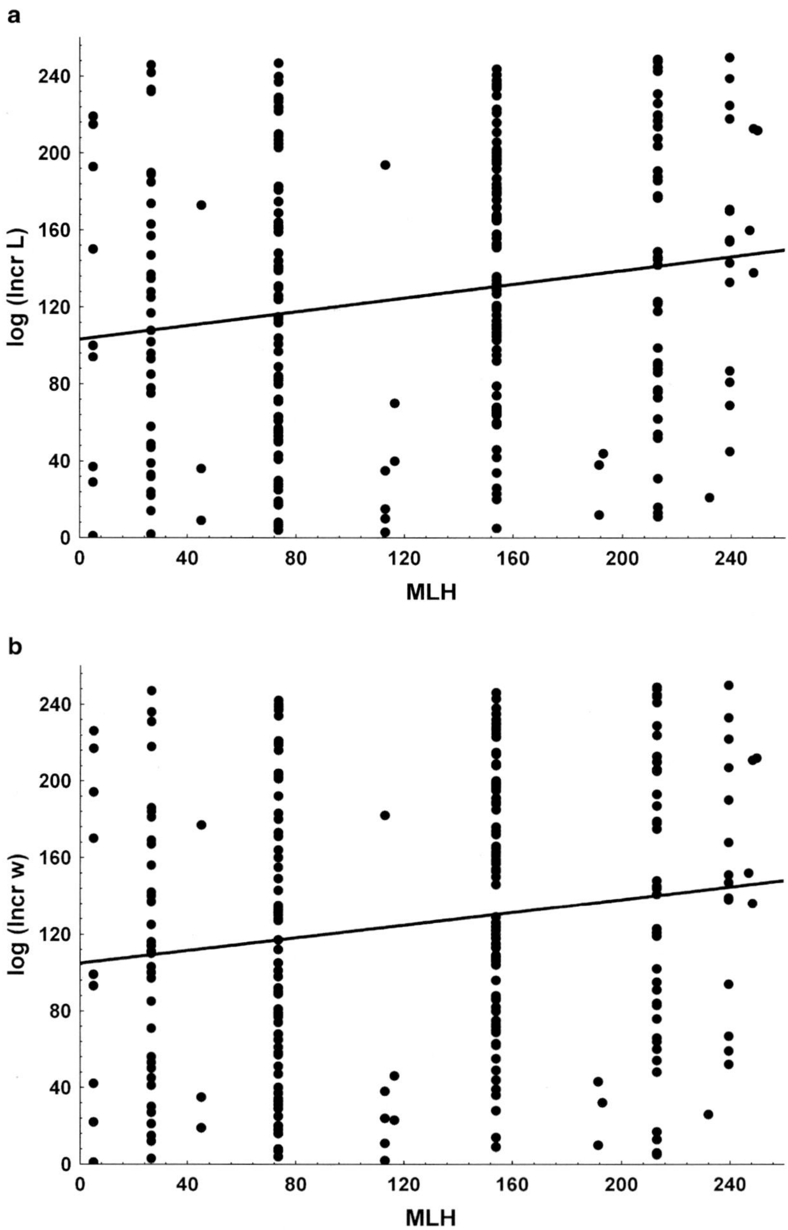

Fig. 17.10 *Anguilla anguilla*. Spearman rank correlation of multi-locus heterozygosity (MLH) versus (a) length (cm) increase (r = 0.177; p = 0.005*) and (b) weight (g) increase (r = 0.164; p = 0.009*) at 12 allozyme loci in all individuals (Modified from Pujolar et al. 2005)

17.5.1 Heterozygosity-Fitness Correlations and Growth

For the first time, Pujolar et al. (2005) provided evidence for a positive correlation between genetic variability and growth rate in a catadromous highly migratory fish. They used farmed individuals grown in aquaculture facilities. The farm environment is thought to increase and magnify fitness differences among individuals, possibly increasing the detection threshold of HFC. Multi-locus heterozygosity (MLH) at allozyme loci turned out to be positively correlated with growth; more heterozygous individuals have a significantly higher length and weight increase compared to more homozygous individuals (Fig. 17.11). In contrast, fast-growth individuals presented the lowest heterozygosities at neutral microsatellite loci. Pujolar et al. (2005) suggested that selection affecting some of the allozyme loci may explain the greater strength of the HFCs found at allozymes in comparison with microsatellites.

Koehn et al. (1988) also reported significant effects of heterozygosity on growth at enzymes involved in metabolic functions such as protein catabolism or glycolysis, while genes without significant effect on growth rate code for enzymes with miscellaneous functions, including the pentose shunt, redox balance or digestion. Metabolic enzymes have an important role in growth. In the study with a positive correlation between genetic variability and growth rate (Pujolar et al. 2005), the two most significant loci are involved in important metabolic functions: mannose-6-phosphate isomerase (MPI) is a pre-glycolytic enzyme, which does not have a direct role in glycolysis. As a supplier of carbon skeletons to glycolysis (converting mannose-6-phosphate to fructose-6-phosphate) it might be considered as belonging functionally to the glycolytic group. Glucose phosphate isomerase (GPI) is a main-pathway glycolytic enzyme that catalyses the reversible interconversion of d-fructose-6-phosphate and d-glucose-6-phosphate. By contrast, the smallest differences between homozygotes and heterozygotes in growth rate were observed at phosphogluconate dehydrogenase (*PGDH**), which codes for an enzyme with a secondary metabolic function (pentose shunt). Since PGDH is not involved in protein catabolism or glycolysis, it would have a less important role in growth. However, associative overdominance (where allozyme loci are merely acting as neutral markers of closely linked fitness loci) might provide an alternative explanation for the HFCs if we consider that allozyme loci have a higher chance than microsatellites to be in linkage disequilibrium with fitness loci.

17.5.2 Heterozygosity-Fitness Correlations and Environmental Pollution

Allozyme polymorphism and heterozygosity might be linked to environmental heterogeneity and stress in animals and plants (Nevo 2001). In this sense, understanding the impact of pollutants, such as heavy metals and pesticides, on the genetic diversity and structure of natural populations is essential to preserve the evolutionary potential of natural populations. The effect of pollution usually translates into a reduced genetic variability, survival of specific genotypes, and strong allelic selection or allele frequency shifts at polymorphic loci. The study of the relationship between pollution (bioaccumulation), fitness and genetic variability in the European eel would allow judging the long-term influence of anthropogenic factors on natural populations. The European eel constitutes a prime model to assess the impact of pollutants on condition and genetic variability, as juveniles recruit to European rivers without appreciable pollution load or interfering genetic background. Because of its high intramuscular fat content and benthic feeding behaviour, the feeding stage (named yellow eel) is considered extremely prone to the bioaccumulation of pollutants.

Maes et al. (2005) studied the relation between heavy metal pollution in European eel juveniles and the level of genetic variability. Individuals from three river basins were analyzed for nine heavy metals and genotyped at 12 presumed polymorphic allozyme and eight microsatellite loci. They observed a significant negative correlation between heavy metal pollution load and condition, suggesting an impact of pollution on the health of sub-adult eels. They also clearly showed a reduced genetic variability in highly polluted eels. A comparison between lowly versus highly polluted groups showed a reduction in genetic variability in the latter group at seven out of nine allozyme loci. A negative relation between individual allozymatic MLH and the level of bioaccumulation was also detected. By contrast, microsatellite genetic variability did not show any differences in genetic variability when comparing high and low pollution groups.

The correlations observed at several functional enzymes suggest a direct role of the allozymes in the bioaccumulation of toxicants and possibly direct overdominance of heterozygous individuals. This could indicate that the overall metabolic gain of heterozygotes fits the prediction that metabolic responses are essentially polygenic in nature. We expect to understand better the link between genotype and phenotype thanks to progress in functional genomics.

17.6 Genetic Research Perspectives and Management of the European Eel

More than 100 years of studies of the European eel have resulted in the testing of hypotheses of its population structure with newly developed techniques each time they became available. The most recent genetic results have answered several evolutionary challenges during its life cycle (Fig. 17.10). Several traits increase the chances for panmixia, such as the variable age-at-maturity, the highly mixed spawning cohorts, the protracted spawning migration and the sex-biased latitudinal distribution. Other aspects promote population fragmentation, such as the protracted spawning season and the variable cohort strength related to the unpredictability of the oceanic conditions. A conclusive answer will require further research.

For a more detailed review of the population genetics and the life cycle of eels of the genus *Anguilla*, we refer to Van Ginneken and Maes (2005).

Several management and scientific measures have been proposed to understand and reverse the decline of eel populations. They include assessing and reducing the impact of the fishery, monitoring recruitment, preserving migration routes (e.g., removing migration barriers), assessing the impact of restocking (preserving potential local populations), assessing anthropogenic influences (e.g., pollution and parasites) and estimating the spawning population size (CEC 2005; ICES 2006). The genetic equivalents of these measures involve the assessment of the spatio-temporal population structure by sampling the spawning grounds, an integrated analysis of census population size (*Nc*), the determination of the relationship between historical and current effective population sizes (*Ne*), the analysis of adaptive variation under natural and anthropogenic conditions and of the functional genomics of important traits for aquaculture and artificial reproduction.

The molecular conclusions are not only important to infer the evolutionary status of the European eel, but also to preserve the genetic resources and to define research priorities. Each priority may be accompanied by a specific management objective and a time-frame during which changes or reversal may be achieved. For instance, genetic diversity within populations may be lost rapidly (this is known as genetic erosion), and recovers only slowly (ICES 2006). We propose three lines of research: assessment of the spawning population structure and effective population size, integration of adaptive genetic variation in management plans, and improvement of artificial reproduction through aquaculture genomics.

17.6.1 Spawning Population Structure and Size

The genetic structure of marine populations is best understood by identifying, sampling and analyzing discrete reproductive aggregations (Waples 1998). But a precise localization of the spawning grounds, nurseries and retention zones, along with a greater knowledge of the ecosystem where spawning takes place is of great importance. To date no adult eel has been found in the Sargasso Sea, and their eggs have yet to be identified (Tesch 2003). In the Pacific Ocean, based on the distribution of newly hatched larvae, the spawning grounds of the Japanese eel have been reconfirmed by genetic identification techniques (Tsukamoto 2006).

The continental populations constitute mixed feeding aggregations, complicating the interpretation of genetic patterns (Dannewitz et al. 2005; Maes et al. 2006a; Pujolar et al. 2006). The age of the eels has to be known such that the site-specific age structure can be understood. The most effective approach would be of course to sample spawning eels and newly hatched larvae throughout the Sargasso Sea, and to analyze them with a representative set of genetic markers. The biological material would allow a reassessment of the spatial and temporal segregation observed so far and an estimate of the effective (genetic) population size (*Ne*). There is even the opportunity with highly performing genetic markers (such as Single Nucleotide Polymorphisms) to compare today's genetic patterns with the patterns some 100 years ago, based on the larval samples of Schmidt (1923). However all attempts to catch or even locate eel spawners have been unsuccessful.

Additionally, the successive recruitment waves of glass eel passing a single site carry a wealth of information on the fine-scale genetic composition and discreteness of spawning groups. A sharp break or clinal pattern in relatedness and genetic differentiation might point to reproductively isolated aggregations (Maes et al. 2006b). Alternatively, stochastic variance in genetic composition might infer genetic patchiness, most likely under the influence of annual and seasonal oceanic and climatic fluctuations (such as the North-Atlantic Oscillation; Knights 2003; Friedland et al. 2007). By analogy to the population dynamics of other fishes, these factors are thought to influence the reproductive success of adults and the survival rate of larvae (Dekker 2004; Pujolar et al. 2006).

Estimating *N*e represents another aim for a conservation strategy. *N*e predicts the rate of loss of neutral genetic variation, the fixation rate of deleterious and favorable genetic variants, and the rate of increase of inbreeding experienced by a population (Frankham et al. 2002). Most important, it is often several orders of magnitude smaller than the census size (*N*c) of the population, owing to unequal sex ratios, variance in reproductive success and assortative mating. In marine fishes (including eels) *Ne*/*Nc* ratios may be expected to be more extreme than in most vertebrates because of the high female fecundity. Indirect methods for estimating *N*e based on molecular markers have been developed to facilitate the inference of population size, which is a challenge in the ocean with its absence of distinct geographic boundaries. The serious decline in census size of the European eel throughout most of its range, puts the maintenance of the spawning stock(s), measured as *Ne* as well as *Nc*, above safe levels at a high priority.

European eels are long-lived animals with reproductive ages ranging from 6 to 60 years (Tesch 2003). A long-term analysis over several generations would be ideal to assess the temporal fluctuations in population size (*Ne*). It should cover a representative period to avoid the shifting-baselines trap (Pauly 2007). Realistically, a time frame over the past 200 years should suffice, because anthropogenic impact seems to have been greatest during that period (e.g. endocrine disruption of spawning, overfishing and river management). Such an analysis has become within reach thanks to genetic techniques appropriate for ancient DNA (Nielsen et al. 1997). For example, reliable estimates of population size have been calculated for several fish species in a pre- and post-industrial fishery (Nielsen et al. 1997; Turner et al. 2002; Hauser et al. 2002).

Finally, the accurate interpretation and extrapolation of genetic results requires an assessment of demographic scenarios through the development of new population dynamic models. Similar models have been the basis of fisheries research for a long time, but here we suggest bio-physical modeling through a joint assessment of demographic, hydrodynamic and genetic parameters (e.g., Galindo et al. 2006). Simulating a range of scenarios of reproductive success, migration, survival, dispersal, age structure, maturation, fisheries pressure and anthropogenic stress, preferably in an ecosystem perspective, looks promising. Subsequent validation with empirical genetic and population dynamic data may identify the key factors involved.

17.6.2 Adaptive Genetic Variation for Fisheries Management

There are more than anthropogenic influences, such as heavy fishing, pollution and migration barriers, to impact the genome and phenotype of eels (see Maes and Volckaert 2007). Over time, large declines in recruits and mature adults may trigger phenotypic and adaptive genetic changes (Law 2000). Such changes may include shifts in age- and size-at-maturity, reduced reproductive success, greater mortality, changes in growth patterns of juveniles and adults, lower fecundity and fertility, and changes in the sex ratio. If changes are heritable, this may lead to almost irreversible genetic changes in life-history traits (Law 2000; de Roos et al. 2006). A strategy to quantify fisheries and climatically induced changes in declining marine stocks involves a joint analysis of phenotypic and genetic data from contemporary populations and compared with a historical reference (preferably before the population decrease). Information on adaptive responses in exploited marine organisms is locked up in archival material (Nielsen et al. 1997; Myers and Worm 2003). Although some evidence exists for phenotypic changes in the European eel stock throughout the past 50 years (increasing adult size and decreasing glass eel size since the 1960s), the evolutionary interpretation of overfishing is complex with too few age-specific data, such as age-at-maturation and growth rate, on hand (Dekker 2004). The long-term genetic consequences of intensive fishing at the adaptive molecular level, such as a decrease or shift in genetic variability at important functional genes related to maturity and growth, have not been assessed yet.

A small level of geographical genetic differentiation at neutral molecular markers may have important consequences for the quantitative and adaptive differentiation between populations. There is growing evidence that observable phenotypic differences between stocks are measurable from the geographic and temporal scales of adaptive genetic variation. As genetic selection is a driver of evolution, it is crucial to species conservation (Conover et al. 2006). The relevance of local adaptation to fisheries management can be divided into two main issues, each differing in temporal scale (ICES 2006). First, local adaptation and population structure affect short-term demographics through effects on local recruitment patterns. Second, local adaptation and genetic heterogeneity affect long-term population dynamics, with respect to the connectivity among stocks/populations and their resilience and response to environmental change and harvesting. Local adaptation and the long-term maintenance of biodiversity have yet to be implemented into fisheries management (ICES 2006). Unfortunately, the understanding of these phenomena is particularly complex in marine organisms. The spatial and temporal scale of adaptive divergence has been assumed to be very large. However, evidence of geographically structured local adaptation in physiological, morphological and functional genetic traits has become apparent (Giger et al. 2006; Nielsen et al.

2006). The proportion of quantitative trait variation at the among-population level (Q_{ST}) has repeatedly been shown to be much higher than for neutral markers (F_{ST}) (Cousyn et al. 2001; Conover et al. 2006). As both metrics of genetic variation are poorly correlated, knowledge of neutral variation does not provide much information on adaptive variation (McKay and Latta 2002; see Conover et al. 2006 for a review). Given the important link between population genetics and dynamics, and the strong potential for selection in species with large population sizes, the application of both selected and neutral markers is obviously needed to resolve the stock structure of marine fishes effectively.

Most genotyping of eels has been performed with a limited number of neutral genetic markers. The observation of a temporal genetic component exceeding the geographic component suggests that genetic differences are possibly under selective influences. Markers under selection (such as Expressed Sequence Tags – ESTs, and Single Nucleotide Polymorphisms – SNPs – in candidate genes) may detect the genetic variation underlying environmentally dependent fitness traits in eels. Additionally, genetic markers in functional regions may affect specific life-history traits, such as mortality during recruitment, age-at-maturity (and its relationship to fishing pressure), and the relationship between heterozygosity, fitness and environment (e.g., temperature, salinity, level of pollution and parasite infection) (Maes et al. 2005; Pujolar et al. 2006). We propose a study of the adaptive potential on contemporary populations and archived otoliths of eels, by jointly screening for genetic changes at a phenotypic and a molecular level. With its high levels of gene flow and large expected *Ne*, the potential for selection in the European eel is huge (Conover et al. 2006; Maes et al. 2006a). Partitioning contemporary and historical phenotypic variance in a plasticity and genetic component can be very useful, because heritable genetic changes might be difficult to reverse and could weaken the species even further (Hutchings et al. 2007).

17.6.3 Artificial Reproduction and Aquaculture Genomics

Fishing pressure on recruits of European eel would disappear if artificial reproduction for a sustainable aquaculture were possible (see Palstra et al. 2005 and references therein). At the moment, the fattening of glass eels in large indoor tanks only merits the designation of ranching. Despite numerous attempts over the past 30 years, economically profitable quantities of eels cannot be produced in aquaculture. Methodologies developed to produce larvae of *A. japonica* have been tested in Europe on *A. anguilla*, resulting in fertilized eggs, embryonic development and occasional hatching (Palstra et al. 2005). Success, however, remains low, calling for further study of the husbandry of eels, and of reproductive and general eel biology. There is some hope as high-throughput genomic tools may shed further light on the physiology and endocrinology. For instance, Miyahara et al. (2000) produced 196 ESTs from a spleen library of Japanese eels, and Kalujnaia et al. (2007) was able to identify, through subtractive hybridization and micro-arrays, a large number of genes down-and up-regulated during osmoregulation in gill, kidney, and intestinal tissue of European eel. As new tools become available in related anguillids (e.g. Japanese eel; Nomura et al. 2006) and other fishes rich in genomic knowledge, promising insights in functional and comparative genomics are expected in the near future. EST sequencing and linkage maps may represent the first steps towards identifying important genes and Quantitative Trait Loci (QTL), the basis for marker-assisted selection. Although larvae of Japanese eel have been bred only at great effort, Nomura et al. (2006) have managed to prepare a low-density linkage map based on 43 microsatellite markers. Given the numerous genetic markers known to cross-amplify between *Anguilla* species (Maes et al. 2006b), paternity screening, gene expression and microarray analyses and a linkage map become realistic goals as soon as European eel larvae are available. Quantitative traits such as growth rate, food conversion, postponed maturity, stress tolerance, and parasite resistance strongly correlates with the possibilities of artificial rearing. One long-term issue where QTL may be of great help is in the management of feed supply. Currently, animal protein is an important ingredient of dry feeding pellets, but it is expected to shift to a diet proportionally enriched in plant material.

17.7 Conclusions

Although much progress has been made since Aristotle's theory of spontaneous generation, the challenge to understand fully the fascinating life cycle of the eel remains open. The chances to understand its biology have improved considerably as novel genetic tools such as molecular markers, genomics and increased means for data analysis have appeared. Both North-Atlantic eel species, the American eel and the European eel, are distinct species, although hybrid populations have been identified in Iceland. The high mitochondrial haplotype diversity of the European eel complicates geographical partitioning, but suggests a panmictic status. This has been confirmed with microsatellite markers. A putative Isolation-By-Distance pattern can be attributed to significant temporal genetic variance between cohorts. Population dynamics may be governed by a broad-scale pattern of Isolation-By-Time of the spawning cohorts, and a small-scale variance in reproductive success within cohorts among seasonally separated spawning groups.

There is strong evidence for Heterozygosity-Fitness Correlations. Farmed heterozygous eels have a significantly greater length and weight increase and an above average condition index in comparison with more homozygous individuals. When examining the effect of heavy metal pollutants in sub-adult eels in Belgian river basins, a reduced genetic variability was observed in strongly polluted eels as well as a negative correlation between the level of bioaccumulation and multi-locus heterozygosity at allozymes. Such evidence might be explained by an effect of either direct allozyme overdominance or associative overdominance.

As the European eel has been brought to the brink of extinction, the need for action has become of the utmost importance. Genetic data support a management plan at the scale of the Eastern-Atlantic shelf (including the Mediterranean), rejecting the currently enforced local approach. Future research should focus on (i) the detection of short term spatio-temporally discrete groups, (ii) the establishment of a biological baseline from pre-decline historical collections, (iii) the analysis of adaptive genetic polymorphism, and (iv) routine artificial reproduction to protect natural stocks from heavy exploitation.

We are convinced that more attention has to be paid to the oceanic component of the life cycle, without neglecting the fresh water habitat. Here biophysics (hydrodynamics and climate), biological traits from historical material, ecotoxicology and the contribution of neutral versus adaptive genetic variation are the much awaited research contributions for an effective global management strategy.

Acknowledgements We thank the following people for the lively and inspiring discussions on eel evolution and management throughout the years: L. Bernatchez, A. Crivelli, J. Dannewitz, W. Dekker, G. de Leo, P. Elie, M. Hansen, B. Jonsson, E. Nielsen, R. Poole, J. Van Houdt, G. van den Thillart, V. van Ginneken, T. Wirth, A. Yahyaoui, L. Zane, and the partners of the EU project EELREP (contract number Q5RS-2001-01836). G.E. Maes is a postdoctoral fellow, funded by the Fund for Scientific Research (FWO Vlaanderen).

References

Albert V, Jonsson B, Bernatchez L (2006) Natural hybrids in Atlantic eels (*Anguilla anguilla, A. rostrata*): evidence for successful reproduction and fluctuating abundance in space and time. Molecular Ecology 15:1903–1916

Anderson EC, Thompson EA (2002) A model-based method for identifying species hybrids using multilocus genetic data. Genetics 160:1217–1229

Aoyama J, Tsukamoto K (1997) Evolution of the freshwater eels. Naturwissenschaften 84:17–21

Aoyama J, Nishida M, Tsukamoto K (2001) Molecular phylogeny and evolution of the freshwater eel, genus *Anguilla*. Molecular Phylogenetics and Evolution 20:450–459

Avise JC (2004) Molecular Markers, Natural History and Evolution. Sinauer, Sunderland, MA

Avise JC, Helfman GS, Saunders NC, Hales LS (1986) Mitochondrial DNA differentiation in North-Atlantic eels: Population genetic consequences of an unusual life-history pattern. Proceedings of the National Academy of Sciences of the USA 83:4350–4354

Avise JC, Nelson WS, Arnold J, Koehn RK, Williams GC, Thorsteinsson V (1990) The evolutionary genetic status of Icelandic eels. Evolution 44:1254–1262

Bastrop R, Strehlow B, Jürss K, Sturmbauer C (2000) A new molecular phylogenetic hypothesis for the evolution of freshwater eels. Molecular Phylogenetics and Evolution 14:250–258

Beacham TD, Brattey J, Miller KM, Le KD, Withler RE (2002) Multiple stock structure of Atlantic cod (*Gadus morhua*) off Newfoundland and Labrador determined from genetic variation. ICES Journal of Marine Science 59:650–665

Boëtius J (1980) Atlantic *Anguilla*. A presentation of old and new data of total numbers of vertebrae with special reference to the occurrence of *Anguilla rostrata* in Europe. Dana 1:93–112

Bossart JL, Prowell DP (1998) Genetic estimates of population structure and gene flow: Limitations, lessons and new directions. Trends in Ecology & Evolution 13:202–206

CEC (2005) Proposal for a council regulation: establishing measures for the recovery of the stock of European eel. European Commission COM (2005) 472 final, 2005/0201 (CNS), 11 pp

Chapman RW, Ball AO, Mash LR (2002) Spatial homogeneity and temporal heterogeneity of red drum (Sciaenops ocellatus) microsatellites: Effective population sizes and management implications. Marine Biotechnology 4:589–603

Cheng PW, Tzeng WN (1996) Timing of metamorphosis and estuarine arrival across the dispersal range of the Japanese eel *Anguilla japonica*. Marine Ecology-Progress Series 131: 87–96

Chenoweth SF, Hughes JM, Keenan CP, Lavery S (1998) Concordance between dispersal and mitochondrial gene flow: Isolation by distance in a tropical teleost, *Lates calcarifer* (Australian barramundi). Heredity 80:187–197

Chistiakov DA, Hellemans B, Volckaert FAM (2006) Microsatellites and their genomic distribution, evolution, function and applications: A review with special reference to fish genetics. Aquaculture 255:1–29

Comparini A, Rizzotti M, Rodinò E (1977) Genetic control and variability of phosphoglucose isomerase (PGI) in eels from the Atlantic ocean and the Mediterranean Sea. Marine Biology 43:109–116

Comparini A, Rodinò E (1980) Electrophoretic evidence for 2 species of *Anguilla* leptocephali in the Sargasso Sea. Nature 287:435–437

Comparini A, Schoth M (1982) Comparison of electrophoretic and meristic characters of 0-group eel larvae from the Sargasso Sea. Helgoländer Meeresuntersuchungen 35:289–299

Conover DO, Clarke LM, Munch SB, Wagner GN (2006) Spatial and temporal scales of adaptive divergence in marine fishes and the implications for conservation. Journal of Fish Biology 69 (Suppl C):21–47

Cousyn C, De Meester L, Colbourne JK, Brendonck L, Verschuren D, Volckaert F (2001) Rapid, local adaptation of zooplankton behavior to changes in predation pressure in the absence of neutral genetic changes. Proceedings of the National Academy of Sciences of the USA 98:6256–6260

Cushing DH (1990) Plankton production and year-class strength in fish populations - An update of the match mismatch hypothesis. Advances in Marine Biology 26:249–293

Daemen E, Cross T, Ollevier F, Volckaert FAM (2001) Analysis of the genetic structure of European eel (*Anguilla anguilla*) using microsatellite DNA and MtDNA markers. Marine Biology 139:755–764

Dannewitz J, Maes GE, Johansson L, Wickström H, Volckaert FAM, Jarvi T (2005) Panmixia in the European eel: a matter of time. Proceedings of the Royal Society of London Series B 272:1129–1137

David P, Delay B, Berthou P, Jarne P (1995) Alternative models for allozyme-associated heterosis in the marine bivalve *Spisula ovalis*. Genetics 139:1719–1726

David P (1998) Heterozygosity-fitness correlations: New perspectives on old problems. Heredity 80:531–537

de Ligny W, Pantelouris EM (1973) Origin of European eel. Nature 246:518–519

de Roos AM, Boukal DS, Persson L (2006) Evolutionary regime shifts in age and size at maturation of exploited fish stocks. Proceedings of the Royal Society of London Series B 273:1873–1880

Dekker W (2003) Did lack of spawners cause the collapse of the European eel, *Anguilla anguilla*? Fisheries Management and Ecology 10:365–376

Dekker W (2004) Slipping through our hands: population dynamics of the European eel. Ph.D. thesis, University of Amsterdam, The Netherlands

Drilhon A, Fine JM, Boffa GA, Amouch P, Drouhet J (1966) Les groupes de transferrines chez l'anguille. Différences phénotypiques entre l'anguille de l'Atlantique et les anguilles de mediterrannée. Comptes Rendus de l'Academie de Sciences de France 262:1315–1318

Drilhon A, Fine JM, Amouch P, Boffa GA (1967) Les groupes de transferrines chez *Anguilla anguilla*. Etude de deux populations d'origine géographique différente. Comptes Rendus de l'Academie de Sciences de France 265:1096–1098

Drilhon A, Fine JM (1968) Importance de l'étude des transferines sériques dans la différenciation des espéces: resultats acquis dans genre *Anguilla*. Experientia 24:555–557

Dulvy NK, Sadovy Y, Reynolds, JD (2003) Extinction vulnerability in marine populations. Fish and Fisheries 4(1):25–64

Ege W (1939) A revision of the genus *Anguilla* Shaw, a systematic, phylogenetic and geographical study. Dana Report 16:1–256

Feder M, Mitchell-Olds T (2003) Evolutionary and ecological functional genomics. Nature Reviews Genetics 4:651–657

Fillatre EK, Etherton P, Heath DD (2003) Bimodal run distribution in a northern population of sockeye salmon (*Oncorhynchus nerka*): life history and genetic analysis on a temporal scale. Molecular Ecology 12:1793–1805

Fine JM, Drilhon A, Ridgeway GJ, Amouch PABG (1967) Les groupes de transferrines dans le genre *Anguilla*. différences dans les frequences phénotypiques de transferrines chez *Anguilla anguilla* et *Anguilla rostrata*. Comptes Rendus de l'Academie de Sciences de France 265:58–60

Flowers JM, Schroeter SC, Burton RS (2002) The recruitment sweepstakes has many winners: genetic evidence from the sea urchin *Strongylocentrotus purpuratus*. Evolution 56:1445–1453

Frankham R, Ballou JD, Briscoe DA (2002) Introduction to Conservation Genetics. Cambridge University Press, Cambridge, UK

Friedland KD, Miller MJ, Knights B (2007) Oceanic changes in the Sargasso Sea and declines in recruitment of the European eel. ICES Journal of Marine Science 64:519–530

Galindo HM, Olson DB, Palumbi SR (2006) Seascape genetics: A coupled oceanographic-genetic model predicts population structure of Caribbean corals. Current Biology 16:1622–1626

Giger T, Excoffier L, Day P, Champigneulle A, Hansen M, Powell R, Largiadèr C (2006) Life history shapes gene expression in salmonids. Current Biology 16:281–282

Hansson B, Westerberg L (2002) On the correlation between heterozygosity and fitness in natural populations. Molecular Ecology 11:2467–2474

Hauser L, Adcock GJ, Smith PJ, Ramirez JHB, Carvalho GR (2002) Loss of microsatellite diversity and low effective population size in an overexploited population of New Zealand snapper (*Pagrus auratus*). Proceedings of the National Academy of Sciences of the USA 99:11742–11747

Hedgecock D (1994) Temporal and spatial genetic structure of marine animal populations in the California Current. CalCOFI Report 35:73–81

Hendry AP, Morbey YE, Berg OK, Wenburg JK (2004) Adaptive variation in senescence: reproductive lifespan in a wild salmon population. Proceedings of the Royal Society of London Series B 271:259–266

Hendry AP, Day T (2005) Population structure attributable to reproductive time: isolation by time and adaptation by time. Molecular Ecology 14:901–916

Hoarau G, Rijnsdorp AD, Van der Veer HW, Stam WT, Olsen JL (2002) Population structure of plaice (*Pleuronectes platessa* L.) in northern Europe: microsatellites revealed large-scale spatial and temporal homogeneity. Molecular Ecology 11:1165–1176

Hutchings JA, Myers RA (1993) Effect of age on the seasonality of maturation and spawning of Atlantic cod, *Gadus morhua*, in the northwest Atlantic. Canadian Journal of Fisheries and Aquatic Sciences 50:2468–2474

Hutchings JA, Swain DP, Rowe S, Eddington JD, Puvanendran V, Brown JA (2007) Genetic variation in life-history reaction norms in a marine fish. Proceedings of the Royal Society of London Series B, 274:1693–1699

ICES (2006) Report of the Working Group on the Application of Genetics in Fisheries and Mariculture (WGAGFM), 24–27 March 2006, Newport, Ireland. ICES Document CM 2006/ MCC: 04. 59 pp

Ishikawa S, Tsukamoto K, Nishida M (2004) Genetic evidence for multiple geographic populations of the giant mottled eel *Anguilla marmorata* in the Pacific and Indian oceans. Ichthyological Research 51:343–353

Jerry DR, Baverstock PR (1998) Consequences of a catadromous life-strategy for levels of mitochondrial DNA differentiation among populations of the Australian bass, *Macquaria novemaculeata*. Molecular Ecology 7:1003–1013

Jørgensen HBH, Hansen MM, Loeschcke V (2005) Spring-spawning herring (*Clupea harengus* L.) in the southwestern Baltic Sea: do they form genetically distinct spawning waves? ICES Journal of Marine Sciences 62:1065–1075

Kalujnaia S, McWilliam I, Feilen A, Nicholson J, Hazon N, Cramb G (2007) Novel genes discovered by transcriptomic approach to the salinity study in European eel *Anguilla anguilla*. Comparative Biochemistry and Physiology A – Molecular and Integrative Physiology 146 (Suppl. S):S94–S94

Kettle AJ, Haines K (2006) How does the European eel (*Anguilla anguilla*) retain its population structure during its larval migration across the North Atlantic Ocean? Canadian Journal of Fisheries and Aquatic Sciences 63:90–106

Kleckner RC, McCleave JD (1985) Spatial and temporal distribution of American eel larvae in relation to North Atlantic current systems. Dana 4:67–92

Knights B (2003) A review of the possible impacts of long-term oceanic and climate changes and fishing mortality on recruitment of anguillid eels of the northern hemisphere. Science of the Total Environment 310:237–244

Koehn RK (1972) Genetic variation in the eel: a critique. Marine Biology 14:179–181

Koehn RK, Diehl WJ, Scott TM (1988) The differential contribution by individual enzymes of glycolysis and protein catabolism to the relationship between heterozygosity and growth-rate in the coot clam, *Mulinia lateralis*. Genetics 118:121–130

Law R (2000) Fishing, selection, and phenotypic evolution. ICES Journal of Marine Science 57:659–668

Li G, Hedgecock D (1998) Genetic heterogeneity, detected by PCR-SSCP, among samples of larval Pacific oysters (*Crassostrea gigas*) supports the hypothesis of large variance in reproductive success. Canadian Journal of Fisheries and Aquatic Sciences 55:1025–1033

Lin YS, Poh YP, Tzeng CS (2001) A phylogeny of freshwater eels inferred from mitochondrial genes. Molecular Phylogenetics and Evolution 20:252–261

Lintas C, Hirano J, Archer S (1998) Genetic variation of the European eel (*Anguilla anguilla*). Molecular Marine Biology and Biotechnology 7:263–269

Maes GE (2005) Evolutionary consequences of a catadromous life-strategy on the genetic structure of European eel (*Anguilla Anguilla* L.). Ph.D. thesis, Catholic University of Leuven, Belgium

Maes GE, Volckaert FAM (2002) Clinal genetic variation and isolation by distance in the European eel *Anguilla anguilla* (L.). Biological Journal of the Linnaean Society 77:509–521

Maes GE, Raeymaekers JAM, Pampoulie C, Seynaeve A, Goemans G, Belpaire C, Volckaert, FAM (2005) The catadromous European eel *Anguilla anguilla* (L.) as a model for freshwater evolutionary ecotoxicology: relationship between heavy metal bioaccumulation, condition and genetic variability. Aquatic Toxicology 73:99–114

Maes GE, Pujolar JM, Hellemans B, Volckaert FAM (2006a) Evidence for Isolation-by-Time in the European eel (*Anguilla anguilla* L.). Molecular Ecology 15:2095–2107

Maes GE, Pujolar JM, Raeymaekers JAM, Dannewitz K, Volckaert FAM (2006b) Microsatellite conservation and Bayesian individual assignment in four *Anguilla* species. Marine Ecology-Progress Series 319:251–261

Maes GE, Volckaert FAM (2007) Challenges for genetic research in European eel management. ICES Journal of Marine Science 64:1463–1471

Manel S, Berthier P, Luikart G (2002). Detecting wildlife poaching: identifying the origin of individuals with bayesian assignment tests and multilocus genotypes. Conservation Biology 16:650–659

Mank JE, Avise JC (2003) Microsatellite variation and differentiation in North Atlantic eels. Journal of Heredity 94:310–314

Mank JE, Avise JC (2004) Individual organisms as units of analysis: Bayesian-clustering alternatives in population genetics. *Genetical Research* 84(3):135–143

Marcogliese DJ, Cone DK (1993) What metazoan parasites tell us about the evolution of American and European eels. Evolution 47:1632–1635

McCleave JD (1993) Physical and behavioural control on the oceanic distribution and migration of leptocephali. Journal of Fish Biology 43 (Suppl. A):243–273

McCleave JD, Kleckner RC (1987) Distribution of leptocephali of the catadromous *Anguilla* species in the western Sargasso Sea in relation to water circulation and migration. Bulletin of Marine Science 41:789–806

McDowall RM (1992) Particular problems for the conservation of diadromous fish. Aquatic Conservation – Marine and Freshwater Ecosystems 2:351–355

McDowall RM (1993) Implications of diadromy for the structuring and modeling of riverine fish communities in New-Zealand. New Zealand Journal of Marine and Freshwater Research 27:453–462

McGoldrick D, Hedgecock D (1997) Fixation, segregation and linkage of allozyme loci in inbred families of the Pacific oyster *Crassostrea gigas* (Thunberg): implications for the causes of inbreeding depression. Genetics 146:321–334

McKay JK, Latta RG (2002) Adaptive population divergence: markers, QTL and traits. Trends in Ecology and Evolution 17:285–291

McPherson AA, Stephenson RL, Taggart CT (2003) Genetically different Atlantic herring *Clupea harengus* spawning waves. Marine Ecology-Progress Series 247:303–309

Minegishi Y, Aoyama J, Inoue JG, Miya M, Nishida M, Tsukamoto K (2005) Molecular phylogeny and evolution of the freshwater eels genus *Anguilla* based on the whole mitochondrial genome sequences. Molecular Phylogenetics and Evolution 34:134–146

Mitton JB (1997) Selection in Natural Populations. Oxford University Press, Oxford

Miyahara T, Hirono I, Aoki T (2000) Analysis of expressed sequence tags from a Japanese eel *Anguilla japonica* spleen cDNA library. Fisheries Science 66:257–260

Mullon C, Fréon P, Cury P (2005) The dynamics of collapse in world fisheries. Fish and Fisheries 6:111–120

Myers RA, Worm B (2003) Rapid worldwide depletion of predatory fish communities. Nature 423:280–283

Nevo E (2001) Evolution of genome-phenome diversity under environmental stress. Proceedings of the National Academy of Sciences of the USA 98:6233–6240

Nielsen EE, Hansen MM, Loeschcke V (1997) Analysis of microsatellite DNA from old scale samples of Atlantic salmon *Salmo salar*: a comparison of genetic composition over 60 years. Molecular Ecology 6:487–492

Kenchington E, Heino M, Nielsen EE (2003) Managing marine genetic diversity: time for action? ICES Journal of Marine Science 60:1172–1176

Nielsen EE, Hansen MM, Meldrup D (2006) Evidence of microsatellite hitch-hiking selection in Atlantic cod (*Gadus morhua* L.): implications for inferring population structure in nonmodel organisms. Molecular Ecology 15:3219–3229

Nomura K, Morishima K, Tanaka H, Unuma T, Okuzawa K, Ohta H, Arai K (2006) Microsatellite-centromere mapping in the Japanese eel (*Anguilla japonica*) by half-tetrad analysis using induced triploid families. Aquaculture 257:53–67

O'Connell M, Wright JO (1997) Microsatellite DNA in fishes. Reviews in Fish Biology and Fisheries 7:331–363

Ohno S, Christian L, Romero M, Dofuku R, Ivey C (1973) On the question of American eels, *Anguilla rostrata* versus European eels, *Anguilla anguilla*. Experientia 29:891

Ohta H, Kagawa H, Tanaka H, Okuzawa K, Iinuma N, Hirose K (1997) Artificial induction of maturation and fertilization in the Japanese eel, *Anguilla japonica*. Fish Physiology and Biochemistry 17:163–169

O'Reilly PT, Canino MF, Bailey KM, Bentzen P (2004) Inverse relationship between F_{ST} and microsatellite polymorphism in the marine fish, walleye pollock (*Theragra chalcogramma*): implications for resolving weak population structure. Molecular Ecology 13:1799–1814

Palstra AP, Cohen EGH, Niemantsverdriet PRW, van Ginneken VJT, van den Thillart GEEJM (2005) Artificial maturation and reproduction of European silver eel: Development of oocytes during final maturation. Aquaculture 249:533–547

Palumbi SR (1994) Genetic-divergence, reproductive isolation, and marine speciation. Annual Review of Ecology and Systematics 25:547–572

Pantelouris EM, Arnason A and Tesch FW (1970) Genetic variation in the eel. 11. Transferrins, haemoglobins, and esterases in the Eastern Atlantic. Possible interpretations of phenotypic frequency differences. Genetical Research 16:277–284

Pantelouris EM, Arnason A, Tesch FW (1971) Genetic variation in the eel. 111, Comparisons of Rhode Island and Icelandic populations. Implications for the Atlantic eel problem. Marine. Biology 9:242–249

Park LK, Moran P (1994) Developments in molecular techniques in fisheries. Reviews in Fish Biology and Fisheries 4:272–299

Patarnello T, Volckaert FAM, Castilho R (2007) Pillars of Hercules: Is the Atlantic-Mediterranean transition a phylogeographic break? Molecular Ecology 16(21):4426–4444

Pauly D (2007) Coral: a pessimist in paradise. Nature 447:33–34

Planes S, Lenfant P (2002) Temporal change in the genetic structure between and within cohorts of a marine fish, *Diplodus sargus*, induced by a large variance in individual reproductive success. Molecular Ecology 11:1515–1524

Pogson GH, Taggart CT, Mesa KA, Boutilier RG (2001) Isolation by distance in the Atlantic cod, Gadus morhua, at large and small geographic scales. Evolution 55:131–146

Pujolar JM, Maes GE, Volckaert FAM (2006) Genetic patchiness among recruits in the European eel, *Anguilla anguilla*. Marine Ecology-Progress Series 307:209–217

Pujolar JM, Maes GE, Vancoillie C, Volckaert FAM (2005) Growth rate correlates to individual heterozygosity in European eel, *Anguilla anguilla* L. Evolution 59:189–199

Richardson BJ, Baverstock PR, Adams M (1986) Allozyme Electrophoresis. A Handbook for Animal Systematics and Population Studies. Sydney: Academic Press

Rousset F (1997) Genetic differentiation and estimation of gene flow from F-statistics under isolation by distance. Genetics 145:1219–1228

Ruzzante DE, Taggart CT, Cook D (1996) Spatial and temporal variation in the genetic composition of a larval cod (*Gadus morhua*) aggregation: Cohort contribution and genetic stability. Canadian Journal of Fisheries and Aquatic Sciences 53:2695–2705

Ruzzante DE, Taggart CT, Cook D (1999) A review of the evidence for genetic structure of cod (Gadus morhua) populations in the NW Atlantic and population affinities of larval cod off Newfoundland and the Gulf of St. Lawrence. Fisheries Research 43:79–97

Schmidt J (1923) Breeding places and migration of the eel. Nature 111:51–54

Schmidt J (1935) Danish eel investigations during 25 years (1905–1930). The Carlsberg Foundation's oceanographical expedition round the world 1928–1930 and previous Danish oceanographical expeditions under the leadership of late Professor Johannes Schmidt. Copenhagen. The Carlsberg Foundation

Sinclair M (1988) Marine Populations. An Essay on Population Regulation and Speciation. University of Washington Press, Seattle

Tagliavini J, Gandolfi G, Cau A, Salvadori S, Deiana AM (1995) Mitochondrial-DNA variability in *Anguilla anguilla* and phylogenetic relationships with congeneric species. Bollettino Di Zoologia 62:147–151

Tesch FW (2003) The Eel. Blackwell Science, Oxford, UK

Tesch FW, Wegner G (1990) The distribution of small larvae of *Anguilla* sp. related to hydrographic conditions 1981 between Bermuda and Puerto Rico. International Revue der Gesamten Hydrobiologie 75:845–858

Tsukamoto K, Aoyama J, Miller MJ (2002) Migration, speciation, and the evolution of diadromy in anguillid eels. Canadian Journal of Fisheries and Aquatic Sciences 59:1989–1998

Tsukamoto, K (2006) Spawning of eels near a seamount, Nature, 439(7079):926

Tucker DW (1959) A new solution to the Atlantic eel problem. Nature 183:495–501

Turner TF, Wares JP, Gold JR (2002) Genetic effective size is three orders of magnitude smaller than adult census size in an abundant, estuarine-dependent marine fish (*Sciaenops ocellatus*). Genetics 162:1329–1339

Ryman N, Utter F (1987) Population Genetics and Fisheries Management. University of Washington Press, Seattle

Van Ginneken V, Maes GE (2005) The European eel (*Anguilla anguilla,* Linnaeus), its lifecycle, evolution and reproduction: a literature review. Reviews in Fish Biology and Fisheries 15:367–398

Waples RS (1998) Separating the wheat from the chaff: patterns of genetic differentiation in high gene flow species. Journal of Heredity 89:438–450

Ward RD (2000) Genetics in fisheries management. Hydrobiologia 420:191–201

Watanabe S, Aoyama J, Tsukamoto K (2004) Reexamination of Ege's (1939) use of taxonomic characters of the genus *Anguilla*. Bulletin of Marine Science 74:337–351

Williams GC, Koehn RK (1984) Population genetics of the North Atlantic catadromous eels (*Anguilla*). In: Turner BJ (ed) Evolutionary Genetics of Fishes. Plenum Press, New York, pp 529–560

Wirth T, Bernatchez L (2001) Genetic evidence against panmixia in the European eel. Nature, 409:1037–1040

Wirth T, Bernatchez L (2003) Decline of North Atlantic eels: a fatal synergy? Proceedings of the Royal Society of London Series B – Biological Sciences 270:681–688

Yahyaoui A, Bruslé J, Pasteur N (1983) Etude du polymorphisme biochimique de deux populations naturelles (Maroc Atlantique et Rousillon) de civelles et anguillettes d'*Anguilla anguilla* L. et de deux échantillons d'élevages. IFREMER Actes de Colloques 1:373–390

Chapter 18
Proximate and Ultimate Control of Eel Continental Dispersal

Eric Edeline, Sylvie Dufour, and Pierre Elie

18.1 Introduction

Eels *Anguilla* spp. are fishes belonging to the Superorder Elopomorph, a group of phylogenetically ancient teleosts (Nelson 1994). Eels have ancestrally evolved a continental growth phase and thus migrate between marine breeding and continental feeding areas. In continental waters, eels colonize an extremely wide variety of salt (SW), brackish and freshwater (FW) habitats. Such ubiquity is almost unique among teleost fishes. However, the mechanisms controlling eel continental dispersal, i.e., distribution in different growth habitats, remain largely unknown. Dispersal is here understood in ecological terms and is thus referred to as movements leading to habitat colonization in general. Dispersal is a pivotal process for both species persistence and evolution, and involves a great diversity of ecological phenomena (Clobert et al. 2001). In the eel, dispersal in different habitats may influence survival, growth, sex differentiation, age and size at silvering [female size affects fecundity, egg-size and larval viability (Einum et al. 2004)], swimming ability during the spawning migration, and finally capacity to reproduce. In turn, decision-makings by individual dispersers for migration to, immigration in and emigration from different habitats depend on genetic, physiological, morphological and social attributes and are affected by a number of environmental parameters (Clobert et al. 2001).

In this chapter, we provide material to understand how (proximate control) and why (ultimate control) dispersal patterns by eels in continental habitats are what they

---

E. Edeline(✉)
Centre for Ecological and Evolutionary Synthesis (CEES), University of Oslo, Dept. of Biology, P.O. Box 1066 Blindern, N-0316 Oslo, Norway

S. Dufour
Muséum National d'Histoire Naturelle, USM 0401, UMR CNRS/MNHN/UPMC 5178 "Biologie des Organismes Marins et Ecosystèmes", 7 rue Cuvier, CP 32, 75231 Paris Cedex 05, France

P. Elie
Cemagref, Unité "Ecosystèmes Estuariens et Poissons Migrateurs Amphihalins" (EPBX), 50 Avenue de Verdun, 33612 Cestas Cedex, France

G. van den Thillart et al. (eds.), *Spawning Migration of the European Eel*,

are. In the first section, we describe the process of continental dispersal, focusing on the behavioural changes that occur during ontogeny. We emphasize the importance of considering separately *migration*, which is mainly an endogenously-controlled behaviour expressed by glass eels and elvers, from *ranging*, which is mainly an environmentally-controlled behaviour expressed by yellow eels. Based on this dichotomy, we review in the second and third sections the internal and external (environmental) drivers of movements. In the fourth section, we investigate the evolutionary forces acting on eel movements and we propose an evolutionarily stable strategy (ESS) model explaining how decision-making for movement by individual dispersers may be ultimately controlled. We further use this ESS model for qualitative predictions on the evolution of diadromy (i.e., colonisation of freshwater habitats) in response to anthropogenic changes in selective pressures. Finally, in the conclusion, we point out lack of knowledge and suggest future research directions.

18.2 Ontogeny of Eel Continental Dispersal

18.2.1 Metamorphosis

Metamorphosis is a drastic developmental strategy consisting of a suite of morphological, physiological and behavioural adaptive changes. In the eel, metamorphosis of translucent leptocephalus larvae, drifting from the oceanic spawning grounds, into late-metamorphic, unpigmented glass eels occurs on the slope of the continental shelf (Schmidt 1909; Tesch 1977). This transformation is an adaptive shift from oceanic drift to river colonization, and marks the start of the continental dispersal phase. The larva's body changes from willow-leaf to eel-shape, undergoes a reduction in length and weight, and an 80% drop in water content (Bertin 1951; Otake 2003). In addition, the brain structure is profoundly remodelled, with the external brain shape of the leptocephali gradually changing from laterally compressed to depressed elongated due to biased growth of the telencephalon and optic tectum (Tomoda and Uematsu 1996). Also, feeding activity stops due to the shedding of larval teeth and reorganization of the digestive system. Then, as shown in *A. anguilla* glass eels, the gut specializes for osmoregulation in FW, independent of the environmental salinity (Ciccoti et al. 1993). This indicates that, although they keep the ability to live and grow in SW, eels are ontogenetically programmed for FW residency. Body pigmentation progressively develops throughout the glass eel phase over different pigment stages VA, VB, VIA_0, VIA_1, VIA_2, VIA_3, VIA_4 and VIB, as defined by Elie et al. (1982). In this chapter, we will not use the term "elver" but instead term "glass eel" for all the development stages between the leptocephalus larvae and the newly transformed yellow eel (stage VII of Elie et al. 1982). Feeding resumption occurs around the pigment stage VIA_3 for most of individuals, after sufficient rearrangements of digestive organs and acquisition of a new set of teeth (Elie 1979). At feeding resumption, stomachs contain mainly a mixture of plant and algal detritus that may be assimilated by the developing gut (Bardonnet and Riera 2005). The true end of metamorphosis, marking the start of the juvenile growth phase (yellow phase), occurs at completion of body pigmentation (stage VII) and matches the full development of gut and teeth (Vilter 1945; Elie 1979; Elie et al. 1982; Jegstrup and Rosenkilde 2003). At that stage, eels start feeding on macro-invertebrates and will progressively shift towards piscivory (Tesch 1977). The colonization of continental habitats is carried out by both the glass and yellow eels, but with quite different modalities.

18.2.2 Glass Eel Dispersal

Glass eels invade coastal and estuarine waters using selective tidal stream transport (STST), a saltatory transport mechanism with alternations of flow-carried swimming during flood tide and benthic sheltering behaviour during ebb tides (Creutzberg 1958; McCleave and Kleckner 1982; Elie and Rochard 1994). STST allows important energy saving compared to constant counter-current swimming (Weihs 1978). Glass eels orientate towards the river mouth and up-estuary following decreasing salinity gradients (Tosi et al. 1990) and river water odours (Creutzberg 1961). Thermal gradients could also play a role in glass eel orientation (Tosi et al. 1990). As they reach the tidal limit (limit of flow reversals), migrating glass eels lose tidal advection for transport and have to switch from STST to constant counter-current swimming (McCleave and Wippelhauser 1987). We have investigated glass eel migration at the obstacle-free tidal limit area of the Dordogne River (France) during a 2-year field study (Edeline et al. 2007). This tidal limit is located about 50 km upstream from the salinity front, and thus provides a rare opportunity to separate the effects of salinity and hydrodynamics on the migration process. We have monitored the distributions of glass eels (pigment stages VB to VIB) vs. newly transformed yellow eels (stage VII, body length <15 cm) at five sites (coded A, B, C, D and E). Site C was located at the limit of flow reversals, i.e., at the point of the watershed where glass eels lose tidal advection for STST. The results indicate that glass eels arriving from the sea rapidly migrated up-estuary. Then, despite the absence of any osmotic barrier, glass eels accumulated at the limit of flow reversals (site C, Fig. 18.1). Upstream of this accumulation point, migration speed dropped and glass eels transformed into small yellow eels before reaching the non-tidal river area, where only small yellow eels and no glass eels were found (site E, Fig. 18.1). This finding is in accordance with other data indicating almost complete absence of glass eels among inland river colonizers (Haro and Krueger 1988; Michaud et al. 1988; Naismith and Knights 1988; Dutil et al. 1989; McGovern and McCarthy 1992; White and Knights 1997b). Together, these results indicate that glass eels are strongly dependent on tidal streams for upstream movements and poorly adapted to colonization of non-tidal rivers. Therefore, the late-metamorphic glass eel should be considered as a transitory developmental stage between the oceanic larva and the continental juvenile, adapted to the colonization of tidal interface habitats by means of STST.

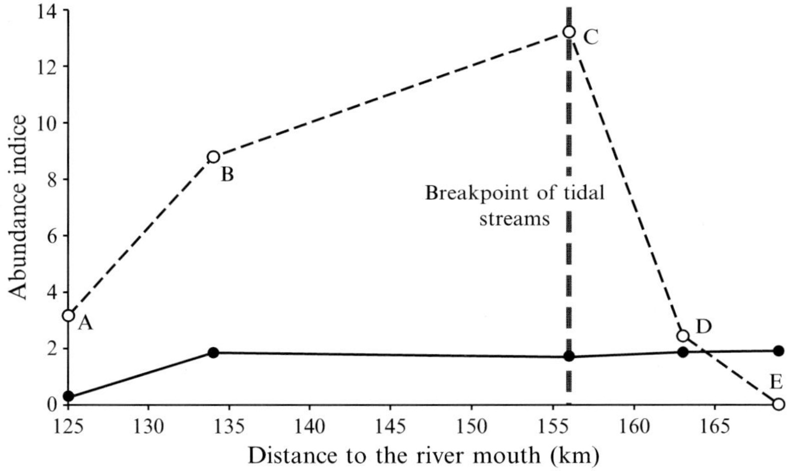

Fig. 18.1 Ontogeny of eel dispersal behaviour, adapted from Edeline et al. (2007). Distributions of glass eels (dotted line) and newly transformed yellow eels (body length <15 cm, solid line) around the tidal limit of the Dordogne River (France). Eels were sampled at five sites (A, B, C, D and E) located at increasing distance from the river mouth. Site C was located at the breakpoint of tidal streams. Glass eels accumulated at site C, most likely due to the loss of tidal streams available for selective tidal stream transport (STST). In contrast, newly transformed yellow eels were homogeneously distributed around site C. This result is most parsimoniously interpreted as a behavioural shift from upstream migration to settlement and food search during the completion of larval metamorphosis (see Section 18.2.3). Abundance indices (y axis) were obtained by multiplying the estimated probability of presence of a fish at each sampling site, by the estimated abundance of fish when fish was present, taking into account the effects of several important environmental variables in generalized linear models (For further details, see Edeline et al. 2007)

18.2.3 Yellow Eel Dispersal

The distribution pattern of small yellow eels in the tidal limit area of the Dordogne River was quite different from that of glass eels (Fig. 18.1), suggesting that completion of larval metamorphosis was related to a pronounced change in dispersal behaviour (Edeline et al. 2007). Upstream migratory movements by newly transformed yellow eels should have induced increased abundance upstream of the site where they accumulated as glass eels (site C, Fig. 18.1). Instead, abundance remained almost constant between sites B and E, suggesting a loss of the upstream-oriented migratory behaviour at the onset of the growth phase. This result matches with experimental aquarium data indicating a shift from pelagic towards benthic behaviour during the transformation of glass eels into small yellow eels (Jegstrup and Rosenkilde 2003). Therefore, termination of metamorphosis and full development of feeding capacity seem to mark the end of the larval migratory period in the European eel, as commonly observed in benthic marine fishes (Moran 1994). Homogenous distribution of newly transformed small yellow eels after initial accumulation at the glass eel stage further suggests density dependent dispersal (Edeline et al. 2007). This hypothesis is supported by recent studies suggesting that upriver movements by yellow eel are largely density-dependent (Smogor et al. 1995; Ibbotson et al. 2002; Feunteun et al. 2003; Briand et al. 2005a). Based on the aforementioned results, we may propose a typical ontogenetic pattern for eel dispersal behaviour. According to Dingle (2006), during migration the individual is not distracted from movement by resource-based stimuli like food or living space. This definition clearly fits glass eel movements because glass eels move upstream even if resources are more abundant in the estuary or on the coastline. In contrast, at the yellow stage, movements are driven by the search for food or other resources, and also aim at avoiding competitors and predators (see Section 18.4.7). These movements are clearly not migratory because they cease when a resource is encountered (Dingle 2006). We here choose to refer to these non-migratory yellow eel movements as "ranging", because they occur at the home range scale-either within or between home ranges [but see Dingle (2006) for further development]. Finally, we define settlement as the behavioural shift from migration to ranging.

18.2.4 Variability of the Eel Dispersal Phenotype

We have now defined the typical ontogenetic pattern for European eel continental dispersal: eels migrate upstream at the glass eel stage, settle at the onset of the yellow stage and then shift to ranging behaviour. However, there is probably a large variability around this typical (or average) ontogenetic pattern. Indeed, settlement is a complex trait involving a number of morphological, physiological and behavioural changes, and complex traits generally show a large variability in populations. In accordance with the view that settlement timing is highly variable, otolith microchemistry shows that some eels never enter freshwater but settle at the glass eel stage at sea or in the estuary (Tsukamoto and Arai 2001; Daverat et al. 2006). Settlement may also occur after the onset of the yellow eel stage. Indeed, small yellow eels may show intense upriver migratory behaviour during spring "runs", and some larger ("pioneer") yellow eels may maintain density-independent, upstream migration (Feunteun et al. 2003).

After settlement, ranging may be aperiodic or instead occur on a periodic basis, and include movements within and between home ranges that may cross salinity boundaries. For instance, seasonal ranging movements by yellow eels may occur between the river and the estuary (Tesch 1977; Daverat et al. 2006). The combination of variability in settlement timing and ranging movements leads to a very large diversity of continental dispersal patterns - this diversity is stressed by otolith microchemistry studies. How the variability of pre-and post-settlement movements is controlled is discussed in the three following sections. The definition of migration given above - that a migrant is not responsive to resource - based stimuli - implies that migration is primarily driven endogenously. However, it should be kept in mind that environmental factors may still affect many internal drivers. Reciprocally, ranging behaviour is mainly controlled by the environment, but final decisions for moving or staying remain under endogenous control.

18.3 Internal Drivers of Eel Continental Dispersal

18.3.1 Genetic Factors

Despite its central importance for both ecological and conservation purposes, the genetic control of eel life history has been historically neglected. Indeed, many authors implicitly consider that phenotypic diversity results from a single genotype showing different norms of reaction in response to various environmental effects. However, assuming no genetic variation (i.e., one genotype) for migratory behaviour simply implies that migration cannot evolve, because evolution results from sorting of genotypes by selection (Futuyma 1998). Instead, in both vertebrates and invertebrates, migration is an heritable syndrome (i.e., a complex trait) that requires genetic programming for morphological, physiological and behavioural adaptations (Berthold 1991; Pulido et al. 1996; Futuyma 1998; Roff and Fairbairn 2001; Alerstam et al. 2003; Dingle 2006). Typically, these adaptations are controlled by many loci each with a small additive effect (polygenic model) together bearing considerable genetic variation (Berthold and Pulido 1994; Futuyma 1998; Roff and Fairbairn 2001). Accordingly, recent studies have shown significant genetic variation in *A. anguilla* (Daemen et al. 2001; Wirth and Bernatchez 2001; Maes and Volckaert 2002; Dannewitz et al. 2005), that may be correlated to fitness-related traits (Pujolar et al. 2005). In the genus *Anguilla*, genetic diversity for migratory traits is necessary to explain evolution of diadromy from ancestral marine residency (Gross et al. 1988), as well as spatiotemporal shifts of migration loops (migration route and life cycle) that caused separation into subpopulations and speciation (Tsukamoto et al. 2002).

18.3.2 Body Condition

Body condition, i.e., energetic status, is a parameter of major importance that influences dispersal in birds and mammals (Dufty and Belthoff 2001). In anadromous fishes, limitations in energy reserves constitute a major constraint for migration (Bernatchez and Dodson 1987; Jonsson et al. 1997; Forseth et al. 1999). Migrating glass eels, due to both their small size and non-trophic state, are likely prone to energetic constraints. Indeed, decreased body condition triggers a swap from FW- to SW-preference and a shift from migration to settlement in glass eels (Bardonnet et al. 2003; Edeline et al. 2004, 2006; Bureau du Colombier et al. 2007). This energetically-controlled migratory plasticity likely reflects an adaptive threshold (see Section 18.5). Indeed, high body condition fishes will gain the highest fitness return by migrating to low-density river habitats (see Sections 18.4.7 and 18.5), while low body condition individuals increase their fitness by stopping migration-related energy expenditure and settling precociously in saline habitats. Glass eels are recruited into estuaries during migration peaks lasting for several months (Tesch 1977). The body condition of estuarine recruits decreases over this migration period in *A. anguilla* (Elie 1979; Charlon and Blanc 1982), *A. rostrata* (Jessop 1998), *A. japonica* (Kawakami et al. 1999), *A. reinhardtii*, *A.*

australis (Sloane 1984) and *A. dieffenbachii* (Jellyman and Lambert 2003), possibly due to a decreased productivity of oceanic ecosystems during the larval migration period (Désaunay and Guerault 1997). Hence, the propensity to colonize river rather than marine and estuarine habitats is probably higher in early than in late recruits. Also, longer term oscillations in oceanic productivity (Désaunay and Guerault 1997) might plastically affect the proportion of diadromous individuals among recruits.

18.3.3 Endocrine Factors

Because they integrate both the genetic and environmental influences and, in turn, regulate gene transcription, neuronal and metabolic activities, hormones are key behavioural mediators. Diadromous fish migrations are under endocrine control (Fontaine 1975). During salmonid smoltification, pre-adaptation to SW residency is controlled in synergy by growth hormone (GH) and cortisol (Boeuf 1993; McCormick 2001), while thyroid hormones (THs, thyroxine T_4 and triiodothyronine T_3) are involved in many adaptive processes including olfactory imprinting (Lema and Nevitt 2004), changes in muscle physiology (Katzman and Cech 2001), and rheotaxis (Specker et al. 2000). In the eel also, THs have a crucial role in controlling leptocephalus metamorphosis (Yamano et al. 1991; Ozaki et al. 2000) body pigmentation and gut transformations (Vilter 1946; Jegstrup and Rosenkilde 2003). In glass eels, THs further stimulate migratory behaviour (Edeline et al. 2004, 2005b). Therefore, THs play a key role in the regulation of morphological, physiological and behavioural adaptations leading to the colonization of FW habitats by the eel. Thyroid gland activity decreases at the yellow stage (Callamand and Fontaine 1942), likely promoting the observed switch from migration to settlement during transformation of glass eels into small yellow eels (see Section 18.2.3). In yellow eels, individuals caught climbing water falls also have higher plasma T_4 levels and locomotor activity compared to sedentary individuals (Castonguay et al. 1990), suggesting that maintenance of a high thyroid gland activity at the yellow stage delays settlement.

It seems likely that cortisol and GH are also involved in the regulation of glass eel dispersal (Fig. 18.2). Indeed, in both glass and yellow eels, cortisol promotes gill and intestine Na^+/K^+-ATPase activity, i.e., adaptation to SW residency, probably in synergy with GH (Epstein et al. 1971; Butler and Carmichael 1972; Wilson et al. 2004). In addition, cortisol injections decrease plasma THs levels (Redding et al. 1986), while THs exert a negative feedback on GH production by the pituitary, an eel-specific regulation (Rousseau et al. 2002). These crossed negative feedbacks between THs vs. cortisol-GH could mediate the swap of glass eels from FW-oriented migration to settlement induced by a lowered body condition (Fig. 18.2). Indeed, in teleosts, a lowered caloric status induces decreased thyroid gland activity (Eales 1988). In glass eels, a low body condition correlates to low thyroid gland activity (Edeline et al. 2004) and high GH levels (Lambert et al. 2003), while physiological stress and fasting in yellow eels stimulate GH and cortisol secretion through brain production of corticotropin-releasing hormone (Marchelidon et al. 1996; Rousseau et al. 1999; Dufour et al. 2001).

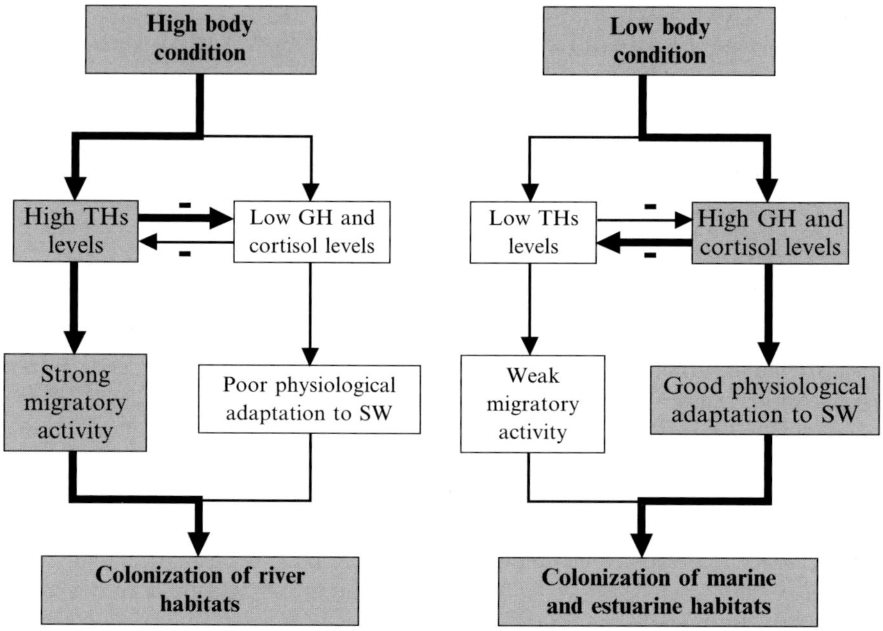

Fig. 18.2 Hypothetical endocrine mechanism for the control of glass eel migratory plasticity based on the results provided in Section 18.3.3. Thyroid hormones (THs) induce a river-oriented migratory behaviour, but also exert a reciprocal negative feedback on cortisol and growth hormone (GH) production, which are both involved in saltwater (SW) acclimation. Left panel: individual's physiology is "THs-dominated". We suggest that a satisfying energetic status (high body condition, see Section 18.3.2) may stimulate thyroid gland activity. Then, high THs levels maintain low GH and cortisol levels and promote colonization of river habitats. Right panel: we depict a possible endocrine switch to a "GH-cortisol-dominated" physiology (bold vs. thin arrows). Decreasing energetic status lowers THs production and increases GH and cortisol secretion, inducing low migratory activity and good hypoosmoregulatory ability. We suggest that shift from a "THs-dominated" to a "GH-cortisol-dominated" physiology occur when individual's energetic status reaches a switch point described in Section 18.5

18.4 External Drivers of Eel Continental Dispersal

18.4.1 Tidal Streams: The Importance of STST

As indicated by their behavioural adaptation to tidal streams (the STST), glass eels are strongly constrained in their movements by the water current, that may either represent a transport vector or a migration barrier depending on its direction. At the tidal limit, migrating glass eels accumulate because tidal advection is no longer available to support upstream movements (see Section 18.2.2). In estuaries, glass eels may be found swimming in the water column during ebb-tide, while the use of STST for migration implies sheltering close to the bottom during ebb-tide. These ebb-tide glass eels present signs of physiological stress, similar to settling glass eels that remain on the bottom of the estuary during flood tide (Edeline et al. 2004). These results suggest that an inefficient use of tidal streams (by ebb-tide glass eels) is one of the mechanisms that could lead precocious settlement in saline habitats at the glass eel stage through a sequence presented in Fig. 18.3.

The rhythmic activity during STST is triggered by both exogenous olfactory cues (Creutzberg 1959; Barbin et al. 1998) and endogenous cues produced by an internal circatidal clock, which rhythm is synchronized by flow reversals (Wippelhauser and McCleave 1988). Disruption of these phasing mechanisms could hinder effective utilization of tidal streams (Fig. 18.3). For instance, pollutants could affect fine sensory processing of water currents and odours necessary for synchronization of the internal clock. Indeed, pollutants may damage olfactory neurons (Halpern 1982) and may also affect a variety of behaviours through the upsetting of sensory, hormonal, neurological, and metabolic systems (Scott and Sloman 2004). Also, in poikilotherms, temperature is a very strong synchronizer of internal oscillators. Slight temperature changes may induce phase shifting of endogenous rhythms

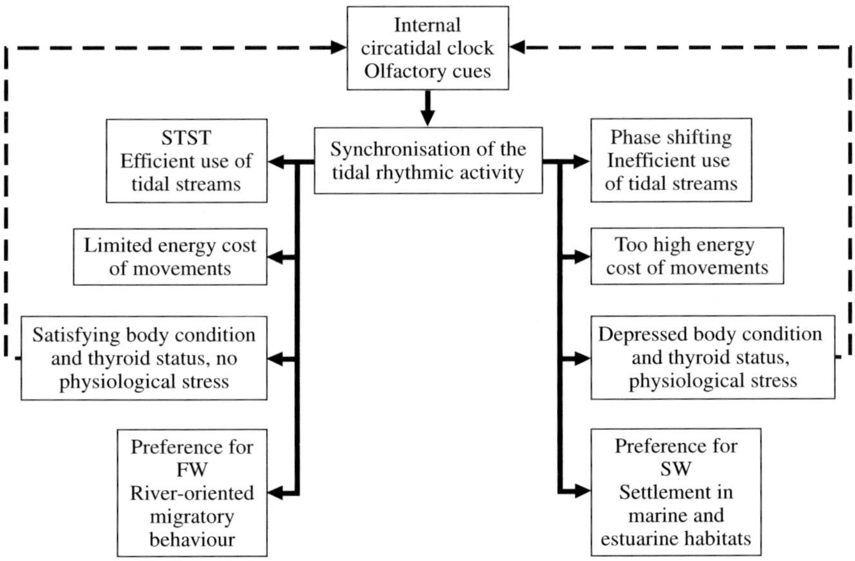

Fig. 18.3 Proximate mechanism for the control of glass eel dispersal by tidal streams based on the results provided in Section 18.4.1. The rhythmic activity during selective tidal stream transport (STST) is triggered by both endogenous (internal clock) and exogenous (olfactory) cues. A rhythmic swimming activity in phase with flood tides allows energy saving and maintenance of satisfying energetic status and high thyroid gland activity, which promote preference for freshwater (FW) and a river-oriented migratory behaviour. In contrast, a phase shifting in activity rhythm, preventing glass eels from efficient using of tidal streams as transport means, leads to physiological stress through the resultant high energy cost of movements. This decreases thyroid gland activity and promotes preference for saltwater (SW), loss of migratory behaviour and finally settlement. We suggest that the physiological condition could possibly exert a feedback on the synchronization of rhythmic activity (dotted arrows) because THs could be involved in olfactory sensitivity (see Section 18.3.3)

through alterations of the clock molecular machinery (Rensing and Ruoff 2002). Accordingly, in *A. japonica* glass eels kept under free-running conditions, a slight temperature change may shift the rhythm of the internal circatidal clock (Kim et al. 2002). Such temperature-mediated phase shifting in the internal tidal clock may be one of the processes leading to an inefficient use of tidal streams by glass eels (see Fig. 18.3).

18.4.2 Salinity

Salinity is a major environmental factor that affects fish distribution (Jung and Houde 2003) and growth (Boeuf and Payan 2001). In the eel, salinity tolerance varies during development. Both glass eels and silver eels show an extraordinary capacity to cope with abrupt salinity transfers (Fontaine and Raffy 1932; Wilson et al. 2004) but osmotic tolerance decreases in small yellow eels (Boucher-Firly 1935) which may suffer high mortalities during acute transfer from FW (0 ppm) to SW (35 ppm) (E. Edeline, unpublished data (2002): 100% mortality among 197 newly transformed small yellow eels previously held and fed in FW for 2 months). Salinity is likely to influence eel movements during both migration and ranging. Indeed, despite exceptional osmotic tolerance, glass eels arriving from the sea need a delay period before voluntary entry into FW (Petit and Vilter 1944; Deelder 1958), indicating that salinity preference may reflect subtle processes that are not revealed by osmoregulation studies. Salinity preference tests show that, at arrival from the sea, the proportion of FW-seeking glass eels varies between 50% and 70% (Tosi et al. 1988; Tosi et al. 1989, 1990), suggesting inter-individual variation in salinity preference that could possibly influence migration and/or habitat selection.

In order to investigate the role of glass eel salinity preference in the control of eel migration, Edeline et al. (2005a) sorted groups of *A. anguilla* glass eels (hereafter termed "contingents") over two consecutive salinity preference tests. This allowed us to study the link between salinity preference and locomotor activity (i.e., positive rheotaxis), and to separate contingents of glass eels that were either plastic or fixed in their preference for FW or SW. During the first trial series, 3,193 glass eels were tested for salinity preference 1, 2 and 3 days after capture; 864 were active (i.e., actively swam towards either a FW or a SW flow). All the active glass eels, as well as a batch (n = 543) of inactive fish, were kept to perform the second behavioural test. The other fish (the rest of the inactive glass eels) were released. During the second trial series, we tested these two batches (543 inactive and 864 active glass eels) for their salinity preference, 9 and 10 days after capture. After the second trial, the 543 glass eels that had been inactive during the first trial were released. Among the 864 glass eels that had been active during the first trial series, 526 were again active during the second trial series. These 526 fish were used to carry out a growth experiment (see below), while the others were released. We classified the 526 glass eels that were active during the two trials into contingents according to their salinity preference: FWC (n = 240) for FW contingent (double preference for FW), SWC (n = 47) for SW contingent (double preference for SW), and PCC (n = 239) for plastic contingent (shift in salinity preference over the acclimatization period: from SW to FW, or from FW to SW). Somatic growth of these three contingents was then monitored during 2 months under excess feeding in controlled FW and SW conditions.

During the two behavioural tests, locomotor activity and salinity preference were significantly linked. Indeed, the glass eels that were active during the first trial (preferring either FW or SW) showed a high locomotor activity during the second trial but also a sharp preference for FW, even if SW had been preferred first. This result indicated that a high locomotor activity was associated with FW preference in glass eels, a behavioural syndrome likely promoting migration to the river (Edeline et al. 2005a). In contrast, inactivity during the first trial was associated with a low locomotor activity during the second trial, but also to a preference for SW if the fish were active. This result indicated that low locomotor activity was associated with SW preference in glass eels, a behavioural syndrome probably promoting an early settlement in marine and estuarine habitats (Edeline et al. 2005a).

Additionally, the behavioural syndromes observed during the behavioural tests were associated with growth patterns that fitted with the observed growth patterns of yellow eels in the wild. Indeed, the FWC (i.e. glass eels that preferred FW twice) had the lowest growth rates in both FW and SW rearing conditions. In contrast, the SWC (i.e. glass eels that preferred SW twice) had the highest growth rates in SW rearing conditions. These results suggest a trade-off (negative correlation) between migration and growth, in accordance with data from the wild showing that freshwater eels grow more slowly than estuarine eels (Tzeng et al. 2003; Jessop et al. 2004; Daverat and Tomás 2006). Additionally, growth was higher in SW than in FW in all contingents, indicating that, in addition to the migration/growth trade-off, habitat salinity may directly affect growth (Edeline et al. 2005a). The PCC (i.e. glass eels that swapped their salinity preference) had an intermediary growth status that could be related to the nomad life style of eels moving between different habitat types during their life (Feunteun et al. 2003; Daverat and Tomás 2006). The results from this behavioural – growth experiment provide a comprehensive ecological mechanism for the control of habitat distribution and growth patterns in glass and yellow eels, as presented in Fig. 18.4.

Trade-offs between migration and other fitness-related traits are frequent because migration incurs heavy fitness costs (Roff and Fairbairn 2001), and may involve various ecological, physiological, and genetic pathways (Zera and Harshman 2001). In juvenile eels, the links between locomotor activity, salinity preference and growth probably involve complex interactions of several mediators, including metabolic hormones and genes (Fig. 18.4). The speculations presented in Fig. 18.4 are based on the facts that: (1) THs promote locomotor activity in glass eels (Edeline et al. 2005b) and have also been shown to affect salinity preference (Baggerman 1960, 1962; Iwata 1995) and growth (Higgs et al. 1982) in other teleosts, (2) GH affects growth, acclimation to SW (see Section 18.3.3) and locomotor activity in salmonids (Øverli et al. 2002; Johansson et al. 2004), and (3) genes

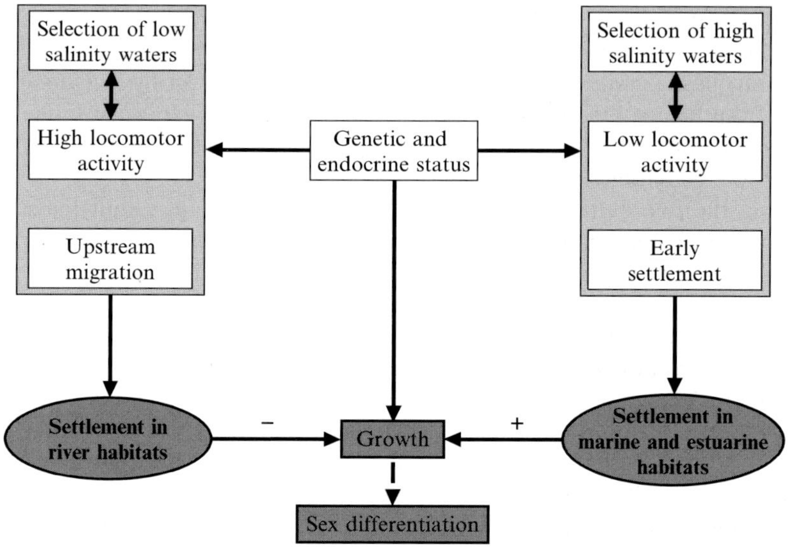

Fig. 18.4 A simple mechanism for the proximate control of eel life history in continental habitats based on the results provided in Section 18.4.2. Non-filled boxes refer to processes occurring in glass eels and elvers (hereafter simply termed glass eels), while grey boxes refer to processes occurring in yellow eels. In glass eels, locomotor activity and salinity preference are correlated (double arrows) and participate in driving upstream migratory behaviour. High locomotor activity and preference for FW lead to upstream migration and diadromy, while a low locomotor activity and preference for SW favour an early settlement and non-diadromy. At the yellow stage, habitat salinity and primary productivity directly influence somatic growth (+ and –). However, we further suggest occurrence of a trade-off between migration and somatic growth, presumably through endocrine and genetic pathways. High migratory propensity at the glass eel stage seems to be correlated with low juvenile growth rates, while a low migratory propensity by glass eels seems to be related to high juvenile growth rates. Following other authors, it is suggested that growth rate during the first year of continental life could affect sex differentiation (dotted line)

(allozyme heterozygosity) affects both salinity tolerance in teleosts (Shikano et al. 2000) and growth performance in *A. anguilla* (Pujolar et al. 2005). Finally, migratory behaviour could also be correlated to sex differentiation through growth (Fig. 18.4). Indeed, recent data suggest that slow growth rates during the first year of continental life may favour female sex differentiation (Holmgren and Mosegaard 1996; Holmgren et al. 1997; Davey and Jellyman 2005)

18.4.3 Water Temperature

Water temperature profoundly affects the whole metabolic machinery and is thus a crucial driver for eel dispersal plasticity. In glass eels, both the estuarine migration and river recruitment are conditioned by temperature thresholds

(Vøllestad and Jonsson 1988; McGovern and McCarthy 1992; Elie and Rochard 1994; Jessop 2003). Accordingly, under experimental conditions, a temperature increase from 10°C to 18°C enhances both locomotor activity (i.e. positive rheotaxis) and FW-preference (Edeline et al. 2006). Runs of small yellow eels in rivers are also triggered by a rise in water temperature; the temperature-dependency of movements decreases with increasing body size (Moriarty 1986; Naismith and Knights 1988; White and Knights 1997b). There are several physiological mechanisms by which temperature may affect locomotion. Decreased environmental temperatures primarily act in reducing enzymatic activities and fluidity of both membranes and internal liquids, causing a drop in the power output from muscles (Johnston and Temple 2002). Water temperature also affects endocrine secretions. In *A. anguilla* yellow eels, thyroid gland activity decreases with water temperature (Leloup 1958; Leloup and De Luze 1985) and, in glass eels, a fall in water temperature slows down the process of metamorphosis (Briand et al. 2005b). Finally, habitat selection at the patch scale may be affected by water temperature (Chen and Chen 1991; Richardson et al. 1994). Thermal preferences match with optimal temperatures for growth (Elie and Daguzan 1976), suggesting that preference is related to optimal functioning of the enzymatic machinery.

18.4.4 Water Odour, Migration and Habitat Selection

Odours are widely known to guide salmon to their native stream during spawning migration (Døving and Stabell 2002). According to the definition of Harden-Jones (1984), migratory fishes may use water odour as a "cue" triggering movements (i.e., shift from inactivity to activity) above sensitivity thresholds, and/or as a "clue"orienting movements through odorous gradients once the fish is in motion (i.e., after the cue triggered onset of movement). In glass eels, the smell of inland waters is thought to provide a cue triggering the onset of swimming during STST (see Section 18.4.1), but also clues orienting movements towards rivers (Creutzberg 1961). Attraction is due to several earthy compounds (Sorensen 1986; Tosi and Sola 1993; Sola 1995), that may either attract or repel glass eels depending on environmental salinity (Sola and Tongiorgi 1996). It has been hypothesized that orientation towards earthy odorants during migration evolved because these compounds are indicative of habitat primary productivity (Sorensen 1986). Odours also transmit social information. Both glass eels (Briand et al. 2002) and yellow eels (Saglio 1982) are attracted by the odour of conspecifics, possibly through epidermal mucous compounds (Saglio 1982), bile salts (Sola and Tosi 1993) and amino acids (Sola et al. 1993; Sola and Tongiorgi 1998). Chemoattraction could promote glass eel's grouping behaviour during migration, mediate conspecific recognition and allow food location during ranging (Sola and Tongiorgi 1998). Therefore, odorous signals may be seen as mediators of optimal foraging and optimal habitat selection by dispersers during ranging.

18.4.5 Light and Photoperiod

In both plants and animals, light is one of the most important synchronizers of internal rhythms (Hastings et al. 1991). Photoperiodic (seasonal) information is transformed into a melatonin secretory rhythm that, through hypothalamic stimulation, regulates many biological functions such as growth and reproduction (Bolliet et al. 1996; Boeuf and Le Bail 1999; Falcon et al. 2003). In teleosts, photoperiod affects circulating levels of numerous hormones, including GH (Björnsson et al. 1994), THs (Leiner and McKenzie 2001) and reproductive hormones (Blázquez et al. 1998). In salmonids, the spring increase in daylength is one of the factors triggering smoltification (Hoar 1988). In the eel, the effects of photoperiod variations on physiology and behaviour remain unexplored, but it is well known that light inhibits eel movements. Glass eels are strongly photonegative (Bardonnet et al. 2003; Dou and Tsukamoto 2003), and light avoidance increases with body pigmentation (Bardonnet et al. 2005). In estuaries, glass eels avoid both the daylight and moonlight by remaining on the bottom or swimming deeper in the water column (Creutzberg 1961; Elie and Rochard 1994; De Casamajor et al. 2000). Yellow eels also show nocturnal foraging activity (Bertin 1951; Tesch 1977; Baisez 2001), reduced during full moon periods (Adam and Elie 1994). This light inhibition of eel movements is mediated by increased secretion of brain catecholamines (Le Bras 1978, 1984). Finally, both glass eels (Wippelhauser and McCleave 1988) and silver eels (Edel 1976) have circadian activity rhythms in free-running conditions, suggesting occurrence of an internal circadian clock.

18.4.6 Anthropogenic Pressures

Anthropogenic pressures on aquatic environments may be considered critical to eel continental dispersal (Feunteun 2002). First, global warming affects oceanic gyres, reduces nutrient availability and slows down oceanic migration, and could thus increase larval mortality through enhanced starvation and predation (Knights 2003). On the other hand, continental influences should not be neglected. Overharvesting, habitat fragmentation and loss, and introduction of exotic species are strong agents of selection and may drive species to extinction (Stockwell et al. 2003). For instance, estuarine fisheries may deplete a large proportion of the migrating glass eels. In the Vilaine River, where a dam blocks upstream movement, the fishery removes up to 99% of the glass eel stock (Briand et al. 2003). In barrier-free estuaries, exploitation rates are probably sharply reduced but glass eel fisheries, by selectively targeting migrants using STST, could possibly favour evolution towards reduced migratory behaviour in the population on a decadal time scale (see Section 18.5). Indeed, fisheries may induce rapid life history evolution in fish populations (Law 2000). Furthermore, dams dramatically decrease habitat accessibility (White and Knights 1997a; Lafaille et al. 2005), and could possibly increase mortality in migrants that accumulate below the obstacle. Introduction of strong FW predators such as *Silurus glanis* could also seriously impair the benefits from colonizing river habitats. Finally, pollution impacts fitness through decreased individual condition and genetic erosion (Maes et al. 2005), and probably also disrupts complex behaviours through various endocrine and metabolic pathways (Robinet and Feunteun 2002; Scott and Sloman 2004).

18.4.7 Social Interactions

Social interactions are critical to animal dispersal (Clobert et al. 2001). In the eel, social entrainment to migration (i.e., positive density-dependent migration) is likely to occur through schooling behaviour (Tesch 1977). However, it is probably during ranging that social interactions play their major role in driving eel continental dispersal patterns. Indeed, yellow eels are strongly agonistic and confrontations may induce severe injuries or even death of the defeated individual (Bertin 1951; Tesch 1977; Peters et al. 1980). This agonistic behaviour is coupled to cannibalism (Degani and Levanon 1983; Knights 1987) and results in negative density-dependent survival (Vøllestad and Jonsson 1988; Briand et al. 2005a) and growth (Degani et al. 1988; Beentjes and Jellyman 2003). In mammals, competition for food and space and agonistic interactions are known to induce positive density-dependent dispersal (Sutherland et al. 2002). The same is likely to be true for the eel. In other words, how far yellow eels disperse from the habitat where they settled (the tidal limit area for many individuals) should be positively related to the density of dominant eels present in this habitat. Indeed, several recent studies indicate that river colonization by yellow eels is positively density-dependent, i.e., is a process equivalent to random diffusion of particles driven by density at the point source (Smogor et al. 1995; Ibbotson et al. 2002; Feunteun et al. 2003; Briand et al. 2005a).

Under density-dependent ranging, the many environmental (seasonal or not) variations that may alter the social structure of local populations will promote a reorganization of distributions, following a model of a patchy "fluid mosaic" (Feunteun et al. 2003). Depending on availability of habitat resources (space, water temperature, food availability, dissolved oxygen, etc.) and/or demographic parameters (density, proportion of dominant eels, etc.), subordinate eels shift towards habitats where survival and growth conditions are better (see also Section 18.5.2). This density-dependent ranging period is by far the longest of the continental phase, giving opportunity for generation of a very large variety of dispersal patterns. How often an individual will cross salinity boundaries during this period may partly depend on its osmotic capacities (see Section 18.4.2). In conclusion, social interactions combined with habitat variations represent major cues inducing plastic dispersal responses through ranging movements. However, we suggest that social interactions further represent a major selective force for the evolution of migration and timing of settlement because they induce habitat-specific survival and growth. This point is developed in the following section.

18.5 Ultimate Control of Eel Continental Dispersal

Game theory and associated concepts such as the Evolutionarily Stable Strategy (ESS) provide a framework for studying eel alternative phenotypes in terms of their costs and benefits to evolutionary fitness. We have analysed eel continental dispersal in the context of evolutionarily stable strategies, considering movement as a threshold trait with two alternative values: moving or staying. For the sake of clarity we have equated moving with diadromy (i.e. colonization of FW habitats) and staying with non-diadromy (i.e. colonization of SW habitats), but we suggest that the approach is valid for any type of movement.

18.5.1 The Cost/Benefits Fitness Ratio of Diadromy

Evolution of diadromy should be seen as driven by a ratio between its costs and benefits in terms of fitness (Gross et al. 1988; Edeline 2007). Diadromy provides evolutionary benefits because it may give access to increased food resources (Gross et al. 1988). In eels, diadromy may be considered as an ancestral trait. Indeed, eels appeared in tropical oceans where they probably evolved diadromy from a marine ancestor because primary productivity is higher in FW than in SW at low latitudes (Aoyama et al. 2001; Tsukamoto et al. 2002). Colonization of temperate oceans by several eel species has promoted evolution of a facultative diadromy (i.e., a fraction of the population never enter FWD presumably because relative productivity of rivers decreases compared to the sea, estuaries or deltas (Tsukamoto and Arai 2001). However, diadromy remains the predominant migratory tactic in temperate eels, as suggested by the ontogenetic programming of *A. anguilla* for osmoregulation in FW (Ciccoti et al. 1993). Most likely, maintenance of diadromy by temperate eels is selected for because it allows reduction of both inter- and intraspecific competition, two pivotal components of dispersal evolution (Clobert et al. 2001). Indeed, a shift to FW permits temperate eels to escape both strong marine predators (Jonsson and Jonsson 1993), such as conger eels (Moriarty 2003), and high conspecific densities (Tesch 1977). In that sense, social interactions may be considered as major drivers for the evolution of migration in the eel.

However, migration and diadromy incur a strong energetic cost (Roff 1991), that results in increased mortality. Indeed, depletion of energy stores by migration induces exhaustion that could lead to death, either directly or through reduced competitive ability and increased sensitivity to predation and infections (Bernatchez and Dodson 1987; Gandon and Michalakis 2001). In addition, migration probably incurs growth costs as already discussed in Section 18.4.2. Indeed, in the eel, energy requirements of digestion and locomotor activity are conflicting (Owen 2001), and the glass eel's migratory propensity is negatively correlated to juvenile growth performance (Edeline et al. 2005a). Therefore, the evolution of eel diadromy may be driven by a trade-off between search for the most productive habitats and competition avoidance favouring shift to less productive areas, while costs of dispersal hinder movements. We now provide an ESS model that may explain how eels cope with these conflicting selective pressures.

18.5.2 Evolutionarily Stable Strategy Model

Conditional strategies are ESSs that are commonly invoked to explain the occurrence of alternative phenotypes (i.e., threshold traits) among populations because they allow individuals to cope with physiological- and environment-dependent fitness trade-offs (Hazel et al. 1990; Gross 1996; Roff 1996; Hazel et al. 2004). Conditional strategies depict a special case of phenotypic plasticity in which the reaction norm is not continuous but instead takes a number of discrete values - most often two values - constituting the threshold trait under scrutiny (Schlichting and Pigliucci 1995). Classical examples of conditional strategies include predator - induced protected morphs, trophic polymorphisms or male alternative reproductive behaviours (but see Roff 1996 for review). The alternative phenotypes may be cued by internal (physiological) cues, and/or by external (environmental) cues if they are reliable indicators of environment quality (Hazel et al. 2004). Following Gross (1996), we hereafter term "tactic" the value taken by the phenotype (or decision taken by the individual), and "status" the value of the internal cue triggering the tactic ("status" is not convenient for environmental cues). Tactics change at the status at which fitness benefits switch from favouring one tactic to favouring the alternative: the ESS switch point (Gross 1996).

Edeline (2005, 2007) suggested that glass eel diadromy is a conditional strategy cued by individual energetic status, based on the observation that shift from migration to settlement by glass eels as their body condition decreases has an adaptive significance (Edeline et al. 2006) (see also Section 18.3.2). Edeline (2007) considered it parsimonious to consider a model in which both the switch point and the status (energy stores) of the conditional strategy as polygenic traits influenced by environmental factors. Hence, in an individual, both traits (the switch point and the status) are genetically fixed but they vary with environmental conditions. For instance, energy stores may possibly be influenced by the amount of food encountered during the larval stage (Désaunay and Guerault 1997). Also, the switch point may change with temperature, as indicated by the fact that low water temperature increases preference for SW in *A. anguilla* glass eels (Edeline et al. 2006). At the population level however, variations for the cue and the switch point result from both genetic and environmental variations, resulting in normal distributions for both traits (Hazel et al. 1990; Roff 1996; Hazel et al. 2004) (Fig. 18.5).

In Fig. 18.5, we depict this conditional strategy model in glass eels. Individual glass eels switch from migration to settlement if their energetic status decreases below their switch point. The proportion of migrants in the population is thus determined by the distribution of switch points relative to the distribution of energetic status (Fig. 18.5). In this model THs vs. GH and cortisol are presumably the

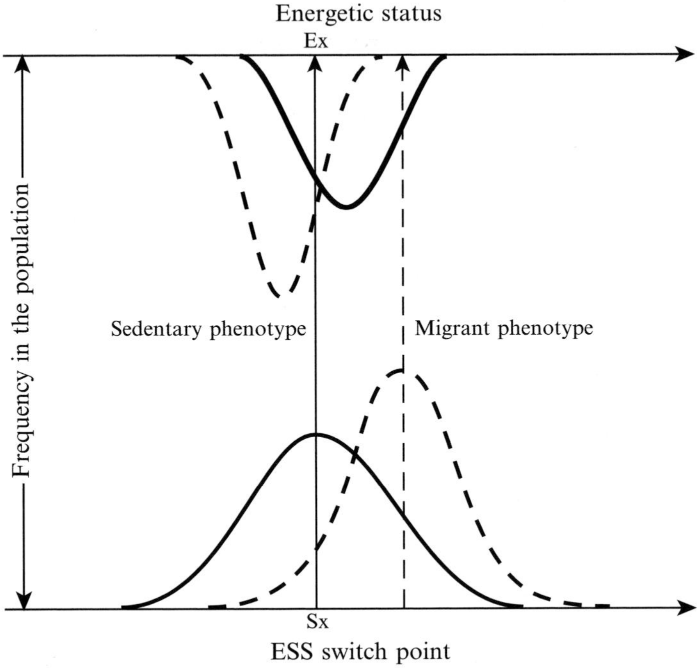

Fig. 18.5 Ultimate control of eel migratory behaviour. Conditional evolutionarily stable strategy (ESS) with alternative migratory tactics depicted as a threshold trait, after Edeline (2007) and adapted from the quantitative genetic models of Hazel et al. (1990, 2004). Consider first the solid curves and arrow. The distribution of the ESS switch points in the population (lower axis) is scaled with exposure to the cueing trait (energetic status, upper axis). An individual x (mean individual with regards to the switch points) will be sedentary if its energetic status Ex is lower than its switch point Sx, and migrant if its energetic status is higher than Sx. The proportion of migrants in the population therefore depends on the distribution of the switch points relative to the distribution of energetic status. For distributions corresponding to solid curves, the majority of the population presents a migratory phenotype (the mean switch point is lower than the mean energetic status). Note that variation for the switch point is larger than variation for energetic status, leading to the existence of unconditional migrants and unconditional sedentary individuals in the population (Hazel et al. 2004). Now, consider the effects of increased selection against migrants due to anthropogenic pressure on FW habitats (see Section 18.4.6). Individuals with a low switch point and/or a high energetic status are counter-selected, resulting in new distributions (dashed curves). This displacement of the curves induces a drop in the proportion of migrants in the population (dashed arrow). We suggest that such evolutionary shift contributes to explain the collapse of eel recruitment into FW during the last decades. Increased selection against migrants strengthens stabilizing selection and thus results in reduced variance in distributions (Futuyma 1998). Note that because the surface below a curve is held constant, reduced variance induces increased peak values

proximate mediators of the alternative tactics, as described in Section 18.3.3. This conditional ESS model, initially designed for glass eel migration, may be adapted to ranging behaviour by yellow eels provided that the cue (internal or environmental) is a reliable proxy for fitness gain from dispersal decisions. For instance, individual body size relative to that of the opponent (hierarchical status) is probably a reliable cue for risk of injury during agonistic interactions. Hence, the relationship between fitness of the alternative dispersal phenotypes and status in yellow eels will be the opposite of that found in glass eels: residency will be favoured in (large, high status) dominant eels, while (small, low status) subordinate eels will derive higher fitness from dispersal to lower density habitats. In that case, variation in the switch point for fighting or avoiding confrontation is related to individual "personality" [aggressiveness or boldness for instance (Stamps 2007)]. In other words, not all subordinate eels of the same size will skip confrontation with a given larger opponent. The view that relative body size is a reliable cue for post-settlement dispersal decisions is supported by data showing that the intensity of movements by yellow eels is inversely proportional to their size (Feunteun et al. 2003).

Ecological events may change selective pressures acting on migrants and sedentary individuals and thus generate equilibrium displacements in the distributions of both switch points and status. For instance, in Fig. 18.5, we have depicted the effects of an anthropogenic increase in selection against migrants (see Section 18.4.6). Enhanced mortality, acting selectively on migrants (individuals with a low switch point and/or a high body condition), pushes distributions away from each other and then dramatically reduces the proportion of migrants in the population. Concurrently, global population collapse will decrease densities in SW habitats and thus favour non-migrant individuals (individuals with a high switch point and/or a low body condition), further contributing to separate the two distributions. As underlined in Section 18.4.6, such evolutionary shifts may be rapid. Hence, we suggest that the collapse of FW eel stocks monitored during the last decades (Dekker et al. 2003; Stone 2003) may have anthropogenic sources. This collapse would not only result from decreased population size, but also from a decrease in the proportion of diadromous individuals in the population (Edeline 2007).

18.6 Conclusion and Suggested Future Research Directions

Eel dispersal in different growth habitats influences pivotal life history traits such as growth, sex differentiation, age and size at silvering and finally capacity to reproduce. Therefore, understanding the mechanisms controlling continental dispersal is crucial to eel biology and conservation. The ontogeny of dispersal behaviour reveals that most of the glass eels switch from upstream migration to food-search and density-dependent dispersal during their transformation to yellow eels. However, the timing of settlement is variable and some individuals settle precociously at the glass eel stage while others continue upstream migration at the yellow stage. These different migratory tactics are underlaid by genetic variation that allows the evolution of migratory behaviour and induces correlations between migratory behaviour and other life history traits such as salinity preference, locomotor activity and growth. Individual energetic status is a major cue for eel migratory plasticity. The endocrine system, processing both genetic and non-genetic influences, may be considered as the key mediator of the alternative migratory tactics. A host of environmental parameters may influence migration reaction norms, among which are water currents, temperature, salinity and odours. Post-settlement, eel movements are driven by search for resources and are thus more appropriately termed "ranging". Ranging is density-dependent and often results from aggressive interactions whose output is size-dependent. During this phase of continental life, variations in demography and resource availability generate a wide array of dispersal patterns. Ultimately, eels have evolved a conditional ESS to regulate dispersal decisions during continental life, i.e. a form of phenotypic plasticity that allows individuals to be either sedentary or dispersing, depending on their environment, and on their physiological and social status. This ESS model suggests that the worldwide collapse of river recruitment during the last decades may result not only from decreased abundance, but also from an adaptive decrease in the proportion of migrants in populations.

There still remain many gaps in our knowledge, as shown by the many speculative or purely theoretical parts of this review. Future works should aim at investigating the genetic basis of the alternative migratory tactics, such as for instance correlations between migratory behaviour and fitness-related genetic markers. More generally, mapping correlations among fitness-related traits could greatly help in dissecting both the proximate and ultimate constraints shaping the eel life history. At the endocrine level, the roles of THs, GH and cortisol in regulating the morphological, physiological and behavioural processes governing dispersal should be explored more deeply. Also, the mechanisms controlling glass eel rhythmic activity are poorly understood, but additional studies could bring fascinating insights about the fine tuning of the glass eel's coastal and estuarine movements. At the environmental level, our understanding of the social control of dispersal remains limited, and the importance of olfaction in conspecific interactions could be underestimated. Finally, a better evaluation of the anthropogenic impacts on eel life history is critical for better management of the endangered eel populations.

Acknowledgments Dr. F. A. Weltzien (University of Oslo) provided comments on an early version of the manuscript. E. Edeline was supported by research grants from the Cemagref, Région Aquitaine and the French Ministère de l'Ecologie.

References

Adam G, Elie P (1994) Mise en évidence des déplacements d'anguilles sédentaires (*Anguilla anguilla* L.) en relation avec le cycle lunaire dans le lac de Grand-Lieu (Loire-Atlantique). Bulletin Français de la Pêche et de la Pisciculture 335:123–132

Alerstam T, Hedenström A, Åkesson S (2003) Long-distance migration: evolution and determinants. Oikos 103:247–260

Aoyama J, Nishida M, Tsukamoto K (2001) Molecular phylogeny and evolution of the freshwater eel, genus *Anguilla*. Molecular Phylogenetics and Evolution 20:459–459

Baggerman B (1960) Salinity preference, thyroid activity and the seaward migration of four species of Pacific salmon (*Oncorhynchus*). Journal of Fisheries Research Board of Canada 17:295–322

Baggerman B (1962) Some endocrine aspects of fish migration. General and Comparative Endocrinology, Suppl. 1:188–205

Baisez A (2001) Optimisation des suivis des indices d'abondances et des structures de taille de l'anguille européenne (*Anguilla anguilla*) dans un marais endigué de la côte Atlantique: relation espèce-habitat. Ph.D. thesis, CEMAGREF/University of Toulouse III

Barbin GP, Parker SJ, McCleave JD (1998) Olfactory cues play a critical role in the estuarine migration of silver-phase American eels. Environmental Biology of Fishes 53:283–291

Bardonnet A, Riera P (2005) Feeding of glass eels (*Anguilla anguilla*) in the course of their estuarine migration: new insights from stable isotope analysis. Estuarine Coastal and Shelf Science 63:201–209

Bardonnet A, Bolliet V, Belon V (2005) Recruitment abundance estimation: role of glass eel (*Anguilla anguilla* L.) response to light. Journal of Experimental Marine Biology and Ecology 321:181–190

Bardonnet A, Dasse S, Parade M, Heland M (2003) Study of glass-eels movements in a flume in relation to nycthemeral changes. Bulletin Français de la Pêche et de la Pisciculture 368:9–19

Beentjes MP, Jellyman DJ (2003) Enhanced growth of longfin eels, *Anguilla dieffenbachii*, transplanted into Lake Hawea, a high country lake in South Island, New Zealand. New Zealand Journal of Marine and Freshwater Research 37:1–11

Bernatchez L, Dodson JJ (1987) Relationship between bioenergetics and behavior in anadromous fish populations. Canadian Journal of Fisheries and Aquatic Sciences 44:399–407

Berthold P (1991) Genetic control of migratory behaviour in birds. Trends in Ecology and Evolution 6:254–257

Berthold P, Pulido F (1994) Heritability of migratory activity in a natural bird population. Proceedings of the Royal Society B-Biological Sciences 257:311–315

Bertin L (1951) Les anguilles. Payot, Paris

Björnsson BT, Taranger GL, Hansen T, Stefansson SO, Haux C (1994) The interrelation between photoperiod, growth hormone, and sexual maturation of adult Atlantic salmon (*Salmo salar*). General and Comparative Endocrinology 93:70–81

Blázquez M, Bosma PT, Fraser EJ, Van Look KJW, Trudeau VL (1998) Fish as models for the neuroendocrine regulation of reproduction and growth. Comparative Biochemistry and Physiology Part C: Pharmacology, Toxicology and Endocrinology 119:345–364

Boeuf G (1993) Salmonid smoltification: a pre-adaptation to the oceanic environment. In: Rankin JC, Jensen FB (eds) Fish Ecophysiology. Chapman & Hall, London, pp 105–135

Boeuf G, Le Bail PY (1999) Does light have an influence on fish growth? Aquaculture 177:129–152

Boeuf G, Payan P (2001) How should salinity influence fish growth? Comparative Biochemistry and Physiology Part C: Toxicology & Pharmacology 130:411–423

Bolliet V, Ali MA, Lapointe FJ, Falcon J (1996) Rhythmic melatonin secretion in different teleost species: an *in vitro* study. Journal of Comparative Physiology B: Biochemical Systemic and Environmental Physiology 165:677–683

Boucher-Firly S (1935) Recherches biochimiques sur les Téléostéens Apodes (Anguille, Congre, Murène). Annales de l'Institut Océanographique 15:220–327

Briand C, Fatin D, Legault A (2002) Role of eel odour on the efficiency of an eel, *Anguilla anguilla*, ladder and trap. Environmental Biology of Fishes 65:473–477

Briand C, Fatin D, Fontenelle G, Feunteun E (2003) Estuarine and fluvial recruitment of the European glass eel, *Anguilla anguilla*, in an exploited Atlantic estuary. Fisheries Management and Ecology 10:377–384

Briand C, Fatin D, Fontenelle G, Feunteun E (2005a) Effect of re-opening of a migratory pathway for eel (*Anguilla anguilla*, L.) at a watershed scale. Bulletin Français de la Pêche et de la Pisciculture 378–379:67–86

Briand C, Fatin D, Ciccoti E, Lambert P (2005b) A stage-structured model to predict the effect of temperature and salinity on glass eel *Anguilla anguilla* pigmentation development. Journal of Fish Biology 67:993–1009

Bureau du Colombier S, Bolliet V, Lambert P, Bardonnet A (2007) Energy and migratory behavior in glass eels (*Anguilla anguilla*). Physiology and Behavior, 92:684–690

Butler DG, Carmichael FJ (1972) (Na^+/K^+)-ATPase activity in eel (*Anguilla rostrata*) gills in relation to changes in environmental salinity: role of adrenocortical steroids. General and Comparative Endocrinology 19:421–427

Callamand O, Fontaine M (1942) L'activité thyroïdienne de l'anguille au cours de son développement. Archives de Zoologie Expérimentale et Générale 82:129–136

Castonguay M, Dutil JD, Audet C, Miller R (1990) Locomotor activity and concentration of thyroid hormones in migratory and sedentary juvenile American eels. Transactions of the American Fisheries Society 119:946–956

Charlon N, Blanc JM (1982) Etude des civelles d'*Anguilla anguilla* L. dans la région du bassin de l'Adour. 1. Caractéristiques biométriques de longueur et de poids en fonction de la pigmentation. Archiv fuer Hydrobiologie 93:238–255

Chen YL, Chen HY (1991) Temperature selections of *Anguilla japonica* (L.) elvers, and their implications for migration. Australian Journal of Marine and Freshwater Research 42:743–750

Ciccoti E, Macchi E, Rossi A, Cataldi E, Cataudella S (1993) Glass eel (*Anguilla anguilla*) acclimation to freshwater and seawater: morphological changes of the digestive tract. Journal of Applied Ichthyology 9:74–81

Clobert J, Danchin E, Dhondt AA, Nichols JD (eds) (2001) Dispersal. Oxford University Press, Oxford

Creutzberg F (1958) Use of tidal streams by migrating elvers (*Anguilla vulgaris* Turt.). Nature 181:357–358

Creutzberg F (1959) Discrimination between ebb and flood tide in migrating elvers (*Anguilla vulgaris* Turt.) by means of olfactory perception. Nature 184:1961–1962

Creutzberg F (1961) On the orientation of migrating elvers (*Anguilla vulgaris* Turt.) in a tidal area. Netherlands Journal of Sea Research 1:257–338

Daemen E, Cross T, Ollevier F, Volckaert FA (2001) Analysis of the genetic structure of the European eel (*Anguilla anguilla*) using microsatellites DNA and mtDNA markers. Marine Biology 139:755–764

Dannewitz J, Maes GE, Johansson L, Wickström H, Volckaert FA, Järvi T (2005) Panmixia in the European eel: a matter of time. Proceedings of the Royal Society B-Biological Sciences 272:1129–1137

Daverat F, Tomás J (2006) Tactics and demographic attributes in the European eel *Anguilla anguilla* in the Gironde watershed, SW France. Marine Ecology Progress Series 307:247–257

Daverat F, Limburg KE, Thibault I, Shiao JC, Dodson JJ, Caron F, Tzeng WN, Lizuka Y, Wickström H (2006) Phenotypic plasticity of habitat use by three temperate eel species, *Anguilla anguilla, A. japonica* and *A. rostrata*. Marine Ecology Progress Series 308:231–241

Davey AJH, Jellyman DJ (2005) Sex determination in freshwater eels and management options for manipulation of sex. Reviews in Fish Biology and Fisheries 15:37–52

De Casamajor MN, Bru N, Prouzet P (2000) Fluctuation des captures de civelles (*Anguilla anguilla* L.) et analyse de la variation de leur capturabilité dans l'estuaire de l'Adour. Bulletin Français de la Pêche et de la Pisciculture 357–358:387–404

Deelder CL (1958) On the behaviour of elvers (*Anguilla vulgaris* Turt.) migrating from the sea into freshwater. Journal du Conseil International pour l'Exploration de la Mer 24:135–146

Degani G, Levanon D (1983) The influence of low density on food adaptation, cannibalism and growth of eels (*Anguilla anguilla* (L.)). Bamidgeh 35:53–60

Degani G, Levanon D, Meltzer A (1988) Influence of high loading density on growth and survival of European glass eels. Progressive Fish Culturist 50:178–181

Dekker W, Casselman JM, Cairns DK, Tsukamoto K, Jellyman D, Lickers H (2003) Worldwide decline of eel resources necessitates immediate action. Fisheries 28:28–30

Désaunay Y, Guerault D (1997) Seasonal and long term changes in biometrics of eel larvae: a possible relationship between recruitment and North Atlantic ecosystem productivity. Journal of Fish Biology 51 Suppl. A:317–339

Dingle H (2006) Animal migration: is there a common migratory syndrome? Journal of Ornithology 147:212–220

Dou SZ, Tsukamoto K (2003) Observations on the nocturnal activity and feeding behavior of *Anguilla japonica* glass eels under laboratory conditions. Environmental Biology of Fishes 67:389–395

Døving KB, Stabell OB (2002) Trails in open waters: sensory cues in salmon migration. In: Collin SP, Marshall NJ (eds) Sensory Processing in Aquatic Environments. Springer, New York, pp 39–52

Dufour S, Rousseau K, Sbaihi M, Le Belle N, Vidal B, Marchelidon J, Schmitz M (2001) Multiple neuroendocrine regulation of growth hormone in the European eel *Anguilla anguilla*. Journal of Taiwan Fisheries Research 9:77–92

Dufty AM, Belthoff JR (2001) Proximate mechanisms of natal dispersal: the role of body condition and hormones. In: Clobert J, Danchin E, Dhondt AA, Nichols JD (eds) Dispersal. Oxford University Press, Oxford, pp 217–229

Dutil JD, Michaud M, Giroux A (1989) Seasonal and diel patterns of stream invasion by American eels (*Anguilla rostrata*) in the Northern Gulf of St. Lawrence. Canadian Journal of Zoology 67:182–188

Eales JG (1988) The influence of nutritional state on thyroid function in various vertebrates. American Zoologist 28:351–362

Edel RK (1976) Activity rhythms of maturing American eels (*Anguilla rostrata*). Marine Biology 36:283–289

Edeline E (2005) Facteurs du contrôle de la dispersion continentale chez l'anguille. Ph.D. thesis, CEMAGREF/University of Toulouse III

Edeline E (2007) Adaptive phenotypic plasticity of eel diadromy. Marine Ecology Progress Series 341:229–232

Edeline E, Dufour S, Elie P (2005a) Role of glass eel salinity preference in the control of habitat selection and growth plasticity in *Anguilla anguilla*. Marine Ecology Progress Series 304:191–199

Edeline E, Lambert P, Rigaud C, Elie P (2006) Effects of body condition and water temperature on *Anguilla anguilla* glass eels migratory behavior. Journal of Experimental Marine Biology and Ecology 331:217–225

Edeline E, Beaulaton L, Le Barh R, Elie P (2007) Dispersal in metamorphosing *Anguilla anguilla* juvenile eel. Marine Ecology Progress Series 344:213–218

Edeline E, Dufour S, Briand C, Fatin D, Elie P (2004) Thyroid status is related to migratory behavior in *Anguilla anguilla* glass eels. Marine Ecology Progress Series 282:261–270

Edeline E, Bardonnet A, Bolliet V, Dufour S, Elie P (2005b) Endocrine control of *Anguilla anguilla* glass eel dispersal: effect of thyroid hormones on locomotor activity and rheotactic behavior. Hormones and Behavior 48:53–63

Einum S, Kinnison MT, Hendry AP (2004) Evolution of egg size and number. In: Hendry AP, Stearns SC (eds) Evolution Illuminated. Oxford University Press, New York, p 510

Elie P (1979) Contribution à l'étude des montées de civelles d'*Anguilla anguilla* Linné (Poisson, Téléostéen, Anguilliforme), dans l'estuaire de la Loire: pêche, écologie, écophysiologie et élevage. Ph.D. thesis, University of Rennes I

Elie P, Daguzan J (1976) Alimentation et croissance des civelles d'*Anguilla anguilla* L. (Poisson Téléostéen Anguilliforme) élevées expérimentalement, à diverses températures, au laboratoire. Annales de la Nutrition et de l'Alimentation 30:95–114

Elie P, Rochard E (1994) Migration des civelles d'anguille dans les estuaires, modalités du phénomène et caractéristiques des individus. Bulletin Français de la Pêche et de la Pisciculture 335:81–98

Elie P, Lecomte-Finiger R, Cantrelle I, Charlon N (1982) Définition des limites des différents stades pigmentaires durant la phase civelle d'*Anguilla anguilla* L. (poisson téléostéen anguilliforme). Vie et Milieu 32:149–157

Epstein FH, Cynamon M, McKay W (1971) Endocrine control of Na-K-ATPase and seawater adaptation in *Anguilla rostrata*. General and Comparative Endocrinology 16:323–328

Falcon J, Besseau L, Fazzari D, Attia J, Gaildrat P, Beauchaud M, Boeuf G (2003) Melatonin modulates secretion of growth hormone and prolactin by trout pituitary glands and cells in culture. Endocrinology 144:4648–4658

Feunteun E (2002) Management and restoration of European eel population (*Anguilla anguilla*): an impossible bargain. Ecological Engineering 18:575–591

Feunteun E, Laffaille P, Robinet T, Briand C, Baisez A, Olivier JM, Acou A (2003) A review of upstream migration and movements in inland waters by anguillid eels. Towards a general theory. In: Aida K, Tsukamoto K, Yamauchi K (eds) Eel Biology. Springer, Tokyo, pp 191–213

Fontaine M (1975) Physiological mechanisms in the migration of marine and amphihaline fish. Advances in Marine Biology 13:241–355

Fontaine M, Raffy A (1932) Recherches physiologiques et biologiques sur les civelles. Bulletin de l'Institut Océanographique 603:1–18

Forseth T, Naesje TF, Jonsson B, Harsaker K (1999) Juvenile migration in brown trout: a consequence of energetic state. Journal of Animal Ecology 68:783–793

Futuyma DJ (1998) Evolutionary biology. Sinauer, Sunderland, Massachussets

Gandon S, Michalakis Y (2001) Multiple causes of the evolution of dispersal. In: Clobert J, Danchin E, Dhondt AA, Nichols JD (eds) Dispersal. Oxford University Press, Oxford, pp 155–167

Gross MR (1996) Alternative reproductive strategies and tactics: diversity within sexes. Trends in Ecology and Evolution 11:92–98

Gross MR, Coleman RM, McDowall RM (1988) Aquatic productivity and the evolution of diadromous fish migration. Science 139:1291–1293

Halpern BP (1982) Environmental factors affecting chemoreceptors: an overview. Environmental Health Perspectives 44:101–105

Harden-Jones FR (1984) A view from the ocean. In: McCleave JD, Arnold GP, Dodson JJ, Neill WH (eds) Mechanisms of Migration in Fishes. Plenum, New York, pp 1–26

Haro A, Krueger WH (1988) Pigmentation, size, and migration of elvers (*Anguilla rostrata* (Lesueur)) in a coastal Rhode Island stream. Canadian Journal of Zoology 66:2528–2533

Hastings JW, Rusak B, Boulos Z (1991) Circadian rhythms: the physiology of biological timing. In: Prosser CL (ed) Neural and Integrative Animal Physiology. Wiley-Liss, New York, pp 435–546

Hazel W, Smock R, Lively CM (2004) The ecological genetics of conditional strategies. American Naturalist 163:888–900

Hazel WN, Smock R, Johnson MD (1990) A polygenic model for the evolution and maintenance of conditional strategies. Proceedings of the Royal Society B-Biological Sciences 242:181–187

Higgs DA, Fagerlund UHM, Eales JG, McBride JR (1982) Application of thyroid and steroid hormones as anabolic agents in fish culture. Comparative Biochemistry and Physiology Part B: Biochemistry and Molecular Biology 73:143–176

Hoar WS (1988) The physiology of smolting salmonids. In: Hoar WS, Randall DJ (eds) Fish Physiology. Academic, London, pp 275–343

Holmgren K, Mosegaard H (1996) Implications of individual growth status on the future sex of the European eel. Journal of Fish Biology 49:910–925

Holmgren K, Wickström H, Clevestam P (1997) Sex-related growth of European eel, *Anguilla anguilla*, with focus on median silver eel age. Canadian Journal of Fisheries and Aquatic Sciences 54:2775–2781

Ibbotson A, Smith J, Scarlett P, Aprahamian MW (2002) Colonisation of freshwater habitats by the European eel *Anguilla anguilla*. Freshwater Biology 47:1696–1706

Iwata M (1995) Downstream migratory behavior of salmonids and its relationship with cortisol and thyroid hormones: A review. Aquaculture 135:131–139

Jegstrup IM, Rosenkilde P (2003) Regulation of post-larval development in the European eel: thyroid hormone level, progress of pigmentation and changes in behaviour. Journal of Fish Biology 63:168–175

Jellyman DJ, Lambert PW (2003) Factors affecting recruitment of glass eels into the Grey River, New Zealand. Journal of Fish Biology 63:1067–1079

Jessop BM (1998) Geographic and seasonal variation in biological characteristics of American eel elvers in the Bay of Fundy area and on the Atlantic coast of Nova Scotia. Canadian Journal of Zoology 76:2172–2185

Jessop BM (2003) Annual variability in the effects of water temperature, discharge, and tidal stage on migration of American eel elvers from estuary to river. American Fisheries Society Symposium 33:3–16

Jessop BM, Shiao JC, Lizuka Y, Tzeng CS (2004) Variation in the annual growth, by sex and migration history, of silver American eels *Anguilla rostrata*. Marine Ecology Progress Series 272:231–244

Johansson V, Winberg S, Jönsson E, Hall D, Björnsson BT (2004) Peripherally administered growth hormone increases brain dopaminergic activity and swimming in rainbow trout. Hormones and Behavior 46:436–443

Johnston IA, Temple GK (2002) Thermal plasticity of skeletal muscle phenotype in ectothermic vertebrates and its significance for locomotory behaviour. Journal of Experimental Biology 205:2305–2322

Jonsson B, Jonsson N (1993) Partial migration: niche shift versus sexual maturation in fishes. Reviews in Fish Biology and Fisheries 3:348–365

Jonsson N, Jonsson B, Hansen LP (1997) Changes in proximate composition and estimates of energetic costs during upstream migration and spawning in Atlantic salmon *Salmo salar*. Journal of Animal Ecology 66:425–436

Jung S, Houde ED (2003) Spatial and temporal variabilities of pelagic fish community structure and distribution in Chesapeake Bay, USA. Estuarine Coastal and Shelf Science 58:335–351

Katzman S, Cech JJ (2001) Juvenile coho salmon locomotion and mosaic muscle are modified by 3′,3′,5′-tri-iodo-l-thyronine (T(3)). Journal of Experimental Biology 204:1711–1717

Kawakami Y, Mochioka N, Kimura R, Nakazono A (1999) Seasonal changes of the RNA/DNA ratio, size and lipid contents and immigration adaptability of Japanese glass eels, *Anguilla japonica*, collected in northern Kyushu, Japan. Journal of Experimental Marine Biology and Ecology 238:1–19

Kim WS, Yoon SJ, Moon HT, Lee TW (2002) Effects of water temperature changes on the endogenous and exogenous rhythms of oxygen consumption in glass eels *Anguilla japonica*. Marine Ecology Progress Series 243:209–216

Knights B (1987) Agonistic behaviour and growth in the European eel, *Anguilla anguilla* L., in relation to warm-water aquaculture. Journal of Fish Biology 31:265–276

Knights B (2003) A review of the possible impacts of long term oceanic and climate changes and fishing mortality on recruitment of anguillid eels of the northern Hemisphere. Science of the Total Environment 310:237–244

Lafaille P, Acou A, Guillouët J, Legault A (2005) Temporal changes in European eel, *Anguilla anguilla*, stocks in a small catchment after installation of fish passes. Fisheries Management and Ecology 12:123–129

Lambert P, Sbaihi M, Rochard E, Marchelidon J, Dufour S, Elie P (2003) Morphological and growth hormone variabilities of European glass eel (*Anguilla anguilla*) in Gironde estuary during 1997–1998 season. Bulletin Français de la Pêche et de la Pisciculture 368:69–83

Law R (2000) Fishing, selection, and phenotypic evolution. ICES Journal of Marine Sciences 57:659–668

Le Bras YM (1978) Circadian rhythm in brain catecholamine concentrations in the teleost: *Anguilla anguilla*. Comparative Biochemistry and Physiology Part C: Pharmacology, Toxicology and Endocrinology 62:115–117

Le Bras YM (1984) Circadian variations of catecholamine levels in brain, heart, and plasma in the eel, *Anguilla anguilla* L., at three different times of year. General and Comparative Endocrinology 55:472–479

Leiner KA, McKenzie DS (2001) The effects of photoperiod on growth rate and circulating thyroid hormone levels in the red drum, *Sciaenops ocellatus*: evidence for a free running circadian rhythm of T_4 secretion. Comparative Biochemistry and Physiology Part A: Molecular & Integrative Physiology 130:141–149

Leloup J (1958) Influence de la température sur le fonctionnement thyroïdien de l'anguille normale. Comptes Rendus des Séances de l'Académie des Sciences 247:2454–2456

Leloup J, De Luze A (1985) Environmental effects of temperature and salinity on thyroid function in Teleost fishes. In: Follet BK, Ishii S, Chandola A (eds) The Endocrine System and the Environment. Springer, Berlin, pp 23–32

Lema SC, Nevitt GA (2004) Evidence that thyroid hormone induces olfactory cellular proliferation in salmon during a sensitive period for imprinting. Journal of Experimental Biology 207:3317–3327

Maes GE, Volckaert FA (2002) Clinal genetic variation and isolation by distance in the European eel *Anguilla anguilla* (L.). Biological Journal of the Linnean Society 77:509–521

Maes GE, Raeymaekers JAM, Pampoulie C, Seynaeve A, Goemans G, Belpaire C, Volckaert FA (2005) The catadromous European eel *Anguilla anguilla* (L.) as a model for freshwater evolutionary ecotoxicology: relationship between heavy metal bioaccumulation, condition and genetic variability. Aquatic Toxicology 73:99–114

Marchelidon J, Schmitz M, Houdebine LM, Vidal B, Le Belle N, Dufour S (1996) Development of a radioimmunoassay for European eel growth hormone and application to the study of silvering and experimental fasting. General and Comparative Endocrinology 102:360–369

McCleave JD, Kleckner RC (1982) Selective tidal stream transport in the estuarine migration of glass eels of the American eel (*Anguilla rostrata*). Journal du Conseil International pour l'Exploration de la Mer 40:262–271

McCleave JD, Wippelhauser GS (1987) Behavioural aspects of selective tidal stream transport in juvenile American eels. American Fisheries Society Symposium 1:138–150

McCormick SD (2001) Endocrine control of osmoregulation in teleost fish. American Zoologist 41:781–794

McGovern P, McCarthy TK (1992) Elver migration in the River Corrib system, western Ireland. Irish Fisheries Investigations Series A (Freshwater) 36:25–32

Michaud M, Dutil JD, Dodson JJ (1988) Determination of the age of young American eels, *Anguilla rostrata*, in freshwater, based on otolith surface area and microstructure. Journal of Fish Biology 32:179–189

Moran NA (1994) Adaptation and constraint in the complex life cycles of animals. Annual Review of Ecology and Systematics 25:573–600

Moriarty C (1986) Riverine migrations of young eels *Anguilla anguilla* (L.). Fisheries Research 4:43–58

Moriarty C (2003) The yellow eel. In: Aida K, Tsukamoto K, Yamauchi K (eds) Eel Biology. Springer, Tokyo, pp 89–105

Naismith IA, Knights B (1988) Migrations of elvers and juvenile European eels, *Anguilla anguilla* L., in the River Thames. Journal of Fish Biology 33:161–177

Nelson JS (1994) Fishes of the World. Wiley, New York

Otake T (2003) Metamorphosis. In: Aida K, Tsukamoto K, Yamauchi K (eds) Eel Biology. Springer, Tokyo, pp 61–74

Øverli Ø, Kotzian S, Winberg S (2002) Effects of cortisol on aggression and locomotor activity in rainbow trout. Hormones and Behavior 42:53–61

Owen SF (2001) Meeting energy budgets by modulation of behaviour and physiology in the eel (*Anguilla anguilla* L.). Comparative Biochemistry and Physiology Part A: Molecular & Integrative Physiology 128:629–642

Ozaki Y, Okamura H, Kazeto Y, Ikeuchi T, Ijiri S, Nagae M, Adachi S, Yamauchi K (2000) Developmental changes in pituitary thyroid axis, and formation of gonads in leptocephali and glass eels of *Anguilla* spp. Fisheries Science 66:1115–1122

Peters G, Delventhal H, Klinger H (1980) Physiological and morphological effects of social stress in the eel, (*Anguilla anguila* L.). Archiv fuer Fischereiwissenschaft 30:157–180

Petit G, Vilter V (1944) Stabulation des civelles à l'embouchure d'un fleuve du golfe de Marseille. Comptes Rendus des Séances de la Société de Biologie et de ses Filiales 138:632–634

Pujolar JM, Maes GE, Vancoillie C, Volckaert FA (2005) Growth rate correlates to individual heterozygosity in the European eel, *Anguilla anguilla* L. Evolution 59:189–199

Pulido F, Berthold P, van Noordwijk AJ (1996) Frequency of migrants and migratory activity are genetically correlated in a bird population: evolutionary implications. Proceedings of the National Academy of Sciences of the USA 93:14642–14647

Redding JM, deLuze A, Leloup-Hâtey J, Leloup J (1986) Suppression of plasma thyroid hormone concentrations by cortisol in the European eel *Anguilla anguilla*. Comparative Biochemistry and Physiology Part A: Physiology 83:409–413

Rensing L, Ruoff P (2002) Temperature effect on entrainment, phase shifting, and amplitude of circadian clocks and its molecular bases. Chronobiology International 19:807–864

Richardson J, Boubée JAT, West DW (1994) Thermal tolerance and preference of some native New Zealand freshwater fish. New Zealand Journal of Marine and Freshwater Research 28:399–407

Robinet T, Feunteun E (2002) Sublethal effects of exposure to chemical compounds: a cause for the decline in Atlantic eels? Ecotoxicology 11:265–277

Roff DA (1991) Life history consequences of bioenergetic and biomechanical constraints on migration. American Zoologist 31:205–216

Roff DA (1996) The evolution of threshold traits in animals. Quarterly Review of Biology 71:3–35

Roff DA, Fairbairn DJ (2001) The genetic basis of dispersal and migration, and its consequences for the evolution of correlated traits. In: Clobert J, Danchin E, Dhondt AA, Nichols JD (eds) Dispersal. Oxford University Press, Oxford, pp 191–202

Rousseau K, Le Belle N, Marchelidon J, Dufour S (1999) Evidence that corticotropin releasing hormone acts as a growth hormone releasing factor in a primitive teleost, the European eel (*Anguilla anguilla*). Journal of Neuroendocrinology 11:385–392

Rousseau K, Le Belle N, Sbaihi M, Marchelidon J, Schmitz M, Dufour S (2002) Evidence for a negative feedback in the control of eel growth hormone by thyroid hormones. Journal of Endocrinology 175:605–613

Saglio P (1982) Piégeage d'anguilles (*Anguilla anguilla* L.) dans le milieu naturel au moyen d'extraits biologiques d'origine intraspécifique. Mise en évidence de l'activité phéromonale du mucus épidermique. Acta Oecologica 3:223–231

Schlichting CD, Pigliucci M (1995) Gene regulation, quantitative genetics and the evolution of reaction norms. Evolutionary Ecology 9:154–168

Schmidt J (1909) Remarks on the metamorphosis and distribution of the larvae of the eel. Meddelelser fra Kommissionen for Havundersøgelser Serie: Fiskeri 3:1–17

Scott GR, Sloman KA (2004) The effects of environmental pollutants on complex fish behaviour: integrating behavioural and physiological indicators of toxicity. Aquatic Toxicology 68:369–392

Shikano T, Chiyokubo T, Nakadate M, Fujio Y (2000) The relationship between allozyme heterozygosity and salinity tolerance in wild and domestic populations of the guppy (*Poecilia reticulata*). Aquaculture 184:233–245

Sloane RD (1984) Upstream migration by young pigmented freshwater eels (*Anguilla australis* Richardson) in Tasmania. Australian Journal of Marine and Freshwater Research 35:61–73

Smogor RA, Angermeier PL, Gaylord CK (1995) Distribution and abundance of American eels in Virginia streams: test of null models across spatial scales. Transactions of the American Fisheries Society 124:789–803

Sola C (1995) Chemoattraction of upstream migrating glass eels *Anguilla anguilla* to earthy and green odorants. Environmental Biology of Fishes 43:179–185

Sola C, Tosi L (1993) Bile salts and taurine as chemical stimuli for glass eels, *Anguilla anguilla*: a behavioural study. Environmental Biology of Fishes 37:197–204

Sola C, Tongiorgi P (1996) The effects of salinity on the chemotaxis of glass eels, *Anguilla anguilla*, to organic earthy and green odorants. Environmental Biology of Fishes 47:213–218

Sola C, Tongiorgi P (1998) Behavioural responses of glass eels of *Anguilla anguilla* to non-protein amino acids. Journal of Fish Biology 53:1253–1262

Sola C, Spampanato A, Tosi L (1993) Behavioural responses of glass eels (*Anguilla anguilla*) toward amino acids. Journal of Fish Biology 42:683–691

Sorensen PW (1986) Origins of the freshwater attractant(s) of migrating elvers of the American eel, *Anguilla rostrata*. Environmental Biology of Fishes 17:185–200

Specker JL, Eales JG, Tagawa M, Tyler III WA (2000) Parr smolt transformation in Atlantic salmon: thyroid deiodination in liver and brain and endocrine correlates of change in rheotactic behavior. Canadian Journal of Zoology 78:696–705

Stamps JA (2007) Growth mortality tradeoffs and 'personality traits' in animals. Ecology Letters 10:355–363

Stockwell CA, Hendry AP, Kinnison MT (2003) Contemporary evolution meets conservation biology. Trends in Ecology and Evolution 18:94–101

Stone R (2003) Freshwater eels are slip sliding away. Science 302:221–222

Sutherland WJ, Gill JA, Norris K (2002) Density dependent dispersal in animals: concepts, evidence, mechanisms and consequences. In: Bollock JM, Kenward RE, Hails RS (eds) Dispersal Ecology. Blackwell, Oxford, pp 134–151

Tesch FW (1977) The Eel. Chapman & Hall, London

Tomoda H, Uematsu K (1996) Morphogenesis of the brain in larval and juvenile Japanese eels, *Anguilla japonica*. Brain Behavior and Evolution 47:33–41

Tosi L, Sola C (1993) Role of geosmin, a typical inland water odour, in guiding glass eel *Anguilla anguilla* (L.) migration. Ethology 95:177–195

Tosi L, Sola C, Spampanato A, Tongiorgi P (1989) The behaviour of glass eel of *Anguilla anguilla* toward salinity: discrimination and preferences. Rivista Italiana di Acquacoltura 24:219–223

Tosi L, Spampanato A, Sola C, Tongiorgi P (1990) Relation of water odour, salinity and temperature to ascent of glass eels, *Anguilla anguilla* (L.): a laboratory study. Journal of Fish Biology 36:327–340

Tosi L, Sala L, Sola C, Spampanato A, Tongiorgi P (1988) Experimental analysis of the thermal and salinity preferences of glass eels, *Anguilla anguilla* (L.), before and during the upstream migration. Journal of Fish Biology 33:721–733

Tsukamoto K, Arai T (2001) Facultative catadromy of the eel *Anguilla japonica* between freshwater and seawater habitats. Marine Ecology Progress Series 220:265–276

Tsukamoto K, Aoyama J, Miller JM (2002) Migration, speciation, and the evolution of diadromy in anguillid eels. Canadian Journal of Fisheries and Aquatic Sciences 59:1989–1998

Tzeng WN, Lizuka Y, Shiao JC, Yamada Y, Oka HP (2003) Identification and growth rates comparison of divergent migratory contingents of Japanese eel (*Anguilla japonica*). Aquaculture 216:77–86

Vilter V (1945) Développement du tractus gastro intestinal au cours de la métamorphose de larves d'anguille (civelles). Comptes Rendus des Séances de la Société de Biologie et de ses Filiales 139:223–224

Vilter V (1946) Action de la thyroxine sur la métamorphose larvaire de l'anguille. Comptes Rendus des Séances de la Société de Biologie et de ses Filiales 140:783–785

Vøllestad LA, Jonsson B (1988) A 13-year study of the population dynamics and growth of the European eel *Anguilla anguilla* in a Norwegian river: evidence for density-dependent mortality, and development of a model for predicting yield. Journal of Animal Ecology 57:983–997

Weihs D (1978) Tidal stream transport as an efficient method for migration. Journal du Conseil International pour l'Exploration de la Mer 38:92–99

White E, Knights B (1997a) Dynamics of upstream migration of the European eel, *Anguilla anguilla* (L.), in the Rivers Severn and Avon, England, with special reference to the effects of man made barriers. Fisheries Management and Ecology 4:311–324

White EM, Knights B (1997b) Environmental factors affecting the migration of the European eel in the Rivers Severn and Avon, England. Journal of Fish Biology 50:1104–1116

Wilson JM, Antunes JC, Buça PD, Coimbra J (2004) Osmoregulatory plasticity of the glass eel of *Anguilla anguilla*: freshwater entry and changes in branchial ion-transport protein expression. Canadian Journal of Fisheries and Aquatic Sciences 61:432–442

Wippelhauser GS, McCleave JD (1988) Rhythmic activity of migrating juvenile American eels *Anguilla rostrata*. Journal of the Marine Biological Association of the United Kingdom 68:81–91

Wirth T, Bernatchez L (2001) Genetic evidence against panmixia in the European eel. Nature 409:1037–1040

Yamano K, Tagawa M, De Jesus EG, Hirano T, Miwa S, Inui Y (1991) Changes in whole body concentrations of thyroid hormones and cortisol in metamorphosing conger eel. Journal of Comparative Physiology, B: Biochemical, Systematic and Environmental Physiology 161:371–375

Zera AJ, Harshman LG (2001) The physiology of life history trade offs in animals. Annual Review of Ecology and Systematics 32:95–126

Part VI
Synthesis

Chapter 19
Reproduction Capacity of European Eels

Sylvie Dufour and Guido van den Thillart

Keywords: silvering, spawning migration, maturation, swimbladder parasite, PCBs, silver eel, quality criteria

19.1 Introduction

Silver eels, which represent the last known stage in eels under natural conditions, are still sexually immature. Their gonad development is heavily depressed, which is likely related to the very long distance they have to swim before reaching the spawning site, which is about 6,000-km from the European west coast. So, the capacity of silver eels to cope with the many different environments during their journey, the capacity to perform extremely long swimming activity and ultimately to find the spawning site and to reproduce, all these factors should be taken together in order to determine the individual's reproduction capacity.

Ecological studies on eel stock survey and management used to focus on the size of the eel population in various hydrosystems. However, not only numbers but also physiological qualities of silver eels are required to predict their potential contribution to stock renewal. Furthermore, quality assessment of wild silver eels is also a prerequisite for future success of ongoing research on European eel artificial reproduction. In the forgoing chapters we have discussed most recent information relating to the individual quality of European silver eels, which in this last chapter is used to develop a semi quantitative indicator for the reproduction capacity of the European silver eel.

S. Dufour
Museum d'Histoire Naturelle, DMPA, UMR CNRS 5178 "Biology of Marine Organisms and Ecosystems", 7 rue Cuvier, CP 32, 75231 Paris cedex 05, France

G. van den Thillart
Institute Biology, Leiden University, P.O. Box 9516, 2300 RA Leiden, The Netherlands

G. van den Thillart et al. (eds.), *Spawning Migration of the European Eel,*

19.2 Development of Criteria to Evaluate Reproduction Capacity of the European Eels

The reproduction capacity of individual silver eels must relate to three different aspects in the last phase of their life cycle: (a) silvering, (b) spawning migration, and (c) maturation. The capacity to complete each of these steps determines the actual reproduction capacity. Obviously these requirements make the eels vulnerable, because any deficiency in any one step would render them automatically incapable of reproduction. Eels are panmictic, which means that the gene flow among groups is too high to establish separate populations. It also means that eels must be extreme polymorphic in their genetic makeup, as they have to be able to live in many different environments from Iceland to Egypt. So, these are the two major factors which determine the phenotypic variability in performance of animals: the environment and the genetic make-up.

19.2.1 Silvering Process

Silvering is a requirement for downstream migration and reproduction. It marks the end of the growth phase and the onset of sexual maturation. This metamorphosis involves a number of different physiological functions (osmoregulatory, reproductive), which prepare the eel for the long return trip to the Sargasso Sea. Silver eels have distinct features that differentiate them from the yellow resident eels. However little is known about the dynamics of the silvering process. Chapter 2 presents new information on the triggers, duration and succession of events up to the silver migratory stage. To analyse the process of silvering, many biometric and physiological parameters were measured in eels from different European locations and compared in a multidimensional correlation matrix. Five different stages for female eels and two different stages for male eels could be distinguished. The further outcome of the PCA analyses showed that 82% of all variability can be described by only four easy to measure biometric parameters: weight, length, pectoral fin length and eye diameter (Chapter 2). Such a "silvering scale", which can be applied to any European eel population, provides a remarkable tool for further basic research as well as for field studies and population survey. The silver index is in fact an extension of the generally used eye index developed by Pankhurst (see Chapter 2). It should be stressed however, that Pankhurst's index was based on changes of the eye diameter during hormonal stimulation of female eels. The silver index on the other hand is based on many morphological and physiological parameters of wild male and female silver eels, and has therefore a better predictive value with respect to the real silver stage.

Concerning the endocrine regulation of silvering, it has always been assumed that silvering was a type of metamorphosis and therefore, would be under the control of thyroid hormones, such as for amphibian metamorphosis. However, recent endocrine studies revealed that, the gonadotropic axis is largely activated during the silvering process, leading to the challenging conclusion that silvering should not be considered as a true "metamorphosis" but rather as a "pubertal" event (Chapter 3).

19.2.2 Migration Performance

Seawater tolerance: Migrating eels have to adapt to a relative fast transition from fresh to seawater, therefore the tolerance for this transition was measured from eels from different locations. All eels, whether they are yellow or silver, are remarkably tolerant to the transfer from fresh to sea water. Their tolerance is for instance much higher than in salmonids, in which it is restricted to the short window of the smolt phase. Eels are able to adapt to acute changes from fresh to sea water. In this respect eels appear quite unique, within 36h the initial osmotic disturbance is stabilised. Using a seawater challenge test it was shown that silver eels have a higher capacity to osmoregulate than yellow eels. The positive correlation between the seawater tolerance and the silver index indicates that silver eels are preparing themselves for life at sea (Chapter 6).

Pressure tolerance: Oceanic migration is believed to occur at great depths; therefore migrating eels must be able to cope with high hydrostatic pressure. The high pressure tolerance was measured in eels from different locations, using a hyperbaric chamber. All eels, whether they are yellow or silver, appear to be remarkably tolerant to pressure changes (2–10 atm min^{-1}) up to 101 atm (equivalent to 1,000 km depth). They are for example considerably more tolerant to pressure than salmonids. In addition, a positive correlation was found between pressure tolerance and silver index, indicating an improved tolerance of silver eels to high pressure, as compared to yellow eels (Chapter 5). Furthermore, ongoing studies suggest that high pressure could be one of the environmental factors of the migration, involved in the activation of eel gonadotropic axis.

Swim fitness: Reaching the spawning grounds, some 6,000-km from the West coast of Europe, requires a very long swimming period of about 6 months. This means that the animals need to have both an impressive endurance and an extensive energy reserve. An extensive study was performed on the swim performance, i.e. the swim fitness and endurance, of eels from different locations. It was observed that all eels swim with impressive low cost of transport (COT), some six times more efficient than salmonids. Most eels, whether they are yellow or silver, are good endurance swimmers. There was no correlation between swim fitness and silver index, in contrast to seawater tolerance and pressure tolerance. Still, the COT shows a positive correlation with size, critical and optimal swim speeds. Therefore, the bigger eels have a significant advantage over the smaller eels. Also the bigger eels have usually larger lipid stores. This suggests that these eels arrive at the spawning site with higher remaining fat reserves for final maturation, resulting in a higher quality and quantity of offspring (Chapter 8).

19.2.3 Maturation Performance

We evaluated the capacity of eels from different locations to mature by measuring their short term and long term responses to hormonal treatment with carp pituitary extracts. Yellow eels are, in any case, not able to complete maturation, indicating that

Table 19.1 Identification table for the reproductive capacity of female European silver eels

Silver stage	I	II	III	IV	Va	Vb
Body length	–	–	–	–	<70 cm	>70 cm
Fat content	–	–	<13%	<16%	16–20%	>20%
Repro capacity	0	0	0	*	***	******

Fat content as % of wet weight; silver stages of female eels are based on silver index scores; reproductive capacity is indicated from very likely (******) to absent (0).

silvering is a strict prerequisite for maturation. Furthermore, among silver eels, the best performance was correlated with the highest silver index. In addition, we observed that eels combining a high silver index, high condition factor, and large body size had the best maturation response in terms of gonad yield (Chapters 2 and 4).

Our studies showed that several performance indices are linked to silvering, including maturation performance. We observed that silvering is an absolute prerequisite for maturation, i.e. the first two stages (Chapter 2) were found incapable of reproduction. Furthermore, among stages 3 to 5, a positive correlation was found between silver index and maturation performance (Table 19.1). Indications were also found that silvering improves the eel's capacity to cope with the environmental conditions of oceanic migration, as demonstrated by the correlation with seawater and pressure tolerance. However, silver index is not the only criterion for discriminating reproduction capacity. Indeed, swimming fitness was not correlated with the silver index. Although, endurance swimming induced silvering and the endurance performance is higher for fatty fish. The same applies for maturation performance which was further improved by body length. So, we can limit the number of discriminating parameters for reproductive capacity to only four external biometric parameters, which are actually the same as those used for the silver index: body length, body weight, eye diameter, and pectoral fin length. For the highest maturation performance we need in addition to a high silver index also a high body length and a high condition factor. For the highest endurance we also need a high fat content. All of these parameters are very easy to measure and applicable to monitoring programs for eel population management. In Table 19.1, the reproductive capacity is expressed in a semi-quantitative way. Female silver eels require at least 60-g fat per kg for a distance of 6,000-km. As the egg production requires an additional 60-g fat per kg, the minimal fat content of successful female spawner should be 120-g fat per kg (Chapter 8). Gonad yield and cost of transport are strongly depending on the length and condition factor of the eels, i.e. the larger and fatter eels have a much higher chance of success to reach the spawning site and when they reach it, they also have a higher percentage of energy stores left for reproduction. Obviously those eels are the best future genitors for the eel population.

19.2.4 Genetic Variability

The major result from the genetic analyses of eels from various European and North African locations is the clear demonstration that global spatial genetic variability in the European eel is extremely weak. This variability is a strong indicator of panmixia. Furthermore the temporal variability was always higher than the spatial one, which further suggests that eels do not return to a predetermined location. Farm studies demonstrated that multilocus heterozygosity is positively correlated with growth. The more heterozygous individuals show a significantly greater length and weight increase and a higher condition factor in comparison with more homozygous individuals. The pioneer data indicate that selection is likely based on heterozygosity (Chapter 17). The experimental data also suggest that heterozygosity may be an important factor for eel reproduction capacity, since eels with large body length and high body condition would represent the best genitors (see above).

19.3 Impact of Environmental Factors (Quality of Continental Habitats) on the Reproduction Capacity of the European Eels

19.3.1 Habitat Trophic Quality

Based on our observations we can conclude that due to the overall high energy costs of migration, the energy reserves may become a limitation to reach the Sargasso Sea when fat levels drop below 12% of the body weight. In addition to the requirements for swimming, migrating silver eels need energy for gonadal growth which suggests that eels with the highest fat stores have the highest recruitment capacity. From our maturation studies it also appears that the large silver eels with a high condition factor have the highest reproductive capacity. The genetic results from the eel farm growth experiments suggest that in natural waters a high body length and a high condition factor are linked to heterozygosity. Therefore heterozygosity together with favourable trophic conditions will produce eels with a high reproductive capacity. Together with the population and hydrosystem structure this may be used for predicting the total recruitment. So, these results provide a strong incentive for studies aiming at correlating habitat quality with production of silver eels with high reproductive capacity.

19.3.2 Habitat Availability

Since eel reproductive capacity depends on favourable habitats during its growth phase, factors controlling eel continental dispersal and habitat selection should have a major impact on eel reproductive capacity. Recent advances on eel continental dispersal are reviewed (Chapter 18). Eels exhibit a large plasticity in their continental dispersal behaviour. Early upstream migration would depend on body condition (levels of energy stores) and thyroid function would play a pivotal role in integrating internal and environmental cues and mediating alternative behaviours. Water management is increasingly focused on maintaining water levels constant in semi-urban areas. This leads to the construction of large numbers of dams and sluices in particularly in wetlands. Drainage in these areas is carried out by high frequency pumps, which do not allow exits out of the enclosed areas. Thus, large areas will in fact become huge traps for many aquatic animals, including eels. Most individuals will be able to enter the area, going down with the water flow, but the return way will not be possible. This not only limits the natural feeding area of eels, but particularly blocks all escapement possibilities of silver eels. So, it is obvious that dams, sluices and other physical barriers have to be equipped with eel ladders to allow (local) upstream migration. Similarly, the many obstacles to silver eel downstream migration such as dams, sluices, and power stations should be prevented. The decline in river recruitment during the last decades, reflecting a drop in the proportion of migrants, may likely interfere with the reproductive capacity profile of the population.

In addition to the general conclusion that reproductive performance can be assessed from only four different biometric parameters, we have to consider the interference of specific harmful environmental factors. In this book we have reviewed the effects of chemical pollutants and the occurrence of virus and parasite infections.

19.3.3 Habitat Contamination

PBCs accumulate in the fat stores of the eels. The short term effect of PCBs is the reduction of energy consumption of swimming- as well as resting eels. This effect is due to suppressed protein synthesis, which in the end will be harmful for the animal. Our preliminary studies also indicate that PCBs have deleterious effects on the fertility by impairing egg quality and embryonic development. Furthermore, long term swimming will increase the levels of PCBs due to usage of the fat as fuel for swimming. New indications for the destructive effects of these water pollutants on reproduction of eel are presented in Chapter 15. So, in areas with high PCB levels, even those eels that have very high scores for the reproduction capacity, can in fact not participate in successful production of vital offspring.

Human activities, including aquaculture practices, have greatly enhanced global transport of fish species and consequently the spreading of pathogens by transfer of infected animals. With the growing transport of fishes and fish products, the transfer of diseases has increased. Widespread infection of the eel-population with the EVEX virus may be the result from the unlimited and uncontrolled transports. Also *Herpesvirus anguillae* has now been isolated and identified in eel populations all over the world, including cultured eels in Taiwan, as well as cultured and wild eels in the Netherlands. Our swim experiments demonstrate (in one occasion) that EVEX infections are particularly destructive for swimming silver eels, as these infected eels developed a severe viral pathology that lead to anemia (Chapter 16). The EVEX infection is normally not visible from the outside, it looks harmless, as the infected animals appear healthy, however it is known, as with other virus infection, to become virulent during stress. The occurrence of EVEX infection was not correlated with the silver index or with the condition factor. The outbreak of the infection takes place only after 1–2 months of continued swimming. So, EVEX infection may become virulent during the long journey to the Sargasso Sea, reducing the possibility to contribute to reproduction.

The infections with *Anguillicola crassus*, also recently brought to Europe due to eel transport, are long lasting. Light and moderate infections already reduce the swim capacity (fitness index). The infection causes deterioration of the swimbladder often with severe permanent damage; heavy infections result in failure during long term swimming (Chapter 9). In addition, heavy infections and severely damaged swimbladders impair buoyancy control, resulting in poor or absent vertical navigation capacity in the open ocean, which obviously interferes with the completion of the spawning migration. So, in the case of heavy swimbladder infection and/or damage, even those eels that have very high scores for the reproductive capacity will in fact never reach the spawning grounds and can not contribute to recruitment.

In addition to the above mentioned water contaminants and pathogens, other water pollutants are likely to impair eel reproductive capacity. For example, ongoing studies on heavy metals indicate that such environmental contaminants accumulate during the eel growth phase and may dramatically disrupt eel sexual maturation. Also, the increasing number of endocrine disruptors (in addition to PCBs), generated by various human activities (pesticides, fungicides, hormones and pharmaceutical compounds), which end up in hydrosystems, represents as many potential threats to eel reproductive capacity. Indeed they may disrupt the complex endocrine controls of eel life cycle, including larval metamorphosis and silvering, continental dispersal and downstream migration, ion regulation, sexual maturation and reproduction.

19.4 General Conclusion

Our conclusions on the reproductive capacity are based on migration performance (seawater and pressure tolerance, swim fitness) and maturation performance (sensitivity to hormonal treatment). The highest reproductive capacity can be described by a combination of a high silver index, high body length and a high condition factor (Fig. 19.1). The latter two parameters can be linked to the quality of the trophic condition of the hydrosystem, as well as to heterozygosity.

Unfortunately this does not preclude the impact of negative factors in the environment, such as obstacles to migration (dams, hydropower stations), water chemical contaminants (PCBs, heavy metals and other endocrine disruptors), and pathogens (viruses, parasites). These environmental threats have to be taken into account in the evaluation of habitat quality and eel reproductive capacity. Action should be taken to protect and restore migration routes and water quality, in the framework of European regulations.

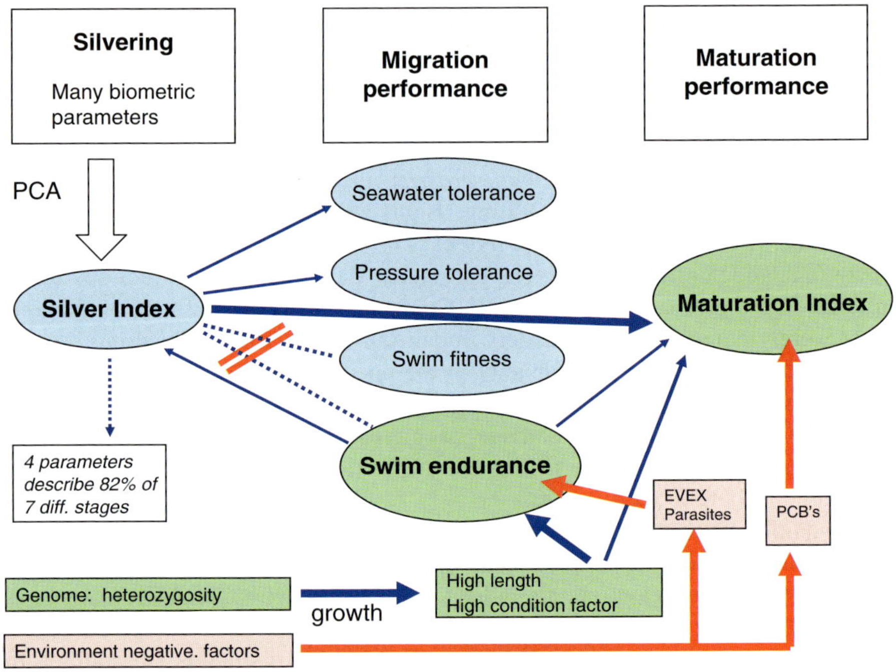

Fig. 19.1 Inter-relationships between the silvering and migration/maturation performance. High silver index correlates with an improved sea water and high pressure tolerance, however, no effect is observed on the swim fitness and endurance. High silver index correlates with high maturation index (more sensitive to hormonal stimulation). Furthermore high length and high fat content (which correlates with heterozygosity) improves maturation index but also swim endurance. Negative environmental factors interfere particularly with swim endurance and maturation index. Infections with EVEX/parasites are devastating for swimming, while PCBs impair fertility

The combination of positive and negative actors on reproduction potential is summarized in Fig. 19.1. (a) Silvering is still a black box, however we know at least the parameters describing the process; (b) Silvering changes the physiology of the eel in terms of osmoregulation and high pressure resistance. It has no effect on swimming performance, but has a strong effect on maturation of both males and females. (c) Swimming does have an effect on silvering, possibly because it stimulates maturation. (d) The positive environmental effects deal with growth rate. The negative environmental effects are parasites, which interfere mainly with swim performance, and pollutants which in turn interfere with maturation and reproduction. Insight in natural responses of eels to biotic and abiotic environmental factors is of importance for further reproduction and ecological studies which hopefully will contribute to the restoration of the European eel population.

As regards the perspectives of restoration of the eel populations, the recent developments in eel reproduction are very hopeful indeed. Apart from the Japanese papers in the 1990s showing mass hatching of Japanese eel larvae, in the last year

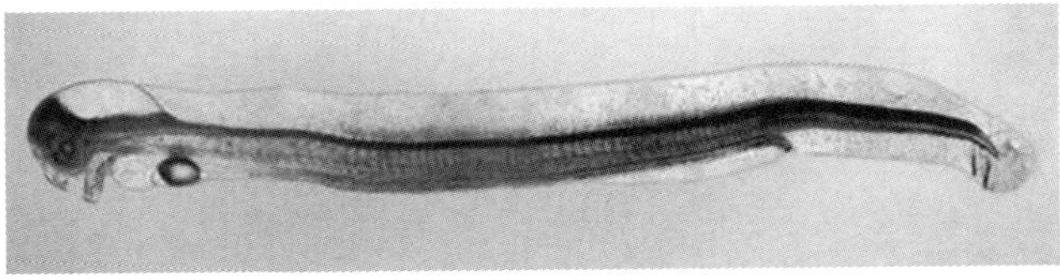

Fig. 19.2 A 10-day-old larva of the New Zealand short finned eel (*Anguilla australis*). The yolk sack has been almost consumed, only a small oil droplet remains. The body is stretched, but the head is not yet turned and ready for feeding. Picture with permission from Tagried Kurwie (Mahurangi Technical Institute, New Zealand). Grant no. EELCO501/TBG/Technology NZ/ Foundation for Research Science and Technology

two new breakthroughs have been presented. In October 2007 at the EAS meeting in Istanbul Jonna Tomkiewicz (DTU-aqua, Kopenhagen) showed results from mass hatchings of European eels, the larvae remained alive for several days. Also in New Zealand the group of Tagried Kurwie from the Mahurangi Technical Institute succeeded in spring 2007 to produce (non-feeding) larvae of short finned eels, *Anguilla australis* (Fig. 19.2). Considering the efforts and collaboration between scientists in this field we expect that the problems of eel reproduction will be resolved in the coming decade.

Index

L

M

N

O

P

R

S